Lecture Notes in Computer Science 11485

Commenced Publication in 1973
Founding and Former Series Editors:
Gerhard Goos, Juris Hartmanis, and Jan van Leeuwen

More information about this series at http://www.springer.com/series/7407

Pinar Heggernes (Ed.)

Algorithms
and Complexity

11th International Conference, CIAC 2019
Rome, Italy, May 27–29, 2019
Proceedings

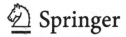 Springer

Editor
Pinar Heggernes
University of Bergen
Bergen, Norway

ISSN 0302-9743   ISSN 1611-3349 (electronic)
Lecture Notes in Computer Science
ISBN 978-3-030-17401-9   ISBN 978-3-030-17402-6 (eBook)
https://doi.org/10.1007/978-3-030-17402-6

LNCS Sublibrary: SL1 – Theoretical Computer Science and General Issues

This Springer imprint is published by the registered company Springer Nature Switzerland AG
The registered company address is: Gewerbestrasse 11, 6330 Cham, Switzerland

Preface

This volume contains the papers selected for presentation at CIAC 2019: the 11th International Conference on Algorithms and Complexity held during May 27–29, 2019 in Rome. This series of conferences presents original research contributions in the theory and applications of algorithms and computational complexity. The papers in this volume are arranged alphabetically by the last names of their authors. We received 95 submissions, and these were handled through the EasyChair system. Each submission was reviewed by three Program Committee members. The committee decided to accept 30 papers.

I would like to thank all the authors who submitted papers, the members of the Program Committee, and the external reviewers who assisted the Program Committee in the evaluation process. The organizers and I are grateful to the three invited speakers, Edith Elkind (University of Oxford), Peter Rossmanith (Aachen University), and Luca Trevisan (U.C. Berkeley), who kindly accepted our invitation to give plenary lectures at CIAC 2019. We are grateful to Springer for sponsoring the CIAC 2019 best paper award. I would like to extend my deep gratitude to Tiziana Calamoneri and Irene Finocchi from the Sapienza University of Rome, who formed the Organizing Committee, for organizing the conference in a very efficient and pleasant way, taking care of administrative and public relation details, and being extremely helpful throughout the whole procedure.

February 2019
Pinar Heggernes

Organization

Program Committee

Faisal Abu-Khzam	Lebanese American University, Lebanon
Isolde Adler	University of Leeds, UK
Cristina Bazgan	LAMSADE, University of Paris-Dauphine, France
Vincenzo Bonifaci	IASI-CNR, Rome, Italy
Tiziana Calamoneri	Sapienza University of Rome, Italy
Pierluigi Crescenzi	Università degli Studi di Firenze, Italy
Konrad K. Dabrowski	Durham University, UK
Lene Favrholdt	University of Southern Denmark, Denmark
Henning Fernau	University of Trier, Germany
Paolo Franciosa	Sapienze University of Rome, Italy
Pinar Heggernes (Chair)	University of Bergen, Norway
Danny Hermelin	Ben-Gurion University of the Negev, Israel
Cecilia Holmgren	Uppsala University, Finland
Giuseppe F. Italiano	Tor Vergata University of Rome, Italy
Mamadou M. Kanté	Université Clermont Auvergne, LIMOS, CNRS, France
Telikepalli Kavitha	Tata Institute of Fundamental Research, Mumbai, India
Jesper Nederlof	Eindhoven University of Technology, The Netherlands
Naomi Nishimura	University of Waterloo, Canada
Nicolas Nisse	Inria, France
Yota Otachi	Kumamoto University, Japan
Sang-Il Oum	Institute for Basic Science (IBS) and KAIST, South Korea
Charis Papadopoulos	University of Ioannina, Greece
Marie-France Sagot	Inria Grenoble Rhône-Alpes and Université de Lyon 1, France
Saket Saurabh	The Institute of Mathematical Sciences, Chennai, India and University of Bergen, Norway
Erik Jan van Leeuwen	Utrecht University, The Netherlands

Steering Committee

Giorgio Ausiello	Sapienza University of Rome, Italy
Vangelis Paschos	University of Paris-Dauphine, France
Rossella Petreschi	Sapienza University of Rome, Italy
Paul Spirakis	University of Liverpool, UK
Peter Widmayer	ETH Zurich, Switzerland

Organizing Committee

Tiziana Calamoneri   Sapienza University of Rome, Italy
Irene Finocchi   Sapienza University of Rome, Italy

Additional Reviewers

Agrawal, Akanksha
Alexandrescu, Andrei
Antoniadis, Antonios
Apollonio, Nicola
Bacci, Tiziano
Banerjee, Niranka
Becchetti, Luca
Bekos, Michael
Belmonte, Rémy
Bernt, Matthias
Berzunza, Gabriel
Bhore, Sujoy
Bonnet, Edouard
Borassi, Michele
Bousquet, Nicolas
Boyar, Joan
Brunetti, Sara
Burghart, Fabian
Cai, Xing Shi
Caillouet, Christelle
Cao, Yixin
Capelli, Florent
Chakraborty, Sankardeep
Cheung, Yun Kuen
Chitnis, Rajesh
Colini-Baldeschi,
    Riccardo
Coudert, David
D'Amore, Fabrizio
Das, Shantanu
de Lima, Paloma
Desmarais, Colin
Devroye, Luc
Di Luna, Giuseppe
    Antonio
Drmota, Michael
Dross, François
Ducoffe, Guillaume

Dyer, Martin
Elbassioni, Khaled
Engels, Christian
Esfandiari, Hossein
Ferraioli, Diodato
Ferraro Petrillo, Umberto
Fournier, Hervé
Fujita, Shinya
Fulek, Radoslav
Georgiadis, Loukas
Gobbert, Moritz
Golin, Mordecai J.
Grossi, Roberto
Gupta, Sushmita
Hanaka, Tesshu
Haque, Sajed
Huang, Chien-Chung
Jakovac, Marko
Jelínková, Eva
Johansson, Tony
Kagaris, Dimitri
Kamiyama, Naoyuki
Katsikarelis, Ioannis
Kim, Eunjung
Klasing, Ralf
Knop, Dušan
Komusiewicz, Christian
Konstantinidis,
    Athanasios
Koutecky, Martin
Krenn, Daniel
Kucherov, Gregory
Kumar, Nirman
Kwon, O-Joung
Lampis, Michael
Lari, Isabella
Lauria, Massimo
Levin, Asaf

Lokshtanov, Daniel
M. S., Ramanujan
Majumdar, Diptapriyo
Manoussakis, Georges
Marchetti-Spaccamela,
    Alberto
Marino, Andrea
Mary, Arnaud
Mattia, Sara
Mc Inerney, Fionn
Migler-Vondollen,
    Theresa
Mitsou, Valia
Miyauchi, Atsushi
Mnich, Matthias
Moscardelli, Luca
Müller, Norbert Th.
Mütze, Torsten
N. S. Pedersen, Christian
Natale, Emanuele
Nichterlein, André
Nilsson, Bengt J.
Nomikos, Christos
Nusser, André
Okamoto, Yoshio
Pagli, Linda
Pandey, Prashant
Panolan, Fahad
Parotsidis, Nikos
Pasquale, Francesco
Penna, Paolo
Perarnau, Guillem
Petreschi, Rossella
Roy, Sanjukta
Roy, Sudip
Sadakane, Kunihiko
Salehkaleybar, Saber
Sampaio Rocha, Leonardo

Sandlund, Bryce
Satti, Srinivasa Rao
Saurabh, Nitin
Schmid, Markus L.
Schweitzer, Pascal
Sharma, Vikram
Shinohara, Ayumi
Sikora, Florian
Sinaimeri, Blerina
Sintos, Stavros
Sitters, Rene

Skerman, Fiona
Spirakis, Paul
Suchý, Ondřej
Talmon, Nimrod
Tardella, Luca
Thacker, Debleena
Tootaghaj, Diman Z.
Trotignon, Nicolas
Truszkowski, Jakub
Tzimas, Spyridon
Ullman, Jeffrey

Velaj, Yllka
Vinyals, Marc
Vocca, Paola
Wasa, Kunihiro
Watrigant, Rémi
Weimann, Oren
Wiederrecht, Sebastian
Woeginger, Gerhard J.
Yedidsion, Harel
Zehavi, Meirav

Contents

Quadratic Vertex Kernel for Split Vertex Deletion

Akanksha Agrawal[1], Sushmita Gupta[2], Pallavi Jain[3(✉)], and R. Krithika[4]

[1] Hungarian Academy of Sciences (MTA SZTAKI), Budapest, Hungary
akanksha@sztaki.hu
[2] National Institute of Science Education and Research, Bhubneswar, India
sushmita.gupta@ii.uib.no
[3] The Institute of Mathematical Sciences, HBNI, Chennai, India
pallavij@imsc.res.in
[4] Indian Institute of Technology Palakkad, Palakkad, India
krithika@iitpkd.ac.in

Abstract. A graph is called a split graph if its vertex set can be partitioned into a clique and an independent set. Split graphs have rich mathematical structure and interesting algorithmic properties making it one of the most well-studied special graph classes. In the SPLIT VERTEX DELETION(SVD) problem, given a graph and a positive integer k, the objective is to test whether there exists a subset of at most k vertices whose deletion results in a split graph. In this paper, we design a kernel for this problem with $\mathcal{O}(k^2)$ vertices, improving upon the previous cubic bound known. Also, by giving a simple reduction from the VERTEX COVER problem, we establish that SVD does not admit a kernel with $\mathcal{O}(k^{2-\epsilon})$ edges, for any $\epsilon > 0$, unless NP $\subseteq$ coNP/poly.

Keywords: Split Vertex Deletion · Kernelization ·
Kernel lower bound · Parameterized Complexity

1 Introduction

The problem of graph editing–adding or deleting vertices or edges–to ensure that the resulting graph has a desirable property is a well-studied problem in graph algorithms. Lewis and Yannakakis [15] showed that if the objective is for the resulting graph to have a non-trivial *hereditary property* (one that exists in every induced subgraph), then the optimization version of the corresponding graph editing problem is known to be NP-hard. Consequently this problem has received attention both within the world of approximation algorithms [9,14,17] as well as parameterized complexity [2,4,11]. A class $\mathcal{G}$ of graphs is called a

The first three authors are supported by the ERC Consolidator Grant SYSTEMATIC-GRAPH (No. 725978) of the European Research Council; Research Council of Norway, Toppforsk project (No. 274526); and SERB-NPDF fellowship (PDF/2016/003508) of DST, India, respectively.

P. Heggernes (Ed.): CIAC 2019, LNCS 11485, pp. 1–12, 2019.
https://doi.org/10.1007/978-3-030-17402-6_1

hereditary class if membership in $\mathcal{G}$ is a hereditary property. Extensive work has been done on the topic of hereditary classes such as chordal graphs and planar graphs [8,12,13,18,19]. In this paper we continue the study initiated by Ghosh *et al.* [10] of the kernelization complexity of the vertex deletion problem of the hereditary class consisting of all *split graphs*. A graph $G = (V, E)$ is called a *split graph* if there exists a partition of the vertex set V into two sets such that the induced subgraph on one of the sets is a complete graph (*i.e.*, a clique) and the other set is an independent set.

Split graphs were introduced by Földes and Hammer [7] and were studied decades later by Tyshkevich and Chernyak [20]. Split graphs have a rich structure and several interesting algorithmic properties. For instance, they can be recognized in polynomial time and they admit polynomial-time algorithms for several classical problems like MAXIMUM INDEPENDENT SET, MAXIMUM CLIQUE and MINIMUM COLORING. They can also be characterised by a finite set of forbidden induced subgraphs (Proposition 1).

Proposition 1 ([7]). *A graph is a split graph if and only if it contains no induced subgraph isomorphic to $2K_2$, C_4, or C_5. Here, $2K_2$ is the graph that is disjoint union of two edges and C_i is a cycle on i vertices.*

In this article we study the kernelization complexity of the following problem.

SPLIT VERTEX DELETION(SVD)   **Parameter:** k
Input: A graph $G = (V, E)$ and an integer k.
Question: Does there exist a set of vertices $S \subseteq V$ such that $|S| \leq k$ and $G[V \setminus S]$ is a split graph?

For a parameterized problem, a *kernelization algorithm* is a polynomial-time algorithm that transforms an arbitrary instance of the problem to an equivalent instance of the same problem whose size is bounded by some computable function g of the parameter of the original instance. The resulting instance is called a *kernel* and if g is a polynomial function, then it is called a *polynomial kernel* and we say that the problem admits a polynomial kernel. Kernelization typically involves applying a set of rules (called *reduction rules*) to the given instance to produce another instance. We say that a reduction rule is applicable on an instance if the output instance is different from the input instance. A reduction rule is said to be *safe* if applying it to the given instance produces an equivalent instance. For more details on kernelization algorithms, we refer the reader to the book by Cygan et al. [3].

From Proposition 1, it is clear that SVD is related to the 5-HITTING SET problem that has a kernel containing $\mathcal{O}(k^4)$ elements [1]. Using the ideas of this kernel, we can also obtain a kernel for SVD containing $\mathcal{O}(k^4)$ vertices. Ghosh et al. [10] improved this result by giving a kernel with $\mathcal{O}(k^3)$ vertices.

Our Results and Techniques. In this paper we improve Ghosh *et al.*'s work on the kernel for SVD by exhibiting a kernel containing $\mathcal{O}(k^2)$ vertices. We complement this result by showing that no kernel with $\mathcal{O}(k^{2-\epsilon})$ edges is possible for any $\epsilon > 0$ (unless $\mathsf{NP} \subseteq \mathsf{coNP/poly}$) by a reduction from VERTEX COVER. Our approach to the quadratic vertex kernel is as follows. Consider an input instance (G, k). First, we compute a maximal set $\mathcal{S}$ of pairwise vertex disjoint induced subgraphs of G isomorphic to $2K_2$, C_4 or C_5. Define the set S to be the vertices of G that are in some element of $\mathcal{S}$. Clearly, $G - S$ is a split graph by Proposition 1 and therefore has a partition of its vertex set into C and I where C is a clique and I is an independent set. Then, we identify certain vertices $C_{\mathsf{fix}} \subseteq C$ and $I_{\mathsf{fix}} \subseteq I$ that have to be in the clique side and independent side, respectively, after the deletion of any solution to (G, k). Using the reduction rules described in [10], we observe that $G - C_{\mathsf{fix}}$ has $\mathcal{O}(k^2)$ vertices. To obtain the claimed kernel, it is necessary and sufficient to bound $|C_{\mathsf{fix}}|$ by $\mathcal{O}(k^2)$. We describe new reduction rules and modify the instance appropriately (while preserving equivalence) to delete redundant vertices in C_{fix}. Essentially, we crucially use the fact that in any split partition of the resulting split graph (after deleting the solution), the clique side can contain at most one vertex from I, and in particular from $I \setminus I_{\mathsf{fix}}$. This property helps us to forget the actual non-neighbourhood in C_{fix} of vertices in $I \setminus I_{\mathsf{fix}}$, and remember only the sizes of the non-neighbourhood of vertices in $I \setminus I_{\mathsf{fix}}$. We show that this information can be encoded using a set of $\mathcal{O}(k^2)$ new vertices whose addition to the graph enables us to delete most of the vertices in C_{fix}. We conclude by showing that the resulting instance has $\mathcal{O}(k^2)$ vertices.

Related Work. In addition to kernels for SVD, there has been quite a bit of work on designing *fixed-parameter tractable* algorithms that beat the trivial bound implied by the kernels. Lokshtanov *et al.* [16] exhibits an $\mathcal{O}^*(2.32^k)$ algorithm[1] for SVD by using the fixed-parameter tractable algorithm for the ABOVE GUARANTEE VERTEX COVER problem. Ghosh *et al.* [10] improve this time-complexity to $\mathcal{O}^*(2^k)$ by combining iterative compression with a bound on the number of split partitions of a split graph. The FPT algorithm with running time $\mathcal{O}^*(1.2738^k k^{\mathcal{O}(\log k)})$ designed by Cygan and Pilipczuk [4] is the current fastest algorithm for SVD, which uses the fastest known algorithm for VERTEX COVER and clique-independent set families. Among all the graph editing problems, the one that most closely resembles SVD is, perhaps, the SPLIT EDGE DELETION problem, where the goal is to decide if there exists a subset of edges of size at most k whose deletion results in a split graph. Guo [11] gave a kernelization algorithm of size $\mathcal{O}(k^4)$. Ghosh *et al.* [10] improved that bound to $\mathcal{O}(k^2)$. In the same paper, the authors gave an $\mathcal{O}^*(2^{\sqrt{k} \log k})$ time algorithm for the problem. It was posited in that paper that this might be the second problem exhibited to have a subexponential-time algorithm on general graphs which does not use bidimensionality theory (besides MINIMUM FILL-IN [8]).

2 Preliminaries

We denote the set of natural numbers by $\mathbb{N}$. For $n \in \mathbb{N}$, by $[n]$ we denote the set $\{1, 2, \ldots, n\}$. We use standard terminology from the book of Diestel [6] for the

---

[1] The $\mathcal{O}^*(.)$ notation suppresses polynomial factors in the input size.

graph related terminologies which are not explicitly defined here. For a graph G, by $V(G)$ and $E(G)$ we denote the vertex and edge sets of the graph G, respectively. Consider a graph G. For $v \in V(G)$, its neighbourhood (denoted by $N_G(v)$) is the set $\{u \in V(G) \mid (v, u) \in E(G)\}$ and its non-neighbourhood (denoted by $\overline{N}_G(v)$) is the set $V(G) \setminus (N_G(v) \cup \{v\})$. For any non-empty subset $S \subseteq V(G)$, the subgraph of G induced by S is denoted by $G[S]$; its vertex set is S and its edge set is $\{(u, v) \in E(G) \mid u, v \in S\}$. For $S \subseteq V(G)$, by $G - S$ we denote the graph $G[V(G) \setminus S]$. The complement of G, denoted by $\overline{G}$, is the graph defined as $V(\overline{G}) = V(G)$ and $E(\overline{G}) = \{(u, v) \mid u, v \in V(G),$ where $u \neq v,$ and $(u, v) \notin E(G)\}$.

A *cycle* $C = (v_1, v_2, \ldots, v_\ell)$ in a graph G is a subgraph of G, with vertex set $\{v_1, v_2, \ldots, v_\ell\} \subseteq V(G)$ and edge set $\{(v_i, v_{(i+1)}) \mid i \in [\ell]\} \subseteq E(G)$. Here, the indices in the edge set are computed modulo ℓ. The length of a cycle is the number of vertices in it. The cycle on i vertices is denoted by C_i. For $q \in \mathbb{N} \setminus \{0\}$, by K_q we denote the clique (complete graph) on q vertices. The graph pK_q is defined as the graph obtained by taking the disjoint union of p copies of K_q.

A *split graph* is a graph G whose vertex set can be partitioned into two sets, C and I, such that C is a clique while I is an independent set. The partition (C, I) of the vertex set of a split graph G into a clique C and an independent set I is called a *split partition* of G. By the definition of split graphs, it is easy to observe the following.

Observation 1. *G is a split graph with split partition (C, I) if and only if $\overline{G}$ is a split graph with split partition (I, C).*

Given a graph G, a *split deletion set* $S \subseteq V(G)$, is a set such that $G - S$ is a split graph. We end this section by stating the following observation that is implied by Proposition 1.

Observation 2. *Given a graph G, there is a polynomial time algorithm that returns a split deletion set S of G such that for any split deletion set S^* of G, we have $|S| \leq 5|S^*|$.*

3   The Quadratic Kernel

In this section, we show that SVD parameterized by k has a kernel with $\mathcal{O}(k^2)$ vertices. Consider an instance (G, k) of SVD. Let S be the split deletion set of G obtained using Observation 2. Observe that if $|S| > 5k$, then (G, k) is a no-instance. Otherwise, let (C, I) be a split partition of $G - S$. As we have $|S| = \mathcal{O}(k)$, to obtain the required kernel, it suffices to bound $|C|$ and $|I|$. To achieve this, we will define a sequence of reduction rules. All reduction rules are applied in the sequence stated. That is, a rule is applied only when none of the preceding rules can be applied.

First, we apply the rules described in [10]. To this end, we first define the subset $S_Y = \{v \in S \mid |N_G(v) \cap I| \geq k + 2$ and $|\overline{N}_G(v) \cap C| \geq k + 2\}$. The use of such a definition is to identify vertices of S that have to be in any solution

S^* to (G, k). Suppose that (C^*, I^*) is a split partition of $G - S^*$. Assume for contradiction that there is a vertex $v \in S_Y \setminus S^*$. Then $v \in C^*$ or $v \in I^*$. Consider the case when v is in C^*. As v has at least $k + 2$ non-neighbours in C and $|S^*| \leq k$, it follows that at least 2 such non-neighbours of v, say x and y, are in $V(G - S^*)$. As x and y are adjacent, at least one of them is in C^*. This leads to a contradiction as $v \in C^*$. A similar argument holds if $v \in I^*$. This is stated as the following lemma that leads to Reduction Rule 1.

Lemma 1 (Lemma 5 of [10]). *For a solution S^* to (G, k), we have $S_Y \subseteq S^*$.*

Reduction Rule 1. *If there is $v \in S_Y$, then delete v and decrease k by 1.*

The safeness of this reduction rule follows from Lemma 1. If our algorithm proceeds to the further steps, we can assume that Reduction Rule 1 is not applicable, which means, $S_Y = \emptyset$. The reasoning behind the safeness of Reduction Rule 1 hints towards identifying vertices that have to be in the "clique side" and vertices that have to be in the "independent side" of the resultant split graph (if the input instance is a *yes*-instance). For this purpose, we define the following subsets of S: $S_C = \{v \in S \mid |N_G(v) \cap I| \geq k + 2\}, S_I = \{v \in S \mid |\overline{N}_G(v) \cap C| \geq k + 2\}$, and $S_Z = S \setminus (S_C \cup S_I)$.

Intuitively, as every vertex v in S_C has a large neighbourhood in I, either v has to be in the solution split deletion set or v has to end up in the "clique side" of the resultant split graph. Similarly, as every vertex u in S_I has a large non-neighbourhood in C, either u has to be in the solution split deletion set or u has to end up in the "independent set side" of the resultant split graph. Note that as $S_Y = \emptyset$, we have $S_C \cap S_I = \emptyset$. In the above we identified some special vertices of S which must necessarily belong to the clique side or belong to the independent set side. Next we identify such "special vertices" in I and C.

$$C_{\text{fix}} = \{v \in C \mid |N_G(v) \cap I| \geq k + 2\} \text{ and } C_Z = C \setminus C_{\text{fix}}$$

$$I_{\text{fix}} = \{v \in I \mid |\overline{N}_G(v) \cap C| \geq k + 2\} \text{ and } I_Z = I \setminus I_{\text{fix}}$$

Next, let C_Y be the set of vertices in C that have no non-neighbours in $S_C \cup S_Z \cup I_Z$. Let I_Y be the set of vertices in I that have no neighbours in $S_I \cup S_Z \cup C_Z$. We will now explain the idea behind partitioning $V(G)$ into these sets. First, we have the following property on C_{fix} and I_{fix}.

Lemma 2 (Lemma 4 of [10]). *Suppose that S^* is a solution to (G, k) and (C^*, I^*) is a split partition of $G - S^*$. Then, for each vertex v in $C_{\text{fix}} \cup S_C$, either $v \in S^*$ or $v \in C^*$ and for each vertex v in $I_{\text{fix}} \cup S_I$, either $v \in S^*$ or $v \in I^*$.*

Next, let us consider the set C_Y. Consider a hypothetical solution S^* to (G, k) and let (C^*, I^*) denote a split partition of $G - S^*$. Observe that $C^* \cap I_{\text{fix}} = \emptyset$ by Lemma 2. In other words, we have $C^* \subseteq S_C \cup S_Z \cup C_{\text{fix}} \cup C_Z \cup I_Z$. By definition, every vertex in C_Y is adjacent to every vertex in $S_C \cup S_Z \cup C_{\text{fix}} \cup C_Z \cup I_Z$. That is, every vertex in C_Y is adjacent to every vertex that has the potential to end up in C^*. Therefore, if $|C_Y|$ is "too large", we can afford to delete a vertex from it as we can always "put it back" into the clique, as described in the following.

Reduction Rule 2 (Reduction Rules 3.3 of [10]). *If* $|C_Y \cap C_{\text{fix}}| > k + 2$, *then delete all edges between* $C_Y \cap C_{\text{fix}}$ *and* $(I_{\text{fix}} \cup S_I)$ *and delete all but* $k + 2$ *vertices of* $C_Y \cap C_{\text{fix}}$.

Reduction Rule 3 (Reduction Rules 3.4 of [10]). *If* $|C_Y \cap C_Z| > k + 2$, *then delete all edges between* $C_Y \cap C_Z$ *and* $(I_{\text{fix}} \cup S_I)$ *and delete all but* $k + 2$ *vertices of* $C_Y \cap C_Z$.

Along similar lines, let us now consider the set I_Y. Once again, consider a hypothetical solution S^* to (G, k) and let (C^*, I^*) denote a split partition of $G - S^*$. Observe that $I^* \cap C_{\text{fix}} = \emptyset$ by Lemma 2. In other words, we have $I^* \subseteq S_I \cup S_Z \cup I_{\text{fix}} \cup I_Z \cup C_Z$. By definition, every vertex in I_Y is non-adjacent to every vertex in $S_I \cup S_Z \cup I_{\text{fix}} \cup I_Z \cup C_Z$. That is, every vertex in I_Y is non-adjacent to every vertex that has the potential to end up in I^*. Therefore, if $|I_Y|$ is too large, we can afford to delete a vertex from it as we can always "put it back" into the independent set side. This justifies the following reduction rules.

Reduction Rule 4 (Reduction Rule 3.5 of [10]). *If* $|I_Y \cap I_{\text{fix}}| > k + 2$, *then delete all edges between* $I_Y \cap I_{\text{fix}}$ *and* $(C_{\text{fix}} \cup S_C)$ *and delete all but* $k + 2$ *vertices of* $I_Y \cap I_{\text{fix}}$.

Reduction Rule 5 (Reduction Rule 3.6 of [10]). *If* $|I_Y \cap I_Z| > k+2$, *then delete all edges between* $I_Y \cap I_Z$ *and* $(C_{\text{fix}} \cup S_C)$ *and delete all but* $k + 2$ *vertices of* $I_Y \cap I_Z$.

The formal safeness proofs of Reduction Rule 1 to 5 can be found in [10]. When none of Reduction Rule 1 to 5 is applicable, we have the following bound.

Lemma 3 (Lemma 8 of [10]). *When neither of Reduction Rule 1 to 5 are applicable, either* $|C_Z| \leq 2k + 2$ *or* $|I_Z| \leq 2k + 2$.

This result leads to the following bound.

Lemma 4 (Lemma 9 of [10]). *When neither of the Reduction Rule 1 to 5 are applicable,* $|C_Z \cup I_Z| = \mathcal{O}(k^2)$. *Moreover, if* $|C_Z| \leq 2k + 2$, *then* $|I_{\text{fix}}|$ *is* $\mathcal{O}(k^2)$, *and if* $|I_Z| \leq 2k + 2$, *then* $|C_{\text{fix}}|$ *is* $\mathcal{O}(k^2)$.

In the following, we describe how we obtain the kernel when $|C_Z \cup I_Z| = \mathcal{O}(k^2)$, $|C_Z| \leq 2k + 2$, and $|I_{\text{fix}}| = \mathcal{O}(k^2)$. Note that if we show that the size of C_{fix} by $\mathcal{O}(k^2)$, then we will obtain the desired kernel. First, we give an intuitive description of the steps that the kernelization algorithm applies to reduce the instance size. The algorithm will not delete any vertices in $S \cup C_Z \cup I_{\text{fix}} \cup I_Z$. From Lemma 2, we know that each vertex in C_{fix} is either in the solution that we are looking for or is in the clique side of the split partition of the resulting split graph. Similarly, each vertex in I_{fix} is either in the solution or is in the independent set side of the split partition of the resulting split graph. Using some gadgets, we will ensure that these properties hold irrespective of any subsequent modification to the instance. After this we will crucially use the fact that in

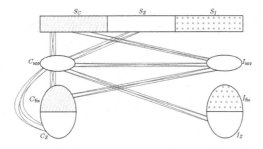

Fig. 1. An illustration of G_{app} which shows only the new edges added. Here, the vertices (remaining) in the squiggled and the starred set always belong to the clique and the independent set side, respectively, after removal of any solution.

any split partition of the resulting split graph, the clique side can contain at most one vertex from I, and in particular from I_Z. This property helps us to forget the actual non-neighbourhood in C_{fix} of vertices in I_Z, and remember only the sizes of the non-neighbourhood of vertices in I_Z. Here, we will also rely on the property that each vertex in I_Z has at most $k + 1$ non-neighbours in C (in particular, in C_{fix}). Next, we formally describe the operations that are performed to obtain the desired kernel.

First, we construct an (equivalent) instance (G_{app}, k) of SVD (in polynomial time) as follows. Roughly speaking, G_{app} is obtained from G by appending a large enough clique and a large enough independent set. The addition of these sets will (roughly) ensure that in the resulting graph, after we remove a solution, the (remaining) vertices in $C_{\mathsf{fix}} \cup S_C$ always belong to the clique side of any split partition. Also, it is assured that the (remaining) vertices in $I_{\mathsf{fix}} \cup S_I$ always belong to the independent side of any split partition. The graph G_{app} is constructed as follows. Initially, $G_{\mathsf{app}} = G$. We add two sets C_{app} and I_{app} each of $k + 2$ (new) vertices to $V(G_{\mathsf{app}})$. The vertices in C_{app} and I_{app} induce a clique and an independent set in G_{app}, respectively. We add all the edges between vertices in C_{app} and vertices in $C \cup S_C \cup S_Z \cup I_Z$. Also, we add all the edges between vertices in I_{app} and $C_{\mathsf{fix}} \cup S_C \cup C_{\mathsf{app}}$. An illustration of the graph G_{app} is depicted in Fig. 1.

Next, we show the equivalence between the instances (G, k) and (G_{app}, k).

Lemma 5. $(\clubsuit)^2$ (G, k) *is a yes-instance of* SVD *if and only if* (G_{app}, k) *is a yes-instance of* SVD.

The following observation follows from the construction of (G_{app}, k).

Observation 3. *The number of vertices in* G_{app} *is* $|V(G)| + 2(k + 2)$.

In the following we construct a (marked) set M, of vertices in G_{app}. Initially, we have $M = C_{\mathsf{app}} \cup I_{\mathsf{app}} \cup S \cup I \cup C_Z$. For each $s \in S_C \cup S_Z$, we add to M,

---
[2] The proofs of results marked with $\star$ will appear in the full version of the paper.

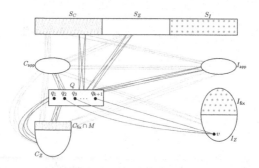

Fig. 2. An illustration of the graph G_ρ, where $\rho(v) = 1$.

all the vertices in $\overline{N}_G(s) \cap C$. That is, we add all the non-neighbours in C of vertices in $S_C \cup S_Z$ to the set M. By our assumption and the definition of S_C and S_Z, it follows that $|M| = \mathcal{O}(k^2)$.

Next, we define some notations that will be useful in obtaining the desired kernel. Let $C^U = C_{\text{fix}} \setminus M$. We define a function $\rho : I_Z \to \mathbb{N}$ as follows. For each $v \in I_Z$, we set $\rho(v) = |\overline{N}_G(v) \cap C^U|$. That is, $\rho(v)$ is the number of non-neighbours of v in C^U. Now we are ready to reduce the size of the instance (G_{app}, k). This step will rely on the fact that in a split partition after deletion of any solution from the graph G_{app}, the clique side can contain at most one vertex from I (and in particular, from I_Z).

We construct another (equivalent) instance (G_ρ, k) of SVD as follows. Initially, $G_\rho = G_{\text{app}} - C^U$. We add a set $Q = \{q_1, q_2, \cdots, q_{k+1}\}$ of (new) vertices inducing a clique, to G_ρ. We add all the edges in $\{(q, v) \mid q \in Q, v \in C_{\text{app}} \cup I_{\text{app}} \cup (C \setminus C^U) \cup S_C \cup S_Z\}$. For each $v \in I_Z$, we add all the edges in $\{(v, q_i) \mid i \in [k+1] \setminus [\rho(v)]\}$ to $E(G_\rho)$. That is, we make v non-adjacent to the first $\rho(v)$ vertices in Q and adjacent to all the remaining vertices in Q. Here, we rely on the fact that each vertex in I_Z has at most $k+1$ non-neighbours in C (and in particular, in C^U). This completes the description of the graph G_ρ; see Fig. 2 for illustration.

From the construction of (G_ρ, k), and the fact that $|C_{\text{app}} \cup I_{\text{app}} \cup Q| = \mathcal{O}(k)$ and $V(G) \setminus C^U = \mathcal{O}(k^2)$, we obtain the following observation.

Observation 4. *The number of vertices in G_ρ is $\mathcal{O}(k^2)$.*

Our next goal is to establish that (G, k) is a *yes*-instance of SVD if and only if (G_ρ, k) is a *yes*-instance of SVD. By Lemma 5, it is enough to show that (G_{app}, k) and (G_ρ, k) are equivalent instances. We start by proving the following.

Lemma 6. (♣) *Suppose that S^* is a solution to (G_{app}, k) and (C^*, I^*) is a split partition of $G_{\text{app}} - S^*$. Then, $(C_{\text{fix}} \cup S_C) \setminus S^* \subseteq C^*$ and $(I_{\text{fix}} \cup S_I) \setminus S^* \subseteq I^*$.*

Similarly, the following claim holds for (G_ρ, k).

Lemma 7. *Suppose that S^* is a solution to (G_ρ, k) and (C^*, I^*) is a split partition of $G_\rho - S^*$. Then, $(Q \cup S_C) \setminus S^* \subseteq C^*$ and $(I_{\text{fix}} \cup S_I) \setminus S^* \subseteq I^*$.*

Lemma 8. (G_{app}, k) *is a yes-instance of* SVD *if and only if* (G_ρ, k) *is a yes-instance of* SVD.

Proof. In the forward direction, let S^* be a solution to SVD in (G_{app}, k), and let (C^*, I^*) be a split partition of $G_{\mathsf{app}} - S^*$.

1. Suppose $I_Z \cap C^* = \emptyset$. In this case, consider the sets $C^*_\rho = (C^* \setminus C^U) \cup Q$ and $I^*_\rho = I^* \setminus C^U$. We will show that (C^*_ρ, I^*_ρ) is a split partition of $G_\rho - S^*$. It is easy to see that C^*_ρ and I^*_ρ partitions the set $V(G_\rho) \setminus S^*$. By construction, we have that I^*_ρ is an independent set in G_ρ. We will show that C^*_ρ is a clique in $G_\rho - S^*$. To show this, we only need to show that Q is adjacent to each vertex in C^*. Consider a vertex $q \in Q$. By construction, $S_I \cup I_{\mathsf{fix}} \subseteq \overline{N}_{G_\rho}(q) \subseteq S_I \cup I_{\mathsf{fix}} \cup I_Z$, and from Lemma 6 we have $(I_{\mathsf{fix}} \cup S_I) \setminus S^* \subseteq I^*$. Moreover, by our assumption, $I_Z \cap C^* = \emptyset$. From the above discussions we can conclude that q is adjacent to every vertex in C^*.

2. Suppose $I_Z \cap C^* \neq \emptyset$. Since I is an independent set in G_{app} (where $I_Z \subseteq I$), we have $|I_Z \cap C^*| = 1$. Let v be the unique vertex in $I_Z \cap C^*$ and $C^U(v) = \overline{N}_{G_{\mathsf{app}}}(v) \cap C^U$. Note that $\rho(v) = |C^U(v)|$. Since $v \notin S^*$, from Lemma 6 we have $C^U(v) \subseteq S^*$. Let $S_\rho = (S^* \setminus C^U) \cup \{q_i \mid i \in [\rho(v)]\}$. Observe that $|S_\rho| \leq |S^*| \leq k$. We will show that S_ρ is a solution to (G_ρ, k). To this end, consider the sets $C^*_\rho = (C^* \setminus C^U) \cup (Q \setminus \{q_i \mid i \in [\rho(v)]\})$ and $I^*_\rho = I^* \setminus C^U$. Notice that C^*_ρ and I^*_ρ forms a partition of $V(G_\rho) \setminus S_\rho$. We will show that (C^*_ρ, I^*_ρ) is a split partition of $G_\rho - S_\rho$. By construction, we have that I^*_ρ is an independent set in G_ρ. Now we will show that C^*_ρ is a clique in G_ρ. Notice that $C^* \setminus C^U$ is a clique in G_ρ, by the construction of G_ρ. We will show that each $q \in Q \setminus \{q_i \mid i \in [\rho(v)]\}$ is adjacent to each vertex in $C^* \setminus C^U$. From Lemma 6 we have $(I_{\mathsf{fix}} \cup S_I) \setminus S^* \subseteq I^*$. Furthermore, $(v, q) \in E(G_\rho)$, $I_Z \cap C^* = \{v\}$, $S_C \cup S_Z \cup (C \setminus C^U) \cup C_{\mathsf{app}} \subseteq N_{G_\rho}(q)$. Thus q is adjacent to each vertex in $C^* \setminus C^U$. This together with the fact that $G_\rho[Q]$ is a clique, implies that C^*_ρ is a clique in G_ρ.

In the reverse direction, let S_ρ be a solution to (G_ρ, k), and (C^*_ρ, I^*_ρ) be a split partition of $G_\rho - S_\rho$. We again have cases similar to the proof of the forward direction, based on the intersection of I_Z with C^*_ρ.

1. Suppose $I_Z \cap C^*_\rho = \emptyset$. In this case, consider the sets $C^* = (C^*_\rho \setminus Q) \cup C^U$ and $I^* = I^*_\rho \setminus Q$. We will show that (C^*, I^*) is a split partition of $G_{\mathsf{app}} - S_\rho$. It is easy to see that C^* and I^* partitions the set $V(G_{\mathsf{app}}) \setminus S_\rho$. By construction, we have that I^* is an independent set in G_{app}. We will show that C^* is a clique in $G_{\mathsf{app}} - S_\rho$. To show this, we only need to show that each vertex in C^U is adjacent to each vertex in $C^*_\rho \setminus Q$. Consider a vertex $v \in C^U$. By the construction of C^U, $S_C \cup S_Z \cup C_{\mathsf{app}} \cup I_{\mathsf{app}} \subseteq N_{G_\rho}(v)$, and from Lemma 7 we have $(I_{\mathsf{fix}} \cup S_I) \setminus S_\rho \subseteq I^*_\rho$. Thus, C^* is a clique in G_{app}.

2. Suppose $I_Z \cap C^*_\rho \neq \emptyset$. Since I is an independent set in G_ρ, we have $|I \cap C^*_\rho| = 1$. Let v be the unique vertex in $I_Z \cap C^*_\rho$ and $Q(v) = \overline{N}_{G_\rho}(v) \cap Q$. Note that $\rho(v) = |Q(v)|$. Since $v \notin S_\rho$, from Lemma 7 we have $Q(v) \subseteq S_\rho$. Let $S^* = (S_\rho \setminus Q) \cup \{w \in C^U \mid (v, w) \notin E(G_{\mathsf{app}})\}$. Observe that $|S^*| \leq |S_\rho| \leq k$.

We will show that S^* is a solution to (G_{app}, k). To this end, consider the sets $C^* = (C^*_\rho \setminus Q) \cup (C^U \setminus \{w \in C^U \mid (v, w) \notin E(G_{\mathsf{app}})\})$ and $I^* = I^*_\rho \setminus Q$. Notice that C^* and I^* partitions $V(G_{\mathsf{app}}) \setminus S^*$. We will show that (C^*, I^*) is a split partition of $G_{\mathsf{app}} - S^*$. By construction, we have that I^* is an independent set in G_{app}. Now we will show that C^* is a clique in G_{app}. Notice that $C^*_\rho \setminus Q$ is a clique in G_{app}. We will show that each $w \in C^U \setminus \{w \in C^U \mid (v, w) \notin E(G_{\mathsf{app}})\}$ is adjacent to every vertex in $C^*_\rho \setminus Q$. From Lemma 7 we have $(I_{\mathsf{fix}} \cup S_I) \setminus S^* \subseteq I^*$. Furthermore, $(v, w) \in E(G_{\mathsf{app}})$, $I_Z \cap C^*_\rho = \{v\}$, $S_C \cup S_Z \cup (C \setminus Q) \cup C_{\mathsf{app}} \cup I_{\mathsf{app}} \subseteq N_{G_{\mathsf{app}}}(w)$. Thus w is adjacent to each vertex in $C^*_\rho \setminus Q$. This implies that C^* is a clique in G_{app}. □

Now, we define the following rule whose safeness follows from Lemmas 5 and 8.

Reduction Rule 6. *Let (G, k) denote the instance on which none of Reduction Rule 1 to 5 are applicable. If $|C_Z| \leq 2k + 2$, then return (G_ρ, k).*

Once this rule is applied, the resulting instance has $\mathcal{O}(k^2)$ vertices by Observation 4. From Lemma 3, we know that when none of Reduction Rule 1 to 6 are applicable on (G, k), we have $|I_Z| \leq 2k + 2$. Using Observation 1 and the fact that (G, k) is a *yes*-instance if and only if $(\overline{G}, k)$ is a *yes*-instance, we define the final reduction rule to handle this situation.

Reduction Rule 7. *Let (G, k) denote the instance on which none of Reduction Rule 1 to 6 are applicable. Then return $(\overline{G}, k)$.*

Observe that after Reduction Rule 7 is applicable on an instance, one of the earlier reduction rules become applicable. This is due to the fact that an independent set (clique) in G becomes a clique (independent set) in $\overline{G}$. This leads to the main result of this paper.

Theorem 1. (♣) SVD *admits a kernel with $\mathcal{O}(k^2)$ vertices.*

4   Kernelization Lower Bound

In Sect. 3, we obtained a kernel for SVD with $\mathcal{O}(k^2)$ vertices. A natural question is to determine if this is the best possible. Though, we do not fully answer this question, we obtain that SVD does not admit a kernel with $\mathcal{O}(k^{2-\epsilon})$ edges, for any $\epsilon > 0$ (assuming NP $\not\subseteq$ coNP/poly). We obtain our lower bound result by giving a simple (linearly) parameter-preserving reduction from the VERTEX COVER problem.

VERTEX COVER   **Parameter:** k
Input: A graph G and an integer k.
Question: Does there exist a set of vertices $S \subseteq V(G)$ such that $|S| \leq k$ and $G - S$ has no edges?

Dell and van Melkebeek [5] proved that VERTEX COVER does not admit a compression of bit-size $\mathcal{O}(k^{2-\epsilon})$, for any $\epsilon > 0$, unless NP $\subseteq$ coNP/poly.

Definition 1. A *compression* of a parameterized problem $\Pi \subseteq \Sigma^* \times \mathbb{N}$ into a problem $\Gamma \subseteq \Sigma^*$ is an algorithm that takes as input an instance (I, k) of Π, and in time polynomial in $|I| + k$ returns an instance I' of Γ such that $|I'| \leq f(k)$ and (I, k) is a *yes*-instance of Π if and only if I' is a *yes*-instance of Γ. The function f is called the *bit-size* of the compression.

Note that if a parameterized problem does not admit a kernel of size $f(k)$, then it does not have a compression of size $f(k)$, as the output kernel can be treated as an instance of the un-parameterized version of the problem.

Theorem 2. SVD *does not admit a kernel with* $\mathcal{O}(k^{2-\epsilon})$ *edges, for any* $\epsilon > 0$, *unless* NP $\subseteq$ coNP/poly.

Proof. Suppose that SVD admits a kernel $\mathcal{A}$, with $\mathcal{O}(k^{2-\epsilon})$ edges, for some $\epsilon > 0$. Now we will design a compression for VERTEX COVER of bit-size $\mathcal{O}(k^{2-\epsilon})$, for some $\epsilon > 0$. Consider an instance (G, k) of VERTEX COVER and let n denote the number of vertices in G. Let $\widehat{G}$ be the graph obtained from G by adding a set Q of $k + 2$ new vertices that induces a clique. We let $(\widehat{G}, k)$ be the instance of SVD. Now we use the algorithm $\mathcal{A}$ to obtain an instance (G^*, k^*) of SVD in time polynomial in $n + k$, such that $(\widehat{G}, k)$ is a *yes*-instance of SVD if and only if (G^*, k^*) is a *yes*-instance of SVD and $|E(G^*)| + k^* = \mathcal{O}(k^{2-\epsilon})$. We can assume that G^* has no isolated vertices (otherwise, we can safely delete them to obtain a smaller equivalent instance), thus we can assume that $|V(G^*)| + |E(G^*)| + k^* = \mathcal{O}(k^{2-\epsilon})$. We output the instance (G^*, k^*) of SVD as a compression for the instance (G, k) of VERTEX COVER.

To obtain the proof, we only need to show that (G, k) is a *yes*-instance of VERTEX COVER if and only if $(\widehat{G}, k)$ a *yes*-instance of SVD. Suppose that S is a vertex cover of size at most k in G. Then, $V(G) \setminus S$ is an independent set in G (and hence in $\widehat{G}$). Thus, $\widehat{G} - S$ is a split graph with split partition $(Q, V(G) \setminus S)$. In other words, S is a split deletion set of size at most k in $\widehat{G}$. Conversely, suppose that S^* is a split deletion set of size at most k in $\widehat{G}$. Let (C^*, I^*) denote a split partition of $\widehat{G} - S^*$. Then, $|Q \setminus S^*| \geq 2$, because $|Q| = k + 2$ and $|S^*| \leq k$. Since Q is a clique in $\widehat{G}$, at most one vertex from $Q \setminus S^*$ can belong to I^*. Thus, there is a vertex $q \in C^* \cap Q$. As no vertex of Q is adjacent to a vertex outside Q in $\widehat{G}$, we have that $C^* \subseteq Q \setminus S^*$. From the above discussions it follows that $S^* \setminus Q$ is a vertex cover of size at most k in G. Hence, (G, k) is a *yes*-instance of VERTEX COVER if and only if $(\widehat{G}, k)$ a *yes*-instance of SVD.

5   Concluding Remarks

We have shown that SVD has a kernel with $\mathcal{O}(k^2)$ vertices. A natural next direction of research is to show that this is the best possible bound (under standard complexity theoretic assumptions) or explore if the problem admits a linear vertex kernel. In fact, even determining the existence of an $\mathcal{O}(k^3)$ edges kernel is an interesting line of work.

Acknowledgement. We are thankful to the reading course headed by Saket Saurabh, held at Institute of Mathematical Sciences, during which the problem was suggested for research.

References

1. Abu-Khzam, F.N.: A kernelization algorithm for d-Hitting Set. J. Comput. Syst. Sci. **76**(7), 524–531 (2010)
2. Cai, L.: Fixed-parameter tractability of graph modification problems for hereditary properties. Inf. Process. Lett. **58**(4), 171–176 (1996)
3. Cygan, M., et al.: Parameterized Algorithms. Springer, Cham (2015). https://doi.org/10.1007/978-3-319-21275-3
4. Cygan, M., Pilipczuk, M.: Split vertex deletion meets vertex cover: new fixed-parameter and exact exponential-time algorithms. Inf. Process. Lett. **113**(5), 179–182 (2013)
5. Dell, H., van Melkebeek, D.: Satisfiability allows no nontrivial sparsification unless the polynomial-time hierarchy collapses. JACM **61**(4), 23:1–23:27 (2014)
6. Diestel, R.: Graph Theory. Springer, Heidelberg (2006)
7. Földes, S., Hammer, P.: Split graph. Congressus Numerantium **19**, 311–315 (1977)
8. Fomin, F.V., Villanger, Y.: Subexponential parameterized algorithm for minimum fill-in. In: Proceedings of SODA, pp. 1737–1746 (2012)
9. Fujito, T.: A unified approximation algorithm for node-deletion problems. Discrete Appl. Math. **86**(2), 213–231 (1998)
10. Ghosh, E., et al.: Faster parameterized algorithms for deletion to split graphs. Algorithmica **71**(4), 989–1006 (2015)
11. Guo, J.: Problem kernels for NP-complete edge deletion problems: split and related graphs. In: Tokuyama, T. (ed.) ISAAC 2007. LNCS, vol. 4835, pp. 915–926. Springer, Heidelberg (2007). https://doi.org/10.1007/978-3-540-77120-3_79
12. Heggernes, P., van't Hof, P., Jansen, B.M.P., Kratsch, S., Villanger, Y.: Parameterized complexity of vertex deletion into perfect graph classes. Theor. Comput. Sci. **511**, 172–180 (2013)
13. Jansen, B.M.P., Pilipczuk, M.: Approximation and kernelization for chordal vertex deletion. In: Proceedings of SODA, pp. 1399–1418 (2017)
14. Kumar, M., Mishra, S., Devi, N.S., Saurabh, S.: Approximation algorithms for node deletion problems on bipartite graphs with finite forbidden subgraph characterization. Theor. Comput. Sci. **526**, 90–96 (2014)
15. Lewis, J.M., Yannakakis, M.: The node-deletion problem for hereditary properties is NP-Complete. J. Comput. Syst. Sci. **20**(2), 219–230 (1980)
16. Lokshtanov, D., Narayanaswamy, N.S., Raman, V., Ramanujan, M.S., Saurabh, S.: Faster parameterized algorithms using linear programming. ACM Trans. Algorithms **11**(2), 15:1–15:31 (2014)
17. Lund, C., Yannakakis, M.: On the hardness of approximating minimization problems. JACM **41**(5), 960–981 (1994)
18. Marx, D.: Chordal deletion is fixed-parameter tractable. Algorithmica **57**(4), 747–768 (2010)
19. Marx, D., Schlotter, I.: Obtaining a planar graph by vertex deletion. Algorithmica **62**(3), 807–822 (2012)
20. Tyshkevich, R.I., Chernyak, A.A.: Yet another method of enumerating unmarked combinatorial objects. Math. Notes Acad. Sci. USSR **48**(6), 1239–1245 (1990)

The Temporal Explorer Who Returns to the Base

Eleni C. Akrida[1]([✉]), George B. Mertzios[2], and Paul G. Spirakis[1,3]

[1] Department of Computer Science, University of Liverpool, Liverpool, UK
{eleni.akrida,p.spirakis}@liverpool.ac.uk
[2] Department of Computer Science, Durham University, Durham, UK
george.mertzios@durham.ac.uk
[3] Computer Engineering and Informatics Department, University of Patras,
Patras, Greece

Abstract. In this paper we study the problem of exploring a temporal graph (i.e. a graph that changes over time), in the fundamental case where the underlying static graph is a star on n vertices. The aim of the exploration problem in a temporal star is to find a temporal walk which starts at the center of the star, visits all leaves, and eventually returns back to the center. We present here a systematic study of the computational complexity of this problem, depending on the number k of time-labels that every edge is allowed to have; that is, on the number k of time points where each edge can be present in the graph. To do so, we distinguish between the decision version STAREXP(k), asking whether a complete exploration of the instance exists, and the maximization version MAXSTAREXP(k) of the problem, asking for an exploration schedule of the greatest possible number of edges in the star. We fully characterize MAXSTAREXP(k) and show a dichotomy in terms of its complexity: on one hand, we show that for both $k = 2$ and $k = 3$, it can be efficiently solved in $O(n \log n)$ time; on the other hand, we show that it is APX-complete, for every $k \geq 4$ (does not admit a PTAS, unless P = NP, but admits a polynomial-time 1.582-approximation algorithm). We also partially characterize STAREXP(k) in terms of complexity: we show that it can be efficiently solved in $O(n \log n)$ time for $k \in \{2, 3\}$ (as a corollary of the solution to MAXSTAREXP(k), for $k \in \{2, 3\}$), but is NP-complete, for every $k \geq 6$.

1 Introduction and Motivation

A temporal graph is, roughly speaking, a graph that changes over time. Several networks, both modern and traditional, including social networks, transportation networks, information and communication networks, can be modeled as temporal graphs. The common characteristic in all the above examples is that the network structure, i.e. the underlying graph topology, is subject to discrete

Partially supported by the NeST initiative of the School of EEE and CS at the University of Liverpool and by the EPSRC Grants EP/P020372/1 and EP/P02002X/1.

P. Heggernes (Ed.): CIAC 2019, LNCS 11485, pp. 13–24, 2019.
https://doi.org/10.1007/978-3-030-17402-6_2

changes over time. Temporal graphs naturally model such time-varying networks using time-labels on the edges of a graph to indicate moments of existence of those edges, while the vertex set remains unchanged. This formalism originates in the foundational work of Kempe et al. [15].

In this work, we focus in particular on temporal graphs where the underlying graph is a star graph and we consider the problem of exploring such a temporal graph starting and finishing at the center of the star. The motivation behind this is inspired from the well known Traveling Salesperson Problem (TSP). The latter asks the following question: "Given a list of cities and the distances between each pair of cities, what is the shortest possible route that visits each city and returns to the origin one?". In other words, given an undirected graph with edge weights where vertices represent cities and edges represent the corresponding distances, find a minimum-cost Hamiltonian cycle. However, what happens when the traveling salesperson has particular temporal constraints that need to be satisfied, e.g. (s)he can only go from city A to city B on Mondays or Tuesdays, or (s)he can only travel by train and, hence, needs to schedule his/her visit based on the train timetables? In particular, consider a traveling salesperson who, starting from his/her home town, has to visit $n-1$ other towns via train, always *returning to his/her own home town* after visiting each city. There are trains between each town and the home town only on specific times/days, possibly different for different towns, and the salesperson knows those times in advance. Can the salesperson decide whether (s)he can visit all towns and return to his/her own home town by a certain day?

Previous Work. Recent years have seen a growing interest in dynamic network studies. Due to its vast applicability in many areas, the notion of temporal graphs has been studied from different perspectives under various names such as time-varying [1], evolving [10], dynamic [8,11,23,24]; for a recent attempt to integrate existing models, concepts, and results from the distributed computing perspective see the survey papers [7] and the references therein. Various temporal analogues of known static graph concepts have also been studied in [2,3,5,13, 17,22].

Notably, temporal graph exploration has been studied before; Erlebach et al. [12] define the problem of computing a foremost exploration of all vertices in a temporal graph (TEXP), without the requirement of returning to the starting vertex. They show that it is NP-hard to approximate TEXP with ratio $O(n^{1-\varepsilon})$ for any $\varepsilon > 0$, and give explicit construction of graphs that need $\Theta(n^2)$ steps for TEXP. They also consider special classes of underlying graphs, such as the grid, as well as the case of random temporal graphs where edges appear in every step with independent probabilities. Michail and Spirakis [18] study a temporal analogue of TSP(1,2) where the objective is to explore the vertices of a complete directed temporal graph with edge weights from $\{1,2\}$ with the minimum total cost. Ilcinkas et al. [14] study the exploration of constantly connected dynamic graphs on an underlying cactus graph. Bodlaender and van der Zanden [6] show that exploring temporal graphs of small pathwidth is NP-complete; they start from the problem that we define in this paper[1], which we prove is NP-complete, and give a reduction to the problem of exploring temporal graphs of small pathwidth.

We focus here on the exploration of temporal stars, inspired by the TRAVELING SALESPERSON PROBLEM (TSP) where the salesperson returns to his/her base after visiting every city. TSP is one of the most well-known combinatorial optimization problems, which still poses great challenges despite having been intensively studied for more than sixty years.

The Model and Definitions. It is generally accepted to describe a network topology using a graph, the vertices and edges of which represent the communicating entities and the communication opportunities between them, respectively. Unless otherwise stated, we denote by n and m the number of vertices and edges of the graph, respectively. We consider graphs whose edge availabilities are described by sets of positive integers (labels), one set per edge.

Definition 1 (Temporal Graph). *Let $G = (V, E)$ be a graph. A temporal graph on G is a pair (G, L), where $L : E \to 2^{\mathbb{N}}$ is a time-labeling function, called a labeling of G, which assigns to every edge of G a set of discrete-time labels. The labels of an edge are the discrete time instances ("days") at which it is available.*

More specifically, we focus on temporal graphs whose underlying graph is an undirected star, i.e. a connected graph of $m = n - 1$ edges which has $n - 1$ leaves, i.e. vertices of degree 1.

Definition 2 (Temporal Star). *A temporal star is a temporal graph (G_s, L) on a star graph $G_s = (V, E)$. Henceforth, we denote by c the center of G_s, i.e. the vertex of degree $n - 1$.*

Definition 3 (Time edge). *Let $e = \{u, v\}$ be an edge of the underlying graph of a temporal graph and consider a label $l \in L(e)$. The ordered triplet (u, v, l) is called* time edge.[2]

A basic assumption that we follow here is that when a message or an entity passes through an available link at time (day) t, then it can pass through a subsequent link only at some time (day) $t' > t$ and only at a time at which that link is available.

Definition 4 (Journey). *A temporal path or journey j from a vertex u to a vertex v $((u, v)$-journey) is a sequence of time edges (u, u_1, l_1), (u_1, u_2, l_2), ..., (u_{k-1}, v, l_k), such that $l_i < l_{i+1}$, for each $1 \le i \le k - 1$. We call the last time label, l_k, arrival time of the journey.*

---

[1] A preliminary version of this paper appeared publicly in ArXiv on 12th May 2018 (https://arxiv.org/pdf/1805.04713.pdf).

[2] Note that an undirected edge $e = \{u, v\}$ is associated with $2 \cdot |L(e)|$ time edges, namely both (u, v, l) and (v, u, l) for every $l \in L(e)$.

Given a temporal star (G_s, L), on the one hand we investigate the complexity of deciding whether G_s is *explorable*: we say that (G_s, L) is explorable if there is a journey starting and ending at the center of G_s that visits every node of G_s. Equivalently, we say that there is an *exploration* that *visits* every node, and *explores* every edge, of G_s; an edge of G_s is explored by crossing it from the center to the leaf at some time t and then from the leaf to the center at some time $t' > t$. On the other hand, we investigate the complexity of computing an exploration schedule that explores the greatest number of edges. A (partial) exploration of a temporal star is a journey J that starts and ends at the center of G_s which visits some nodes of G_s; its size $|J|$ is the number of nodes of G_s that are visited by J, where the centre is only accounted for once even if it is visited multiple times. We, therefore, identify the following problems:

STARExp(k)

Input: A temporal star (G_s, L) such that every edge has at most k labels.
Question: Is (G_s, L) explorable?

MAXSTARExp(k)

Input: A temporal star (G_s, L) such that every edge has at most k labels.
Output: A (partial) exploration of (G_s, L) of *maximum* size.

Note that the case where one edge e of the input temporal star has only one label is degenerate. Indeed, in the decision variant (i.e. STARExp(k)) we can immediately conclude that (G_s, L) is a no-instance as this edge cannot be explored; similarly, in the maximization version (i.e. MAXSTARExp(k)) we can just ignore edge e for the same reason. We say that we "enter" an edge $e = \{c, v\}$ of (G_s, L) when we cross the edge from c to v at a time on which the edge is available. We say that we "exit" e when we cross it from v to c at a time on which the edge is available. Without loss of generality we can assume that, in an exploration of (G_s, L), the entry to any edge e is followed by the exit from e at the earliest possible time (after the entry). That is, if the labels of an edge e are $l_1, l_2, \ldots, l_k$ and we enter e at time l_i, we exit at time l_{i+1}. The reason is that, waiting at a leaf (instead of exiting as soon as possible) does not help in exploring more edges; we are better off returning to the center c as soon as possible.

In order to solve the problem of exploring as many edges of a temporal star as possible, we define here the JOB INTERVAL SELECTION PROBLEM where each job has at most k associated intervals (JISP(k)), $k \geq 1$.

JOB INTERVAL SELECTION PROBLEM - JISP(k) [21]

Input: n jobs, each described as a set of at most k intervals on the real line.
Output: A schedule that executes as many jobs as possible; to execute a job one needs to select one interval associated with the job.

Notice that every edge e with labels $l_1, l_2, \ldots, l_k$ can be seen as a job to be scheduled where the corresponding intervals are $[l_1, l_2], [l_2, l_3], \ldots, [l_{k-1}, l_k]$, hence MAXSTAREXP($k$) is a special case of JISP($k - 1$); in the general JISP($k - 1$), the intervals associated with each job are not necessarily consecutive. JISP(k) is a well-studied problem in the Scheduling community, with several known complexity results. In particular, Spieksma [21] showed a 2-approximation for the problem, later improved to a 1.582-approximation by Chuzhoy et al. [9]. This immediately implies a 1.582-approximation algorithm for MAXSTAREXP(k); we use the latter to conclude on the APX-completeness[3] of MAXSTAREXP(k). JISP(k) was also shown [21] to be APX-hard for any $k \geq 2$, but since MAXSTAREXP(k) is a special case of JISP($k-1$), its hardness does not follow from the already known results. In fact, we show that MAXSTAREXP(3)- which is a special case of JISP(2)- is polynomially solvable.

Our Contribution. In this paper we do a systematic study of the computational complexity landscape of the temporal star exploration problems STAREXP(k) and MAXSTAREXP(k), depending on the maximum number k of labels allowed per edge. As a warm-up, we first prove in Sect. 2 that the maximization problem MAXSTAREXP(2) and MAXSTAREXP(3), i.e. when every edge has at most three labels per edge, can be efficiently solved in $O(n \log n)$ time; sorting the labels of the edges is the dominant part in the running time.

In Sect. 3 we prove that, for every $k \geq 6$, the decision problem STAREXP(k) is NP-complete and, for every $k \geq 4$, the maximization problem MAXSTAREXP(k) is APX-hard, and thus it does not admit a Polynomial-Time Approximation Scheme (PTAS), unless P = NP. These results are proved by reductions from special cases of the satisfiability problem, namely 3SAT(3) and MAX2SAT(3). The APX-hardness result is complemented by a 1.582-approximation algorithm for MAXSTAREXP(k) for any k, which concludes that MAXSTAREXP(k) is APX-complete for $k \geq 4$. This approximation algorithm carries over from an approximation for the JOB INTERVAL SELECTION PROBLEM [9], which we show is generalization of MAXSTAREXP(k).

The table below summarizes the results presented in this paper regarding the complexity of the two studied problems, shows the clear dichotomy in the complexity of MAXSTAREXP(k), as well as the open problem regarding the complexity of STAREXP(k) for $k \in \{4, 5\}$. The entry NP-c (resp. APX-c) denotes NP-completeness (resp. APX-completeness). Where $k = 1$, any instance of either problem is clearly a NO-instance, since one can explore no edge (i.e. by also returning to the center) with a single label:

	Maximum number of labels per edge					
	$k = 1$	$k = 2$	$k = 3$	$k = 4$	$k = 5$	$k \geq 6$
STAREXP(k)	No	$O(n \log n)$	$O(n \log n)$	?	?	NP-c
MAXSTAREXP(k)	No	$O(n \log n)$	$O(n \log n)$	APX-c	APX-c	APX-c

---

[3] APX is the complexity class of optimization problems that allow constant-factor approximation algorithms.

2   Efficient Algorithm for $k \leq 3$ Labels per Edge

In this section we show that, when every edge has two or three labels, a maximum size exploration in (G_s, L) can be efficiently solved in $O(n \log n)$ time. Thus, clearly, the decision variation of the problem, i.e. STARExp(2) and STARExp(3), can also be solved within the same time bound. We give here the proof for $k = 3$ labels, which also covers the case of $k = 2$.

Theorem 1. MAXSTARExp(3) *can be solved in* $O(n \log n)$ *time.*

Proof. We show that MAXSTARExp(3) is reducible to the Interval Scheduling Maximization Problem (ISMP).

Interval Scheduling Maximization Problem (ISMP)

Input: A set of intervals, each with a start and a finish time.
Output: Find a set of non-overlapping intervals of maximum size.

Given (G_s, L) we construct a set I of at most $2(n-1)$ intervals as follows:

All edges of (G_s, L) with a single label can be ignored as they can not be explored in any exploration of (G_s, L); for every edge e of (G_s, L) with labels $l_e < l'_e$ we create a single *closed* time interval, $[l_e, l'_e]$; for every edge e of (G_s, L) with labels $l_e < l'_e < l''_e$ we create two *closed* time intervals, $[l_e, l'_e]$ and $[l'_e, l''_e]$.

We may now compute a maximum size subset I' of I of non-conflicting (i.e., disjoint) time intervals, using the greedy algorithm that can find an optimal solution for ISMP [16]. It suffices to observe that no two intervals associated with the same edge will ever be selected in I', as any two such intervals are non-disjoint; indeed, two intervals associated with the same edge e are of the form $[l_e, l'_e]$ and $[l'_e, l''_e]$, hence they overlap at the single time point l'_e.

So a maximum-size set I' of non-overlapping intervals corresponds to a maximum-size exploration of (G_s, L) (in fact, of the same size as the size of I'). Also, we may indeed solve STARExp(3) by checking whether $|I'| = n - 1$ or not. The above works in $O(n \log n)$ time [16].   □

3   Hardness for $k \geq 4$ Labels per Edge

In this section we show that, whenever $k \geq 6$, STARExp(k) is NP-complete. Furthermore, we show that MAXSTARExp(k) is APX-hard for $k \geq 4$. Thus, in particular, MAXSTARExp(k) does not admit a Polynomial-Time Approximation Scheme (PTAS), unless P = NP. In fact, due to a known polynomial-time constant-factor approximation algorithm for JISP(k) [9], it follows that MAXSTARExp(k) is also APX-complete.

3.1   StarExp(k) is NP-complete for $k \geq 6$ Labels per Edge

We prove our NP-completeness result through a reduction from a special case of 3SAT, namely 3SAT(3), which is known to be NP-complete [19].

3SAT(3)

Input: A boolean formula in CNF with variables $x_1, x_2, \ldots, x_p$ and clauses $c_1, c_2, \ldots, c_q$, such that each clause has at most 3 literals, and each variable appears in at most 3 clauses.

Output: Decision on whether the formula is satisfiable.

Intuition and Overview of the Reduction: Given an instance F of 3SAT(3), we shall create an instance (G_s, L) of STAREXP(k) such that F is satisfiable if and only if (G_s, L) is explorable. Henceforth, we denote by $|\tau(F)|$ the number of clauses of F that are satisfied by a truth assignment τ of F. Without loss of generality we make the following assumptions on F. Firstly, if a variable occurs only with positive (resp. only with negative) literals, then we trivially set it to *true* (resp. *false*) and remove the associated clauses. Furthermore, without loss of generality, if a variable x_i appears three times in F, we assume that it appears once as a negative literal $\neg x_i$ and two times as a positive literal x_i; otherwise we rename the negation with a new variable. Similarly, if x_i appears two times in F, then it appears once as a negative literal $\neg x_i$ and once as a positive literal x_i.

We introduce here the intuition behind the reduction. (G_s, L) will have one edge corresponding to each clause of F, and three edges (one "primary" and two "auxiliary" edges) corresponding to each variable of F. We shall assign labels in pairs to those edges so that it is possible to explore an edge only by using labels from the same pair to enter and exit the edge; for example, if an edge e is assigned the pairs of labels l_1, l_2 and l_3, l_4, with $l_1 < l_2 < l_3 < l_4$, we shall ensure that one cannot enter e with, say, label l_2 and exit with, say, label l_3. In particular, for the "primary" edge corresponding to a variable x_i we will assign to it two pairs of labels, namely $(\alpha i - \beta, \alpha i - \beta + \gamma)$ and $(\alpha i + \beta, \alpha i + \beta + \gamma)$, for some $\alpha, \beta, \gamma \in \mathbb{N}$. The first (entry, exit) pair corresponds to setting x_i to false, while the second pair corresponds to setting x_i to true. We shall choose α, β, γ so that the entry and exit from the edge using the first pair is not conflicting with the entry and exit using the second pair.

Then, to any edge corresponding to a clause c_j that contains x_i unnegated for the first time (resp. second time[4]), we shall assign an (entry, exit) pair of labels $(\alpha i - \delta, \alpha i - \delta + \varepsilon)$ (resp. $(\alpha i - \delta', \alpha i - \delta' + \varepsilon')$), choosing $\delta, \varepsilon \in \mathbb{N}$ (resp. $\delta', \varepsilon' \in \mathbb{N}$) so that $(\alpha i - \delta, \alpha i - \delta + \varepsilon)$ (resp. $(\alpha i - \delta', \alpha i - \delta' + \varepsilon')$) is in conflict with the $(\alpha i - \beta, \alpha i - \beta + \gamma)$ pair of labels of the edge corresponding to x_i, which is associated with $x_i = false$ but not in conflict with the $(\alpha i + \beta, \alpha i + \beta + \gamma)$ pair. If x_i is false in F then c_j cannot be satisfied through x_i so we should not be able to explore a corresponding edge via a pair of labels associated with x_i. If c_j contains x_i negated, we shall assign to its corresponding edge an (entry, exit) pair of labels $(\alpha i + \zeta, \alpha i + \zeta + \theta)$, choosing $\zeta, \theta \in \mathbb{N}$ so that the latter is in conflict with the $(\alpha i + \beta, \alpha i + \beta + \gamma)$ pair of labels of the edge corresponding to x_i, which is associated with $x_i = true$ but not in conflict with the $(\alpha i - \beta, \alpha i - \beta + \gamma)$ pair.

---

[4] We consider here the order $c_1, c_2, \ldots, c_q$ of the clauses of C; we say that x_i appears unnegated for the *first* time in some clause c_μ if $x_i \notin c_m$, $m < \mu$.

If x_i is true in F then c_j cannot be satisfied through $\neg x_i$ so we should not be able to explore a corresponding edge via a pair of labels associated with $\neg x_i$.

Finally, for every variable x_i we also introduce two additional "auxiliary" edges: the first one will be assigned the pair of labels $(\alpha i, \alpha i + \xi)$, $\xi \in \mathbb{N}$, so that it is not conflicting with any of the above pairs – the reason for introducing this first auxiliary edge is to avoid entering and exiting an edge corresponding to some variable x_i using labels from different pairs. The second auxiliary edge for variable x_i will be assigned the pair of labels $(\alpha i + \chi, \alpha i + \chi + \psi)$, $\chi, \psi \in \mathbb{N}$, so that it is not conflicting with any of the above pairs – the reason for introducing this edge is to avoid entering an edge that corresponds to some clause c_j using a label associated with some variable x_i and exiting using a label associated with a *different* variable $x_{i'}$. The reader is referred to Fig. 1 for an example construction, where the specific choices of the constants $\alpha, \beta, \gamma, \delta, \varepsilon, \delta', \varepsilon', \zeta, \theta, \xi, \chi, \psi$ are $\alpha = 50, \beta = 10, \gamma = 3, \delta = 12, \varepsilon = 3, \delta' = 8, \varepsilon' = 3, \zeta = 8, \theta = 3, \xi = 1, \chi = 15, \psi = 1$.

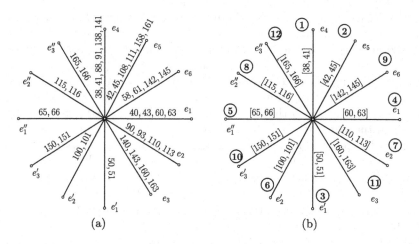

Fig. 1. The temporal star constructed for the formula $(x_1 \lor x_2 \lor x_3) \land (x_1 \lor \neg x_2 \lor \neg x_3) \land (\neg x_1 \lor x_3)$. Setting x_1 to true, x_2 to true and x_3 to true yields a satisfying truth assignment whose corresponding exploration is indicated in (b), where the numbers in the circles indicate the order over time of the exploration of each edge.

The following lemmas are needed for the NP-completeness proof (Theorem 2).

Lemma 1. *There exists a (partial) exploration J of (G_s, L) of maximum size which explores all $3p$ edges associated with the variables of F.*

Lemma 2. *There exists a truth assignment τ of F with $|\tau(F)| \geq \beta$ if and only if there exists a (partial) exploration J of (G_s, L) of size $|J| \geq 3p + \beta$.*

Theorem 2. STAREXP(k) *is NP-complete for every $k \geq 6$.*

3.2 MaxStarExp(k) is APX-complete for $k \geq 4$ Labels per Edge

It can be shown that the reduction of Sect. 3.1 linearly preserves approximability features; this would in turn prove that MAXSTAREXP(k) is APX-hard for $k \geq 6$, since MAX3SAT(3), i.e. the maximization version of 3SAT(3), is APX-complete [4]. However, this leaves a gap in the complexity of the problem for $k \in \{4, 5\}$. To close this gap we instead give an *L-reduction* [20] from the MAX2SAT(3) problem, i.e. an approximation preserving reduction which linearly preserves approximability features. MAX2SAT(3) is known to be APX-complete [20].

MAX2SAT(3)

Input: A boolean formula in CNF with variables $x_1, x_2, \ldots, x_p$ and clauses $c_1, c_2, \ldots, c_q$, such that each clause has at most 2 literals, and each variable appears in at most 3 clauses.
Output: Maximum number of satisfiable clauses in the formula.

The reduction: Given an instance F of MAX2SAT(3) we shall create an instance (G_s, L) of MAXSTAREXP(k) such that F has β satisfiable clauses if and only if (G_s, L) has $\beta + 3p$ explorable edges. As previously, we assume without loss of generality that every variable appears once as a negative literal and once or twice as a positive literal.

The reduction is the same as the one presented in Sect. 3.1, with the edges of (G_s, L) being assigned the same labels as in the previous reduction to appropriately introduce conflicts between exploration windows of edges. The only difference in the construction is that now we start from a 2-CNF formula F (instead of a 3-CNF formula in Sect. 3.1). Thus every edge of (G_s, L) that corresponds to a clause of F now receives four labels instead of six, i.e. two labels for every literal that appears in the clause.

The following lemmas are needed for the APX-hardness proof (Theorem 3).

Lemma 3. *There exists a (partial) exploration J of (G_s, L) of maximum size which explores all $3p$ edges associated with the variables of F.*

Lemma 4. *There exists a truth assignment τ of F with $|\tau(F)| \geq \beta$ if and only if there exists a (partial) exploration J of (G_s, L) of size $|J| \geq 3p + \beta$.*

Theorem 3. MAXSTAREXP(k) *is APX-hard, for $k \geq 4$.*

Proof. Denote by $OPT_{\mathrm{MAX2SAT(3)}}(F)$ the greatest number of clauses that can be simultaneously satisfied by a truth assignment of F. The proof is done by an *L-reduction* [20] from the MAX2SAT(3) problem, i.e. by an approximation preserving reduction which linearly preserves approximability features. For such a reduction, it suffices to provide a polynomial-time computable function g and two constants $\gamma, \delta > 0$ such that:

- $OPT_{\mathrm{MAXSTAREXP}}((G_s, L)) \leq \gamma \cdot OPT_{\mathrm{MAX2SAT(3)}}(F)$, for any boolean formula F, and

- for any (partial) exploration J' of (G_s, L), $g(J')$ is a truth assignment for F and $OPT_{\text{Max2SAT}(3)}(F) - |g(J')| \leq \delta(OPT_{\text{MaxStarExp}}((G_s, L)) - |J'|)$, where $|g(J')|$ is the number of clauses of F that are satisfied by $g(J')$.

We will prove the first condition for $\gamma = 13$. Recall that p and q are the numbers of variables and clauses of F, respectively. Note that a random truth assignment satisfies each clause of F with probability at least $\frac{1}{2}$ (if each clause had exactly 2 literals, then it would be satisfied with probability $\frac{3}{4}$, but we have to account also for single-literal clauses), and thus there exists an assignment τ that satisfies at least $\frac{q}{2}$ clauses of F. Furthermore, since every clause has at most 2 literals and every variable appears at least once, it follows that $q \geq \frac{p}{2}$. Therefore $OPT_{\text{Max2SAT}(3)}(F) \geq \frac{q}{2} \geq \frac{p}{4}$, and thus $p \leq 4 \cdot OPT_{\text{Max2SAT}(3)}(F)$. Now Lemma 4 implies that:

$$
\begin{aligned}
OPT_{\text{MaxStarExp}}((G_s, L)) &= 3p + OPT_{MAX2SAT(3)}(F) \\
&\leq 3 \cdot 4 \cdot OPT_{MAX2SAT(3)}(F) + OPT_{MAX2SAT(3)}(F) \\
&= 13 \cdot OPT_{\text{Max2SAT}(3)}(F)
\end{aligned}
$$

To prove the second condition for $\delta = 1$, consider an arbitrary partial exploration J' of $G_s(L)$ of maximum size. In the proof of Lemma 4, we describe how one can start from any such J' and construct in polynomial time a truth assignment $g(J') = \tau$ that satisfies at least $OPT_{\text{MaxStarExp}}((G_s, L)) - 3p$ clauses of F, i.e. $|g(J')| = |\tau(F)| \geq |J'| - 3p$. Then:

$$
\begin{aligned}
OPT_{\text{Max2SAT}(3)}(F) - |g(J')| &\leq OPT_{\text{Max2SAT}(3)}(F) - |J'| + 3p \\
&= OPT_{\text{MaxStarExp}}((G_s, L)) - 3p - |J'| + 3p \\
&= OPT_{\text{MaxStarExp}}((G_s, L)) - |J'|
\end{aligned}
$$

This completes the proof of the theorem.   □

Corollary 1. MaxStarExp(k) is APX-complete, for $k \geq 4$.

Now we prove a correlation between the inapproximability bounds for the MaxStarExp(k) problem and Max2SAT(3), as a result of the L-reduction presented in the proof of Theorem 3. Note that, since Max2SAT(3) is APX-hard [4], there exists a constant $\varepsilon_0 > 0$ such that there exists no polynomial-time constant-factor approximation algorithm for Max2SAT(3) with approximation ratio greater than $(1 - \varepsilon_0)$, unless P = NP.

Theorem 4. Let $\varepsilon_0 > 0$ be the constant such that, unless $P = NP$, there exists no polynomial-time constant-factor approximation algorithm for Max2SAT(3) with approximation ratio greater than $(1 - \varepsilon_0)$. Then, unless $P = NP$, there exists no polynomial-time constant-factor approximation algorithm for MaxStarExp(k) with approximation ratio greater than $(1 - \frac{\varepsilon_0}{13})$.

Proof. Let $\varepsilon > 0$ be a constant such that there exists a polynomial-time approximation algorithm $\mathcal{A}$ for MAXSTAREXP(k) with ratio $(1-\varepsilon)$. Let F be an instance of MAX2SAT(3) with p variables and q clauses. We construct the instance (G_s, L) of MAXSTAREXP(k) corresponding to F, as described in the L-reduction (see Theorem 3). Then we apply the approximation algorithm $\mathcal{A}$ to (G_s, L), which returns a (partial) exploration J. Note that $|J| \geq (1 - \varepsilon) \cdot OPT_{\text{MAXSTAREXP}}$. We construct from J in polynomial time a truth assignment τ; we denote by $|\tau|$ the number of clauses in F that are satisfied by the truth assignment τ. It now follows from the proof of Theorem 3 that:

$$OPT_{\text{MAX2SAT(3)}}(F) - |\tau| \leq OPT_{\text{MAXSTAREXP}}((G_s, L)) - |J|$$
$$\leq 13\varepsilon \cdot OPT_{\text{MAX2SAT(3)}}(F)$$

Therefore $|\tau| \geq (1 - 13\varepsilon) \cdot OPT_{\text{MAX2SAT(3)}}(F)$. That is, using algorithm $\mathcal{A}$, we can devise a polynomial-time algorithm for MAX2SAT(3) with approximation ratio $(1 - 13\varepsilon)$. Therefore, due to the assumptions of the theorem it follows that $\varepsilon \geq \frac{\varepsilon_0}{13}$, unless P = NP. This completes the proof of the theorem. $\square$

Note that we have fully characterized MAXSTAREXP(k) in terms of complexity, for all values of $k \in \mathbb{N}$. However, the reduction that shows APX-hardness for MAXSTAREXP(k) cannot be employed to show NP-hardness of the decision version STAREXP(k), since the decision problem 2SAT is polynomially solvable.

Open Problem. *What is the complexity of* STAREXP(k), *for* $k \in \{4, 5\}$?

References

1. Aaron, E., Krizanc, D., Meyerson, E.: DMVP: foremost waypoint coverage of time-varying graphs. In: International Workshop on Graph-Theoretic Concepts in Computer Science (WG), pp. 29–41 (2014)
2. Akrida, E.C., Gasieniec, L., Mertzios, G.B., Spirakis, P.G.: The complexity of optimal design of temporally connected graphs. Theory Comput. Syst. **61**(3), 907–944 (2017)
3. Akrida, E.C., Mertzios, G., Spirakis, P.G., Zamaraev, V.: Temporal vertex covers and sliding time windows. In: International Colloquium on Automata, Languages and Programming (ICALP) (2018)
4. Ausiello, G., Protasi, M., Marchetti-Spaccamela, A., Gambosi, G., Crescenzi, P., Kann, V.: Complexity and Approximation: Combinatorial Optimization Problems and Their Approximability Properties. Springer, Heidelberg (1999). https://doi.org/10.1007/978-3-642-58412-1
5. Biswas, S., Ganguly, A., Shah, R.: Restricted shortest path in temporal graphs. In: Chen, Q., Hameurlain, A., Toumani, F., Wagner, R., Decker, H. (eds.) DEXA 2015. LNCS, vol. 9261, pp. 13–27. Springer, Cham (2015). https://doi.org/10.1007/978-3-319-22849-5_2
6. Bodlaender, H.L., van der Zanden, T.C.: On exploring temporal graphs of small pathwidth. CoRR, abs/1807.11869 (2018)
7. Casteigts, A., Flocchini, P.: Deterministic algorithms in dynamic networks: formal models and metrics. Technical report, Defence R&D Canada, April 2013

8. Chan, T.-H.H., Ning, L.: Fast convergence for consensus in dynamic networks. ACM Trans. Algorithms **10**, 15-1 (2014)
9. Chuzhoy, J., Ostrovsky, R., Rabani, Y.: Approximation algorithms for the job interval selection problem and related scheduling problems. Math. Oper. Res. **31**, 730–738 (2006)
10. Clementi, A.E.F., Macci, C., Monti, A., Pasquale, F., Silvestri, R.: Flooding time of edge-Markovian evolving graphs. SIAM J. Discrete Math. (SIDMA) **24**(4), 1694–1712 (2010)
11. Demetrescu, C., Italiano, G.F.: Algorithmic techniques for maintaining shortest routes in dynamic networks. Electron. Notes Theor. Comput. Sci. **171**, 3–15 (2007)
12. Erlebach, T., Hoffmann, M., Kammer, F.: On temporal graph exploration. In: Halldórsson, M.M., Iwama, K., Kobayashi, N., Speckmann, B. (eds.) ICALP 2015. LNCS, vol. 9134, pp. 444–455. Springer, Heidelberg (2015). https://doi.org/10.1007/978-3-662-47672-7_36
13. Himmel, A.-S., Molter, H., Niedermeier, R., Sorge, M.: Adapting the Bron-Kerbosch algorithm for enumerating maximal cliques in temporal graphs. Soc. Netw. Anal. Min. **7**(1), 35:1–35:16 (2017)
14. Ilcinkas, D., Klasing, R., Wade, A.M.: Exploration of constantly connected dynamic graphs based on cactuses. In: Halldórsson, M.M. (ed.) SIROCCO 2014. LNCS, vol. 8576, pp. 250–262. Springer, Cham (2014). https://doi.org/10.1007/978-3-319-09620-9_20
15. Kempe, D., Kleinberg, J.M., Kumar, A.: Connectivity and inference problems for temporal networks. In: ACM Symposium on Theory of Computing (STOC), pp. 504–513 (2000)
16. Kleinberg, J., Tardos, E.: Algorithm Design. Addison-Wesley Longman, Boston (2005)
17. Mertzios, G.B., Michail, O., Chatzigiannakis, I., Spirakis, P.G.: Temporal network optimization subject to connectivity constraints. In: Fomin, F.V., Freivalds, R., Kwiatkowska, M., Peleg, D. (eds.) ICALP 2013. LNCS, vol. 7966, pp. 657–668. Springer, Heidelberg (2013). https://doi.org/10.1007/978-3-642-39212-2_57
18. Michail, O., Spirakis, P.G.: Traveling salesman problems in temporal graphs. Theor. Comput. Sci. **634**, 1–23 (2016)
19. Papadimitriou, C.H., Steiglitz, K.: Combinatorial Optimization: Algorithms and Complexity. Prentice-Hall, Upper Saddle River (1982)
20. Papadimitriou, C.H., Yannakakis, M.: Optimization, approximation, and complexity classes. J. Comput. Syst. Sci. **43**(3), 425–440 (1991)
21. Spieksma, F.C.R.: On the approximability of an interval scheduling problem. J. Sched. **2**, 215–227 (1999)
22. Viard, T., Latapy, M., Magnien, C.: Computing maximal cliques in link streams. Theor. Comput. Sci. **609**, 245–252 (2016)
23. Wagner, D., Willhalm, T., Zaroliagis, C.D.: Dynamic shortest paths containers. Electron. Notes Theor. Comput. Sci. **92**, 65–84 (2004)
24. Zhuang, H., Sun, Y., Tang, J., Zhang, J., Sun, X.: Influence maximization in dynamic social networks. In: International Conference on Data Mining (2013)

Minimum Convex Partition of Point Sets

Allan S. Barboza, Cid C. de Souza$^{(\boxtimes)}$, and Pedro J. de Rezende

Institute of Computing, University of Campinas, Campinas, Brazil
allansapucaia@gmail.com, {cid,rezende}@ic.unicamp.br

Abstract. A convex partition of a point set P in the plane is a planar subdivision of the convex hull of P whose edges are segments with both endpoints in P and such that all internal faces are empty convex polygons. In the Minimum Convex Partition Problem (MCPP) one seeks to find a convex partition with the least number of faces. The complexity of the problem is still open and so far no computational tests have been reported. In this paper, we formulate the MCPP as an integer program that is used both to solve the problem exactly and to design heuristics. Thorough experiments are conducted to compare these algorithms in terms of solution quality and runtime, showing that the duality gap is decidedly small and grows quite slowly with the instance size.

1 Introduction

Let P be a set of n points in the plane in general position, i.e., with no three points being collinear. We say that a simple polygon is *empty*, w.r.t. P, if it contains no points of P in its interior. Denote by $H(P)$ the convex hull of P. A convex *partition* (or *decomposition*) of P is a planar subdivision of $H(P)$ into non overlapping empty convex polygons whose vertices are the points of P. The Minimum Convex Partition Problem (MCPP) asks to find a convex partition of P minimizing the number of faces. These concepts are illustrated in Fig. 1.

A practical application of the MCPP in the area of network design is described in [7]. The goal is to form a communication network connecting the points of P. When the edges used as links in the network form a convex partition of P, a simple randomized algorithm can be used for routing packages [2]. Hence, one way to build a low-cost network, i.e., one with few links that still enables the application of that routing algorithm, is to solve the MCPP for P.

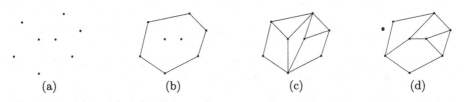

Fig. 1. (a) Point set P; (b) $H(P)$; (c) minimal and (d) optimal partition.

P. Heggernes (Ed.): CIAC 2019, LNCS 11485, pp. 25–37, 2019.
https://doi.org/10.1007/978-3-030-17402-6_3

Besides the actual application outlined above, the MCPP lies in the broader context of polygon decomposition, an important topic of study in computational geometry [6]. A frequently used approach for designing divide-and-conquer algorithms for geometric problems for general polygons is to decompose these into simpler component parts, solve the original problem on each component using a specialized algorithm and then combine the partial solutions. Convex polygons are often the best choice for the role of the smaller components. This is, in fact, a reason why triangulations of sets of points, and of polygons in general, have been so extensively studied and became a central problem in computational geometry. Since triangulations are special cases of convex partitions, a deeper understanding of the MCPP gains importance.

Literature Overview. To the best of our knowledge, the complexity of the MCPP remains unknown. Moreover, even though some articles describe non-polynomial algorithms to solve the problem, no implementations or results were found that make an empirical assessment of the efficiency of those algorithms.

Fevens et al. [3] proposed an exact algorithm for the MCPP using dynamic programming. If h denotes the onion depth of P, i.e. the number of nested convex hulls that need to be removed before P becomes empty [8], the complexity of this algorithm can be expressed as $O(n^{3h+3})$ and is, therefore, exponential in h, which can actually be as large as $\Theta(n)$. Spillner [10] designed another exact algorithm whose complexity is $O(2^k k^4 n^3 + n \log n)$, where $k = |I(P)|$ is the number of points of P in the interior of $H(P)$. Moreover, Spillner et al. [7] present a $\frac{30}{11}$-factor approximation algorithm for the MCPP.

Other papers investigate the MCPP in an attempt to find theoretical bounds for the optimal value. Let $Inst(n)$ be the set of all MCPP instances of size n. Denoting the optimal value of an instance $i \in Inst(n)$ by $\text{OPT}(i)$, define $F(n) = \max_{i \in Inst(n)} \text{OPT}(i)$, that is, $F(n)$ is the maximum among all optima for instances of size n. The best known lower bound for $F(n)$ is $\frac{12}{11}n - 2$ as shown in [4], for $n \geq 4$, whereas the best upper bound proven to date is $\lceil \frac{7(n-3)}{5} \rceil$, see [5]. It is shown in [9] that any minimal convex partition has at most $\frac{3(n-2)}{2}$ faces, which also serves as an upper bound on $F(n)$.

Our Contribution. In this paper, we introduce the first known integer linear programming formulation of the MCPP. Through extensive experimentation, we show that the solutions of the linear relaxation of this model provide invaluable information on which segments are likely to be in an optimal solution of the problem. From this observation, we derive a powerful heuristic for the MCPP that is capable of producing high quality solutions for instances with up to one hundred points in no more than a few minutes of computation on a currently standard machine. We hope that the present work will spearhead efforts of researchers in the field to publish computational evaluation of algorithms for the MCPP. To this end, all instances we used and their solutions, are made available [1].

Organization of the Text. The next sections of the paper are organized as follows. Section 2 describes the computational environment in which our tests were carried out. Section 3 presents an integer programming formulation for the

MCPP and reports on experiments with this model. Section 4 is devoted to the discussion of the heuristics developed in this work, whereas in Sect. 5 we analyze the results yielded by these algorithms. Section 6 contains a few conclusions and points out future research directions.

2   Experimental Environment

In the following sections, we propose exact and heuristic algorithms for the MCPP and report on their experimental evaluation. Since observations made from initial empirical results had an impact on the design of these algorithms, it is necessary to first introduce the computational environment where tests were carried out and to describe how the benchmark instances were created.

Software and Hardware. All experiments were run on an Intel Xeon E5-2420 at 1.9 GHz, and 32 GB of RAM running Ubuntu 14.04. The algorithms were implement in C++ 11 and compiled with GCC 7.2.0. Geometrical structures and procedure were implement using CGAL 4.2-5, library Gmpq was used to represent rational numbers exactly. To solve integer linear programs and relaxations, we used CPLEX 12.8.0, in single-thread mode, with default configurations, except for a time limit of 1200 s.

Instances. Instances consist of points whose x and y coordinates were randomly generated in the interval $[0, 1]$ according to a uniform distribution. For an instance of size n, points were created in sequence until n points in general position were included. To that end, when a newly spawned point resulted collinear with a previously generated pair the former was rejected, otherwise it was accepted.

Two sets of instances were generated, with different instance sizes. For each size, 30 instances were created. The set $T1$ is comprised of instances of sizes 30, 32, 34, . . . , 50. For each size, we chose the first 30 found to be optimally solvable within 20 min. The second set, $T2$, simply contains instances of sizes 55, 60, 65, . . . , 110 for which optimal solutions are still not known.

3   A Mathematical Model for the MCPP

In this section, we propose an Integer Linear Programming (ILP) formulation for the MCPP and discuss its correctness and performance on the randomly generated instances described in Sect. 2.

Model Description. Before describing the model itself, we need to introduce some terminology and notation. Recall that P denotes a set of n points in general position. The set of *internal points* of P, denoted $I(P)$, is the subset of P formed by the points that are not vertices of the convex hull of P, $H(P)$. Let S denote the set of $\Theta(n^2)$ line segments whose endpoints belong to P. Given a pair of segments $\overline{ij}, \overline{k\ell} \in S$, we say that $\overline{ij}$ and $\overline{k\ell}$ *cross* when $\overline{ij} \cap \overline{k\ell} \setminus \{i, j, k, \ell\} \neq \emptyset$.

Consider the complete (geometric) undirected graph $G = (P, E(P))$, induced by P, where $E(P) = \{\{i, j\} \mid \overline{ij} \in S\}$. We will refer to an edge $\{i, j\}$ of $E(P)$ and its corresponding segment $\overline{ij}$ in S interchangeably. We will also need to allude to the set of pairs of crossing edges (segments), denoted S^C. These pairs, $(\overline{ij}, \overline{k\ell}) \mid \overline{ij}$ and $\overline{k\ell}$ cross, may easily be identified in $O(n^4)$ time using simple geometric procedures. Similarly, a complete directed graph $\overrightarrow{G} = (P, A(P))$ can be defined, whose arcs correspond to the segments of S with either one of the two possible orientations: $(i, j) \in A(P)$ iff $\overline{ij} \in S$.

A few additional terminologies are needed before we can present the model and argue about its correctness. Given two points a, b in the plane, we denote by CCW(ab) (CW(ab)) the set of points c in the plane such that the triple abc is positively (negatively) oriented, i.e., $0° < \sphericalangle abc < 180°$ ($-180° < \sphericalangle abc < 0°$). Let a be a point in the plane and L be a list of non collinear segments, all sharing a as one of their endpoints. Assume that $|L| \geq 2$ and that its segments are given in a *clockwise circular order* around a. Point a is called *reflex with respect to L* if there are two consecutive segments in L, say ba and ca, so that $c \in$ CW(ba). Notice that this is equivalent to say that CCW(ac) contains no endpoints of segments in L. With respect to L, when a is not reflex, we call it *convex*.

Lastly, to each edge $\{i, j\} \in E(P)$ we associate a binary variable x_{ij} whose value is one iff $\{i, j\}$ is in the solution. In the formulation below, the unique variable associated to the segment $\overline{ij}$ is naturally referred to as x_{ij} and as x_{ji}. Accordingly, the proposed ILP model, referred to as BASIC, reads:

$$z = \min \sum_{\{i,j\} \in E(P)} x_{ij} \tag{1a}$$

$$\text{s.t.} \quad x_{ij} + x_{k\ell} \leq 1 \qquad\qquad \forall \{\{i, j\}, \{k, \ell\}\} \in S^c \tag{1b}$$

$$x_{ij} = 1 \qquad\qquad \forall \{i, j\} \in H(P) \tag{1c}$$

$$\sum_{k \in \text{CCW}(ij) \cap P} x_{ik} \geq 1 \qquad\qquad \forall (i, j) \in A(P), i \in I(P) \tag{1d}$$

$$\sum_{j \in P} x_{ij} \geq 3 \qquad\qquad \forall i \in I(P) \tag{1e}$$

$$x_{ij} \in \{0, 1\} \qquad\qquad \forall \{i, j\} \in E(P) \tag{1f}$$

The objective function (1a) can be expressed in terms of the number of edges in a solution because Euler's formula implies that minimizing the number of edges in a connected planar graph (subdivision) is equivalent to minimizing the number of faces of any planar embedding of that graph. That is, if f, e and v denote, respectively, the number of faces, edges and vertices of an embedding of a planar graph, then $f = e - v + 2$. Since we seek to build a planar connected subdivision and $v = |P|$ is given, minimizing e is equivalent to minimizing f.

Constraints (1b) guarantee planarity since they prevent that both edges of a crossing pair be included in a solution. Constraints (1c) establish that the edges of the convex hull of P are included in the solution. Constraints (1d) ensure that any point i in $I(P)$ is convex with respect to the set of segments that are in the

Fig. 2. An instance of the MCPP whose solution of the RELAX model without constraints (1e) has a vertex p with fractional degree 2.5. Dashed edges have value 0.5 and solid edges have value 1.

solution. As shown in Proposition 1, the degree constraints (1e) are redundant when we consider integer solutions and constraints (1d). However, we opted to keep them in the formulation because we verified that they improve the value of the relaxation. Figure 2 depicts an instance whose solution of the relaxation of the BASIC model without constraints (1e) has an internal point with *fractional* degree less than three. Finally, constraints (1f) require all variables to be binary.

Proposition 1. *Let x^L be the incidence vector of a subset L of segments in S satisfying all constraints (1d). Then, x^L also satisfies constraints (1e).*

Proof. Denote by $G(L)$ the subgraph induced by L in G. Due to constraints (1d), the degree in $G(L)$ of any point $i \in I(P)$ is at least one, meaning that, at the minimum, there is one segment $\overline{ia}$ in L for some $a \in P$. Now, taking $j = a$ in constraint (1d), we obtain that there must be another point b in CCW$(ia) \cap P$ for which $\overline{ib}$ is also in L. Hence, i has degree at least two. By construction, a must be in CW(ib). Together with constraint (1d) for (i, b), this implies that there must be a third point c in CCW$(ib) \cap P$ so that the segment $\overline{ic}$ is in L. □

Theorem 1. *The BASIC model is a correct formulation for the MCPP.*

Proof. First, we show that the incidence vector x^L of any set L of segments in S that corresponds to a feasible solution of the MCPP satisfies all the constraints of the BASIC model. By definition, x^L is a binary vector and hence, constraint (1f) holds. The feasibility of L implies planarity, so x^L satisfies all constraints (1b). Besides, as the segments in the convex hull of P are all present in any feasible solution, constraints (1c) are also verified by x^L. Proposition 1 implies the fulfillment of constraints (1e). It remains to prove that x^L satisfies constraints (1d) as well. Let L_i be the clockwise ordered subset of segments of L that are incident to $i \in I(P)$. The feasibility of L requires i to be convex with respect to L_i, or else i would be a reflex vertex of a face in the planar subdivision corresponding to L, contradicting the hypothesis that the set is a feasible solution for the MCPP. As a consequence, for any direction $\overrightarrow{u}$, the set CCW$(i(i + \overrightarrow{u}))^1$ contains at least one endpoint of a segment in L_i. Hence, for any $(i, j) \in A(P)$, at least one variable x_{ik}^L in the summation on the left of constraint (1d) has value one, ensuring that the inequality is satisfied by x^L.

---

[1] Here, $(i + \overrightarrow{u})$ denotes any point obtained by a translation of i in the direction $\overrightarrow{u}$.

We now focus on the proof that the set L associated to any incidence vector x^L satisfying the linear system (1b)–(1f) is a convex partition of P. Let $G(L)$ denote the subgraph of G induced by L. Because of (1b), L is planar and, due to (1c), it contains all the segments in $H(P)$. From Proposition 1, the degree of any point $i \in I(P)$ in $G(L)$ is at least three. For L_i defined as before, we claim that i is convex with respect to L_i. To see this, suppose by contradiction that i is reflex. So, w.l.o.g, there is a segment $\overline{ia}$ in L_i such that, for any b with $\overline{ib} \in L_i$, b is in $\mathrm{CW}(ia)$. This implies that there is no $k \in \mathrm{CCW}(ia) \cap P$ so that $\overline{ik}$ is in L. Consequently, x^L does not satisfy constraint (1d) for (i,a), contradicting the feasibility of x^L. We next show that $G(L)$ is connected.

By contradiction, suppose that $G(L)$ has two or more connected components. Since L contains the segments in $H(P)$, the points in at least one of these connected components must all be internal. Let i be the rightmost point in this component. Since i is not connected to any point to its right and has degree at least three, i must be reflex relative to L. But then, constraint (1d) for (i, j) is violated by x^L, where $\overline{ij}$ is the first segment visited when L_i is traversed starting from a vertical line that goes through i.

On the other hand, we can also prove that $G(L)$ contains no articulation points (and hence, no bridges) by applying arguments similar to those employed above. It then follows that each segment in $L \backslash H(P)$ is incident to exactly two (distinct) faces of the planar subdivision defined by L and that all faces in this subdivision are empty polygons. Moreover, as all points in $I(P)$ are convex w.r.t. L, these polygons are convex. The minimization condition is ensured by (1a). $\square$

Empirical Evaluation. We now describe the empirical evaluation of the BASIC model for the benchmark set $T1$. Recall that to construct the 330 instances in $T1$, we sought to have only instances for which an optimal solution could be found. To achieve that, a sequence of instances was generated for each of the 11 intended sizes and we attempted to solve them to optimality within the time limit of 20 min of computation. Those instances for which this process was successful were kept and the failed ones were discarded and replaced.

As a form of probing the space of possible instances in regard to how much harder it gets, as sizes increase, for finding instances that are solvable within our time limit, consider Fig. 3a. It shows how many instances had to be generated until a set of 30 instances could be found that satisfied our criteria. One can see that the first 30 instances of sizes between 30 and 42 were all solvable. Differently, from size 44 onwards, some instances timed out and had to be replaced. Clearly, solvable instances within our time frame become rarer as the size increases.

One of the major drawbacks for the performance of the BASIC model is the huge number of crossing constraints (1b), which is $O(n^4)$. For instances with 100 points, the formulation is too big even to be loaded into memory. A conceivable approach in this case is to implement a branch-and-cut algorithm to compute the BASIC model where the crossing constraints are added as they are needed. Initially, no constraints (1b) are included in the model. The processing of a node in the enumeration tree involves the execution a cutting-plane algorithm that is composed of the following steps. In the first one, the current linear relaxation is

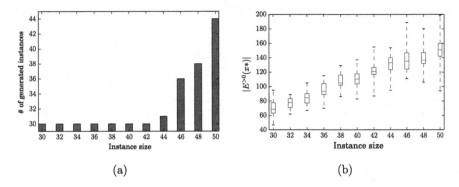

Fig. 3. (a) Number of instances generated until 30 were solved exactly, and (b) cardinality of the support of $x^\star$, as functions of the instance size.

computed. Then, a separation routine is run that checks for crossing constraints that are violated by the optimal solution. The latter can be done by inspection in $O(n^4)$ time and, if violated constraints are found, they are added to the current relaxation and the previous steps are repeated; if not, the processing of the node halts and branching takes place provided that the current solution is not integral.

Having implemented and tested this algorithm, we noticed that the computation of the linear relaxation via the cutting-plane algorithm was faster than solving the complete relaxed model. This proved helpful for developing our heuristics. However, considering the imposed time limit and the size of the tested instances, the branch-and-cut algorithm as a whole was slower than the standard branch-and-bound algorithm used by CPLEX when applied to the full model.

The difficulty in computing optimal solutions motivated us to design heuristics for the MCPP. The starting point for the development of such algorithms came from observations on the solutions of the relaxation of the BASIC model.

Given a vector $y \in \mathbb{R}^{|E(P)|}$, the *support* of y is the set of edges of $E(P)$ for which the corresponding variables have positive values in y. Now, let RELAX be the linear relaxation of the BASIC model, $x^\star$ an optimal solution of RELAX and $E^{>0}(x^\star)$ the support of $x^\star$. Similarly, let $E^{=1}(x^\star)$ be the subset of edges of $E(P)$ corresponding to the variables of value one in $x^\star$.

With these definitions, we are ready to analyze the quality of optimal solutions of the linear relaxation of the BASIC model computed by CPLEX. We consider the support of $x^\star$ to be *good* if it is small-sized and there exists an optimal (integer) solution of BASIC whose support intersects $E^{>0}(x^\star)$ for a large number of edges. In that vein, Fig. 3b shows the average number of edges in $E^{>0}(x^\star)$ for instances in the set $T1$. Notice that this value grows linearly in n, even though there are $\Theta(n^2)$ edges in $E(P)$. Since any convex partition of P has $\Omega(n)$ edges, $E^{>0}(x^\star)$ is indeed small-sized.

To estimate the quality of the support of the solutions found by CPLEX, we modify the BASIC model to look for an optimal solution with the largest intersection with $E^{>0}(x^\star)$. To achieve this, the weights of the variables in the

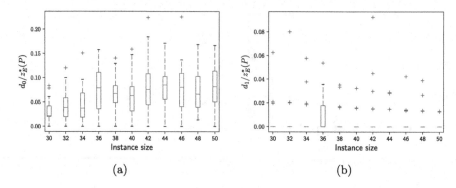

Fig. 4. Ratio between the minimum number of (a) zeros in $x^\star$ that changed to ones in an optimal solution of value d_0; (b) ones in $x^\star$ that changed to zeros in an optimal solution of value d_1, and the optimum $(z_E^*(P))$ of the MCPP.

objective function are changed. A null weight is assigned to edges in $E^{>0}(x^\star)$ while a weight of one is assigned to all remaining variables. To ascertain that the solution found is optimal for the MCPP, we add the following constraint: $\sum_{e \in E(P)} x_e = z_E^*(P)$, where $z_E^*(P)$ is the number of edges in a minimum convex partition of P. Objectively, we seek an optimal solution having as few edges not in $E^{>0}(x^\star)$ as possible. Let $d_0(P)$ be the optimum of the modified model.

Figure 4a shows the ratio between $d_0(P)$ and $z_E^*(P)$ obtained for each size for the instances in $T1$. Note that this value is quite small, with an average of less than 10%. This suggests that the support of $x^\star$ contains most of the edges present in some optimal solution.

Another important observation for the development of heuristics for the MCPP concerns the number of edges in $E^{=1}(x^\star)$ that do not appear in an optimal solution. To assess this quantity, we modify the BASIC model again but this time to find a minimum convex partition of P that uses as many ones from $x^\star$ as possible. As before, we add the constraint $\sum_{e \in E(P)} x_e = z_E^*(P)$ to ensure that the solution found is an optimal solution for the MCPP. Also, we alter the objective function by replacing it with $|E^{=1}(x^\star)| - \sum_{e \in E^{=1}(x^\star)} x_e$. This function computes the number of variables that have value one in $x^\star$ but not in the solution of the new model. Hence, minimizing its value is tantamount to obtaining an optimal convex partition with as many edges in $E^{=1}(x^\star)$ as possible.

Now, let $d_1(P)$ be the optimal value of the latter model. Figure 4b displays the ratio between $d_1(P)$ and $z_E^*(P)$ obtained for each size for the instances in $T1$. As can be seen, the ratios are minute, meaning that a very small fraction of the edges in $E^{=1}(x^\star)$ are not present in some minimum convex partition.

Inspired by the two previous remarks, we decided to develop heuristics for the MCPP based on the optimal solutions of the RELAX model. Accordingly, we explored such solutions prioritizing the use of edges in their support, noticeably those at value one. The next section explains how this is done.

4 Heuristics

We now describe our proposed heuristics for the MCPP. The general framework of the heuristics is summarized by the following steps:

Step 1: Solve the RELAX model and let $x^\star$ be the optimal solution found;

Step 2: Using $x^\star$, build a subset B of $E(P)$ that induces at least one convex partition of P and, for the sake of efficiency, having $O(n)$ crossings;

Step 3: Find a smallest convex partition S^h of P in B and return this solution. From the discussion in the previous section, in **Step 2** we try to use the information conveyed by $x^\star$ to select edges that have a high probability of belonging to a minimum convex partition. We devise different procedures to construct the set B containing these edges, which results in the heuristics that are detailed below. The main strategy here is to pack B with more edges than are really needed to obtain a convex partition of P and then, in **Step 3** to select among them a subset of minimum size that is a feasible solution for the MCPP.

Steps 1 and **3** are addressed in the same way in all heuristics developed in this work. **Step 1** corresponds to solving the RELAX model, in our case using CPLEX, to obtain $x^\star$. As for **Step 3**, S^h is computed by solving a restricted version of the BASIC model where the variables x_{ij} for (i, j) not in B are removed from the model (equivalently, one could set them to zero *a priori*). It only remains to decide how the set B is to be built in **Step 2**. Alternatives are proposed below to accomplish this task giving rise to four distinct heuristics.

The GREEDY Heuristic. In this heuristic, the set B in **Step 2** is constructed as follows. Initially, B is empty and the edges in $E(P)$ are organized in a sorted list σ in non increasing order of their corresponding value in $x^\star$. Ties are broken by the number of times the edge crosses the support of the relaxation; edges with less crossings appear earlier in σ. Then, a triangulation of P is constructed iteratively in a greedy fashion. At each iteration, the next edge e in σ is considered. If it does not cross any of the edges already in B, the set is updated to $B \cup \{e\}$; otherwise, B remains unchanged. Clearly, the greedy strategy prioritizes the edges in the support of $x^\star$, which is in consonance with the results seen in Sect. 3. In the end, since all edges in $E(P)$ have been considered, B must determine a triangulation of P. Thus, the BASIC model relative to B computed in **Step 3** has no constraints of the form (1b). In the experiments reported at the end of this section, it is shown that the computation of this restricted version of the BASIC model does not compromise the efficiency of the GREEDY heuristic. In fact, the algorithm turned out to be remarkably fast.

The MAXSUP Heuristic. The previous heuristic constructs B as the set of edges of a greedy triangulation of P by favoring edges in the support of $x^\star$ which, as seen, are more likely to be in a minimum convex partition. In a similar mode, the MAXSUP heuristic provides an alternative way for obtaining B. The set is initialized with a convex partition of P having the largest possible intersection with the support of $x^\star$. Next, edges are added to B in a greedy fashion until a triangulation is formed whose associated BASIC model is computed in **Step 3**.

The difficulty with this approach lies on obtaining a convex partition of P that maximizes the number of edges that are in the support of $x^\star$. This can be accomplished by modifying the costs of the variables in the objective function of the BASIC model and changing the problem into one of maximization. In the new formulation, the costs of the variables associated to edges in the support of $x^\star$ are set to one and all others to zero. Besides, variables related to edges in $E^{=1}(x^\star)$ are fixed to one. Since this ILP has fewer variables with positive costs, in practice, it can be computed faster than the original model.

Adding Flips to a Triangulation. In an attempt to improve the results from the heuristics even further, we decided to forgo the planarity requirements for B in **Step 2**. By this strategy, we expect to increase the search space in **Step 3**, potentially leading to better solutions. Nonetheless, some of the constraints (1b) will have to be brought back into the model in **Step 3**, which could severely impact its runtime. To avert this, we strive to keep its size linear in n.

With that intent in mind, assume that initially B determines a triangulation of P. We say that an edge $e \in B \setminus H(P)$ *admits a flip* if the two triangles incident to e form a convex quadrilateral. If e admits a flip, its *flip edge* f_e, which is not in B, corresponds to the other diagonal of this quadrilateral. Clearly, each edge in B has at most one flip edge and, hence, there are $O(n)$ flip edges in total. Let $F(B)$ be the set of flip edges of the edges in B (that admit one). Evidently, $B \cup F(B)$ has $O(n)$ edges and the total number of pairs of edges in this set that cross each other is also $O(n)$. Thus, in **Step 2**, if B is initialized as a triangulation of P and is later extended with its flip edges, in **Step 3** we have to solve a shorter version of the BASIC model with $O(n)$ variables and constraints.

Notice that interchanging an edge of a triangulation and its flip edge gives us an alternative triangulation. Therefore, one can think of the effect of **Step 3** when B is replaced with $B \cup F(B)$ as the computation of the smallest convex partition that can be obtained from a triangulation generated from B by a sequence of flips involving the edges of $B \cup F(B)$. Observe that this amounts to a considerable growth in the search space when compared to the use of a single triangulation. Of course, this is only worthwhile when the computing time does not increase too much while the solution quality improves. As we shall see later, the expected benefit is confirmed in practice.

Motivated by the preceding discussion, we created one new heuristic from each of the previous ones by aggregating flip edges. The enhanced version of GREEDY (MaxSup) using flip edges is called GREEDYF (MaxSupF).

5   Computational Results

We now assess the results from the four heuristics discussed above comparing them in regard to solution quality and running time. Firstly, we analyze the performance of the heuristics for the instances with known optimum $(T1)$. The first row of the table below displays statistics on the mean values, per instance size, of the average relative gap given by $100(z_E^h - z_E^*)/z_E^*$, where z_E^* and z_E^h are the cost of an optimal and of the heuristic solution, respectively. The minimum, average, standard deviation and maximum values are given for each heuristic.

	GREEDY	MAXSUP	GREEDYF	MAXSUPF
mean gap (min/avg±std/max)	0.0/4.2±2.4/12.5	0.0/3.0±1.9/9.0	0.0/1.9±1.6/7.5	0.0/1.1±1.2/6.0
% of instances solved to optimality	7.58	12.12	27.58	43.33

The improvements caused by considering flips are evident. The version of each heuristic including flips reduces the gap by more than half when compared to their original counterparts. When flips are taken into consideration, no gap exceeds 8%. As expected, MAXSUP has a better overall performance relative to GREEDY. The second row of the table above shows the percentage of instances for which the optimum was found in each case. Again, the benefit of including flips and the superiority of MAXSUP over GREEDY are clear. In fact, there was no instance where any other heuristic found a better solution than MAXSUPF.

The mean values of the average ratio, per instance size, between the time spent by each heuristic and the BASIC model is given in the next table. On average, the additional computational cost incurred by considering flips is quite small both for GREEDY and for MAXSUP. Together with the enhancement in solution quality this favors even more the latter strategy. Although both heuristics took just a fraction of the time spent by the BASIC model, one can see that the GREEDY versions used about half the time needed by their MAXSUP equivalents. Also, as revealed by the maximum values, the time required by the MAXSUP heuristics can surpass that of the BASIC model. This occurs because these heuristics compute two ILPs, and the instances in $T1$ are small.

time ratio relative to BASIC	GREEDY	MAXSUP	GREEDYF	MAXSUPF
mean (min/avg±std/max)	0.3/13.2±13.4/60.6	0.5/28.7±28.7/146.8	0.3/15.5±15.9/82.2	0.5/31.1±31.3/169.3

Next, the heuristics are evaluated for the instances in $T2$. While the optimum in this case is unknown, we observed that in all instances of $T2$ the best heuristic solution was again found by MAXSUPF. Therefore, for comparison purposes, MAXSUPF is used as reference, playing a role similar to that of the BASIC model in the analysis of the instances in $T1$. The table below exhibits, per instance size, the mean of the average gap between each heuristic and MAXSUPF. As before, the minimum, average, standard deviation and maximum values are given.

	GREEDY	MAXSUP	GREEDYF	MAXSUPF
mean gap (min/avg±std/max)	0.0/6.0±2.3/14.5	0.0/3.6±1.5/9.0	0.0/1.4±1.4/6.4	0.0/0.0±0.0/0.0

The same pattern perceived for $T1$ is observed here with GREEDYF performing slightly worse than MAXSUPF, while the versions with flips outstrip by far their original counterparts. Mean solving times relative to MAXSUPF for all heuristics are shown in the next table. Once again, we observe the insubstantial impact of the inclusion of flips in the running time of the algorithms. The fact that the MAXSUP versions are almost four times slower on average than their GREEDY equivalents is remarkable. This is due to the fact that the computation of the additional ILP in MAXSUP to find the triangulation with the largest intersection with x^* consumes too much time as the instance size grows.

time ratio relative to MAXSUPF	GREEDY	MAXSUP	GREEDYF	MAXSUPF
mean (min/avg±std/max)	1.7/25.3±16.2/55.2	44.0/98.4±14.1/264.5	1.7/27.0±17.5/59.4	100.0/100.0±0.0/100.0

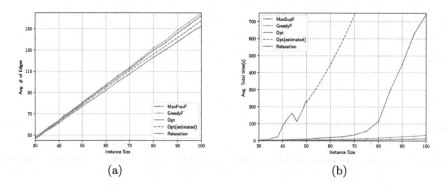

Fig. 5. (a) Average number of edges in the solution; (b) Runtime (seconds)

From the previous discussions, we are left with two competitive heuristics: GREEDYF and MAXSUPF. In Fig. 5a, we compare the solutions found by MAXSUPF and GREEDYF with the dual bound given by the RELAX model and the optimal solution, when available. We also extrapolate the partial average function corresponding to the optimal solution for the instances with sizes equal to those in $T2$ and, as expected, the extrapolation corresponds to a linear function since the optimum of the MCPP for a point set P is in $\Theta(|P|)$. The graph reveals that the quality of the heuristic solutions deteriorates quite slowly as the instance sizes increase. In particular, the distances between the values of the MAXSUPF and GREEDYF solutions and between the dual bound and the (estimated) optimum seem to grow at no more than a constant rate, as the instance sizes get larger. With this in mind, one can estimate the error incurred by the algorithm when executed on an instance of a given size.

Similarly, we analyze the total time spend by MAXSUPF and GREEDYF in Fig. 5b, comparing them to the total time to solve the model and to find optimal solutions for the instances in $T1$. We notice an erratic behavior for the BASIC model at around size 45. This correlates to the difficulty of finding harder instances that are still solvable to optimality within our preset time limit above size 42, as shown in Fig. 3a. This suggests that the estimating curve should actually grow even faster if no instances had to be discarded.

The MAXSUPF heuristic eventually becomes less efficient since it involves the solution of two ILPs. However, it still remains much faster than computing a true optimum. Thus, if more running time is allowed, and solution quality is a prime concern, MAXSUP should be considered an effective alternative when solving the MCPP. On the other hand, we can see that, on average, running RELAX and GREEDYF are both very fast. In fact, at the expense of a small loss in quality, the GREEDY approach takes less than one minute, on average, to find solutions for random instances of up to one hundred points. Thus, it is also a viable option to obtain high quality solutions quickly. Furthermore, both heuristics endorse that using the support of linear relaxations to guide heuristics is a powerful strategy.

6 Final Remarks

In this paper, we investigated the problem of finding a minimum convex partition of a set of points. An ILP model was developed that enabled the computation of exact solutions for small-sized instances. The linear relaxation of this formulation served as a guide for the design of heuristics that lead to demonstrably high quality solutions within reasonable runtimes. To the best of our knowledge, both the ILP modeling and the extensive computational experimentation with MCPP algorithms done here are novelties on the study of this problem and may lead to further practical developments. Moreover, research directions currently being pursued include the development of new mathematical models for this problem aiming at solving larger instances to optimality.

Acknowledgments. This research was financed in part by grants from Conselho Nacional de Desenvolvimento Científico e Tecnológico (CNPq) #311140/2014-9, #304727/2014-8, Coordenação de Aperfeiçoamento do Pessoal do Ensino Superior - Brasil (Capes) #001 and Fundação de Amparo à Pesquisa do Estado de São Paulo (FAPESP) #2014/12236-1, #2017/12523-9, #2018/14883-5.

References

1. Barboza, A.S., de Souza, C.C., de Rezende, P.J.: Minimum Convex Partition of Point Sets - Benchmark Instances and Solutions (2018). www.ic.unicamp.br/~cid/Problem-instances/Convex-Partition
2. Bose, P., et al.: Online routing in convex subdivisions. Int. J. Comput. Geom. Appl. **12**(4), 283–296 (2002)
3. Fevens, T., Meijer, H., Rappaport, D.: Minimum convex partition of a constrained point set. Discrete Appl. Math. **109**(1–2), 95–107 (2001)
4. García-López, J., Nicolás, M.: Planar point sets with large minimum convex partitions. In: Abstracts 22nd European Workshop on Computational Geometry, pp. 51–54 (2006)
5. Hosono, K.: On convex decompositions of a planar point set. Discrete Math. **309**(6), 1714–1717 (2009)
6. Keil, J.M.: Polygon decomposition. In: Sack, J.R., Urrutia, J. (eds.) Handbook of Computational Geometry, pp. 491–518. North-Holland, Amsterdam (2000)
7. Knauer, C., Spillner, A.: Approximation algorithms for the minimum convex partition problem. In: Arge, L., Freivalds, R. (eds.) SWAT 2006. LNCS, vol. 4059, pp. 232–241. Springer, Heidelberg (2006). https://doi.org/10.1007/11785293_23
8. Löffler, M., Mulzer, W.: Unions of onions: preprocessing imprecise points for fast onion decomposition. J. Comput. Geom. **5**(1), 1–13 (2014)
9. Neumann-Lara, V., Rivera-Campo, E., Urrutia, J.: A note on convex decompositions of a set of points in the plane. Graphs Comb. **20**(2), 223–231 (2004)
10. Spillner, A.: A fixed parameter algorithm for optimal convex partitions. J. Discrete Algorithms **6**(4), 561–569 (2008)

Parameterized Complexity of Safe Set

Rémy Belmonte[1] , Tesshu Hanaka[2] , Ioannis Katsikarelis[3]([✉]) ,
Michael Lampis[3] , Hirotaka Ono[4] , and Yota Otachi[5]

[1] The University of Electro-Communications, Chofu, Tokyo 182-8585, Japan
remy.belmonte@uec.ac.jp
[2] Chuo University, Bunkyo-ku, Tokyo 112-8551, Japan
hanaka.91t@g.chuo-u.ac.jp
[3] Université Paris-Dauphine, PSL University, CNRS, LAMSADE, Paris, France
{ioannis.katsikarelis,michail.lampis}@dauphine.fr
[4] Nagoya University, Nagoya 464-8601, Japan
ono@nagoya-u.jp
[5] Kumamoto University, Kumamoto 860-8555, Japan
otachi@cs.kumamoto-u.ac.jp

Abstract. In this paper we study the problem of finding a small safe
set S in a graph G, i.e. a non-empty set of vertices such that no connected component of $G[S]$ is adjacent to a larger component in $G - S$.
We enhance our understanding of the problem from the viewpoint of
parameterized complexity by showing that (1) the problem is W[2]-hard
when parameterized by the pathwidth pw and cannot be solved in time
$n^{o(\mathsf{pw})}$ unless the ETH is false, (2) it admits no polynomial kernel parameterized by the vertex cover number vc unless PH $= \Sigma_3^p$, but (3) it is
fixed-parameter tractable (FPT) when parameterized by the neighbor-
hood diversity nd, and (4) it can be solved in time $n^{f(\mathsf{cw})}$ for some double
exponential function f where cw is the clique-width. We also present (5)
a faster FPT algorithm when parameterized by solution size.

Keywords: Safe set · Parameterized complexity ·
Vulnerability parameter · Pathwidth · Clique-width

1 Introduction

Let $G = (V, E)$ be a graph. For a vertex set $S \subseteq V(G)$, we denote by $G[S]$ the
subgraph of G induced by S, and by $G - S$ the subgraph induced by $V \setminus S$.
If $G[S]$ is connected, we also say that S is connected. A vertex set $C \subseteq V$ is
a *component* of G if C is an inclusion-wise maximal connected set. Two vertex
sets $A, B \subseteq V$ are *adjacent* if there is an edge $\{a, b\} \in E$ such that $a \in A$ and

Partially supported by JSPS and MAEDI under the Japan-France Integrated Action
Program (SAKURA) Project GRAPA 38593YJ, and by JSPS/MEXT KAKENHI
Grant Numbers JP24106004, JP17H01698, JP18K11157, JP18K11168, JP18K11169,
JP18H04091, 18H06469.

P. Heggernes (Ed.): CIAC 2019, LNCS 11485, pp. 38–49, 2019.
https://doi.org/10.1007/978-3-030-17402-6_4

$b \in B$. Now, a non-empty vertex set $S \subseteq V$ of a graph $G = (V, E)$ is a *safe set* if no connected component C of $G[S]$ has an adjacent connected component D of $G - S$ with $|C| < |D|$. A safe set S of G is a *connected safe set* if $G[S]$ is connected. The *safe number* $\mathsf{s}(G)$ of G is the size of a minimum safe set of G, and the *connected safe number* $\mathsf{cs}(G)$ of G is the size of a minimum connected safe set of G. It is known [13] that $\mathsf{s}(G) \leq \mathsf{cs}(G) \leq 2 \cdot \mathsf{s}(G) - 1$.

Fujita et al. [13] introduced the concept of safe sets motivated by a facility location problem that can be used to design a safe evacuation plan. Subsequently, Bapat et al. [2] observed that a safe set can control the consensus of the underlying network with a majority of each part of the subnetwork induced by a component in the safe set and an adjacent component in the complement, thus the minimum size of a safe set can be used as a vulnerability measure of the network. That is, contrary to its name, having a small safe set could be unsafe for a network. Both the combinatorial and algorithmic aspects of the safe set problem have already been extensively studied [1,8,11,12].

In this paper, we study the problem of finding a small safe set mainly from the parameterized-complexity point of view. We show a number of both tractability and intractability results, that highlight the difference in complexity that this parameter exhibits compared to similarly defined vulnerability parameters.

Our Results

Our main results are the following (see also Fig. 1).

1. Both problems are W[2]-hard parameterized by the pathwidth pw and cannot be solved in time $n^{o(\mathsf{pw})}$ unless the Exponential Time Hypothesis (ETH) fails, where n is the number of vertices.
2. They do not admit kernels of polynomial size when parameterized by vertex cover number even for connected graphs unless $\mathrm{PH} = \Sigma_3^{\mathrm{p}}$.
3. Both problems are fixed-parameter tractable (FPT) when parameterized by neighborhood diversity.
4. Both problems can be solved in XP-time when parameterized by clique-width.
5. Both problems can be solved in $O^*(k^{O(k)})$ time[1] when parameterized by the solution size k.

The W[2]-hardness parameterized by pathwidth complements the known FPT result when parameterized by the solution size [1], since for every graph the size of the solution is at least half of the graph's pathwidth (see Sect. 2). The $n^{o(\mathsf{pw})}$-time lower bound is tight since there is an $n^{O(\mathsf{tw})}$-time algorithm [1], where tw is the treewidth. The second result also implies that there is no polynomial kernel parameterized by solution size, as the vertex cover number is an upper bound on the size of the solution. The third result marks the first FPT algorithm by a parameter that is incomparable to the solution size. The fourth result implies XP-time solvability for all the parameters mentioned in this paper and extends the result for treewidth from [1]. The fifth result improves the known algorithm [1] that uses Courcelle's theorem.

---

[1] The $O^*(\cdot)$ notation omits the polynomial dependency on the input size.

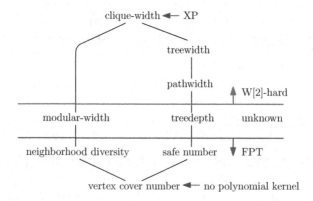

Fig. 1. Graph parameters and an overview of the parameterized complexity landscape for SS and CSS. Connections between two parameters imply the existence of a function in the one above (being in this sense more general) that lower bounds the one below.

Previous Work

In the first paper on this topic, Fujita et al. [13] showed that the problems are NP-complete in general. Their hardness proof implies that the parameters are hard to approximate within a factor of 1.3606 ([1]). They also showed that a minimum connected safe set in a tree can be found in linear time.

Bapat et al. [2] considered the problems on vertex-weighted graphs, where the problems are naturally generalized. They showed that these are weakly NP-complete even for weighted stars (thus for all parameters generalizing vertex cover number and graph classes like interval and split graphs). On the other hand, they showed that the problems can be solved in $O(n^3)$ time for weighted paths. Ehard and Rautenbach [8] presented a PTAS for the connected safe number of a weighted tree. Fujita et al. [12] showed among other results that the problems can be solved in linear time for weighted cycles.

Águeda et al. [1] studied their unweighted versions. They presented an XP algorithm for graphs of bounded treewidth and showed that the problems can be solved in polynomial time for interval graphs, while they are NP-complete for split and bipartite planar graphs of maximum degree at most 7. Observing that the treewidth of a graph is bounded by a function of its safe number, they also showed that the problems are FPT parameterized by solution size.

2   Preliminaries

We assume that the reader is familiar with the concepts relating to fixed-parameter tractability. Due to space restrictions, we omit some formal definitions and most of our proofs. See [5] (and references therein) for the definitions of relevant notions in parameterized complexity theory. We let $N[v]$ denote the *closed neighborhood* of vertex v, i.e. the set containing v and all vertices adjacent to it. Furthermore, for a positive integer k, we denote the set $\{1, \ldots, k\}$ by $[k]$.

Graph Parameters

We recall relationships among some graph parameters used in this paper, and give a map of the parameters with the results.

The graph parameters we explicitly use in this paper are vertex cover number vc, pathwidth pw, neighborhood diversity nd, and clique-width cw. They are situated in the hierarchy of well-studied graph parameters, along with safe number s, as depicted in Fig. 1.

The treewidth $tw(G)$, pathwidth $pw(G)$, treedepth $td(G)$, and vertex cover number $vc(G)$ of a graph G can be defined as the minimum of the maximum clique-size (-1 for $tw(G)$, $pw(G)$, and $vc(G)$) among all supergraphs of G that are of type chordal, interval, trivially perfect and threshold, respectively. This gives us $tw(G) \leq pw(G) \leq td(G) - 1 \leq vc(G)$ for every graph G. One can easily see that $s(G) \leq vc(G)$ and $td(G) \leq 2s(G)$. (See also the discussion in [1].) This justifies the hierarchical relationships among them in Fig. 1.

Although modular-width $mw(G)$ and neighborhood diversity $nd(G)$ are incomparable to most of the parameters mentioned above, all these parameters are generalized by clique-width, while the vertex cover number is their specialization.

3 W[2]-Hardness Parameterized by Pathwidth

In this section we show that SAFE SET is W[2]-hard parameterized by pathwidth, via a reduction from DOMINATING SET. Given an instance $[G = (V, E), k]$ of DOMINATING SET, we will construct an instance $[G' = (V', E'), pw(G')]$ of SAFE SET parameterized by pathwidth. Let $V = \{v_1, \ldots, v_n\}$ and $k' = 1 + kn + \sum_{v \in V} k(\delta(v) + 1)$ be the target size of the safe set in the new instance, where $\delta(v)$ is the degree of v in G.

Before proceeding, let us give a high-level description of some of the key ideas of our reduction (an overview is given in Fig. 2). First, we note that our new instance will include a universal vertex. This simplifies things, as such a vertex must be included in any safe set (of reasonable size) and ensures that the safe set is connected. The problem then becomes: can we select $k' - 1$ additional vertices so that their deletion disconnects the graph into components of size at most k'.

The main part of our construction consists of a collection of k cycles of length n^2. By attaching an appropriate number of leaves to every n-th vertex of such a cycle we can ensure that any safe set that uses exactly n vertices from each cycle must space them evenly, that is, if it selects the i-th vertex of the cycle, it also selects the $(n + i)$-th vertex, the $(2n + i)$-th vertex, etc. As a result, we expect the solution to invest kn vertices in the cycles, in a way that encodes k choices from the set $[n]$.

We must now check that these choices form a dominating set of the original graph G. For each vertex of G we construct a gadget and connect this gadget to a different length-n section of the cycles. This type of connection ensures that the construction will in the end have small pathwidth, as the different gadgets are only connected through the highly-structured "choice" part that consists of the k cycles. We then construct a gadget for each vertex v_i that can be broken down into small enough components by deleting $k(\delta(v_i) + 1)$ vertices, if and only if we have already selected from the cycles a vertex corresponding to a member of $N[v]$ in the original graph.

Domination Gadget: Before we go on to describe in detail the full construction, we describe a *domination gadget* $\hat{D}_i$. This gadget refers to the vertex $v_i \in V$ and its purpose is to model the domination of v_i in G and determine the member of $N[v_i]$ that belongs to the dominating set. We construct $\hat{D}_i$ as follows:

- We make a *central* vertex z^i. We attach to this vertex $k' - k(\delta(v_i) + 1)$ leaves. Call this set of leaves W^i.
- We make k independent sets $X_1^i, \ldots, X_k^i$ of size $|N[v_i]|$. For each $j \in [1, k]$ we associate each $x \in X_j^i$ with a distinct member of $N[v_i]$. We attach to each vertex x of each $X_j^i, j \in [1, k]$ an independent set of $k' - 1$ vertices. Call this independent set Q_x.
- We then make another k independent sets $Y_1^i, \ldots, Y_k^i$ of size $|N[v_i]|$. For each $j \in [1, k]$ we construct a perfect matching from X_j^i to Y_j^i. We then connect z to all vertices of Y_j^i for all $j \in [1, k]$.

The intuition behind this construction is the following: we will connect the vertices of X_j^i to the j-th selection cycle, linking each vertex with the element of $N[v_i]$ it represents. In order to construct a safe set, we need to select at least one vertex $y \in Y_j^i$ for some $j \in [1, k]$, otherwise the component containing z will be too large. This gives us the opportunity to *not* place the neighbor $x \in X_j^i$ of y in the safe set. This can only happen, however, if the neighbor of x in the main part is also in the safe set, i.e. if our selection dominates v_i.

Construction: Graph G' is constructed as follows:

- We first make n copies of V and serially connect them in a long cycle, conceptually divided into n *blocks*: $v_1, v_2, \ldots, v_n | v_{n+1}, \ldots, v_{2n} | v_{2n+1}, \ldots | \ldots, v_{n^2} | v_1$. Each block corresponds to one vertex of V.
- We make k copies $V^1, \ldots, V^k$ of this cycle, where $V^i = \{v_1^i, \ldots, v_{n^2}^i\}, \forall i \in [1, k]$ and refer to each as a *line*. Each such line will correspond to one vertex of a dominating set in G.
- We add a set B_j^i of $k' - n + 1$ neighbors to each vertex $v_{(j-1)n+1}^i, \forall j \in [1, n]$, i.e. the first vertex of every block of every line. We refer to these sets collectively as the *guards*.
- Then, for each *column* of blocks (i.e. the i-th block of all k lines for $i \in [1, n]$), we make a domination gadget $\hat{D}_i$ that refers to vertex $v_i \in V$. As described above, the gadget contains k copies $X_1^i, \ldots, X_k^i$ of $N[v_i] \subseteq V$.
- For $i \in [1, n]$ and $j \in [1, k]$, we add an edge between each vertex in X_j^i and its corresponding vertex in the i-th block of V^j, i.e. for the given i-th column, we connect a vertex from the j-th line (V^j) to the vertex from the j-th copy of $N[v_i]$ (X_j^i) if they correspond to the same original vertex from V.

– We add a *universal vertex* u and connect it to every other vertex in the graph. This concludes our construction (see also Fig. 2).

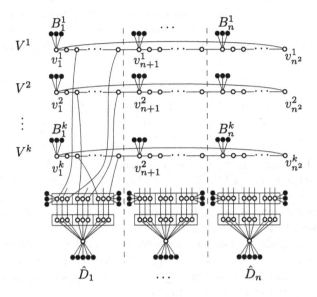

Fig. 2. A simplified picture of our construction. Note sets Q are only shown for two vertices per gadget, exemplary connections between corresponding vertices in sets X and lines V are only shown for the first gadget, while the universal vertex u is omitted.

Theorem 3.1. SAFE SET *and* CONNECTED SAFE SET *are W[2]-hard parameterized by the pathwidth of the input graph. Furthermore, both problems cannot be solved in time* $n^{o(\mathsf{pw})}$ *unless the ETH is false.*

4 No Polynomial Kernel Parameterized by Vertex Cover Number

A set $X \subseteq V$ is a *vertex cover* of $G = (V, E)$ if each edge $e \in E$ has at least one endpoint in X. The minimum size of a vertex cover in G is the *vertex cover number* of G, denoted by $\mathsf{vc}(G)$. Parameterized by vertex cover number vc, both problems are FPT (see Fig. 1) and in this section we show the following kernelization hardness of SS and CSS.

Theorem 4.1. SAFE SET *and* CONNECTED SAFE SET *parameterized by the vertex cover number do not admit polynomial kernels even for connected graphs unless* $\mathrm{PH} = \Sigma_3^{\mathrm{p}}$.

Since for every graph G, it is $\mathsf{cs}(G)/2 \leq \mathsf{s}(G) \leq \mathsf{vc}(G)$ ([1]), the above theorem implies that SS and CSS parameterized by the natural parameters do not admit a polynomial kernel.

Corollary 4.2. *SS and CSS parameterized by solution size do not admit polynomial kernels even for connected graphs unless* $\mathrm{PH} = \Sigma_3^{\mathrm{p}}$.

Let P and Q be parameterized problems. A polynomial-time computable function $f\colon \Sigma^* \times N \to \Sigma^* \times N$ is a *polynomial parameter transformation* from P to Q if there is a polynomial p such that for all $(x, k) \in \Sigma^* \times N$, it is: $(x, k) \in P$ if and only if $(x', k') = f(x, k) \in Q$, and $k' \le p(k)$. If such a function exits, then P is *polynomial-parameter reducible* to Q.

Proposition 4.3 ([3]). *Let P and Q be parameterized problems, and P' and Q' be unparameterized versions of P and Q, respectively. Suppose P' is NP-hard, Q' is in NP, and P is polynomial-parameter reducible to Q. If Q has a polynomial kernel, then P also has a polynomial kernel.*

To prove Theorem 4.1, we present a polynomial-parameter transformation from the well-known RED-BLUE DOMINATING SET problem (RDBS, see [5]) to SS (and CSS) parameterized by vertex cover number.

RDBS becomes trivial when $k \ge |R|$ and thus we assume that $k < |R|$ in what follows. It is known that RBDS parameterized simultaneously by k and $|R|$ does not admit a polynomial kernel unless $\mathrm{PH} = \Sigma_3^{\mathrm{p}}$ [6]. Since RBDS is NP-hard and SS and CSS are in NP, it suffices to present a polynomial-parameter transformation from RBDS parameterized by $k + |R|$ to SS and CSS parameterized by the vertex cover number.

From an instance $[G, k]$ of RDBS with $G = (R, B; E)$, we construct an instance $[H, s]$ of SS as follows (see Fig. 3). Let $s = k + |R| + 1$. We add a vertex u to G and make it adjacent to all vertices in B. We then attach $2s$ pendant vertices to each vertex in R and to u. Finally, for each $r \in R$, we make a star $K_{1,s-1}$ and add an edge between r and the center of the star. We call the resultant graph H. Observe that $\mathsf{vc}(H) \le 2|R| + 1$ since $\{u\} \cup R \cup C$ is a vertex cover of H, where C is the set of centers of stars attached to R. This reduction is a polynomial-parameter transformation from RBDS parameterized by $k + |R|$ to SS parameterized by the vertex cover number.

If $D \subseteq V(G)$ is a solution of RBDS of size k, then $S := \{u\} \cup R \cup D$ is a connected safe set of size s. To see this, recall that B is an independent set and $N_H(B) = \{u\} \cup R$. Thus each component in $H - S$ is either an isolated vertex in $D \setminus B$, or a star $K_{1,s-1}$ with s vertices.

Assume that (H, s) is a yes instance of SS and let S be a safe set of H with $|S| \le s$. Observe that $\{u\} \cup R \subseteq S$ since u and all vertices in R have degree at least $2s$. Since $|S \setminus (\{u\} \cup R)| \le k < |R|$, S cannot intersect all stars attached to the vertices in R. Hence $H - S$ has a component of size at least $|V(K_{1,s-1})| = s$. This implies that S is connected. Since R is an independent set and each path from u to a vertex in R passes through B, each vertex in R has to have a neighbor in $B \cap S$. Thus $B \cap S$ dominates R. Since $B \cap S \subseteq S \setminus (\{u\} \cup R)$, it has size at most k. Therefore, $[G, k]$ is a Yes instance of RBDS. This completes the proof of Theorem 4.1.

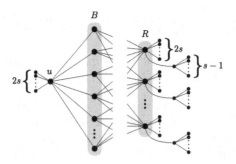

Fig. 3. The graph H in our reduction from RBDS to SS for the proof of Theorem 4.1.

5   FPT Algorithm Parameterized by Neighborhood Diversity

In this section, we present FPT algorithms for SS and CSS parameterized by neighborhood diversity. That is, we prove the following theorem.

Theorem 5.1. SAFE SET *and* CONNECTED SAFE SET *are fixed-parameter tractable when parameterized by the neighborhood diversity.*

In a graph $G = (V, E)$, two vertices $u, v \in V$ are *twins* if $N(u) \setminus \{v\} = N(v) \setminus \{u\}$. The *neighborhood diversity* $\mathsf{nd}(G)$ of $G = (V, E)$ is the minimum integer k such that V can be partitioned into k sets $T_1, \ldots, T_k$ of pairwise twin vertices. It is known that such a minimum partition can be found in linear time using fast modular decomposition algorithms [18,21]. It is also known that $\mathsf{nd}(G) \leq 2^{\mathsf{vc}(G)} + \mathsf{vc}(G)$ for every graph G [16].

Let G be a connected graph such that $\mathsf{nd}(G) = k$. Let $T_1, \ldots, T_k$ be the partition of $V(G)$ into sets of pairwise twin vertices. Note that each T_i is either a clique or an independent set by definition. We assume that $k \geq 2$ since otherwise the problem becomes trivial. Since each T_i is a twin set, the sizes of intersections $|S \cap T_i|$ completely characterize the sizes of the components in $G[S]$ and $G - S$, and the adjacency among them.

Let $S \subseteq V(G)$ and $s_i = |S \cap T_i|$ for $i \in [k]$. We partition $[k]$ into $I_{\mathbf{f}}$, $I_{\mathbf{p}}$, and $I_{\emptyset}$ as follows:

$$i \in \begin{cases} I_{\emptyset} & \text{if } s_i = 0, \\ I_{\mathbf{p}} & \text{if } 1 \leq s_i \leq |T_i| - 1, \\ I_{\mathbf{f}} & \text{otherwise}(s_i = |T_i|). \end{cases} \tag{1}$$

For $i, i' \notin I_{\emptyset}$ (not necessarily distinct), twin sets T_i and $T_{i'}$ are *reachable* in S if either

- $T_i = T_{i'}$ and T_i is a clique, or
- there is a sequence $i_0, \ldots, i_\ell$ of indices such that $\ell \geq 1$, $i_0 = i$, $i_\ell = i'$, $i_j \notin I_{\emptyset}$ for all j, and T_{i_j} and $T_{i_{j+1}}$ are adjacent for $0 \leq j < \ell$.

Lemma 5.2. *If $i \notin I_\emptyset$ and T_i is not reachable to T_i itself in S, then each vertex in $T_i \cap S$ induces a component of size 1.*

Lemma 5.3. *Two vertices $u, v \in S$ are in the same component of $G[S]$ if and only if $u \in T_i$ and $v \in T_{i'}$ for some i and i', and T_i is reachable from $T_{i'}$ in S.*

By Lemmas 5.2 and 5.3, each component of $G[S]$ is either a single vertex, or the intersection of S and the union of a maximal family of pairwise reachable twin sets in S. Observe that the maximal families of pairwise reachable twin sets in S is determined only by the set $I_\emptyset$. Also, if R is a maximal family of pairwise reachable twin sets in S, then the corresponding component of $G[S]$ has size $\sum_{T_i \in R} |T_i \cap S|$.

Now, just by interchanging the roles of S and $V(G) \setminus S$ in Lemmas 5.2 and 5.3, we can show the following counterparts that imply that the maximal families of pairwise reachable twin sets in $V(G) \setminus S$ are determined only by the set $I_\mathbf{f}$, while the size of the component of $G - S$ corresponding to a maximal family R of pairwise reachable twin sets in $V(G) \setminus S$ is $\sum_{T_i \in R} |T_i \setminus S|$.

Lemma 5.4. *If $i \notin I_\mathbf{f}$ and T_i is not reachable to T_i itself in $V(G) \setminus S$, then each vertex in $T_i \setminus S$ induces a component of size 1.*

Lemma 5.5. *Two vertices $u, v \in V(G) \setminus S$ are in the same component of $G - S$ if and only if $u \in T_i$ and $v \in T_{i'}$ for some i and i', and T_i is reachable from $T_{i'}$ in $V(G) \setminus S$.*

ILP Formulation

Now we reduce the problem to an FPT number of integer linear programs with a bounded number of variables. We first divide $[k]$ into the subsets $I_\emptyset$, $I_\mathbf{p}$, $I_\mathbf{f}$ in Eq. (1). There are 3^k candidates for such a partition.

For each $i \in [k]$, we use a variable x_i to represent the size of $T_i \cap S$. To find a minimum safe set satisfying $I_\emptyset$, $I_\mathbf{p}$, $I_\mathbf{f}$, we set the objective function to be $\sum_{i \in [k]} x_i$ and minimize it subject to the following linear constraints. The first set of constraints is to make S consistent with the guess of $I_\emptyset$, $I_\mathbf{p}$, $I_\mathbf{f}$:

$$x_i = 0 \quad \text{for } i \in I_\emptyset,$$
$$1 \le x_i \le |T_i| - 1 \quad \text{for } i \in I_\mathbf{p},$$
$$x_i = |T_i| \quad \text{for } i \in I_\mathbf{f}.$$

As discussed above, the set of sizes x_i completely characterizes the structure of components in $G[S]$ and $G - S$. In particular, we can decide whether $G[S]$ is connected or not at this point. We reject the disconnected case if we are looking for a connected safe set.

Let $\mathcal{C}$ and $\mathcal{D}$ be the sets of maximal families of pairwise reachable twin sets in S and $V(G) \setminus S$, respectively. Note that the twin sets that satisfy the conditions of Lemma 5.2 (Lemma 5.4) are not included in any member of $\mathcal{C}$ ($\mathcal{D}$, respectively).

For each $C_j \in \mathcal{C}$, we use a variable y_j to represent the size of the corresponding component of $G[S]$. Also, for each $D_h \in \mathcal{D}$, we use a variable z_h to represent the size of the corresponding component of $G - S$. This can be stated as follows:

$$y_j = \sum_{T_i \in C_j} x_i \quad \text{for } C_j \in \mathcal{C},$$

$$z_h = \sum_{T_i \in D_h} |T_i| - x_i \quad \text{for } D_h \in \mathcal{D}.$$

We say that $C_j \in \mathcal{C}$ and $D_h \in \mathcal{D}$ are *touching* if there are $T \in C_j$ and $T' \in D_h$ that are adjacent, or the same. We can see that C_j and D_h are touching if and only if the corresponding components are adjacent via an edge from T and T' or an edge completely in $T = T'$. We add the following constraint to guarantee the safeness of S: $y_j \geq z_h$, for each pair of touching $C_j \in \mathcal{C}$ and $D_h \in \mathcal{D}$.

Now we have to deal with the singleton components of $G[S]$ (we can ignore the singleton components of $G - S$ because the components of $G[S]$ adjacent to them have size at least 1). Let T_i be a twin set that satisfies the conditions of Lemma 5.2. That is, $T_i \cap S \neq \emptyset$, T_i is an independent set, and no twin set adjacent to T_i has a non-empty intersection with S. Hence a component of $G - S$ is adjacent to the singleton components in $T_i \cap S$, if and only if the corresponding family $D_h \in \mathcal{D}$ includes a twin set adjacent to T_i. We say that such D_h is *adjacent to* T_i. Therefore, we add the following constraint: $z_h \leq 1$, for each $D_h \in \mathcal{D}$ adjacent to T_i satisfying Lemma 5.2.

Solving the ILP

Lenstra [17] showed that the feasibility of an ILP formula can be decided in FPT time when parameterized by the number of variables (see also [10, 15]). Fellows et al. [9] extended it to the optimization version. More precisely, we define the problem as follows:

p-OPT-ILP
Input: A matrix $A \in \mathbb{Z}^{m \times p}$, and vectors $b \in \mathbb{Z}^m$ and $c \in \mathbb{Z}^p$.
Question: Find a vector $x \in \mathbb{Z}^p$ that minimizes $c^\top x$ and satisfies that $Ax \geq b$.

They then showed the following:

Theorem 5.6 (Fellows et al. [9]). p-OPT-ILP *can be solved using $O(p^{2.5p+o(p)} \cdot L \cdot \log(MN))$ arithmetic operations and space polynomial in L, where L is the number of bits in the input, N is the maximum absolute values any variable can take, and M is an upper bound on the absolute value of the minimum taken by the objective function.*

In the formulation for SS and CSS, we have at most $O(k)$ variables: x_i for $i \in [k]$, y_j for $C_j \in \mathcal{C}$, and z_h for $D_h \in \mathcal{D}$. Observe that the elements of $\mathcal{C}$ (and of $\mathcal{D}$ as well) are pairwise disjoint. We have only $O(k^2)$ constraints and the variables and coefficients can have values at most $|V(G)|$. Therefore, Theorem 5.1 holds.

6   XP Algorithm Parameterized by Clique-Width

This section presents an XP-time algorithm for SS and CSS parameterized by clique-width. The algorithm runs in time $O(g(c) \cdot n^{f(c)})$, where c is the clique-width. It is known that for any constant c, one can compute a $(2^{c+1} - 1)$-expression of a graph of clique-width c in $O(n^3)$ time [14,19,20].

Due to space restrictions the proof is omitted here, but we can compute in a bottom-up manner for each node t of such an expresssion all entries of the DP tables describing our partial solutions (i.e. the sizes of components containing the vertices of each label and their adjacent components) in time $O(n^{18 \cdot 2^c})$, assuming that the entries for the children of t are already computed. Note that there are $O(n^{10 \cdot 2^c + 1})$ such entries. A $\cup$-node requires time $O(n^{18 \cdot 2^c})$, while for ρ- and η-nodes $O(n^{10 \cdot 2^c})$ time will suffice.

Theorem 6.1. *Given an n-vertex graph G and an irredundant c-expression T of G, the values of $\mathsf{s}(G)$ and $\mathsf{cs}(G)$, along with their corresponding sets can be computed in $O(n^{28 \cdot 2^c + 1})$ time.*

Corollary 6.2. *Given an n-vertex graph G, the values of $\mathsf{s}(G)$ and $\mathsf{cs}(G)$, along with their corresponding sets can be computed in time $n^{O(f(\mathsf{cw}(G)))}$, where $f(c) = 2^{2^{c+1}}$.*

7   Faster Algorithms Parameterized by Solution Size

We know that both SS and CSS admit FPT algorithms [1] when parameterized by the solution size. The algorithms in [1] use Courcelle's theorem [4], however, and thus their dependency on the parameter may be gigantic. The natural question would be whether they admit $O^*(k^k)$-time algorithms as is the case for vertex integrity [7].

We answer this question with the following theorems. The first step of our algorithm for SS is a branching procedure to first guess the correct number of components (k choices) and then guess their sizes (at most k^k choices). We complete our solutions (ensuring they are connected) by constructing and solving appropriate STEINER TREE sub-instances. With a simple modification our algorithm also works for CSS.

Theorem 7.1. SAFE SET *can be solved in $O^*(2^k k^{3k})$ time, where k is the solution size.*

Corollary 7.2. CONNECTED SAFE SET *can be solved in $O^*(2^k k^k)$ time, where k is the solution size.*

References

1. Águeda, R., et al.: Safe sets in graphs: graph classes and structural parameters. J. Comb. Optim. **36**(4), 1221–1242 (2018)
2. Bapat, R.B., et al.: Safe sets, network majority on weighted trees. Networks **71**, 81–92 (2018)
3. Bodlaender, H.L., Downey, R.G., Fellows, M.R., Hermelin, D.: On problems without polynomial kernels. J. Comput. Syst. Sci. **75**(8), 423–434 (2009)
4. Courcelle, B.: The monadic second-order logic of graphs III: tree-decompositions, minor and complexity issues. Theor. Inform. Appl. **26**, 257–286 (1992)
5. Cygan, M., et al.: Parameterized Algorithms. Springer, Cham (2015). https://doi.org/10.1007/978-3-319-21275-3
6. Dom, M., Lokshtanov, D., Saurabh, S.: Kernelization lower bounds through colors and ids. ACM Trans. Algorithms **11**(2), 13:1–13:20 (2014)
7. Drange, P.G., Dregi, M.S., van 't Hof, P.: On the computational complexity of vertex integrity and component order connectivity. Algorithmica, **76**(4), 1181–1202 (2016)
8. Ehard, S., Rautenbach, D.: Approximating connected safe sets in weighted trees. CoRR, abs/1711.11412 (2017)
9. Fellows, M.R., Lokshtanov, D., Misra, N., Rosamond, F.A., Saurabh, S.: Graph layout problems parameterized by vertex cover. In: Hong, S.-H., Nagamochi, H., Fukunaga, T. (eds.) ISAAC 2008. LNCS, vol. 5369, pp. 294–305. Springer, Heidelberg (2008). https://doi.org/10.1007/978-3-540-92182-0_28
10. Frank, A., Tardos, É.: An application of simultaneous diophantine approximation in combinatorial optimization. Combinatorica **7**, 49–65 (1987)
11. Fujita, S., Furuya, M.: Safe number and integrity of graphs. Discrete Appl. Math. **247**, 398–406 (2018)
12. Fujita, S., Jensen, T., Park, B., Sakuma, T.: On weighted safe set problem on paths and cycles. J. Comb. Optim. (to appear)
13. Fujita, S., MacGillivray, G., Sakuma, T.: Safe set problem on graphs. Discrete Appl. Math. **215**, 106–111 (2016)
14. Hlinený, P., Oum, S.: Finding branch-decompositions and rank-decompositions. SIAM J. Comput. **38**(3), 1012–1032 (2008)
15. Kannan, R.: Minkowski's convex body theorem and integer programming. Math. Oper. Res. **12**, 415–440 (1987)
16. Lampis, M.: Algorithmic meta-theorems for restrictions of treewidth. Algorithmica **64**(1), 19–37 (2012)
17. Lenstra Jr., H.W.: Integer programming with a fixed number of variables. Math. Oper. Res. **8**, 538–548 (1983)
18. McConnell, R.M., Spinrad, J.P.: Modular decomposition and transitive orientation. Discrete Math. **201**(1–3), 189–241 (1999)
19. Oum, S.: Approximating rank-width and clique-width quickly. ACM Trans. Algorithms **5**, 1 (2008)
20. Oum, S., Seymour, P.D.: Approximating clique-width and branch-width. J. Comb. Theor. Ser. B **96**, 514–528 (2006)
21. Tedder, M., Corneil, D., Habib, M., Paul, C.: Simpler linear-time modular decomposition via recursive factorizing permutations. In: Aceto, L., Damgård, I., Goldberg, L.A., Halldórsson, M.M., Ingólfsdóttir, A., Walukiewicz, I. (eds.) ICALP 2008. LNCS, vol. 5125, pp. 634–645. Springer, Heidelberg (2008). https://doi.org/10.1007/978-3-540-70575-8_52

Parameterized Complexity of Diameter

Matthias Bentert[(⊠)] and André Nichterlein

Algorithmics and Computational Complexity, Faculty IV,
TU Berlin, Berlin, Germany
{matthias.bentert,andre.nichterlein}@tu-berlin.de

Abstract. DIAMETER—the task of computing the length of a longest shortest path—is a fundamental graph problem. Assuming the Strong Exponential Time Hypothesis, there is no $O(n^{1.99})$-time algorithm even in sparse graphs [Roditty and Williams, 2013]. To circumvent this lower bound we aim for algorithms with running time $f(k)(n+m)$ where k is a parameter and f is a function as small as possible. We investigate which parameters allow for such running times. To this end, we systematically explore a hierarchy of structural graph parameters.

1 Introduction

The diameter is arguably among the most fundamental graph parameters. Most known algorithms for determining the diameter first compute the shortest path between each pair of vertices (APSP: ALL-PAIRS SHORTEST PATHS) and then return the maximum [1]. The currently fastest algorithms for APSP in weighted graphs have a running time of $O(n^3/2^{\Omega(\sqrt{\log n})})$ in dense graphs [11] and $O(nm + n^2 \log n)$ in sparse graphs [23], respectively. In this work, we focus on the unweighted case. Formally, we study the following problem:

DIAMETER
Input: An undirected, connected, unweighted graph $G = (V, E)$.
Task: Compute the length of a longest shortest path in G.

The (theoretically) fastest algorithm for DIAMETER runs in $O(n^{2.373})$ time and is based on fast matrix multiplication [33]. This upper bound can (presumably) not be improved by much as Roditty and Williams [32] showed that solving DIAMETER in $O((n+m)^{2-\varepsilon})$ time for any $\varepsilon > 0$ breaks the SETH (Strong Exponential Time Hypothesis [21,22]). Seeking for ways to circumvent this lower bound, we follow the line of "parameterization for polynomial-time solvable problems" [18] (also referred to as "FPT in P"). This approach is recently actively studied and sparked a lot of research [1,4,9,13,15,16,24,25,27]. Given some parameter k we search for an algorithm with a running time of $f(k)(n+m)^{2-\varepsilon}$ that solves DIAMETER. Starting FPT in P for DIAMETER, Abboud et al. [1] proved that, unless the SETH fails, the function f has to be an *exponential* function if k is the treewidth of the graph. We extend their research by systematically exploring the parameter space looking for parameters where f can be a polynomial. If this is

© Springer Nature Switzerland AG 2019
P. Heggernes (Ed.): CIAC 2019, LNCS 11485, pp. 50–61, 2019.
https://doi.org/10.1007/978-3-030-17402-6_5

not possible (due to conditional lower bounds), then we seek for matching upper bounds of the form $f(k)(n + m)^{2-\varepsilon}$ where f is exponential.

In a second step, we combine parameters that are known to be small in many real world graphs. We concentrate on social networks which often have special characteristics, including the "small-world" property and a power-law degree distribution [26,28–31]. We therefore combine parameters related to the diameter with parameters related to the h-index[1]; both parameters can be expected to be orders of magnitude smaller than the number of vertices.

Related Work. Due to its importance, DIAMETER is extensively studied. Algorithms employed in practice have usually a worst-case running time of $O(nm)$, but are much faster in experiments. See e. g. Borassi et al. [5] for a recent example which also yields good performance bounds using average-case analysis [6]. Concerning worst-case analysis, the theoretically fastest algorithms are based on matrix multiplication and run in $O(n^{2.373})$ time [33] and any $O((n+m)^{2-\varepsilon})$-time algorithm refutes the SETH [32].

The following results on approximating DIAMETER are known: It is easy to see that a simple breadth-first search gives a linear-time 2-approximation. Aingworth et al. [2] improved the approximation factor to $3/2$ at the expense of the higher running time of $O(n^2 \log n + m\sqrt{n \log n})$. The lower bound of Williams [32] also implies that approximating DIAMETER within a factor of $3/2 - \delta$ in $O(n^{2-\varepsilon})$ time refutes the SETH. Moreover, a $3/2 - \delta$-approximation in $O(m^{2-\varepsilon})$ time or a $5/3 - \delta$-approximation in $O(m^{3/2-\varepsilon})$ time also refute the SETH [3,10]. On planar graphs, there is an approximation scheme with near linear running time [35]; the fastest exact algorithm for DIAMETER on planar graphs runs in $O(n^{1.667})$ time [17].

Concerning FPT in P, DIAMETER can be solved in $2^{O(k)}n^{1+o(1)}$ time where k is the treewidth of the graph [9]; however, a $2^{o(k)}n^{2-\varepsilon}$-time algorithm refutes the SETH [1]. In fact, the construction actually proves the same running time lower bound with k being the vertex cover number. The reduction for the lower bound of Roditty and Williams [32] also implicitly implies that the SETH is refuted by any $f(k)(n+m)^{2-\varepsilon}$-time algorithm for DIAMETER for any computable function f when k is either the (vertex deletion) distance to chordal graphs or the combined parameter h-index and domination number. Evald and Dahlgaard [14] adapted the reduction by Roditty and Williams and proved that any $f(k)(n + m)^{2-\varepsilon}$-time algorithm for DIAMETER parameterized by the maximum degree k for any computable function f refutes the SETH.

Our Contribution. We make progress towards systematically classifying the complexity of DIAMETER parameterized by structural graph parameters. Figure 1 gives an overview of previously known and new results and their implications (see Brandstädt et al. [7] for definitions of the parameters). In Sect. 3, we follow the "distance from triviality parameterization" [20] aiming to extend

---

[1] The h-index of a graph G is the largest number ℓ such that G contains at least ℓ vertices of degree at least ℓ.

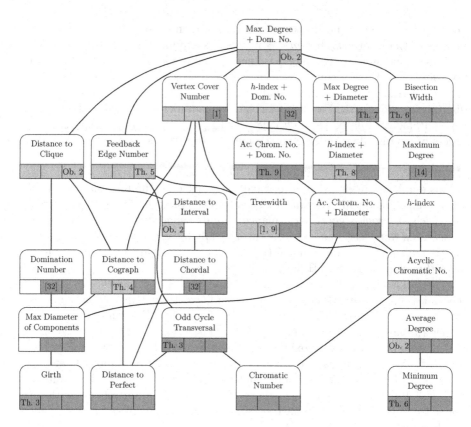

Fig. 1. Overview of the relation between the structural parameters and the respective results for DIAMETER. An edge from a parameter α to a parameter β below of α means that β can be upper-bounded in a polynomial (usually linear) function in α (see also [34]). The three small boxes below each parameter indicate whether there exists (from left to right) an algorithm running in $f(k)n^2$, $f(k)(n \log n + m)$, or $k^{O(1)}(n \log n + m)$ time, respectively. If a small box is green, then a corresponding algorithm exists and the box to the left is also green. Similarly, a red box indicates that a corresponding algorithm is a major breakthrough. More precisely, if a middle box (right box) is red, then an algorithm running in $f(k) \cdot (n + m)^{2-\varepsilon}$ (or $k^{O(1)} \cdot (n + m)^{2-\varepsilon}$) time refutes the SETH. If a left box is red, then an algorithm with running time $f(k)n^2$ implies a faster algorithm for DIAMETER in general. Hardness results for a parameter α imply the same hardness results for the parameters below α. Similarly, algorithms for a parameter β imply algorithms for the parameters above β. (Color figure online)

known tractability results for special graph classes to graphs with small modulators. For example, DIAMETER is linear-time solvable on trees. We obtain for the parameter **feedback edge number** k (edge deletion number to trees) an $O(k \cdot n)$-time algorithm. However, this is our only $k^{O(1)}(n + m)$-time algorithm in this section. For the remaining parameters, it is already known that such algorithms refute the SETH. For the parameter **distance** k **to cographs** we therefore provide a $2^{O(k)}(n + m)$-time algorithm. Finally, for the parameter odd cycle transversal k, we use the recently introduced notion of *General-Problem-hardness* [4] to show that DIAMETER parameterized by k is "as hard" as the unparameterized DIAMETER problem. In Sect. 4, we investigate parameter combinations. We prove that a $k^{O(1)}(n + m)$-time algorithm where k is the combined parameter diameter and maximum degree would refute the SETH. Complementing this lower bound, we provide an $f(k)(n + m)$-time algorithm where k is the combined parameter diameter and h-index.

2   Preliminaries and Basic Observations

For $\ell \in \mathbb{N}$ we set $[\ell] := \{1, 2, \ldots, \ell\}$. Given a graph $G = (V, E)$ set $n := |V|$ and $m := |E|$. A path $P = v_0 \ldots v_a$ is a graph with vertex set $\{v_0, \ldots, v_a\}$ and edge set $\{\{v_i, v_{i+1}\} \mid 0 \le i < a\}$. For $u, v \in V$, we denote with $\mathrm{dist}_G(u, v)$ the distance between u and v in G, that is, the number of edges in a shortest path between u and v. If G is clear from the context, then we omit the subscript. We denote by $d(G)$ the diameter of G. For a vertex subset $V' \subseteq V$, we denote with $G[V']$ the graph induced by V'. We set $G - V' := G[V \setminus V']$.

Parameterized Complexity and GP-Hardness. A language $L \subseteq \Sigma^* \times \mathbb{N}$ is a *parameterized problem* over some finite alphabet Σ, where $(x, k) \in \Sigma^* \times \mathbb{N}$ denotes an instance of L and k is the parameter. Then L is called *fixed-parameter tractable* if there is an algorithm that on input (x, k) decides whether $(x, k) \in L$ in $f(k) \cdot |x|^{O(1)}$ time, where f is some computable function only depending on k and $|x|$ denotes the size of x. For a parameterized problem L, the language $\hat{L} = \{x \mid (x, k) \in L\}$ is called the *unparameterized problem* associated to L. We use the notion of General-Problem-hardness which formalizes the types of reduction that allow us to exclude parameterized algorithms as they would lead to faster algorithms for the general, unparameterized, problem.

Definition 1 ([4, Definition 2]). *Let $P \subseteq \Sigma^* \times \mathbb{N}$ be a parameterized problem, let $\hat{P} \subseteq \Sigma^*$ be the unparameterized decision problem associated to P, and let $g : \mathbb{N} \to \mathbb{N}$ be a polynomial. We call P ℓ-General-Problem-hard(g) 0(ℓ-GP-hard(g)) if there exists an algorithm $\mathcal{A}$ transforming any input instance I of $\hat{P}$ into a new instance (I', k') of P such that*

(G1) $\mathcal{A}$ runs in $O(g(|I|))$ time,   *(G3) $k' \le \ell$, and*
(G2) $I \in \hat{P} \iff (I', k') \in P$,   *(G4) $|I'| \in O(|I|)$.*

We call P General-Problem-hard (g) (GP-hard(g)) if there exists an integer ℓ such that P is ℓ-GP-hard(g). We omit the running time and call P ℓ-General-Problem-hard (ℓ-GP-hard) if g is a linear function.

Showing GP-hardness for some parameter k allows to lift algorithms for the parameterized problem to the unparameterized setting as stated next.

Lemma 2 ([4, Lemma 3]). *Let* $g\colon \mathbb{N} \to \mathbb{N}$ *be a polynomial, let* $P \subseteq \Sigma^* \times \mathbb{N}$ *be a parameterized problem that is GP-hard(g), and let* $\hat{P} \subseteq \Sigma^*$ *be the unparameterized decision problem associated to* P. *If there is an algorithm solving each instance* (I, k) *of* P *in* $O(f(k) \cdot g(|I|))$ *time, then there is an algorithm solving each instance* I' *of* $\hat{P}$ *in* $O(g(|I'|))$ *time.*

Applying Lemma 2 to DIAMETER yields the following. First, having an $f(k)n^{2.3}$ time algorithm with respect to a parameter k for which DIAMETER is GP-hard would yield a faster DIAMETER algorithm. Moreover, from the known SETH-based hardness results [3,10,32] we get the following.

Observation 1. *If the SETH is true and* DIAMETER *is GP-hard($n^{2-\varepsilon}$) with respect to some parameter* k *for some* $\varepsilon > 0$, *then there is no* $f(k) \cdot n^{2-\varepsilon'}$ *time algorithm for any* $\varepsilon' > 0$ *and any function* f.

We next present a simple observation that completes the overview in Fig. 1.

Observation 2 ($\star^2$). DIAMETER *parameterized by distance i to interval graphs, distance c to clique, average degree a, maximum degree Δ, diameter d, and domination number γ is solvable*

1. *in* $O(i \cdot n^2)$ *time provided that the deletion set is given,*
2. *in* $O(c \cdot (n + m))$ *time,*
3. *in* $O(a \cdot n^2)$ *time,*
4. *in* $O(\Delta^{2d+3})$ *time, and,*
5. *in* $O(\gamma^2 \cdot \Delta^3)$ *time, respectively.*

3   Deletion Distance to Special Graph Classes

In this section, we investigate parameterizations that measure the distance to special graph classes. The hope is that when DIAMETER can be solved efficiently in a special graph class Π, then DIAMETER can be solved if the input graph is "almost" in Π. We study the following parameters in this order: odd cycle transversal (which is the same as distance to bipartite graphs), distance to cographs, and feedback edge number. The first two parameters measure the vertex deletion distance to some graph class. Feedback edge number measures the edge deletion distance to trees. Note that the lower bound of Abboud et al. [1] for the parameter vertex cover number already implies that there is no $k^{O(1)}(n + m)^{2-\varepsilon}$-time algorithm for k being one of the first two parameters in our list unless the SETH breaks, since each of these parameters is smaller than vertex cover number (see Fig. 1). This is also the motivation for studying feedback edge number rather than feedback vertex number.

Odd Cycle Transversal. We show that DIAMETER parameterized by odd cycle transversal (from now on called oct) and girth is 4-GP-hard. Consequently solving DIAMETER in $f(k) \cdot n^{2.3}$ for any computable function f, implies an $O(n^{2.3})$-time algorithm for DIAMETER which would improve the currently best (unparameterized) algorithm. The girth of a graph is the length of a shortest cycle in it.

---

[2] Results marked with ($\star$) are deferred to a full version, available under https://arxiv.org/abs/1802.10048.

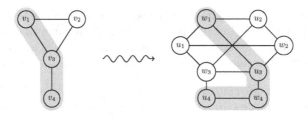

Fig. 2. Example for the construction in the proof of Theorem 3. The input graph given on the left side has diameter two and the constructed graph on the right side has diameter three. In each graph one longest shortest path is highlighted.

Theorem 3. DIAMETER *is 4-GP-hard with respect to the combined parameter* oct *and* girth.

Proof. Let $G = (V, E)$ be an arbitrary undirected graph where $V = \{v_1, v_2, \ldots, v_n\}$. We construct a new graph $G' = (V', E')$ as follows: $V' := \{u_i, w_i \mid v_i \in V\}$ and $E' := \{\{u_i, w_j\}, \{u_j, w_i\} \mid \{v_i, v_j\} \in E\} \cup \{\{u_i, w_i\} \mid v_i \in V\}$. An example of this construction can be seen in Fig. 2. We will now prove that all properties of Definition 1 hold. It is easy to verify that the reduction can be implemented in linear time and therefore the resulting instance is of linear size as well. Observe that $\{u_i \mid v_i \in V\}$ and $\{w_i \mid v_i \in V\}$ are both independent sets and therefore G' is bipartite. Notice further that for any edge $\{v_i, v_j\} \in E$ there is an induced cycle in G' containing the vertices $\{u_i, w_i, u_j, w_j\}$. Since G' is bipartite there is no cycle of length three in G' and the girth of G' is four.

Lastly, we show that $d(G') = d(G) + 1$ by proving that if $\mathrm{dist}(v_i, v_j)$ is odd, then $\mathrm{dist}(u_i, w_j) = \mathrm{dist}(v_i, v_j)$ and $\mathrm{dist}(u_i, u_j) = \mathrm{dist}(v_i, v_j) + 1$, and if $\mathrm{dist}(v_i, v_j)$ is even, then $\mathrm{dist}(u_i, u_j) = \mathrm{dist}(v_i, v_j)$ and $\mathrm{dist}(u_i, w_j) = \mathrm{dist}(v_i, v_j) + 1$. Since $\mathrm{dist}(u_i, w_i) = 1$ and $\mathrm{dist}(u_i, w_j) = \mathrm{dist}(u_j, w_i)$, this will conclude the proof.

Let $P = v_{a_0} v_{a_1} \ldots v_{a_d}$ be a shortest path from v_i to v_j where $v_{a_0} = v_i$ and $v_{a_d} = v_j$. Let $P' = u_{a_0} w_{a_1} u_{a_2} w_{a_3} \ldots$ be a path in G'. Clearly, P' is also a shortest path as there are no edges $\{u_i, w_j\} \in E'$ where $\{v_i, v_j\} \notin E$.

If d is odd, then $u_{a_0} w_{a_1} \ldots w_{a_d}$ is a path of length d from u_i to w_j and $u_{a_0} w_{a_1} \ldots w_{a_d} u_{a_d}$ is a path of length $d + 1$ from u_i to u_j. If d is even, then $u_{a_0} w_{a_1} \ldots w_{a_{d-1}} u_{a_d}$ is a path of length d from u_i to u_j and $u_{a_0} w_{a_1} \ldots w_{a_{d-1}} u_{a_d} w_{a_d}$ is a path of length $d + 1$ from u_i to w_j. Notice that G' is bipartite and thus $\mathrm{dist}(u_i, u_j)$ must be even and $\mathrm{dist}(u_i, w_j)$ must be odd. □

Distance to Cographs. Providing an algorithm that matches the lower bound of Abboud et al. [1], we will show that DIAMETER parameterized by distance k to cographs can be solved in $O(k \cdot (n + m) + 2^{O(k)})$ time.

A graph is a cograph if and only if it does not contain a P_4 as an induced subgraph, where P_4 is the path on four vertices. Given a graph G one can determine in linear time whether G is a cograph and can return an induced P_4 if this is not the case [8,12]. This implies that in $O(k \cdot (n + m))$ time one can compute a set $K \subseteq V$ with $|K| \leq 4k$ such that $G - K$ is a cograph: Iteratively add all four vertices of a returned P_4 into the solution set and delete those vertices from G until it is P_4-free. In the following, we hence assume that such a set K is given. Notice that every cograph has diameter at most two as any graph with diameter at least three contains an induced P_4.

Theorem 4 ($\star$). DIAMETER *can be solved in* $O(k \cdot (n + m) + 2^{16k}k)$ *time when parameterized by* distance k to cographs.

Feedback Edge Number. We will prove that DIAMETER parameterized by feedback edge number k can be solved in $O(k \cdot n)$ time. One can compute a minimum feedback edge set K (with $|K| = k$) in linear time by taking all edges not in a spanning tree. Recently, this parameter was used to speed up algorithms computing maximum matchings [24].

Theorem 5 ($\star$). DIAMETER *parameterized by* feedback edge number k *can be solved in* $O(k \cdot n)$ *time.*

4   Parameters for Social Networks

Here, we study parameters that we expect to be small in social networks. Recall that social networks have the "small-world" property and a power-law degree distribution [26, 28–31]. The "small-world" property directly transfers to the diameter. We capture the power-law degree distribution by the h-index as only few high-degree exist in the network. Thus, we investigate parameters related to the diameter and to the h-index starting with degree-related parameters.

4.1   Degree Related Parameters

We next investigate the parameter minimum degree. Unsurprisingly, the minimum degree is not helpful for parameterized algorithms. In fact, we show that DIAMETER is 2-GP-hard with respect to the combined parameter bisection width and minimum degree. The bisection width of a graph G is the minimum number of edges to delete from G in order to partition G into two connected component whose number of vertices differ by at most one.

Theorem 6 ($\star$). DIAMETER *is 2-GP-hard with respect to* bisection width and minimum degree.

We mention in passing that the constructed graph in the proof behind Theorem 6 contains the original graph as an induced subgraph and if the original graph is bipartite, then so is the constructed graph. Thus, first applying the construction in the proof of Theorem 3 (see also Fig. 2) and then the construction in the proof of Theorem 6 proves that DIAMETER is GP-hard even parameterized by the sum of girth, bisection width, minimum degree, and oct.

4.2   Parameters Related to Both Diameter and h-Index

Here, we will study combinations of two parameters where the first one is related to diameter and the second to h-index (see Fig. 1 for an overview of closely related parameters). We start with the combination maximum **degree** and **diameter**. Interestingly, although the parameter is quite large, the naive algorithm (see Observation 2) cannot be improved to a fully polynomial running time.

Theorem 7. *There is no* $(d + \Delta)^{O(1)}(n + m)^{2-\epsilon}$*-time algorithm that solves* DIAMETER *parameterized by* maximum degree Δ *and diameter* d *unless the SETH is false.*

Proof. We prove a slightly stronger statement excluding $2^{o(\sqrt[c]{d+\Delta})} \cdot (n + m)^{2-\epsilon}$-time algorithms for some constant c. Assume towards a contradiction that for each constant r there is a $2^{o(\sqrt[r]{d+\Delta})} \cdot (n + m)^{2-\epsilon}$-time algorithm that solves DIAMETER parameterized by maximum **degree** Δ and **diameter** d. Evald and Dahlgaard [14] have shown a reduction from CNF-SAT to DIAMETER where the resulting graph has maximum degree three such that for any constant $\epsilon > 0$ an $O((n + m)^{2-\epsilon})$-time algorithm (for DIAMETER) would refute the SETH. A closer look reveals that there is some constant c such that the diameter d in their constructed graph is in $O(\log^c(n + m))$. By assumption we can solve DIAMETER parameterized by maximum **degree** and **diameter** in $2^{o(\sqrt[c]{d+\Delta})} \cdot (n + m)^{2-\epsilon}$ time. Observe that

$$2^{o(\sqrt[c]{d+\Delta})} \cdot (n + m)^{2-\epsilon} = 2^{o(\sqrt[c]{\log^c(n+m)})} \cdot (n + m)^{2-\epsilon}$$
$$= (n + m)^{o(1)} \cdot (n + m)^{2-\epsilon} \subseteq O((n + m)^{2-\epsilon'}) \text{ for some } \varepsilon' > 0.$$

Since we constructed for some $\epsilon' > 0$ an $O((n + m)^{2-\epsilon'})$-time algorithm for DIAMETER the SETH fails and thus we reached a contradiction. Finally, notice that $(d + \Delta)^{O(1)} \subset 2^{o(\sqrt[c]{d+\Delta})}$ for any constant c.   □

h-Index and Diameter. We next investigate in the combined parameter h-index and diameter. The reduction by Roditty and Williams [32] produces instances with constant domination number and logarithmic **vertex cover** number (in the input size). Since the **diameter** d is linearly upper-bounded by the domination number and the h-index is linearly upper-bounded by the **vertex cover** number, any algorithm that solves DIAMETER parameterized by the combined parameter $(d + h)$ in $2^{o(d+h)} \cdot (n + m)^{2-\epsilon}$ time disproves the SETH. We will now present an algorithm for DIAMETER parameterized by h-index and diameter that almost matches the lower bound.

Theorem 8. DIAMETER *parameterized by* diameter d *and* h-Index h *is solvable in* $O(h \cdot (m + n) + n \cdot d \cdot 2^{h \log d + d \log h})$ *time.*

Proof. Let $H = \{x_1, \ldots, x_h\}$ be a set of vertices such that all vertices in $V \setminus H$ have degree at most h in G. Clearly, H can be computed in linear time. We will describe a two-phase algorithm with the following basic idea: In the first phase it performs a breadth-first search from each vertex $v \in H$, stores the distance to each other vertex and uses this to compute the "type" of each vertex, that is, a characterization by the distance to each vertex in H. In the second phase it iteratively increases a value e and verifies whether there is a vertex pair of distance at least e. If at any point no vertex pair is found, then the diameter of G is $e - 1$.

The first phase is straight forward: Compute a BFS from each vertex v in H and store the distance from v to every other vertex w in a table. Then iterate over each vertex $w \in V \setminus H$ and compute a vector of length h where the ith entry represents the distance from w to x_i. Also store the number of vertices of each type containing at least one vertex. Since the distance to any vertex is at most d, there are at most d^h different types. This first phase takes $O(h \cdot (m+n))$ time.

For the second phase, we initialize e with the largest distance found so far, that is, the maximum value stored in the table and compute $G' = G - H$. Iteratively check whether there is a pair of vertices in $V \setminus H$ of distance at least $e + 1$ as follows. We check for each vertex $v \in V \setminus H$ and each type whether there is a path of length at most e from v to each vertex of this type through a vertex in H. This can be done by computing the sum of the two type-vectors in $O(h)$ time and comparing the minimum entry in this sum with e. If all entries are larger than e, then no shortest path from v to some vertex w of the respective type of length at most e can contain any vertex in H. Thus we compute a BFS from v in G' up to depth e and count the number of vertices of the respective type we found. If this number equals the total number of vertices of the respective type, then for all vertices w of this type it holds that $\text{dist}(v, w) \le e$. If the two numbers do not match, then there is a vertex pair of distance at least $e + 1$ so we can increase e by one and start the process again.

There are at most d iterations in which e is increased and the check is done. Recall that the maximum degree in G' is h and therefore each of these iteration takes $O(n \cdot d^h \cdot (h^e + h))$ time as each BFS to depth e takes $O(h^e)$ time. Thus, the overall running time is in $O(h \cdot (m + n) + n \cdot d \cdot 2^{h \log d + d \log h})$. □

Acyclic Chromatic Number and Domination Number. We next analyze the parameterized complexity of DIAMETER parameterized by acyclic chromatic number a and domination number d. The acyclic chromatic number of a graph is the minimum number of colors needed to color each vertex with one of the given colors such that each subgraph induced by all vertices of one color is an independent set and each subgraph induced by all vertices of two colors is acyclic. The acyclic chromatic number upper-bounds the average degree, and therefore the standard $O(n \cdot m)$-time algorithm runs in $O(n^2 \cdot a)$ time. We will show that this is essentially the best one can hope for as we can exclude $f(a, d) \cdot (n + m)^{2-\varepsilon}$-time algorithms under SETH. Our result is based on the reduction by Roditty and Williams [32] and is modified such that the acyclic chromatic number and domination number are both four in the resulting graph.

Theorem 9 (⋆). *There is no $f(a,d) \cdot (n+m)^{2-\epsilon}$-time algorithm for any computable function f that solves* DIAMETER *parameterized by* acyclic chromatic number a *and* domination number d *unless the SETH is false.*

5 Conclusion

We have resolved the complexity status of DIAMETER for most of the parameters in the complexity landscape shown in Fig. 1. However, several open questions remain. For example, is there an $f(k)n^2$-time algorithm with respect to the parameter diameter? Moreover, our algorithms working with parameter combinations have mostly impractical running times which, assuming SETH, cannot be improved by much. So the question arises, whether there are parameters $k_1, \ldots, k_\ell$ that allow for practically relevant running times like $\prod_{i=1}^{\ell} k_i \cdot (n+m)$ or even $(n+m) \cdot \sum_{i=1}^{\ell} k_i$? The list of parameters displayed in Fig. 1 is by no means exhaustive. Hence, the question arises which other parameters are small in typical scenarios? For example, what is a good parameter capturing the special community structures of social networks [19]?

References

1. Abboud, A., Williams, V.V., Wang, J.R.: Approximation and fixed parameter subquadratic algorithms for radius and diameter in sparse graphs. In: Proceedings of the 27th Annual ACM-SIAM Symposium on Discrete Algorithms (SODA 2016), pp. 377–391. SIAM (2016)
2. Aingworth, D., Chekuri, C., Indyk, P., Motwani, R.: Fast estimation of diameter and shortest paths (without matrix multiplication). SIAM J. Comput. **28**(4), 1167–1181 (1999)
3. Backurs, A., Roditty, L., Segal, G., Williams, V.V., Wein, N.: Towards tight approximation bounds for graph diameter and eccentricities. In: Proceedings of the 50th Annual ACM SIGACT Symposium on Theory of Computing (STOC 2018), pp. 267–280. ACM (2018)
4. Bentert, M., Fluschnik, T., Nichterlein, A., Niedermeier, R.: Parameterized aspects of triangle enumeration. In: Klasing, R., Zeitoun, M. (eds.) FCT 2017. LNCS, vol. 10472, pp. 96–110. Springer, Heidelberg (2017). https://doi.org/10.1007/978-3-662-55751-8_9
5. Borassi, M., Crescenzi, P., Habib, M., Kosters, W.A., Marino, A., Takes, F.W.: Fast diameter and radius BFS-based computation in (weakly connected) real-world graphs: with an application to the six degrees of separation games. Theor. Comput. Sci. **586**, 59–80 (2015)
6. Borassi, M., Crescenzi, P., Trevisan, L.: An axiomatic and an average-case analysis of algorithms and heuristics for metric properties of graphs. In: Proceedings of the 28th Annual ACM-SIAM Symposium on Discrete Algorithms (SODA 2017), pp. 920–939. SIAM (2017)
7. Brandstädt, A., Le, V.B., Spinrad, J.P.: Graph Classes: A Survey, SIAM Monographs on Discrete Mathematics and Applications, vol. 3. SIAM, Philadelphia (1999)

8. Bretscher, A., Corneil, D.G., Habib, M., Paul, C.: A simple linear time LexBFS cograph recognition algorithm. SIAM J. Discret. Math. **22**(4), 1277–1296 (2008)

9. Bringmann, K., Husfeldt, T., Magnusson, M.: Multivariate analysis of orthogonal range searching and graph distances parameterized by treewidth. Computing Research Repository abs/1805.07135 (2018). Accepted at IPEC 2018

10. Cairo, M., Grossi, R., Rizzi, R.: New bounds for approximating external distances in undirected graphs. In: Proceedings of the 27th Annual ACM-SIAM Symposium on Discrete Algorithms (SODA 2016), pp. 363–376. SIAM (2016)

11. Chan, T.M., Williams, R.: Deterministic APSP, orthogonal vectors, and more: quickly derandomizing Razborov-Smolensky. In: Proceedings of the 27th Annual ACM-SIAM Symposium on Discrete Algorithms (SODA 2016), pp. 1246–1255. SIAM (2016)

12. Corneil, D.G., Perl, Y., Stewart, L.K.: A linear recognition algorithm for cographs. SIAM J. Comput. **14**(4), 926–934 (1985)

13. Coudert, D., Ducoffe, G., Popa, A.: Fully polynomial FPT algorithms for some classes of bounded clique-width graphs. In: Proceedings of the 29th Annual ACM-SIAM Symposium on Discrete Algorithms (SODA 2018), pp. 2765–2784. SIAM (2018)

14. Evald, J., Dahlgaard, S.: Tight hardness results for distance and centrality problems in constant degree graphs. Computing Research Repository abs/1609.08403 (2016)

15. Fluschnik, T., Komusiewicz, C., Mertzios, G.B., Nichterlein, A., Niedermeier, R., Talmon, N.: When can graph hyperbolicity be computed in linear time? Algorithms and Data Structures. LNCS, vol. 10389, pp. 397–408. Springer, Cham (2017). https://doi.org/10.1007/978-3-319-62127-2_34

16. Fomin, F.V., Lokshtanov, D., Saurabh, S., Pilipczuk, M., Wrochna, M.: Fully polynomial-time parameterized computations for graphs and matrices of low treewidth. ACM Trans. Algorithms **14**(3), 34:1–34:45 (2018)

17. Gawrychowski, P., Kaplan, H., Mozes, S., Sharir, M., Weimann, O.: Voronoi diagrams on planar graphs, and computing the diameter in deterministic $\tilde{O}(n^{5/3})$ time. In: Proceedings of the 29th Annual ACM-SIAM Symposium on Discrete Algorithms (SODA 2018), pp. 495–514. SIAM (2018)

18. Giannopoulou, A.C., Mertzios, G.B., Niedermeier, R.: Polynomial fixed-parameter algorithms: a case study for longest path on interval graphs. Theor. Comput. Sci. **689**, 67–95 (2017)

19. Girvan, M., Newman, M.E.J.: Community structure in social and biological networks. Proc. Natl. Acad. Sci. **99**(12), 7821–7826 (2002)

20. Guo, J., Hüffner, F., Niedermeier, R.: A structural view on parameterizing problems: distance from triviality. In: Downey, R., Fellows, M., Dehne, F. (eds.) IWPEC 2004. LNCS, vol. 3162, pp. 162–173. Springer, Heidelberg (2004). https://doi.org/10.1007/978-3-540-28639-4_15

21. Impagliazzo, R., Paturi, R.: On the complexity of k-SAT. J. Comput. Syst. Sci. **62**(2), 367–375 (2001)

22. Impagliazzo, R., Paturi, R., Zane, F.: Which problems have strongly exponential complexity? J. Comput. Syst. Sci. **63**(4), 512–530 (2001)

23. Johnson, D.B.: Efficient algorithms for shortest paths in sparse networks. J. ACM **24**(1), 1–13 (1977)

24. Korenwein, V., Nichterlein, A., Niedermeier, R., Zschoche, P.: Data reduction for maximum matching on real-world graphs: theory and experiments. In: Proceedings of the 26th Annual European Symposium on Algorithms (ESA 2018). LIPIcs, vol. 112. Schloss Dagstuhl - Leibniz-Zentrum fuer Informatik (2018)

25. Kratsch, S., Nelles, F.: Efficient and adaptive parameterized algorithms on modular decompositions. In: Proceedings of the 26th Annual European Symposium on Algorithms (ESA 2018). LIPIcs, vol. 112, pp. 55:1–55:15. Schloss Dagstuhl - Leibniz-Zentrum fuer Informatik (2018)

26. Leskovec, J., Horvitz, E.: Planetary-scale views on a large instant-messaging network. In: Proceedings of the 17th International World Wide Web Conference (WWW 2008), pp. 915–924. ACM (2008). ISBN 978-1-60558-085-2

27. Mertzios, G.B., Nichterlein, A., Niedermeier, R.: The power of linear-time data reduction for maximum matching. In: Proceedings of the 42nd International Symposium on Mathematical Foundations of Computer Science (MFCS 2017). LIPIcs, vol. 83, pp. 46:1–46:14. Schloss Dagstuhl - Leibniz-Zentrum fuer Informatik (2017)

28. Milgram, S.: The small world problem. Psychol. Today **1**, 61–67 (1967)

29. Newman, M.E.J.: The structure and function of complex networks. SIAM Rev. **45**(2), 167–256 (2003)

30. Newman, M.E.J.: Networks: An Introduction. Oxford University Press, Oxford (2010)

31. Newman, M.E.J., Park, J.: Why social networks are different from other types of networks. Phys. Rev. E **68**(3), 036122 (2003)

32. Roditty, L., Williams, V.V.: Fast approximation algorithms for the diameter and radius of sparse graphs. In: Proceedings of the 45th Symposium on Theory of Computing Conference (STOC 2013), pp. 515–524. ACM (2013)

33. Seidel, R.: On the all-pairs-shortest-path problem in unweighted undirected graphs. J. Comput. Syst. Sci. **51**(3), 400–403 (1995)

34. Sorge, M., Weller, M.: The graph parameter hierarchy (2013, manuscript)

35. Weimann, O., Yuster, R.: Approximating the diameter of planar graphs in near linear time. ACM Trans. Algorithms **12**(1), 12:1–12:13 (2016)

Fixed-Parameter Algorithms
for Maximum-Profit Facility Location
Under Matroid Constraints

René van Bevern[1,2]([✉]), Oxana Yu. Tsidulko[1,2], and Philipp Zschoche[3]

[1] Department of Mechanics and Mathematics, Novosibirsk State University,
Novosibirsk, Russian Federation
rvb@nsu.ru
[2] Sobolev Institute of Mathematics of the Siberian Branch of the Russian Academy
of Sciences, Novosibirsk, Russian Federation
tsidulko@math.nsc.ru
[3] Algorithmics and Computational Complexity, Fakultät IV, TU Berlin,
Berlin, Germany
zschoche@tu-berlin.de

Abstract. We consider an uncapacitated discrete facility location problem where the task is to decide which facilities to open and which clients to serve for maximum profit so that the facilities form an independent set in given facility-side matroids and the clients form an independent set in given client-side matroids. We show that the problem is fixed-parameter tractable parameterized by the number of matroids and the minimum rank among the client-side matroids. To this end, we derive fixed-parameter algorithms for computing representative families for matroid intersections and maximum-weight set packings with multiple matroid constraints. To illustrate the modeling capabilities of the new problem, we use it to obtain algorithms for a problem in social network analysis. We complement our tractability results by lower bounds.

Keywords: Matroid set packing · Matroid parity · Matroid median ·
Representative families · Social network analysis · Strong triadic closure

1 Introduction

The uncapacitated facility location problem (UFLP) is a classical problem studied in operations research [14]: a company has to decide where to open facilities in order to serve its clients. Opening a facility incurs a cost, whereas serving clients yields profit. The task is to decide where to open facilities so as to maximize the profit minus the cost for opening facilities.

Numerous algorithms have been developed for the UFLP [14, Sect. 3.4]. Yet in practice, the required solutions are often subject to additional side constraints [14], which make algorithms for the pure UFLP inapplicable. For example:

© Springer Nature Switzerland AG 2019
P. Heggernes (Ed.): CIAC 2019, LNCS 11485, pp. 62–74, 2019.
https://doi.org/10.1007/978-3-030-17402-6_6

- The number of facilities in each of several, possibly overlapping areas is limited (environmental protection, precaution against terrorism or sabotage).
- Facilities may be subject to move [25].
- A client may be served at one of several locations, which influences the profit.

In order to analyze the influence of such constraints on the complexity of UFLP and to capture such constraints in a single algorithm, we introduce the following uncapacitated facility location problem with matroid constraints (UFLP-MC).

Problem 1.1 (UFLP-MC).

Input: A universe U with $n := |U|$, for each pair $u, v \in U$ a *profit* $p_{uv} \in \mathbb{N}$ obtained when a facility at u serves a client at v, for each $u \in U$ a *cost* $c_u \in \mathbb{N}$ for opening a facility at u, *facility matroids* $\{(U_i, A_i)\}_{i=1}^a$, and *client matroids* $\{(V_i, C_i)\}_{i=1}^c$, where $U_i \cup V_i \subseteq U$.
Task: Find two disjoint sets $A \uplus C \subseteq U$ that maximize the *profit*

$$\sum_{v \in C} \max_{u \in A} p_{uv} - \sum_{u \in A} c_u \quad \text{such that} \quad A \in \bigcap_{i=1}^a A_i \quad \text{and} \quad C \in \bigcap_{i=1}^c C_i.$$

Besides modeling natural facility location scenarios like the ones described above, UFLP-MC generalizes several well-known combinatorial optimization problems.

Example 1.2. Using UFLP-MC with $a = 1$ facility matroid and $c = 0$ client matroids one can model the classical NP-hard problem of covering a maximum number of elements of a set V using at most r sets of a collection $\mathcal{H} \subseteq 2^V$. To this end, choose the universe $U = V \cup \mathcal{H}$, a single facility matroid $(\mathcal{H}, A_1)$ with $A_1 := \{H \subseteq \mathcal{H} \mid |H| \le r\}$, $c_u = 0$ for each $u \in U$, and, for each $u, v \in U$,

$$p_{uv} = \begin{cases} 1 & \text{if } u \in \mathcal{H} \text{ such that } v \in u, \\ 0 & \text{otherwise.} \end{cases}$$

Already UFLP (without matroid constraints) does not allow for polynomial-time approximation schemes unless $P = NP$ [2]. Moreover, from Example 1.2 and the W[2]-hardness of SET COVER [6], it immediately follows that UFLP-MC is W[2]-hard parameterized by r even for zero costs, binary profits, and a single facility matroid of rank r, making the problem of optimally placing a *small* number r of facilities already hard when the set of clients is unconstrained. However, facility location problems also have been studied when the number of *clients* is small [1] and have several applications, for example:

- When clients are not end customers, but in the middle of the supply chain, then their number may indeed be small. For example, there are many possible locations for waste-to-energy plants, yet clients are a few city waste dumps.[1]

---

[1] The State Register of Waste Disposal Facilities of the Russian Federation lists 18 dumps for municipal solid waste in Moscow region—the largest city of Europe.

– In the stop location problem [23], the clients correspond to cities that are served by new train stops to be built along existing rail infrastructure.
– Due to resource constraints or due to the constraint that each client be served at only one of several possible locations, the number of actually served clients may be small compared to the number of possible client and facility locations.

We aim for optimally solving UFLP-MC when the number of served clients is small (not necessarily constant). To this end, we employ *fixed-parameter algorithms* and thus contribute to the still relatively scarce results on the parameterized complexity of problems in operations research.

Parameterized Complexity. The main idea of fixed-parameter algorithms is to accept the exponential running time seemingly inherent to solving NP-hard problems, yet to confine the combinatorial explosion to a parameter of the problem, which can be small in applications [6]. A problem is *fixed-parameter tractable* if it can be solved in $f(k) \cdot \mathrm{poly}(n)$ time on inputs of length n and some function f depending only on some parameter k. This requirement is stronger than an algorithm that merely runs in polynomial time for fixed k, say, in $O(n^k)$ time, which is intractable even for small values of k. The parameterized analog of NP and NP-hardness is the W-hierarchy FPT $\subseteq W[1] \subseteq W[2] \subseteq \ldots W[P] \subseteq$ XP and $W[t]$-hardness, where FPT is the class of fixed-parameter tractable problems and all inclusions are conjectured to be strict. If some $W[t]$-hard problem is in FPT, then FPT $= W[t]$ [6].

1.1   Our Contributions and Organization of this Work

We introduce UFLP-MC and present first algorithmic results (we refer the reader to Sect. 2 for basic definitions from matroid theory).

Theorem 1.3. *UFLP-MC is*

(i) *solvable in $2^{O(r \log r)} \cdot n^2$ time for a single facility matroid given as an independence oracle and a single uniform client matroid of rank r.*
(ii) *fixed-parameter tractable parameterized by $a+c+r$, where r is the minimum rank of the client matroids and representations of all matroids over the same finite field $\mathbb{F}_{p^d}$ are given for some prime p polynomial in the input size.*

We point out that, if our aim is to maximize the profit from serving only k clients, then we can always add a uniform client matroid of rank k and Theorem 1.3(ii) gives a fixed-parameter algorithm for the parameter $a+c+k$. In contrast, Example 1.2 shows that one cannot replace r by the minimum rank of *all* matroids in Theorem 1.3(ii)

To illustrate the modeling capabilities of the newly introduced problem, in Sect. 3, we use Theorem 1.3 to obtain fixed-parameter algorithms for a problem in social network analysis, complementing known approximation results.

In Sect. 4, we present new results of independent interest to matroid optimization. In Sect. 5, we show how to use these results to prove Theorem 1.3.

Using a parameterized reduction from the CLIQUE problem via a construction resembling that of Schrijver [24, Sect. 43.9], we can also show that Theorem 1.3(ii) does not generalize to non-representable client matroids:

Theorem 1.4. *UFLP-MC even with unit costs, binary profits, without facility matroids, and a single (non-representable) client matroid of rank r is W[1]-hard parameterized by r.*

Proofs are deferred to a full version of this article, available on arXiv:1806.11527.

1.2 Related Work

Uncapacitated Facility Location. The literature usually studies the variant where each facility u has an opening cost c_u, serving client v by facility u costs p_{uv}, and one *minimizes* the total cost of serving *all* clients. Fellows and Fernau [7] study the parameterized complexity of this variant. Krishnaswamy et al. [13] and Swamy [25] study approximation algorithms for this variant with one facility matroid and without client matroids, called the *matroid median* problem. In case where the clients also have to obey matroid constraints, the minimization problem is meaningless: the minimization variant of UFLP-MC would simply not serve clients and not open facilities. Thus, we study the problem of *maximizing* total profit minus facility opening costs. Ageev and Sviridenko [2] study approximation algorithms for this variant (yet without matroid constraints).

Combinatorial Optimization with Matroid Constraints. Other classical combinatorial optimization problems such as MAXIMUM COVERAGE [8], submodular function maximization [5], and SET PACKING [16] have been studied with a matroid constraint. Towards proving Theorem 1.3, we obtain a fixed-parameter algorithm WEIGHTED SET PACKING subject to multiple matroid constraints.

Matroids in Parameterized Complexity. Matroids are an important tool in the development of fixed-parameter algorithms [21]. Many of them are based on so-called *representative families* for matroids [9,10,17,19], which we will generalize to representative families for weighted matroid *intersections*. Marx [19] showed that a common independent set of size r in m matroids can be found in $f(r, m) \cdot$ poly(n) time. Bonnet et al. [4] showed that the problem of covering at least p elements of a set V using at most k sets of a given family $\mathcal{H} \subseteq 2^V$ is fixed-parameter tractable parameterized by p. Earlier, Marx [18] showed that PARTIAL VERTEX COVER is fixed-parameter tractable by p. Our Theorem 1.3 generalizes all of these results and, indeed, is based on the color coding approach in Marx's [18] algorithm.

2 Preliminaries

By $\mathbb{N}$, we denote the naturals numbers including zero. By $\mathbb{F}_p$, we denote the field on p elements. By $n = |U|$, we denote the size of the universe (matroid ground set).

Sets and Set Functions. By $A \uplus B$, we denote the union of sets A and B that we require to be disjoint. By convention, the intersection of no sets is the whole universe and the union of no sets is the empty set.

We call $Z_1, \ldots, Z_\ell$ a *partition* of a set A if $Z_1 \uplus \cdots \uplus Z_\ell = A$ and $Z_i \neq \emptyset$ for each $i \in \{1, \ldots, \ell\}$. We call $A \subseteq 2^U$ a γ-*family* if each set in A has cardinality exactly γ. A set function $w \colon 2^U \to \mathbb{R}$ is *additive* if, for any subsets $A \cup B \subseteq U$, one has $w(A \cup B) = w(A) + w(B) - w(A \cap B)$. If "$\leq$" holds instead of "$=$", then w is called *submodular*.

Matroid Fundamentals. For proofs of the following propositions and for illustrative examples of the following definitions, we refer to the book of Oxley [20].

A pair $(U, \mathcal{I})$, where U is the *ground set* and $\mathcal{I} \subseteq 2^U$ is a family of *independent sets*, is a *matroid* if (i) $\emptyset \in \mathcal{I}$, (ii) if $A' \subseteq A$ and $A \in \mathcal{I}$, then $A' \in \mathcal{I}$, and (iii) if $A, B \in \mathcal{I}$ and $|A| < |B|$, then there is an $x \in B \setminus A$ such that $A \cup \{x\} \in \mathcal{I}$. An inclusion-wise maximal independent set $A \in \mathcal{I}$ is a *basis*. The cardinality of the bases of M is called the *rank* of M.

Each matroid M has a *dual matroid* M^* whose bases are the complements of bases of M. The *union* $M_1 \vee M_2 = (U_1 \cup U_2, \{J_1 \cup J_2 \mid J_1 \in I_1, J_2 \in I_2\})$ of two matroids $M_1 = (U_1, I_1)$ and $M_2 = (U_2, I_2)$ is a matroid. If $U_1 \cap U_2 = \emptyset$, we write $M_1 \oplus M_2 := M_1 \vee M_2$ and call their union *direct sum*.

In a *free matroid* $(U, 2^U)$ every set is independent. A *uniform matroid of rank* r is a matroid $(U, \mathcal{I})$ such that $\mathcal{I} := \{S \subseteq U \mid |S| \leq r\}$. The direct sum of uniform matroids is called *partition matroid*. We call the direct sum of uniform matroids of rank one a *multicolored matroid*. The k-*truncation* of a matroid $(U, \mathcal{I})$ is a matroid $(U, \mathcal{I}')$ with $\mathcal{I}' = \{S \subseteq U \mid S \in \mathcal{I} \wedge |S| \leq k\}$.

Matroid Representations. An *independence oracle* is an algorithm that answers in constant time whether a given set is independent in a given matroid. A matroid $M = (U, \mathcal{I})$ is *representable over a field* $\mathbb{F}$ if there is a matrix A with n columns labeled by elements of U such that $S \in \mathcal{I}$ if and only if the columns of A with labels in S are linearly independent over $\mathbb{F}$. Not all matroids are representable over all fields [20, Theorem 6.5.4]. Some are not representable at all [20, Example 1.5.14]. Given a representation of a matroid M over a field $\mathbb{F}$, a representation of the dual matroid M^* over $\mathbb{F}$ is computable in linear time [20, Theorem 2.2.8]. Given representation of two matroids M_1 and M_2 over $\mathbb{F}$, one can easily obtain a representation of $M_1 \oplus M_2$ over $\mathbb{F}$ [20, Exercise 6, p. 132]. Uniform matroids of rank r on a universe of size n are representable over all fields with at least n elements [19, Sect. 3.5]. The uniform matroid of rank one is trivially representable over *all* fields. Thus, so are multicolored matroids.

3   Finding Strong Links in Social Networks

We now illustrate the modeling capabilities of our newly introduced problem UFLP-MC by using it to model a problem from social network analysis. In sociological work, Granovetter [12] stated the *Strong Triadic Closure (STC)* hypothesis: if two agents in a social network have a strong tie to a third agent, then they should have at least a weak tie to each other. An induced path on three vertices, also called an *open triangle*, consisting of strong edges only is an *STC violation*.

Finding the strong ties in social networks helps improving clustering and advertising algorithms. Rozenshtein et al. [22] consider the following scenario: one is given a graph $G = (V, E)$ and communities $X_1, \ldots, X_m \subseteq V$, each of which represents a group with common interests and is thus assumed to be connected via strong ties. They showed that it is NP-hard to check whether one can label the edges of a graph strong or weak so that the subgraphs $G[X_i]$ be connected via strong ties and there be no STC violations. Thus, the number of STC violations is inapproximable within any factor and the problem is not fixed-parameter tractable with respect to the parameter "number of allowed STC violations" unless P = NP. In contrast, the problem of maximizing the number r of non-violated triangles admits a polynomial-time $1/(m+1)$-approximation [22]. Complementing the approximability result for r, we show that the problem is fixed-parameter tractable parameterized by m and r.

Problem 3.1 (STC with tight communities).

Input: A graph $G = (V, E)$, communities $X_1, \ldots, X_m \subseteq V$, and an integer r.
Question: Are there weak edges $A \subseteq E$ (edges in $E \setminus A$ are *strong*) such that each $G[X_i] \setminus A$ is connected and there are at least r non-violated triangles?

We model Problem 3.1 as UFLP-MC: Let the universe $U = E \cup \mathcal{K}$, where E is the set of edges and $\mathcal{K}$ is the set of all open triangles in G. Note that $\mathcal{K}$ can be computed in $O(n^3)$ time. The set $\mathcal{K}$ corresponds to the set of clients, and the set of edges E corresponds to the set of facilities. The cost of opening a facility $u \in E$ (that is, the cost of making an edge weak) is $c_u = 0$. The profit that facility u gets from serving client v, that is, from resolving STC violation v, is

$$p_{uv} = \begin{cases} 1 & \text{if the edge } u \in E \text{ is a part of the open triangle } v \in \mathcal{K} \text{ in } G, \\ 0 & \text{otherwise.} \end{cases}$$

As client matroid we can choose $(\mathcal{K}, C_1)$ with $C_1 = \{\mathcal{K}' \subseteq \mathcal{K} \mid |\mathcal{K}'| \leq r\}$, since resolving r STC violations is sufficient in order to resolve *at least* r of them. Since it is a uniform matroid, we easily get a representation over any field with at least $|\mathcal{K}|$ elements. The facility matroids are $\{M_i\}_{i=1}^m$ with

$$M_i := (E_i, A_i) \oplus (E \setminus E_i, 2^{E \setminus E_i}),$$
$$E_i := \{\{u, v\} \in E \mid \{u, v\} \subseteq X_i\}, \text{ and}$$
$$A_i := \{E' \subseteq E_i \mid G[X_i] \setminus E' \text{ is connected}\}.$$

Herein, (E_i, A_i) is the so-called *bond matroid* of $G[X_i]$ [20, Sect. 2.3]. As the dual of a *graphic matroid*, we easily represent it over *any* field [20, Proposition 5.1.2]. Since the free matroid $(E \setminus E_i, 2^{E \setminus E_i})$ is also representable over any field, so is M_i. Therefore, having represented all matroids over the same field, we can apply Theorem 1.3(ii) and thus complement the known $1/(m+1)$-approximation of the parameter r due to Rozenshtein et al. [22] by the following result:

Proposition 3.2. *Problem 3.1 is fixed-parameter tractable with respect to $r + m$.*

As noted by Rozenshtein et al. [22], if two communities X_i and X_j are disjoint, then the matroids M_i and M_j can be combined into one via a direct sum. Thus, Problem 3.1 is fixed-parameter tractable parameterized by r if the input communities are pairwise disjoint.

Along the lines of Golovach et al. [11], we can also generalize Proposition 3.2 to the problem of maximizing induced subgraphs belonging to some set F and containing at least one weak edge, if all graphs in F have constant size.

4   Representative Families for Matroid Intersections and Set Packing with Multiple Matroid Constraints

In this section, we present generalizations of some known matroid optimization fixed-parameter algorithms for use in our algorithm for UFLP-MC. These results are completely independent from UFLP-MC and of independent interest. Among them, we will see a fixed-parameter algorithm of the following problem.

Problem 4.1 (SET PACKING WITH MATROID CONSTRAINTS *(SPMC))*.

Input: Matroids $\{(U, \mathcal{I}_i)\}_{i=1}^{m}$, a family $\mathcal{H} \subseteq 2^U$, $w \colon \mathcal{H} \to \mathbb{R}$, and $\alpha \in \mathbb{N}$.
Task: Find sets $H_1, \ldots, H_\alpha \in \mathcal{H}$ such that

$$\biguplus_{i=1}^{\alpha} H_i \in \bigcap_{i=1}^{m} \mathcal{I}_i \quad \text{and maximizing} \quad \sum_{i=1}^{\alpha} w(H_i).$$

SPMC is a generalization of the MATROID PARITY and MATROID MATCH-ING problems introduced by Lawler [15]. Lee et al. [16] studied approximation algorithms for the variant MATROID HYPERGRAPH MATCHING with one input matroid and unweighted input sets. Marx [19] and Lokshtanov et al. [17] obtained fixed-parameter tractability results for MATROID γ-PARITY, where one has only one input matroid and pairwise non-intersecting unweighted sets of size γ in the input.

We generalize the fixed-parameter algorithms of Marx [19] and Lokshtanov et al. [17] to SPMC. Both are based on *representative families*: intuitively, a representative $\widehat{S}$ of some family S for a matroid M ensures that, if S contains a set S that can be extended to a basis of M, then $\widehat{S}$ also contains such a set that is "as least as good" as S. To solve SPMC (and later UFLP-MC), we generalize this concept to representative families for matroid *intersections*:

Definition 4.2 (max intersection q-representative family). *Given matroids $\{(U, \mathcal{I}_i)\}_{i=1}^{m}$, a family $S \subseteq 2^U$, and a function $w \colon S \to \mathbb{R}$, we say that a subfamily $\widehat{S} \subseteq S$ is max intersection q-representative for S with respect to w if, for each set $Y \subseteq U$ of size at most q, it holds that:*

  – if there is a set $X \in S$ with $X \uplus Y \in \bigcap_{i=1}^{m} \mathcal{I}_i$,

– *then there is a set* $\widehat{X} \in \widehat{\mathcal{S}}$ *with* $\widehat{X} \uplus Y \in \bigcap_{i=1}^{m} \mathcal{I}_i$ *and* $w(\widehat{X}) \geq w(X)$.

If $m = 1$, *then we call* $\widehat{\mathcal{S}}$ *a* max q-representative family *of* $\mathcal{S}$.

To solve SPMC, we compute a max intersection representative of the family of all feasible solutions to SPMC, yet not only with respect to the goal function of SPMC, but more generally with respect to set functions of the following type.

Definition 4.3 (inductive union maximizing function). *Let* $\mathcal{H} \subseteq 2^U$ *and*

$$\mathcal{B}(\mathcal{H}) := \left\{ \biguplus_{j=1}^{i} H_j \; \middle| \; i \in \mathbb{N}, H_1, \ldots, H_i \in \mathcal{H} \right\}.$$

A set function $\mathbf{w} \colon \mathcal{B}(\mathcal{H}) \to \mathbb{R}$ *is called an* inductive union maximizing function *if there is a* generating function $\mathbf{g} \colon \mathbb{R} \times \mathcal{H} \to \mathbb{R}$ *that is non-decreasing in its first argument and such that, for each* $X \neq \emptyset$,

$$\mathbf{w}(X) = \max_{\substack{H \in \mathcal{H}, S \in \mathcal{B}(\mathcal{H}) \\ S \uplus H = X}} \mathbf{g}(\mathbf{w}(S), H).$$

Note that an inductive union maximizing function $\mathbf{w}$ is fully determined by the value $\mathbf{w}(\emptyset)$ and its generating function $\mathbf{g}$. Inductive union maximizing functions resemble primitive recursive functions on natural numbers, where $S \uplus H$ plays the role of the "successor" of S in primitive recursion.

Example 4.4. Let $\mathcal{H} \subseteq 2^U$ and $w \colon \mathcal{H} \to \mathbb{R}$. For solving SPMC, we compute max intersection representative families with respect to the function $\mathbf{w}$ determined by $\mathbf{w}(\emptyset) = 0$ and $\mathbf{g} \colon (k, H) \mapsto k + w(H)$. Concretely, for $\emptyset \neq X \subseteq \mathcal{B}(\mathcal{H})$, one gets

$$\mathbf{w}(X) = \max_{\substack{X = H_1 \uplus \cdots \uplus H_i \\ H_1, \ldots, H_i \in \mathcal{H}}} \sum_{j=1}^{i} w(H_j)$$

due to the associativity and commutativity of the maximum and sum.

Inductive union maximizing functions generalize additive set functions, yet submodular functions are generally not inductive union maximizing.

To compute max intersection representatives of the families of feasible solutions of SPMC, we can prove the following theorem.

Theorem 4.5. *Let* $\{M_i = (U, \mathcal{I}_i)\}_{i=1}^{m}$ *be linear matroids of rank* $r := (\alpha+\beta)\gamma \geq 1$, $\mathcal{H} \subseteq 2^U$ *be a* γ-family of size t, *and* $\mathbf{w} \colon \mathcal{B}(\mathcal{H}) \to \mathbb{R}$ *be an inductive union maximizing function.*

Given a representation A_i *of* M_i *for each* $i \in \{1, \ldots, m\}$ *over the same field* $\mathbb{F}$ *and the value* $\mathbf{w}(\emptyset)$, *one can compute a max intersection* $\beta\gamma$-*representative of size at most* $\binom{rm}{\alpha\gamma m}$ *of the family*

$$\mathcal{S} = \left\{ S = H_1 \uplus \cdots \uplus H_\alpha \; \middle| \; S \in \bigcap_{i=1}^{m} \mathcal{I}_i \text{ and } H_j \in \mathcal{H} \text{ for } j \in \{1, \ldots, \alpha\} \right\}$$

with respect to $\mathbf{w}$ *in time of* $2^{O(rm)} \cdot t + O(m^2 rn)$ *operations over* $\mathbb{F}$ *and calls to the function* $\mathbf{g}$ *generating* $\mathbf{w}$.

The main feature of Theorem 4.5 is that it allows us to compute max intersection representatives of the family $\mathcal{S}$, whose size may be exponential in the size of $\mathcal{H}$, in time growing merely linearly in the size of $\mathcal{H}$. The literature uses several implicit ad-hoc proofs of variants of Theorem 4.5 with $m = 1$ matroid in algorithms for concrete problems [9,10,19]. These proofs usually use non-negative additive functions in place of $\mathbf{w}$. Our Theorem 4.5 works for but is not limited to additive weight functions, yet does not generalize to coverage functions—a special case of submodular functions.

Proposition 4.6. *If Theorem 4.5 holds for coverage functions* $\mathbf{w}$ *even with* $m = 1$ *matroid, then* $FPT = W[2]$.

Using Theorem 4.5 to compute a max intersection 0-representative of the family of all feasible solutions to SPMC, we can prove the following result.

Theorem 4.7. *SPMC on matroids of rank at most* r *represented over a field* $\mathbb{F} = \mathbb{F}_{p^d}$ *is solvable in time of* $2^{O(\alpha \gamma m)} \cdot |\mathcal{H}|^2 \cdot \mathrm{poly}(r) + m^2 n \cdot \mathrm{poly}(r, \alpha, \gamma, p, d)$ *operations over* $\mathbb{F}$, *where* γ *is an upper bound on the sizes of the sets in* $\mathcal{H}$.

Remark 4.8. Although we define SPMC as a *maximization* problem and prove our results for *max* intersection representative families, all results in this section can also be proved for the *minimization* variants by negating the weight functions.

5   A Fixed-Parameter Algorithm for UFLP-MC

In this section, we sketch our algorithm for Theorem 1.3. One major difficulty in solving UFLP-MC is that the profit from opening a facility depends on which other facilities are already open. To name an extreme example: when opening only facility u, it induces cost c_u and yields profit from serving all the clients. However, when some other facility v is already open, then additionally opening u induces cost c_u yet might not yield any profit if all clients are more profitably already served by v. To avoid such interference between facilities, we use the color coding technique [3] to reduce UFLP-MC to $2^{O(r \log r)}$ instances of the following uncapacitated facility location problem with matroid and *color constraints* (UFLP-MCC).

Problem 5.1 (UFLP-MCC).

Input: A universe U, a coloring col: $U \to \{1, \ldots, k+\ell\}$, a partition $Z_1 \uplus \cdots \uplus Z_\ell = \{\ell + 1, \ldots, \ell + k\}$, for each pair $u, v \in U$ a *profit* $p_{uv} \in \mathbb{N}$ gained when a facility at u serves a client at v, for each $u \in U$ a *cost* $c_u \in \mathbb{N}$ for opening a facility at u, *facility matroids* $\{(U_i, A_i)\}_{i=1}^{a}$, and *client matroids* $\{(V_i, C_i)\}_{i=1}^{c}$, where $U_i \cup V_i \subseteq U$.

Task: Find two sets $A \uplus C \subseteq U$ such that

(i) for each $i \in \{1, \ldots, \ell\}$, there is exactly one $u \in A$ with col$(u) = i$,

(ii) for each $i \in \{\ell + 1, \ldots, \ell + k\}$, there is exactly one $v \in C$ with col$(v) = i$,

(iii) $A \in \bigcap_{i=1}^{a} A_i$ and $C \in \bigcap_{i=1}^{c} C_i$,

and that maximizes

$$\sum_{u \in A} \left(-c_u + \sum_{v \in C \cap Z(u)} p_{uv}\right), \text{ where } Z(u) := \{v \in U \mid \mathrm{col}(v) \in Z_{\mathrm{col}(u)}\}.$$

In case where there is only a single facility matroid and a uniform client matroid, UFLP-MCC can be shown to be solvable in polynomial time by computing a maximum-weight common independent set of the facility matroid and a multicolored matroid, which yields Theorem 1.3(i). To prove Theorem 1.3(ii), we solve UFLP-MCC with multiple linear client and facility matroids. To describe the algorithm, we introduce some notation.

Algorithm 5.1. Algorithm for UFLP-MC with linear matroids

Input: An UFLP-MCC instance: universe $U = \{1, \ldots, n\}$,
partition $Z_1 \uplus \cdots \uplus Z_\ell = \{\ell+1, \ldots, \ell+k\}$, coloring $\mathrm{col} \colon U \to \{1, \ldots, k+\ell\}$,
profits $p_{uv} \in \mathbb{N}$ for each $u, v \in U$, costs $c_u \in \mathbb{N}$ for each $u \in U$, facility
matroids $\mathcal{A} = \{(U_i, A_i)\}_{i=1}^{a}$, client matroids $\mathcal{C} = \{(V_i, C_i)\}_{i=1}^{c}$, all given as
representations over the same finite field, where $U_i, V_i \subseteq U$.
Output: An optimal solution $A \uplus C$ to UFLP-MCC.

1   $M_P \leftarrow (U, \{I \subseteq U \mid I \text{ has at most one element of each color in } \{1, \ldots, k+\ell\}\})$.
2   $\mathcal{M} \leftarrow \{M_P\} \cup \{M \vee (U_C, 2^{U_C}) \mid M \in \mathcal{A}\} \cup \{M \vee (U_A, 2^{U_A}) \mid M \in \mathcal{C}\}$.
3   **if** *some matroid in $\mathcal{M}$ has rank less than $k+\ell$* **then return** *No solution exists.*
4   Truncate all matroids in $\mathcal{M}$ to rank $k+\ell$ (using [17, Theorem 3.15]).
5   **foreach** $u \in U_A$ *and* $i := \mathrm{col}(u)$ **do**
6      $\widehat{\mathcal{F}}(u) \leftarrow$ max intersection $(k+\ell - |Z_i|)$-representative for the family

$$\mathcal{F}(u) := \{I \subseteq Z(u) \mid I \text{ is independent in each of } \mathcal{M} \text{ and } |I| = |Z_i|\}$$

     with respect to weights $w_u \colon 2^U \to \mathbb{N}, I \mapsto \sum_{v \in I} p_{uv}$ (via Theorem 4.5).
7      $\widehat{\mathcal{F}}[u] \leftarrow \{X \cup \{u\} \mid X \in \widehat{\mathcal{F}}(u)\}$.
8   $\widehat{\mathcal{F}} \leftarrow \bigcup_{u \in U_A} \widehat{\mathcal{F}}[u]$.
9   $S_1, \ldots, S_\ell \leftarrow$ solution to SPMC with matroids $\mathcal{M}$, family $\widehat{\mathcal{F}}$, and weights
     $w \colon \widehat{\mathcal{F}} \to \mathbb{Z}, X \mapsto w_u(X \setminus \{u\}) - c_u$, where $\{u\} = X \cap U_A$ (via Theorem 4.7).
10   **if** *not found* **then return** *No solution exists.*
11   $A \leftarrow U_A \cap (S_1 \cup \cdots \cup S_\ell)$.
12   $C \leftarrow U_C \cap (S_1 \cup \cdots \cup S_\ell)$.
13   **return** $A \uplus C$.

Definition 5.2. *For a coloring* $\mathrm{col} \colon U \to \{1, \ldots, k+\ell\}$, *we denote by*

$$
\begin{aligned}
U(i) &:= \{u \in U \mid \mathrm{col}(u) = i\} & &\text{the elements of color } i, \\
U_A &:= \textstyle\bigcup_{i=1}^{\ell} U(i) & &\text{is the set of facilities, and} \\
U_C &:= \textstyle\bigcup_{i=\ell+1}^{\ell+k} U(i) & &\text{is the set of clients.}
\end{aligned}
$$

Algorithm 5.1 now solves UFLP-MCC as follows. In line 1, it constructs a multicolored matroid M_P that will ensure that any independent set of k facilities and ℓ clients fulfills Problem 5.1(i) and (ii). In line 2, it computes a family $\mathcal{M}$ of matroids that contains M_P and all facility and client matroids, which are extended so that a set $A \uplus C \subseteq U$ is independent in all of them if and only if A is independent in all facility matroids and C is independent in all client matroids. Now, if one of the matroids in $\mathcal{M}$ has rank less than $k+\ell$, then there is no common independent set of ℓ facilities and k clients, which is checked in line 3. The truncation in line 4 thus results in each matroid in $\mathcal{M}$ having rank exactly $k + \ell$, which is needed to apply Theorem 4.5 in line 6. In line 6, we construct for each $u \in U_A$ with $\mathrm{col}(u) = i$ a max intersection $(k + \ell - |Z_i|)$-representative $\widehat{\mathcal{F}}(u)$ for the family $\mathcal{F}(u)$ of all sets of clients that could potentially be served by u in a solution. Afterwards, in line 8, we construct a family of sets, each consisting of one facility $u \in U_A$ and a potential client set from $\widehat{\mathcal{F}}(u)$. Finally, in line 9, we will use Theorem 4.7 to combine ℓ of such sets into a set that is independent in all matroids in $\mathcal{M}$ and yields maximum profit.

6   Conclusion

We can show fixed-parameter algorithms for UFLP-MC parameterized by the minimum rank of the client matroids in case when the facility matroid is arbitrary and the client matroid is uniform, or when all matroids are linear. The problem becomes W[1]-hard when the client matroid is *not* linear, even without facility matroids. The complexity of UFLP-MC thus seems to be determined by the client matroids. It would be interesting to settle the complexity of UFLP-MC with one *arbitrary* facility matroid parameterized by the rank of a single linear client matroid. For future research, we point out that our algorithm for Theorem 1.3(i) works in polynomial space, whereas Theorem 1.3(ii) requires exponential space due to Theorems 4.5 and 4.7. It is interesting whether this is avoidable. Moreover, given that approximation algorithms are known for UFLP without matroid constraints [2], for the minimization variant of UFLP with a single facility matroid [13,25], as well as for other optimization problems under matroid constraints [5,8,16], it is canonical to study approximation algorithms for UFLP-MC.

Acknowledgments. This study was initiated at the 7th annual research retreat of the Algorithmics and Computational Complexity group of TU Berlin, Darlingerode, Germany, March 18th–23rd, 2018. R. van Bevern and O. Yu. Tsidulko were supported by the Russian Foundation for Basic Research, grants 16-31-60007 mol_a_dk and 18-31-00470 mol_a, respectively. Both were supported by Russian Science Foundation grant 16-11-10041 while working on Sect. 3.

References

1. Aardal, K., van den Berg, P.L., Gijswijt, D., Li, S.: Approximation algorithms for hard capacitated k-facility location problems. Eur. J. Oper. Res. **242**(2), 358–368 (2015)
2. Ageev, A.A., Sviridenko, M.I.: An 0.828-approximation algorithm for the uncapacitated facility location problem. Discr. Appl. Math. **93**(2), 149–156 (1999)
3. Alon, N., Yuster, R., Zwick, U.: Color-coding. J. ACM **42**(4), 844–856 (1995)
4. Bonnet, É., Paschos, V.T., Sikora, F.: Parameterized exact and approximation algorithms for maximum k-set cover and related satisfiability problems. RAIRO-Inf. Theor. Appl. **50**(3), 227–240 (2016)
5. Calinescu, G., Chekuri, C., Pál, M., Vondrák, J.: Maximizing a monotone submodular function subject to a matroid constraint. SIAM J. Comput. **40**(6), 1740–1766 (2011)
6. Cygan, M., et al.: Parameterized Algorithms. Springer, Heidelberg (2015)
7. Fellows, M.R., Fernau, H.: Facility location problems: a parameterized view. Discr. Appl. Math. **159**(11), 1118–1130 (2011)
8. Filmus, Y., Ward, J.: The power of local search: maximum coverage over a matroid. In: Proceedings of 29th STACS, LIPIcs, vol. 14, pp. 601–612. Schloss Dagstuhl-Leibniz-Zentrum für Informatik, Dagstuhl (2012)
9. Fomin, F.V., Lokshtanov, D., Panolan, F., Saurabh, S.: Efficient computation of representative families with applications in parameterized and exact algorithms. J. ACM **63**(4), 29:1–29:60 (2016)
10. Fomin, F.V., Lokshtanov, D., Panolan, F., Saurabh, S.: Representative families of product families. ACM T. Algorithms **13**(3), 36:1–36:29 (2017)
11. Golovach, P.A., Heggernes, P., Konstantinidis, A.L., Lima, P.T., Papadopoulos, C.: Parameterized aspects of strong subgraph closure. In: Proceedings of 16th SWAT, LIPIcs, vol. 101, pp. 23:1–23:13. Schloss Dagstuhl-Leibniz-Zentrum für Informatik, Dagstuhl (2018)
12. Granovetter, M.S.: The strength of weak ties. Am. J. Sociol. **78**(6), 1360–1380 (1973)
13. Krishnaswamy, R., Kumar, A., Nagarajan, V., Sabharwal, Y., Saha, B.: Facility location with matroid or knapsack constraints. Math. Oper. Res. **40**(2), 446–459 (2015)
14. Laporte, G., Nickel, S., Saldanha da Gama, F. (eds.): Location Science. Springer, Heidelberg (2015). https://doi.org/10.1007/978-3-319-13111-5
15. Lawler, E.: Combinatorial Optimization-Networks and Matroids. Holt, Rinehart and Winston, New York (1976)
16. Lee, J., Sviridenko, M., Vondrák, J.: Matroid matching: the power of local search. SIAM J. Comput. **42**(1), 357–379 (2013)
17. Lokshtanov, D., Misra, P., Panolan, F., Saurabh, S.: Deterministic truncation of linear matroids. ACM Trans. Algorithms **14**(2), 14:1–14:20 (2018)
18. Marx, D.: Parameterized complexity and approximation algorithms. Comput. J. **51**(1), 60–78 (2008)
19. Marx, D.: A parameterized view on matroid optimization problems. Theor. Comput. Sci. **410**(44), 4471–4479 (2009)
20. Oxley, J.G.: Matroid Theory. Oxford University Press, Oxford (1992)
21. Panolan, F., Saurabh, S.: Matroids in parameterized complexity and exact algorithms. In: Kao, M.Y. (ed.) Encyclopedia of Algorithms, pp. 1203–1205. Springer, Heidelberg (2016). https://doi.org/10.1007/978-1-4939-2864-4

22. Rozenshtein, P., Tatti, N., Gionis, A.: Inferring the strength of social ties: a community-driven approach. In: Proceedings of 23rd ACM SIGKDD, pp. 1017–1025. ACM, New York (2017)
23. Schöbel, A., Hamacher, H.W., Liebers, A., Wagner, D.: The continuous stop location problem in public transportation networks. Asia Pac. J. Oper. Res. **26**(1), 13–30 (2009)
24. Schrijver, A.: Combinatorial Optimization: Polyhedra and Efficiency, Algorithms and Combinatorics, vol. 24. Springer, Heidelberg (2003). https://doi.org/10.1007/s10288-004-0035-9
25. Swamy, C.: Improved approximation algorithms for matroid and knapsack median problems and applications. ACM Trans. Algorithms **12**(4), 49:1–49:22 (2016)

Project Games

Vittorio Bilò[1], Laurent Gourvès[2(✉)], and Jérôme Monnot[2]

[1] Department of Mathematics and Physics "Ennio De Giorgi", University of Salento,
Lecce, Italy
`vittorio.bilo@unisalento.it`
[2] CNRS, Université Paris-Dauphine, Université PSL, LAMSADE,
75016 Paris, France
{`laurent.gourves,jerome.monnot`}`@dauphine.fr`

Abstract. We consider a strategic game called project game where each agent has to choose a project among his own list of available projects. The model includes positive weights expressing the capacity of a given agent to contribute to a given project. The realization of a project produces some reward that has to be allocated to the agents. The reward of a realized project is fully allocated to its contributors, according to a simple proportional rule. Existence and computational complexity of pure Nash equilibria is addressed and their efficiency is investigated according to both the utilitarian and the egalitarian social function.

Keywords: Strategic games · Price of anarchy/stability · Congestion

1 Introduction

We introduce and study the PROJECT GAME, a model where some agents take part to some projects. Every agent chooses a single project but several agents can select the same project. This situation happens for example when some scientists decide on which problem they work, when some investors choose the business in which they spend their money, when some benefactors select which artistic project they support, etc. Our model includes positive weights which express the capacity of a given agent to contribute to a given project. By assumption, a project is realized if it is selected by at least one agent. The realization of a project produces some reward that has to be allocated to the agents.

We take a game theoretic perspective, i.e. an agent's strategy is to select, within the projects that are available to her, the one inducing the largest piece of reward. Therefore, the way the rewards are allocated is essential to this game. Here we suppose that the reward of a realized project is fully allocated to its contributors, according to a simple proportional rule based on the aforementioned weights.

Our motivation is to analyze the impact of this simple and natural allocation rule. Do the players reach a Nash equilibrium, that is a stable state in which no

Supported by ANR Project CoCoRICo-CoDec.

P. Heggernes (Ed.): CIAC 2019, LNCS 11485, pp. 75–86, 2019.
https://doi.org/10.1007/978-3-030-17402-6_7

one wants to deviate from the project she is currently contributing? How bad is a Nash equilibrium compared to the situation where a central authority would, at best, decide by which agent(s) a project is conducted? In other words, does the allocation rule incentivize the players to realize projects that optimize the total rewards?

The Model. The PROJECT GAME is a strategic game with a set of n players $N = \{1, \cdots, n\} = [n]$ and a set of m projects $M = \{1, \cdots, m\} = [m]$. The strategy space of every player i, denoted by S_i, is a subset of M. We assume that $\bigcup_{i \in N} S_i = M$ and a strategy for player i is to select a project $j \in S_i$. Each project $j \in M$ has a positive *reward* r_j. We suppose without loss of generality that the minimum reward is always equal to 1. Each player $i \in N$ has a positive weight $w_{i,j}$ when she selects project j.

The *load* of project j under strategy profile σ, denoted by $L(\sigma, j)$, is the total weight of the players who play j. Thus, $L(\sigma, j) = \sum_{\{i \in N \, : \, \sigma_i = j\}} w_{i,j}$.

The utility of player i (that she wants to maximize) under σ is defined as

$$u_i(\sigma) = \frac{w_{i,\sigma_i}}{L(\sigma, \sigma_i)} \, r_{\sigma_i}. \tag{1}$$

A player's utility is defined as a portion of the reward of the realized project that she is contributing to. This portion is proportional to the player's weight.

We will sometimes consider special cases of the PROJECT GAME. An instance of the PROJECT GAME is *symmetric* when $S_i = M$ for every player i. The players' weights are *universal* when, for every player i, $w_{i,j}$ is equal to some positive number w_i for every project j; in particular, they are *identical* when $w_i = 1$ for every player i. The weights are *project-specific* when they are not universal. The projects' rewards are *identical* when the reward is the same for all projects, and this reward is equal to 1 by assumption.

A strategy profile σ is a pure Nash equilibrium if for each $i \in N$ and $j \in S_i$, $u_i(\sigma) \geq u_i(\sigma_{-i}, j)$ where $\sigma' = (\sigma_{-i}, j) =$ is defined by $\sigma'_\ell = \sigma_\ell$ for $\ell \in N \setminus \{i\}$ and $\sigma'_i = j$. For a PROJECT GAME G, denote by $\mathsf{NE}(G)$ its set of pure Nash equilibria.

For a strategy profile σ, $P(\sigma) = \{j \in M : L(\sigma, j) > 0\}$ will denote the set of projects selected by some players in σ. The social utility under strategy profile σ, denoted by $\mathsf{U}(\sigma)$, is defined as the total sum of the rewards of the selected projects (also known as the *utilitarian social welfare*), i.e., $\mathsf{U}(\sigma) = \sum_{j \in P(\sigma)} r_j$. Note that $\mathsf{U}(\sigma) = \sum_{i \in N} u_i(\sigma)$. A social optimum, denoted as σ^*, is a strategy profile maximizing U.

Given a PROJECT GAME G, the *price of anarchy* of G is the worst-case ratio between the social utility of a social optimum and the social utility of a pure Nash equilibrium for G, namely, $\mathsf{PoA}(G) = \sup_{\sigma \in \mathsf{NE}(G)} \frac{\mathsf{U}(\sigma^*)}{\mathsf{U}(\sigma)}$ [1]; the *price of stability* of G is the best-case ratio between the social utility of a social optimum and the social utility of a pure Nash equilibrium for G, namely, $\mathsf{PoS}(G) = \inf_{\sigma \in \mathsf{NE}(G)} \frac{\mathsf{U}(\sigma^*)}{\mathsf{U}(\sigma)}$ [2].

For any two integers $n, m > 1$, let $\mathcal{G}_{n,m}$ denote the set of all PROJECT GAMES with n players and m projects. We define $\mathsf{PoA}(n, m) = \sup_{G \in \mathcal{G}_{n,m}} \mathsf{PoA}(G)$ (resp. $\mathsf{PoS}(n, m) = \sup_{G \in \mathcal{G}_{n,m}} \mathsf{PoS}(G)$) as the price of anarchy (resp. stability) of games with n players and m projects.

Our Contribution. We focus on existence, computational complexity and efficiency of pure Nash equilibria in PROJECT GAMEs. Given the structural simplicity of these games, it will be possible to derive some results from the state of the art of similar classes of games.

For instance, by making use of the notion of better response equivalence [3], we derive that the problem of computing a pure Nash equilibrium in the PROJECT GAME with universal weights belongs to the complexity class PLS and can be solved in polynomial time as long as at least one of the following three conditions holds: the game is symmetric, the rewards are identical, the weights are identical. For the more general case of project-specific weights, instead, we show by means of a potential function argument that the problem is in PLS as long as the rewards are identical. Without this assumption, the problem gets fairly much more complicated and even the existence of pure Nash equilibria remains an open problem.

As to the efficiency of pure Nash equilibria, it is easy to see that the PROJECT GAMEs belong to the class of valid utility games. For these games, Vetta [4] gives an upper bound of 2 on the price of anarchy. We show that this bound is tight only for the case of asymmetric games with non-identical rewards and non-identical weights. In all other cases, we give refined bounds parameterized by both the number of players and projects, also with respect to the price of stability. All these bounds are shown to be tight except for one case involving the price of anarchy of asymmetric games with identical rewards and identical weights. For this particular variant of the game, we also consider an interesting restriction in which all players have exactly two available strategies. These games admit a multigraph representation and we provide some bounds on the price of anarchy as a function of the multigraph topology.

Before concluding, we explore the efficiency of equilibria under an alternative notion of social welfare which focuses on the utility of the poorest player. In this document, some proofs are omitted due to space constraints but they will appear in a journal version.

Related Work. Our PROJECT GAMEs fall within the class of *monotone valid utility games* introduced by Vetta [4] and further considered in [5–11]. In a monotone valid utility game there is a ground set of objects V and a strategy for a player consists in selecting some subset of V. A social function $\gamma : 2^V \mapsto \mathbb{R}$ associates a non-negative value to each strategy profile; γ is assumed to be monotone and submodular. The utility of player i in a strategy profile σ is at least the value $\gamma(\sigma) - \gamma(\sigma_{-i})$. Moreover, the sum of the players' utilities in σ does not exceed the value $\gamma(\sigma)$. Vetta [4] shows that the price of anarchy of these games is at most 2.

Among the special cases of monotone valid utility games considered in the literature, the one that mostly relates to our PROJECT GAMEs is the one studied by Kleinberg and Orel in [9]. They consider a set of projects modeling open problems in scientific research and a set of players/scientists each of which chooses a single problem to work on. However, there are several differences between the two models which make the achieved results non comparable. In fact, in the games studied by Kleinberg and Orel, players may fail in solving a problem, and so the reward associated with each project is not always guaranteed to be realized; moreover, when a problem is solved, its reward is always shared equally among the solving players. This assumption makes these games instances of *congestion games*, whereas this is not the case in our PROJECT GAMEs.

Congestion games [12] is a well known category of strategic games which, by a potential argument [13], always admit a pure Nash equilibrium. In a congestion game, there is a set of resources M and every players' strategy set is a non-empty subset of 2^M. For example, M contains the links of a network from which each player wants to choose a path. Each resource j is endowed with a latency function ℓ_j which depends on the number of players having j in their strategy. A player's cost is the sum of the latencies of the resources that she uses. This model received a lot of attention in the computer science community, see e.g. [14]. Congestion games where generalized to the case where the players have different weights (*weighted congestion games*), or when a resource's latency depends on the identity of the player (*player specific congestion games*) [15]. These extensions still admit a pure Nash equilibrium if the players' strategies are singletons. Nevertheless, a pure Nash equilibrium is not guaranteed when we combine weights and player-specific costs, even with singleton strategies [15]. Singleton congestion games with weighted players are also known as *Load Balancing* games (c.f. [16]): resources and players may represent machines and jobs, respectively. In this context each job goes on the machine that offers her the lowest completion time.

Finally, it is worth mentioning the PROJECT GAME is remotely connected with *hedonic games* [17] and the *group activity selection problem* [18] as the realized projects induce a partition of the player set.

2   Existence of a Pure Strategy Nash Equilibrium

In this section, we focus on the existence and efficient computation of pure Nash equilibria in the PROJECT GAME. We shall show how several positive results can be obtained from the realm of load balancing games and singleton congestion games by making use of the notion of better response equivalence [3]. Intuitively, two games are better response equivalent when, for every pair of strategies, they agree when one is better than the other (i.e., they have the same Nash dynamics graph). By definition, two games which are better response equivalent share the same set of pure Nash equilibria. Thus, existential and computational results for one game can be directly applied to the other.

Fix a PROJECT GAME with universal weights. By (1), we have that, for each strategy profile σ, player $i \in N$ and strategy $j \in S_i$,

$$u_i(\sigma_{-i}, j) > u_i(\sigma) \iff \frac{L(\sigma, j) + w_i}{r_j} < \frac{L(\sigma, \sigma_i)}{r_{\sigma_i}}. \tag{2}$$

If one interprets the set of projects as a set of related machines, where machine j has a speed r_j, and the set of players as a set of tasks, where task i has a processing time w_i, it follows immediately from (2) that any PROJECT GAME with universal weights is better response equivalent to a load balancing game with related machines. Similarly, a PROJECT GAME with universal weights and identical projects is better response equivalent to a load balancing game with identical machines and a PROJECT GAME with identical weights is better response equivalent to a singleton congestion game with linear latency functions.

So, for the PROJECT GAME with universal weights, the existential result for load balancing games with related machines as well as the polynomial time algorithm for the case of symmetric games, both given in [19], can be reused. For asymmetric games with identical rewards, the polynomial time algorithm given in [20] can be applied. For asymmetric games with identical weights, the algorithm given in [14] can be applied. These results are summarized in the following theorem.

Theorem 1 ([14,19,20]). *The* PROJECT GAME *with universal weights admits a potential function. Moreover, a pure Nash equilibrium can be computed in polynomial time when at least one of the following conditions is true: the game is symmetric, the rewards are identical, the weights are identical.*

For the case of project-specific weights, no transformation to other known classes of games are possible (up to our knowledge) and a direct approach needs to be developed. For projects with identical rewards, we show the existence of pure Nash equilibria by providing a potential function argument.

Theorem 2. *For the* PROJECT GAME *with identical rewards, the vector $\langle |P(\sigma)|,$ $\Phi(\sigma)\rangle$, where $\Phi(\sigma) := \Pi_{j \in P(\sigma)} L(\sigma, j)$ lexicographically increases after every profitable unilateral deviation.*

It follows from Theorem 2 that the better response dynamics of the PROJECT GAME with identical rewards never cycles: it always converges to a pure Nash equilibrium. As the potential function given in Theorem 2, as well as the one given in [19] for games with universal weights, can be computed in polynomial time, it follows that the problem of computing a pure Nash equilibrium in games with project-specific weights and identical rewards and in games with universal weights belongs to the complexity class PLS, see, for instance, [21].

For the case of general rewards and project-specific weights, it is easy to see that the PROJECT GAME is better response equivalent to a particular subclass of singleton weighted congestion games with player-specific linear latency functions and resource-specific weights. These games are defined as follows. There is a set of

n players $N = \{1, \cdots, n\} = [n]$ and a set of m resources $R = \{1, \cdots, m\} = [m]$. Each player $i \in N$ can choose a resource from a prescribed set $S_i \subseteq R$ and has a weight $w_{i,j} > 0$ on resource $j \in R$. The load (congestion) of resource j in a strategy profile σ is $L(\sigma, j) = \sum_{i \in N : \sigma_i = j} w_{i,j}$. Each resource $j \in R$ has a player-specific linear latency function $\ell_j^i(x) = \alpha_j^i x$, with $\alpha_j^i \geq 0$, for each $i \in N$. The cost of player i in σ is defined as $c_i(\sigma) = \ell_{\sigma_i}^i(L(\sigma, \sigma_i)) = \alpha_{\sigma_i}^i L(\sigma, \sigma_i)$.

To the best of our knowledge, singleton weighted congestion games with player-specific linear latency functions and resource-specific weights have been considered so far in the literature only under the assumption that the players' weights are not resource-specific, i.e., each player $i \in N$ has a weight $w_i > 0$ for each resource $j \in S_i$. These games have been considered in [22,23]. In particular, [22] shows that they do admit a potential function if and only if $n = 2$, while [23] proves the existence of a pure Nash equilibrium for the cases of either $n = 3$ or $m = 2$; in the latter, a polynomial time algorithm for computing an equilibrium is also provided. However, there is no relationship between these games and our PROJECT GAMES. In fact, if from one perspective PROJECT GAMES are more general than singleton weighted congestion games with player-specific linear latency functions in the definition of the players' weights (which are resource-specific in the former and resource-independent in the latter), on the other hand singleton weighted congestion games with player-specific linear latency functions are more general than PROJECT GAMES in the definition of the latency functions (which are arbitrary in the former and resource-related in the latter).

We close this section with the most general case of the PROJECT GAME, but for a small number of players.

Proposition 1. *The best response dynamics of the* PROJECT GAME *with two players always converges.*

Proposition 2. *The* PROJECT GAME *with three players always admits a pure Nash equilibrium.*

3   Social Utility and the Price of Anarchy/Stability

In this section, we analyze the quality of pure Nash equilibria in the PROJECT GAME in term of price of anarchy and stability. Before presenting our complete characterization of their bounds, note that a social optimum can be computed efficiently.

Proposition 3. *Maximizing the utilitarian social welfare of the* PROJECT GAME *can be done in polynomial time.*

3.1   Games with Identical Rewards

In this subsection, we give results for games with identical rewards. The first result states that there is always a pure Nash equilibrium that is socially optimal.

Theorem 3. *For any two integers $n, m > 1$, $\mathsf{PoS}(n, m) = 1$.*

Next, we show that, under the assumption of symmetric games, all pure Nash equilibria are socially optimal.

Theorem 4. *For any two integers $n, m > 1$, $\mathsf{PoA}(n, m) = 1$ for symmetric games.*

For asymmetric games, instead, next theorem shows that the price of anarchy rises to almost 2 even when considering universal weights.

Theorem 5. *For any two integers $n, m > 1$ and $s := \min(n, m)$, $\mathsf{PoA}(n, m) \geq \frac{2\lfloor \frac{s-1}{2} \rfloor + 1}{\lceil \frac{s-1}{2} \rceil + 1}$ for games with universal weights.*

A matching upper bound, which holds for the more general case of project-specific weights is achieved in the following theorem.

Theorem 6. *For any two integers $n, m > 1$ and $s := \min(n, m)$, $\mathsf{PoA}(n, m) \leq \frac{2\lfloor \frac{s-1}{2} \rfloor + 1}{\lceil \frac{s-1}{2} \rceil + 1}$.*

By Theorems 5 and 6, we get that $\mathsf{PoA}(n, m) = \frac{2\lfloor \frac{s-1}{2} \rfloor + 1}{\lceil \frac{s-1}{2} \rceil + 1}$ for games with both project-specific and universal weights.

Identical Weights. Here, we consider the case of games with identical weights. Games with this property admit an interesting representation via hypergraphs (it becomes multigraphs when $|S_i| \leq 2$ for each $i \in N$).

Theorem 7. *For identical weights and identical rewards and for any two integers $n > 5, m > 1$, $1.582 \approx \frac{e}{e-1} \leq \mathsf{PoA}(n, m) \leq \frac{5}{3} \approx 1.667$.*

3.2   Games with Non Identical Rewards

In this subsection, we address the more general case of general rewards. We start by showing a lower bound on the price of stability which holds even for symmetric games with identical weights.

Proposition 4. *For any two integers $n, m > 1$, $\mathsf{PoS}(n, m) \geq 1 + \frac{\min(n,m)-1}{n}$ for symmetric games with identical weights.*

Proof. For any two integers $n, m > 1$, consider a game with n players of weight 1, one project p with reward $n + \epsilon$, where $\epsilon > 0$ is an arbitrary number, and $m - 1$ projects with reward 1.

As choosing project p is a dominant strategy for each player, this game has only one pure Nash equilibrium in which all the players select p. Under this strategy profile, the social utility is $n + \epsilon$. In a social optimum, a maximum number of $\min(n, m)$ projects can be selected by some player, so that the social utility is at most $n + \epsilon + \min(n, m) - 1$. Thus, by the arbitrariness of ϵ, the price of stability is at least $1 + \frac{\min(n,m)-1}{n}$. □

We now show a matching upper bound that holds even for the price of anarchy of symmetric games with project-specific weights.

Theorem 8. *For any two integers $n, m > 1$, $\mathsf{PoA}(n,m) \leq 1 + \frac{\min(n,m)-1}{n}$ for symmetric games with project-specific weights.*

We now move to the case of asymmetric games. Again, we shall prove a lower bound on the price of stability which holds for universal weights and then provides a matching upper bound on the price of anarchy for the case of project-specific weights. As to the upper bound, from Vetta's result [4], we have that, for any two integers $n, m > 1$, $\mathsf{PoA}(n,m) \leq 2$ for games with project-specific weights. Now, we show the matching lower bound.

Proposition 5. *For any two integers $n, m > 1$, $\mathsf{PoS}(n,m) \geq 2$ for games with universal weights.*

The lower bound for the price of stability given in Proposition 5 does not apply to games with identical weights. This leaves open the possibility to obtain better bounds on both the price of anarchy and the price of stability in this setting. The following two results cover this case. Again, we shall give a lower bound on the price of stability and a matching upper bound on the price of anarchy.

Proposition 6. *For any two integers $n, m > 1$, we have $\mathsf{PoS}(n,m) \geq 2 - 1/n$ if $m \geq n$, $\mathsf{PoS}(n,m) \geq \frac{n+1}{n}$ if $n > m = 2$, and $\mathsf{PoS}(n,m) \geq 2 - \frac{1}{m-1}$ if $n > m > 2$ for games with identical weights.*

We shall prove the upper bounds by exploiting the primal-dual method developed in [24]. Before doing this, we need some additional notation. Given two strategy profiles σ and σ^*, denote as $\alpha(\sigma, \sigma^*) = |P(\sigma^*) \setminus P(\sigma)|$; moreover, for each $j \in M$, denote as $C_j(\sigma, \sigma^*) = \{i \in N : \sigma_i = \sigma_i^* = j\}$ the set of players selecting project j in both σ and σ^* and as $O_j(\sigma, \sigma^*) = \{i \in N \setminus C_j(\sigma, \sigma^*) : \sigma_i^* = j\}$ the set of players selecting project j in σ^* but not in σ. In the application of this method, we shall make use of the following technical lemma.

Lemma 1. *Fix a game with identical weights. For each strategy profile σ and social optimum σ', there exists a social optimum σ^* such that (i) $P(\sigma^*) = P(\sigma')$ and (ii) for each $j \in P(\sigma^*) \cap P(\sigma)$, $|C_j(\sigma, \sigma^*)| \geq L(\sigma, j) - \alpha(\sigma, \sigma^*)$.*

Proof. Fix a strategy profile σ and a social optimum σ' and, for the sake of simplicity, set $\alpha = \alpha(\sigma, \sigma')$. Our aim is to slightly modify σ so as to obtain a social optimum σ^* mimicking the assignment of players to projects realized in σ for as much as possible. To do this, consider the following algorithm operating in three steps.

At step 1, for each $j \in P(\sigma') \setminus P(\sigma)$, choose a unique player $o(j)$ such that $j \in S_{o(j)}$ and define $\sigma^*_{o(j)} = j$. Let T_1 be the set of players chosen at this step; clearly, $|T_1| = \alpha$. At step 2, for each $j \in P(\sigma') \cap P(\sigma)$, choose a unique player $o(j)$ in $N \setminus T_1$ such that $j \in S_{o(j)}$ and define $\sigma^*_{o(j)} = j$. Let T_2 be the set of players chosen at this step. At step 3, for each $i \notin T_1 \cup T_2$, set $\sigma_i^* = \sigma_i$ if $\sigma_i \in P(\sigma') \cap P(\sigma)$ and $\sigma_i^* = j$ otherwise, where j is an arbitrary project in $P(\sigma')$.

The existence of σ' implies that there exists a choice for T_1 and T_2 which guarantees that $P(\sigma^*) = P(\sigma')$. To show part (ii) of the claim, consider a project $j^* \in P(\sigma) \cap P(\sigma^*)$ such that $L(\sigma, j^*) \geq \alpha$ (if no such project exists, then the claim is trivially true). Let

$$\beta = |\{j \in P(\sigma') \cap P(\sigma) : \{i \in N \setminus T_1 : \sigma_i = j\} = \emptyset\}|$$

be the number of projects in $P(\sigma') \cap P(\sigma)$ that lost all of their users in σ after step 1 of the algorithm. We have that step 1 selects at least β players from β different projects in $P(\sigma) \cap P(\sigma^*)$. This implies that j^* loses at most $\alpha - \beta$ users after step 1. At step 2, j^* can lose at most other β additional users for a total of α users. Hence, at least $L(\sigma, j) - \alpha$ players are assigned to j^* in σ^* at step 3 of the algorithm and this shows claim (ii). □

Theorem 9. *For any two integers $n, m > 1$, we have $\mathsf{PoA}(n, m) \leq 2 - 1/n$ if $m \geq n$, $\mathsf{PoA}(n, m) \leq \frac{n+1}{n}$ if $n > m = 2$, and $\mathsf{PoA}(n, m) \leq 2 - \frac{1}{m-1}$ if $n > m > 2$ for games with identical weights.*

Proof. Fix a pure Nash equilibrium σ and a social optimum σ^* and, for the sake of simplicity, set $\alpha = \alpha(\sigma, \sigma^*)$. By Lemma 1, we can assume without loss of generality that, for each $j \in P(\sigma^*) \cap P(\sigma)$, $|C_j(\sigma, \sigma^*)| \geq L(\sigma, j) - \alpha$. We assume $\alpha \geq 1$ as, otherwise, the price of anarchy is trivially equal to 1. By applying the primal-dual method, we get that the inverse of the optimal solution of the following linear program provides an upper bound on $\mathsf{PoA}(n, m)$:

$$\min \sum_{j \in P(\sigma)} r_j$$
$$s.t.$$
$$\frac{r_{\sigma_i}}{L(\sigma, \sigma_i)} - \frac{r_{\sigma_i^*}}{L((\sigma_{-i}, \sigma_i^*), \sigma_i^*)} \geq 0 \quad \forall i \in N,$$
$$\sum_{j \in P(\sigma^*)} r_j = 1$$
$$r_j \geq 0 \qquad \forall j \in M$$

For a strategy profile τ and a project j, denote by $1_j(\tau)$ the indicator function that is equal to 1 if and only if $j \in P(\tau)$. The dual of the above linear program is the following (we associate variable x_i with the first constraint for each $i \in N$ and variable γ with the second one):

$$\max \gamma$$
$$s.t.$$
$$\sum_{i : \sigma_i = j} \frac{x_i}{L(\sigma, j)} - \sum_{i : \sigma_i^*} \frac{x_i}{L((\sigma_{-i}, j), j)} + \gamma 1_j(\sigma^*) \leq 1_j(\sigma) \quad \forall j \in M,$$
$$x_i \geq 0 \qquad \forall i \in N$$

The inverse of the objective value of any feasible solution to this program provides an upper bound on $\mathsf{PoA}(n, m)$.

First of all, we observe that, for any dual solution such that $x_i = x$ for each $i \in N$ and $\gamma = x$, the dual constrain becomes:

$$x \left(1_j(\sigma) - \frac{|C_j(\sigma, \sigma^*)|}{L(\sigma, j)} - \frac{|O_j(\sigma, \sigma^*)|}{L(\sigma, j) + 1} + 1_j(\sigma^*) \right) \leq 1_j(\sigma). \tag{3}$$

If $1_j(\sigma^*) = 0$, (3) is satisfied as long as $x \leq 1$. If $1_j(\sigma^*) = 1$ and $1_j(\sigma) = 0$, which imply $|C_j(\sigma, \sigma^*)| = 0$, $|O_j(\sigma, \sigma^*)| \geq 1$, and $L(\sigma, j) + 1 = 1$, (3) is satisfied independently of the value of x. The case of $1_j(\sigma^*) = 1$ and $1_j(\sigma) = 1$ is then the only one which can cause a price of anarchy higher than 1 and we focus on this case in the remainder on the proof. Note that, in this case, we can always assume $|C_j(\sigma, \sigma^*)| + |O_j(\sigma, \sigma^*)| \geq 1$.

Consider the dual solution such that $x = \frac{n}{2n-1}$. As $L(\sigma, j) + 1 \leq n$, (3) is satisfied. This proves a general upper bound of $2 - 1/n$. However, for the case of $n > m$, better upper bounds can be derived. Note that, in this case, we have $1 \leq \alpha \leq m - 1$.

Assume $\alpha \leq m - 2$ and consider the dual solution such that $x = \frac{m-1}{2m-3}$. If $L(\sigma, j) \leq \alpha$, the term within the parenthesis in the left-hand side of (3) is at most $\frac{2m-3}{m-1}$ and the constraint is satisfied. If $L(\sigma, j) > \alpha$, as $|C_j(\sigma, \sigma^*)| \geq L(\sigma, j) - \alpha$, the term within the parenthesis in the left-hand side of (3) is at most $2 - \frac{L(\sigma, j) - \alpha}{L(\sigma, j)} = \frac{L(\sigma, j) + \alpha}{L(\sigma, j)}$ which is maximized for $\alpha = m - 2$ and $L(\sigma, j) = \alpha + 1 = m - 1$. Again, (3) is satisfied. This proves an upper bound of $2 - \frac{1}{m-1}$. Note that this bound does not apply to the case of $m = 2$, as α cannot be equal to $m - 2$ in this case.

Assume now $\alpha = m - 1$ and consider the dual solution such that $x = \frac{n}{n+m-1}$. The assumption $\alpha = m - 1$ implies that there exists a unique project $j \in P(\sigma) \cap P(\sigma^*)$ and so $L(\sigma, j) = n$ and $C_j(\sigma, \sigma^*) = n - m + 1$. In this case, the term within the parenthesis in the left-hand side of (3) is exactly $\frac{n+m-1}{n}$ and (3) is satisfied. This proves an upper bound of $\frac{n+m-1}{n}$.

As $2 - \frac{1}{m-1} \geq \frac{n+m-1}{n}$ for $n > m > 2$, the claimed upper bounds follow. □

4 Egalitarian Social Welfare

So far we have considered the utilitarian social welfare $U(\sigma) := \sum_{i \in N} u_i(\sigma)$. In this section we use the *egalitarian social welfare* $E(\sigma) := \min_{i \in N} u_i(\sigma)$ (to be maximized). For this section we suppose adapted definitions of the PoA and the PoS which include E instead of U. The motivation for considering E instead of U is fairness among the players.

Proposition 7. *For the egalitarian social welfare, the* PoS *of the* PROJECT GAME *is unbounded even with 4 players, 2 projects, universal weights and identical rewards.*

Since PoA $\geq$ PoS, the PoA of the PROJECT GAME is unbounded as well. One can be tempted to try to enforce a social optimum. However, unlike the utilitarian social welfare (see Proposition 3), the problem is intractable.

Proposition 8. *It is* **NP**-*hard to compute a strategy profile that maximizes the egalitarian social welfare of the* PROJECT GAME *even if there are two projects, identical rewards, and universal weights.*

Nevertheless, we were able to identify a polynomial case.

Proposition 9. *Maximizing the egalitarian social welfare of the* PROJECT GAME *can be done in polynomial time when the players have identical weights.*

5   Conclusion and Open Problems

We introduced a new class of games sharing similarities with valid utility games, singleton congestion games, and hedonic games. We focused on existence, computational complexity and efficiency of pure Nash equilibria under a natural method for sharing the rewards of the projects that are realized.

Though the existence of a pure Nash equilibrium is showed for many important special cases, proving (or disproving) its existence in general is a challenging task. An interesting special case that is left open is when the number of projects is small (e.g. $m = 2$). Other solution concepts (e.g. strong Nash equilibria) deserve attention.

Our upper bounds on PoA and PoS under the utilitarian social welfare never exceed 2, but it does not prevent to explore other sharing methods. Moreover, closing the gap shown in Theorem 7 is an intriguing open problem.

Regarding the computation of an optimal strategy profile with respect to the egalitarian social welfare, there is a gap between hard and polynomial cases (see Propositions 8 and 9). As a first step, it would be interesting to settle the complexity of the symmetric case. As the PoS is unbounded under the egalitarian social welfare, it is natural to ask if a different reward sharing method can provide better results.

References

1. Koutsoupias, E., Papadimitriou, C.H.: Worst-case equilibria. In: Meinel, C., Tison, S. (eds.) STACS 1999. LNCS, vol. 1563, pp. 404–413. Springer, Heidelberg (1999). https://doi.org/10.1007/3-540-49116-3_38
2. Anshelevich, E., Dasgupta, A., Kleinberg, J.M., Tardos, É., Wexler, T., Roughgarden, T.: The price of stability for network design with fair cost allocation. In: Proceedings of the 45th Symposium on Foundations of Computer Science (FOCS), pp. 295–304 (2004)
3. Morris, S., Ui, T.: Best response equivalence. Games Econ. Behav. **49**, 260–287 (2004)
4. Vetta, A.: Nash equilibria in competitive societies, with applications to facility location, traffic routing and auctions. In: Proceedings of the 43rd Symposium on Foundations of Computer Science (FOCS), pp. 416–425 (2002)
5. Augustine, J., Chen, N., Elkind, E., Fanelli, A., Gravin, N., Shiryaev, D.: Dynamics of profit-sharing games. Internet Math. **11**, 1–22 (2015)
6. Bachrach, Y., Syrgkanis, V., Vojnović, M.: Incentives and efficiency in uncertain collaborative environments. In: Chen, Y., Immorlica, N. (eds.) WINE 2013. LNCS, vol. 8289, pp. 26–39. Springer, Heidelberg (2013). https://doi.org/10.1007/978-3-642-45046-4_4
7. Gollapudi, S., Kollias, K., Panigrahi, D., Pliatsika, V.: Profit sharing and efficiency in utility games. In: Proceedings of the 25th Annual European Symposium on Algorithms (ESA), pp. 43:1–43:14 (2017)
8. Goemans, M.X., Li, L., Mirrokni, V.S., Thottan, M.: Market sharing games applied to content distribution in ad hoc networks. IEEE J. Sel. Areas Commun. **24**, 1020–1033 (2006)

9. Kleinberg, J.M., Oren, S.: Mechanisms for (mis)allocating scientific credit. In: Proceedings of the 43rd ACM Symposium on Theory of Computing (STOC), pp. 529–538 (2011)

10. Marden, J.R., Roughgarden, T.: Generalized efficiency bounds in distributed resource allocation. In: Proceedings of the 49th IEEE Conference on Decision and Control (CDC), pp. 2233–2238 (2010)

11. Marden, J.R., Wierman, A.: Distributed welfare games. Oper. Res. **61**, 155–168 (2013)

12. Rosenthal, R.: A class of games possessing pure-strategy Nash equilibria. Int. J. Game Theory **2**, 65–67 (1973)

13. Monderer, D., Shapley, L.S.: Potential games. Games Econ. Behav. **14**, 124–143 (1996)

14. Ieong, S., McGrew, R., Nudelman, E., Shoham, Y., Sun, Q.: Fast and compact: a simple class of congestion games. In: Proceedings of the 20th National Conference on Artificial Intelligence (AAAI), pp. 489–494 (2005)

15. Milchtaich, I.: Congestion games with player-specific payoff functions. Games Econ. Behav. **13**, 111–124 (1996)

16. Vöcking, B.: Selfish load balancing. In: Nisan, N., Roughgarden, T., Tardos, E., Vazirani, V.V. (eds.) Algorithmic Game Theory, pp. 517–542. Cambridge University Press, New York (2007)

17. Drèze, J.H., Greenberg, J.: Hedonic coalitions: optimality and stability. Econometrica **48**, 987–1003 (1980)

18. Darmann, A., Elkind, E., Kurz, S., Lang, J., Schauer, J., Woeginger, G.: Group activity selection problem. In: Goldberg, P.W. (ed.) WINE 2012. LNCS, vol. 7695, pp. 156–169. Springer, Heidelberg (2012). https://doi.org/10.1007/978-3-642-35311-6_12

19. Fotakis, D., Kontogiannis, S., Koutsoupias, E., Mavronicolas, M., Spirakis, P.: The structure and complexity of Nash equilibria for a selfish routing game. In: Widmayer, P., Eidenbenz, S., Triguero, F., Morales, R., Conejo, R., Hennessy, M. (eds.) ICALP 2002. LNCS, vol. 2380, pp. 123–134. Springer, Heidelberg (2002). https://doi.org/10.1007/3-540-45465-9_12

20. Gairing, M., Lucking, T., Mavronicolas, M., Monien, B.: Computing Nash equilibria for scheduling on restricted parallel links. In: Proceedings of the 36th Annual ACM Symposium on Theory of Computing (STOC), pp. 613–622 (2004)

21. Fabrikant, A., Papadimitriou, C.H., Talwar, K.: The complexity of pure Nash equilibria. In: Proceedings of the 36th Annual ACM Symposium on Theory of Computing (STOC), pp. 604–612 (2004)

22. Gairing, M., Monien, B., Tiemann, K.: Routing (un-)splittable flow in games with player-specific linear latency functions. In: Bugliesi, M., Preneel, B., Sassone, V., Wegener, I. (eds.) ICALP 2006. LNCS, vol. 4051, pp. 501–512. Springer, Heidelberg (2006). https://doi.org/10.1007/11786986_44

23. Georgiou, C., Pavlides, T., Philippou, A.: Selfish routing in the presence of network uncertainty. Parallel Process. Lett. **19**, 141–157 (2009)

24. Bilò, V.: A unifying tool for bounding the quality of non-cooperative solutions in weighted congestion games. Theory Comput. Syst. **62**, 1288–1317 (2018)

Subgraph Isomorphism on Graph Classes that Exclude a Substructure

Hans L. Bodlaender[1,2], Tesshu Hanaka[3], Yoshio Okamoto[4,5],
Yota Otachi[6(✉)], and Tom C. van der Zanden[1]

[1] Utrecht University, Utrecht, The Netherlands
{H.L.Bodlaender,T.C.vanderZanden}@uu.nl
[2] University of Technology Eindhoven, Eindhoven, The Netherlands
[3] Chuo University, Bunkyo-ku, Tokyo, Japan
hanaka.91t@g.chuo-u.ac.jp
[4] The University of Electro-Communications, Chofu, Tokyo, Japan
okamotoy@uec.ac.jp
[5] RIKEN Center for Advanced Intelligence Project, Tokyo, Japan
[6] Kumamoto University, Kumamoto 860-8555, Japan
otachi@cs.kumamoto-u.ac.jp

Abstract. We study SUBGRAPH ISOMORPHISM on graph classes defined by a fixed forbidden graph. Although there are several ways for forbidding a graph, we observe that it is reasonable to focus on the minor relation since other well-known relations lead to either trivial or equivalent problems. When the forbidden minor is connected, we present a near dichotomy of the complexity of SUBGRAPH ISOMORPHISM with respect to the forbidden minor, where the only unsettled case is the path of five vertices. We then also consider the general case of possibly disconnected forbidden minors. We show in particular that: the problem is fixed-parameter tractable parameterized by the size of the forbidden minor H when H is a linear forest such that at most one component has four vertices and all other components have three or less vertices; and it is NP-complete if H contains four or more components with at least five vertices each. As a byproduct, we show that SUBGRAPH ISO-MORPHISM is fixed-parameter tractable parameterized by vertex integrity. Using similar techniques, we also observe that SUBGRAPH ISOMORPHISM is fixed-parameter tractable parameterized by neighborhood diversity.

Keywords: Subgraph isomorphism · Minor free graphs ·
Parameterized complexity

Partially supported by NETWORKS (the Networks project, funded by the Netherlands Organization for Scientific Research NWO), the ELC project (the project Exploring the Limits of Computation, funded by MEXT), JSPS/MEXT KAKENHI grant numbers JP24106004, JP18K11168, JP18K11169, JP18H04091, JP18H06469, JP15K00009, JST CREST Grant Number JPMJCR1402, and Kayamori Foundation of Informational Science Advancement. The authors thank Momoko Hayamizu, Kenji Kashiwabara, Hirotaka Ono, Ryuhei Uehara, and Koichi Yamazaki for helpful discussions. The authors are grateful to the anonymous reviewer of an earlier version of this paper who pointed out a gap in a proof.

© Springer Nature Switzerland AG 2019
P. Heggernes (Ed.): CIAC 2019, LNCS 11485, pp. 87–98, 2019.
https://doi.org/10.1007/978-3-030-17402-6_8

1   Introduction

Let Q and G be graphs. A *subgraph isomorphism* η is an injection from $V(Q)$ to $V(G)$ that preserves the adjacency in Q; that is, if $\{u, v\} \in E(Q)$, then $\{\eta(u), \eta(v)\} \in E(G)$. We say that Q is *subgraph-isomorphic* to G if there is a subgraph isomorphism from Q to G, and write $Q \preceq G$. In this paper, we study the following problem of deciding the existence of a subgraph isomorphism.

SUBGRAPH ISOMORPHISM
Input: Two graphs G (the *host* graph) and Q (the *pattern* graph).
Question: $Q \preceq G$?

The problem SUBGRAPH ISOMORPHISM is one of the most general and fundamental graph problems and generalizes many other graph problems such as GRAPH ISOMORPHISM, CLIQUE, HAMILTONIAN PATH/CYCLE, and BANDWIDTH. Obviously, SUBGRAPH ISOMORPHISM is NP-complete in general. When both host and pattern graphs are restricted to be in a graph class $\mathcal{C}$, we call the problem SUBGRAPH ISOMORPHISM *on* $\mathcal{C}$. By slightly modifying known reductions in [7,14], one can easily show that the problem is hard even for very restricted graph classes. Recall that a linear forest is the disjoint union of paths and a cluster graph is the disjoint union of complete graphs. We can show the following hardness of SUBGRAPH ISOMORPHISM by a simple reduction from 3-PARTITION [14].

Proposition 1.1 ($\bigstar^1$). SUBGRAPH ISOMORPHISM *on linear forests and cluster graphs is NP-complete even if both graphs have the same number of vertices.*

Since most of the well-studied graph classes contain all linear forests or all cluster graphs, it is often hopeless to have a polynomial-time algorithm for an interesting graph class. This is sometimes true even if we further assume that the graphs are connected [19,21]. On the other hand, it is polynomial-time solvable for trees [27]. This result was first generalized for 2-connected outerplanar graphs [24], and finally for k-connected partial k-trees [15,26] (where the running time is XP parameterized by k). In [26], a polynomial-time algorithm for partial k-trees of bounded maximum degree is presented as well, which is later generalized to partial k-trees of log-bounded fragmentation [16]. It is also known that for chain graphs, co-chain graphs, and threshold graphs, SUBGRAPH ISOMORPHISM is polynomial-time solvable [19–21]. In the case where only the pattern graph has to be in a restricted graph class that is closed under vertex deletions, a complexity dichotomy with respect to the graph class is known [17].

Because of its unavoidable hardness in the general case, it is often assumed that the pattern graph is small. In such a setting, we can study the parameterized complexity[2] of SUBGRAPH ISOMORPHISM parameterized by the size of the pattern graph. Unfortunately, the W[1]-completeness of CLIQUE [9] implies

---

[1] A black star $\bigstar$ means that the proof is omitted or shortened.
[2] We assume that the readers are familiar with the concept of parameterized complexity. See e.g. [6] for basic definitions omitted here.

that this parameterization does not help in general. Indeed, the existence of a $2^{o(n \log n)}$-time algorithm for SUBGRAPH ISOMORPHISM is ruled out assuming the Exponential Time Hypothesis, where n is the total number of vertices [5]. So we need further restrictions on the considered graph classes even in the parameterized setting. For planar graphs, it is known to be fixed-parameter tractable [8,11]. This result is later generalized to graphs of bounded genus [4]. For several graph parameters, the parameterized complexity of SUBGRAPH ISO-MORPHISM parameterized by combinations of them is determined in [25]. In [3], it is shown that when the pattern graph excludes a fixed graph as a minor, the problem is fixed-parameter tractable parameterized by treewidth and the size of the pattern graph. The result in [3] implies also that SUBGRAPH ISOMORPHISM can be solved in subexponential time when the host graph also excludes a fixed graph as a minor.

1.1   Our Results

As mentioned above, the research on SUBGRAPH ISOMORPHISM has been done mostly when the size of the pattern graph is considered as a parameter. However, in this paper, we are going to study the general case where the pattern graph can be as large as the host graph.

We first observe that forbidding a graph as an induced substructure (an induced subgraph, an induced topological minor, or an induced minor) does not help for making SUBGRAPH ISOMORPHISM tractable unless we make the graph class trivial by forbidding either adjacent vertices or nonadjacent vertices. This can be done just by combining some easy observations and known results.

Observation 1.2 (★). *Let C be the graph class that forbids a fixed graph H as either an induced subgraph, an induced topological minor, or an induced minor. Then, SUBGRAPH ISOMORPHISM on C is polynomial-time solvable if H has at most two vertices; otherwise, it is NP-complete.*

Our main contribution in this paper is the following pair of results on SUB-GRAPH ISOMORPHISM on graph classes forbidding a fixed graph as a substructure. (We prove Theorem 1.3 in Sect. 3 and Theorem 1.4 in Sect. 4.)

Theorem 1.3. *Let C be the graph class that forbids a fixed connected graph $H \neq P_5$ as either a subgraph, a topological minor, or a minor. Then, SUBGRAPH ISOMORPHISM on C is polynomial-time solvable if H is a subgraph of P_4; otherwise, it is NP-complete.*

Theorem 1.4. *Let C be the graph class that forbids a fixed (not necessarily connected) graph H as either a subgraph, a topological minor, or a minor. Then, SUBGRAPH ISOMORPHISM on C is*

- *fixed-parameter tractable parameterized by the order of H if H is a linear forest such that at most one component is of order 4 and all other components are of order at most 3;*

– *NP-complete if either H is not a linear forest, H contains a component with six or more vertices, or H contains four components with five vertices.*

Note that we have some missing cases. We do not know the complexity of the problem when the forbidden linear forest H contains either

– two or more disjoint P_4 subgraphs but no P_5 subgraph, or
– one, two, or three disjoint P_5 subgraphs but no P_6 subgraph.

2 Preliminaries and Basic Observations

We denote the path of n vertices by P_n, the complete graph of n vertices by K_n, and the star with ℓ leaves by $K_{1,\ell}$. For notational convenience, we allow ℓ to be 0; that is, $K_{1,0} = P_1 = K_1$. The disjoint union of graphs X and Y is denoted by $X \cup Y$ and the disjoint union of k copies of a graph Z is denoted by kZ.

A graph Q is a *minor* of G if Q can be obtained from G by removing vertices, removing edges, and contracting edges, where contracting an edge $\{u, v\}$ means adding a new vertex $w_{u,v}$, making the neighbors of u and v adjacent to $w_{u,v}$, and removing u and v. A graph Q is a *topological minor* of G if Q can be obtained by removing vertices, removing edges, and contracting edges, where contraction of an edge is allowed if one of the endpoints of the edge is of degree 2. A graph Q is a *subgraph* of G if Q can be obtained by removing vertices and edges. If we cannot remove edges but can do the other modifications as before, then we get the induced variants *induced minor*, *induced topological minor*, and *induced subgraph*.

Recall that a graph is a linear forest if it is the disjoint union of paths. In other words, a graph is a linear forest if and only if it does not contain a cycle nor a vertex of degree at least 3. Observe that in all graph containment relations mentioned above, if we do not forbid any linear forest from a graph class, then the class includes all linear forests. Thus, by Proposition 1.1, we have the following lemma.

Lemma 2.1. *If H is not a linear forest, then* SUBGRAPH ISOMORPHISM *is NP-complete for graphs that do not contain H as a minor, a topological minor, a subgraph, an induced minor, an induced topological minor, or an induced subgraph.*

2.1 Graphs Forbidding a Short Path as a Minor

By the discussion above, we can focus on a graph class forbidding a linear forest as a minor (or equivalently as a topological minor or a subgraph). We here characterize graph classes forbidding a short path as a minor.

Lemma 2.2 (★). *A connected P_3-minor free graph is isomorphic to K_1 or K_2.*

Lemma 2.3 (★). *A connected P_4-minor free graph is isomorphic to either K_3 or $K_{1,s}$ for some $s \geq 0$.*

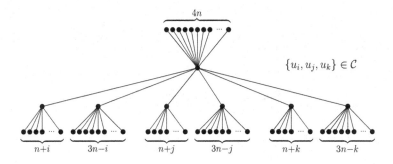

Fig. 1. The tree in G corresponding to $\{u_i, u_j, u_k\} \in \mathcal{C}$.

3   Forbidding a Connected Graph as a Minor

Here we first show that SUBGRAPH ISOMORPHISM on P_k-minor free graphs is linear-time solvable if $k \leq 4$. Note that P_k-minor free graphs include all $P_{k'}$-minor free graphs if $k' \leq k$.

The following result can be easily obtained from Lemma 2.3.

Lemma 3.1 (★). SUBGRAPH ISOMORPHISM on P_4-minor free graphs is linear-time solvable.

The following theorem implies that SUBGRAPH ISOMORPHISM on P_k-minor free graphs is NP-complete for every $k \geq 6$.

Theorem 3.2. SUBGRAPH ISOMORPHISM is NP-complete when the host graph is a forest without paths of length 6 and the pattern is a collection of stars.

Proof. The problem clearly is in NP. To show hardness, we reduce from EXACT 3-COVER [14]:

EXACT 3-COVER
Input: Collection $\mathcal{C}$ of subsets of a set U such that each $c \in \mathcal{C}$ has size 3.
Question: Is there a subcollection $\mathcal{C}' \subseteq \mathcal{C}$ such that $\bigcup_{C \in \mathcal{C}'} C = U$ and $|\mathcal{C}'| = |U|/3$?

Suppose we have an instance $(\mathcal{C}, U)$ of EXACT 3-COVER given, where $U = \{u_0, \ldots, u_{n-1}\}$. From $(\mathcal{C}, U)$, we construct the host graph G and the pattern Q.

The host G consists of the disjoint union of $|\mathcal{C}|$ trees as follows (see Fig. 1). For each set $C \in \mathcal{C}$, we take a tree in G as follows. Take a star $K_{1,4n+6}$. For each $u_i \in C$, do the following: take one of the leaves of the star, and add $n+i$ pendant vertices to it. Take another leaf of the star, and add $3n - i$ pendant vertices to it. I.e., if $C = \{u_i, u_j, u_k\}$, then the corresponding tree has seven vertices of degree more than 1: one vertex with degree $4n+6$, which is also adjacent to each of the other six non-leaf vertices; the non-leaf vertices have degree $n+i+1$, $3n-i+1$, $n+j+1$, $3n-j+1$, $n+k+1$, and $3n-k+1$. Call the vertex of degree $4n+6$ the *central* vertex of the component of C.

The pattern graph Q consists of a number of stars (see Fig. 2):

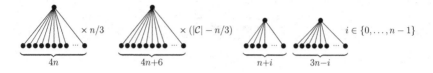

Fig. 2. The pattern graph Q.

- We have $n/3$ stars $K_{1,4n}$.
- We have $|\mathcal{C}| - n/3$ stars $K_{1,4n+6}$.
- For each $i \in \{0, \ldots, n-1\}$, we have stars $K_{1,n+i}$ and $K_{1,3n-i}$. Call these the *element stars*.

From $(\mathcal{C}, U)$, G and Q can be constructed in polynomial time. Now we show that $Q \preceq G$ if and only if $(\mathcal{C}, U)$ is a yes-instance of Exact 3-Cover. We assume that $n > 6$ in the following.

The if direction: Suppose that the Exact 3-Cover instance $(\mathcal{C}, U)$ has a solution $\mathcal{C}' \subseteq \mathcal{C}$.

We map each $K_{1,4n+6}$ of Q into a component M of G corresponding to a set $D \notin \mathcal{C}'$. The center of $K_{1,4n+6}$ is mapped to the central vertex of M and all leaves to its neighbors. The other vertices T are isolated and not used.

Embed each $K_{1,4n}$ of Q into a component L of G corresponding to a set $C \in \mathcal{C}'$, mapping the center of $K_{1,4n}$ to the central vertex of L, and the leaves of $K_{1,4n}$ to leaves neighboring the central vertex of L. After we have done so, we left in this component six stars: if $C = \{u_i, u_j, u_k\}$, then the vertices in L that we did not yet use form stars $K_{1,n+i}$, $K_{1,3n-i}$, $K_{1,n+j}$, $K_{1,3n-j}$, $K_{1,n+k}$, $K_{1,3n-k}$. We thus can embed the element stars corresponding to u_i, u_j, and u_k in these stars, and have embedded the entire pattern in the host graph since $\mathcal{C}'$ is a cover of U.

The only if direction: Suppose that $Q \preceq G$. Note that both Q and G have exactly $|\mathcal{C}|$ vertices of degree at least $4n$. Thus it follows that each vertex of degree at least $4n$ in Q must be mapped to a central vertex of a component in G. We can see that one of the following two cases must hold for the components in the host graph G.

Case 1: A star $K_{1,4n+6}$ is embedded in the component. This "uses up" the central vertex and all its neighbors. The only vertices in the component that are not in the image of the star $K_{1,4n+6}$ are leaves with its neighbor being used: these isolated vertices thus cannot be used for embedding any other stars. So all element stars must be embedded in components for which Case 2 holds.

Case 2: A star $K_{1,4n}$ is embedded in the component. At this point, note that the total number of vertices of element stars in Q equals $4n^2 + 2n$: each of the n elements has in total $4n$ leaves and two high degree vertices in its element stars. Also, the total number of vertices not used by the stars $K_{1,4n}$ in the Case 2-components equals $4n^2 + 2n$: we have $n/3$ components of Case 2 in G and each has $16n + 7$ vertices of which $4n + 1$ are used for embedding the star $K_{1,4n}$. Thus, each vertex in a Case 2-component M must be used for embedding a vertex. This is only possible if we embed in M the element stars of the elements in the set corresponding to M.

So, let $\mathcal{C}'$ be the sets whose component is of Case 2, i.e., where we embedded a $K_{1,4n}$ in its component. This subcollection $\mathcal{C}'$ is a solution for EXACT 3-COVER: for each element u_i, its element stars are embedded in a component that corresponds to a set C that contains u_i, and by the argument above $C \in \mathcal{C}'$. $\square$

By Lemma 2.1, if a connected graph H is not a path, then SUBGRAPH ISO-MORPHISM on H-minor free graphs is NP-complete. Assume that H is a path P_k. If $k \geq 6$, then by Theorem 3.2 the problem is NP-complete. If $k \leq 4$, then by Lemma 3.1 the problem can be solved in polynomial time. This completes the proof of Theorem 1.3.

4   Forbidding a Disconnected Graph as a Minor

In this section, we study the more general cases where the forbidden minor H is not necessarily connected. By Lemma 2.1, we can focus on linear forests H. We already know, by Theorem 3.2, if H contains a component with six or more vertices the problem becomes NP-complete. Thus in the following we consider the case where the components of H have five or less vertices.

Using the results in this section, we can prove Theorem 1.4. Corollary 4.2 implies the positive case of Theorem 1.4. Theorems 3.2 and 4.4 together with Lemma 2.1 imply the negative cases.

4.1   Subgraph Isomorphism on $(P_4 \cup kP_3)$-Minor Free Graphs

We show that SUBGRAPH ISOMORPHISM on $(P_4 \cup kP_3)$-minor free graphs is fixed-parameter tractable when parameterized by k. To this end, we present an algorithm that is parameterized by the vertex integrity, which we think is of independent interest. The *vertex integrity* [1] of a graph is the minimum integer k such that there is a vertex set $S \subseteq V$ such that $|S| \leq k$ and the maximum order of the components of $G - S$ is at most $k - |S|$. We call such S a vi(k) *set* of G. Note that the property of having vertex integrity at most k is closed under the subgraph relation.

This subsection is devoted to the proof of the following theorem.

Theorem 4.1. SUBGRAPH ISOMORPHISM *on graphs of vertex integrity at most k is fixed-parameter tractable when parameterized by k.*

By combining Theorem 4.1, Lemma 3.1, and the fact that kP_3-minor free graphs have vertex integrity at most $3k - 1$, we can prove the following.

Corollary 4.2 (★). SUBGRAPH ISOMORPHISM *on $(P_4 \cup kP_3)$-minor free graphs is fixed-parameter tractable when parameterized by k.*

To prove Theorem 4.1, we start with the following simple fact.

Lemma 4.3 (★). *Let η be a subgraph isomorphism from Q to G. For every* $\mathsf{vi}(k)$ *set T of G, there exists a minimal* $\mathsf{vi}(k)$ *set S of Q such that $\eta(S) \subseteq T$.*

Our algorithm assumes that there is a subgraph isomorphism η from Q to G and proceeds as follows:

1. find a $\mathsf{vi}(k)$ set T of G;
2. guess a minimal $\mathsf{vi}(k)$ set S of Q such that $\eta(S) \subseteq T$;
3. guess the bijection between S and $R := \eta(S)$;
4. guess a subset $F \subseteq E(G - R)$ of the edges "unused" by η such that R is a $\mathsf{vi}(k)$ set of $G - F$;
5. solve the problem of deciding the extendability of the guessed parts as the feasibility problem of an integer linear program with a bounded number of variables.

Proof (Proof of Theorem 4.1). Let G and Q be graphs of vertex integrity at most k. Our task is to find a subgraph isomorphism η from Q to G in FPT time parameterized by k.

We first find a $\mathsf{vi}(k)$ set T of G and then guess a minimal $\mathsf{vi}(k)$ set S of Q such that $\eta(S) \subseteq T$ for some subgraph isomorphism η from Q to G. By Lemma 4.3, such a set S exists if η exists. Finding T can be done in $O(k^{k+1}n)$ time [10], where $n = |V(G)|$. To guess S, it suffices to list all minimal $\mathsf{vi}(k)$ set S of Q. The same algorithm in [10] can be used again: it lists all $O(k^k)$ candidates by branching on $k + 1$ vertices that induce a connected subgraph.

We then guess the subset R of T such that $\eta(S) = R$. We also guess for each $s \in S$, the image $\eta(s) \in R$. That is, we guess an injection from S to T. The number of such injections is $\binom{|T|}{|S|} \cdot |S|! \leq k!$. If there is an edge $\{u, v\} \in E(Q[S])$ such that $\{\eta(u), \eta(v)\} \notin E(G[R])$, then we reject this guess. Otherwise, we try to further extend η.

Observe that R is not necessarily a $\mathsf{vi}(k)$ set of G. In the following, we guess "unnecessary" edges in $G - R$. That is, we guess a subset F of the edges that are not used by η as images of any edges in Q. Furthermore, we select F so that R is a $\mathsf{vi}(k)$ set of $G - F$. Such F exists because η embeds $Q - S$ (and no other things) into $G - R$.

Guessing F: We now show that the number of candidates of F that we need to consider is bounded by some function in k. We partition F into three sets $F_1 = F \cap E(G[T - R])$, $F_2 = F \cap E(V(G) - T, T - R)$, and $F_3 = F \cap E(G - T)$ and then count the numbers of candidates separately.

Guessing F_1: For F_1, we just use all $2^{|E(G[T-R])|} < 2^{k^2}$ subsets of $E(G[T - R])$ as candidates. If R is not a $\mathsf{vi}(k)$ set of $G[T] - F_1$, we reject this F_1.

Guessing F_2: Since we are finding F such that R is a $\mathsf{vi}(k)$ set of $G - F$, each vertex in $T - R$ has less than k edges to $V(G) - T$ in $G - F$. Thus fewer than k^2 components of $V(G) - T$ have edges to $T - R$ in $G - F$. We guess such components $\mathcal{C}$.

Observe that each component in $V(G) - T$ is of order at most k and that each vertex of $V(G) - T$ can be partitioned into at most 2^k types with respect to the adjacency to T. This implies that the components of $V(G) - T$ can be classified into at most 4^{k^2} types (2^{k^2} for the isomorphism type and $(2^k)^k$ for the adjacency to T) in such a way that if two components C_1 and C_2 of $G - T$ are of the same type, then there is an automorphism of G that fixes T and maps C_1 to C_2. Given this classification of the components in $V(G) - T$, we only need to guess how many components of each type are included in $\mathcal{C}$. For this guess, we have at most $\binom{4^{k^2}+k^2-1}{k^2} < 4^{k^4+k^2}$ options.

For each guess $\mathcal{C}$, we guess the edges connecting the components in $\mathcal{C}$ to $T - R$ in $G - F$. Since $|\mathcal{C}| < k^2$ and $|C| \le k$ for each $C \in \mathcal{C}$, there are at most $k^3 \cdot |T - R| \le k^4$ candidate edges. We just try all $O(2^{k^4})$ subsets F_2' of such edges, and set $F_2 = E(V(G) - T, T - R) - F_2'$. In total, we have $O(2^{k^4+k^2} \cdot 2^{k^4})$ options for F_2.

Guessing F_3: Recall that $G - T$ does not contain any component of order more than k. Hence, if $G - R - (F_1 \cup F_2)$ has a component of order more than k, then it consists of some vertices in $T - R$ and some components in $\mathcal{C}$. Thus, we only need to pick some edges of the components in $\mathcal{C}$ for F_3 to make R a $\mathrm{vi}(k)$ set of $G - F$. We use all 2^{k^4} subsets of the edges of the components in $\mathcal{C}$ as a candidate of F_3.

In total, $F = F_1 \cup F_2 \cup F_3$ has at most $2^{k^2} \cdot 4^{k^4+k^2} \cdot 2^{k^4} \cdot 2^{k^4}$ candidates, and each candidate can be found in FPT time. We reject this guess F if R is not a $\mathrm{vi}(k)$ set of $G - F$. In the following, we assume that F is guessed correctly and denote $G - F$ by G'.

Extending η: Recall that we already know how η maps S to R and that each component in $Q - S$ and $G' - R$ is of order at most k. We now extend η by determining how η maps $Q - S$ to $G' - R$. By renaming vertices, we can assume that $S = \{s_1, \ldots, s_q\}$, $R = \{r_1, \ldots, r_q\}$, and $\eta(s_i) = r_i$ for $1 \le i \le q$.

We say that a vertex u in $Q - S$ *matches* a vertex v in $G' - R$ if $\{i \mid s_i \in N_Q(u) \cap S\} \subseteq \{i \mid r_i \in N_{G'}(v) \cap R\}$. A set of components $\{C_1, \ldots, C_h\}$ of $Q - S$ *fits* a component D of $G' - R$ if there is an isomorphism ϕ from the disjoint union of $C_1, \ldots, C_h$ to D such that for all $u \in \bigcup_i V(C_i)$ and $v \in V(D)$, $\phi(u) = v$ holds only if u matches v. Note that if $h > k$, then $\{C_1, \ldots, C_h\}$ can fit no component of $G' - R$.

As we did before for guessing F_2, we classify the components of $Q - S$ and $G' - R$ into at most 4^{k^2} types. Two components C_1 and C_2 of $Q - S$ (or of $G' - R$) are of the same type if and only if there is an isomorphism ϕ from C_1 to C_2 such that $\phi(v_1) = v_2$ implies that $N_Q(v_1) \cap S = N_Q(v_2) \cap S$ (or $N_{G'}(v_1) \cap R = N_{G'}(v_2) \cap R$, respectively). We denote by $t(C)$ the type of a component C and by $t(\{C_1, \ldots, C_h\})$ the multi-set $\{t(C_1), \ldots, t(C_h)\}$. Observe that $\{C_1, \ldots, C_h\}$ fits D if and only if all sets $\{C_1', \ldots, C_h'\}$ with $t(\{C_1', \ldots, C_h'\}) = t(\{C_1, \ldots, C_h\})$ fits D' with $t(D') = t(D)$.

Observe that the guessed part $\eta|_S$ can be extended to a subgraph isomorphism η from Q to G' if and only if there is a partition of the components of $Q - S$ such that each part $\{C_1, \ldots, C_h\}$ in the partition can be injectively mapped to a component D of $G' - R$ where $\{C_1, \ldots, C_h\}$ fits D. To check the existence of such a partition, we only need to find for each pair of a multi-set $\mathcal{T}$ of types of a set of components in $Q - S$ and a type τ of a component in $G' - R$, how many sets of components of type $\mathcal{T}$ the map η embeds to components of type τ. We use the following ILP formulation to solve this problem.

Let n_τ and n'_τ be the numbers of type-τ components in $Q - S$ and $G' - R$, respectively. These numbers can be computed in FPT time parameterized by k.

For each type τ and for each multi-set $\mathcal{T}$ of types such that $\mathcal{T}$ fits τ, we use a variable $x_{\mathcal{T},\tau}$ to represent the number of type-$\mathcal{T}$ multi-sets of components in $Q - S$ that are mapped to type-τ components in $G' - R$. For each type τ of components in $G' - R$, we can embed at most n_τ sets of components in $Q - S$. This constraint is expressed as follows:

$$n_\tau \geq \sum_{\mathcal{T}\,:\,\mathcal{T} \text{ fits } \tau} x_{\mathcal{T},\tau} \quad \text{for each type } \tau. \tag{1}$$

For each type σ of components in $Q - S$, we need to embed all n_σ components of type σ into some components of $G - R'$. We can express this constraint as follows:

$$n_\sigma = \sum_{\mathcal{T},\tau\,:\,\sigma \in \mathcal{T} \text{ and } \mathcal{T} \text{ fits } \tau} \mu_{\mathcal{T},\sigma} \cdot x_{\mathcal{T},\tau} \quad \text{for each type } \sigma, \tag{2}$$

where $\mu_{\mathcal{T},\sigma}$ is the multiplicity of σ in $\mathcal{T}$. This completes the ILP formulation of the problem. We do not have any objective function and just ask for the feasibility. The construction can be done in FPT time parameterized by k.

Observe that there are at most $\binom{4^{k^2}+k-1}{k} < 4^{k^3+k}$ multi-sets $\mathcal{T}$ of types of components. Thus the ILP above has at most $4^{k^2} \cdot 4^{k^3+k}$ variables (the first factor for τ and the second for $\mathcal{T}$) and at most $4^{k^2} \cdot 4^{k^3+k} + 4^{k^2} \cdot 4^{k^2} \cdot 4^{k^3+k}$ constraints (the first term for (1) and the second for (2)) of length $O(4^{k^2} \cdot 4^{k^3+k})$. The coefficients are upper bounded by $|V(G')|$. It is known that the feasibility check of such an ILP can be done in FPT time parameterized by k [12,18,23]. Thus, the problem can be solved in FPT time when parameterized by k. $\square$

4.2   Subgraph Isomorphism on $4P_5$-Minor Free Graphs

For this case, we show the NP-hardness by a reduction from (3, B2)-SAT [2], which is a restricted version of 3-SAT. (The proof is omitted in this version.)

Theorem 4.4 (★). SUBGRAPH ISOMORPHISM on $4P_5$-minor free graphs is NP-complete.

5   Concluding Remarks

As we mentioned before, there are some unsettled cases for SUBGRAPH ISOMORPHISM on H-minor free graphs. If H is connected, then $H = P_5$ is only the unknown case. When H can be disconnected, we do not know the complexity

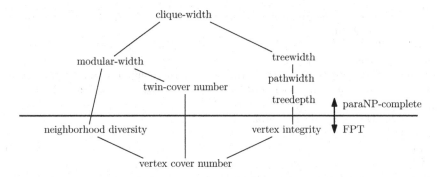

Fig. 3. Graph parameters and SUBGRAPH ISOMORPHISM. For each connection of parameters, there is a function in the parameter above that lower bounds the one below.

when H is a linear forest and either H contains kP_4 as a subgraph for $k \geq 2$ but no P_5; or H contains kP_5 as a subgraph for $k \in \{1, 2, 3\}$ but no P_6.

Our results imply some parameterized results. See Fig. 3. (We omit the definitions of the parameters.) The proof of Theorem 3.2 implies that SUBGRAPH ISOMORPHISM is NP-complete even for graphs of tree-depth [28] at most 3. This bound is tight by Lemma 3.1 since graphs of tree-depth at most 2 does not contain P_4 as a subgraph. Proposition 1.1 implies it is NP-complete even for graphs of constant twin-cover number [13] because cluster graphs have twin-cover number 0. For the parameterization by neighborhood diversity [22], we can use techniques similar to the ones we used in this paper.

Theorem 5.1 (★). SUBGRAPH ISOMORPHISM *on graphs of neighborhood diversity at most k is fixed-parameter tractable parameterized by k.*

References

1. Barefoot, C.A., Entringer, R.C., Swart, H.C.: Vulnerability in graphs–a comparative survey. J. Comb. Math. Comb. Comput. **1**, 13–22 (1987)
2. Berman, P., Karpinski, M., Scott, A.D.: Approximation hardness of short symmetric instances of MAX-3SAT. Technical report TR03-049, Electronic Colloquium on Computational Complexity (ECCC) (2003)
3. Bodlaender, H.L., Nederlof, J., van der Zanden, T.C.: Subexponential time algorithms for embedding H-minor free graphs. In: ICALP 2016. LIPIcs, vol. 55, pp. 9:1–9:14 (2016)
4. Bonsma, P.: Surface split decompositions and subgraph isomorphism in graphs on surfaces. In: STACS 2012. LIPIcs, vol. 14, pp. 531–542 (2012)
5. Cygan, M., et al.: Tight lower bounds on graph embedding problems. J. ACM **64**(3), 18:1–18:22 (2017)
6. Cygan, M., et al.: Parameterized Algorithms. Springer, Cham (2015). https://doi.org/10.1007/978-3-319-21275-3
7. Damaschke, P.: Induced subgraph isomorphism for cographs is NP-complete. In: Möhring, R.H. (ed.) WG 1990. LNCS, vol. 484, pp. 72–78. Springer, Heidelberg (1991). https://doi.org/10.1007/3-540-53832-1_32

8. Dorn, F.: Planar subgraph isomorphism revisited. In: STACS 2010. LIPIcs, vol. 5, pp. 263–274 (2010)
9. Downey, R.G., Fellows, M.R.: Fixed-parameter tractability and completeness II: on completeness for W[1]. Theor. Comput. Sci. **141**(1&2), 109–131 (1995)
10. Drange, P.G., Dregi, M.S., van 't Hof, P.: On the computational complexity of vertex integrity and component order connectivity. Algorithmica **76**(4), 1181–1202 (2016)
11. Eppstein, D.: Subgraph isomorphism in planar graphs and related problems. J. Graph Algorithms Appl. **3**(3), 1–27 (1999)
12. Frank, A., Tardos, É.: An application of simultaneous Diophantine approximation in combinatorial optimization. Combinatorica **7**(1), 49–65 (1987)
13. Ganian, R.: Improving vertex cover as a graph parameter. Discret. Math. Theor. Comput. Sci. **17**(2), 77–100 (2015)
14. Garey, M.R., Johnson, D.S.: Computers and Intractability: A Guide to the Theory of NP-Completeness. W. H. Freeman, New York (1979)
15. Gupta, A., Nishimura, N.: The complexity of subgraph isomorphism for classes of partial k-trees. Theor. Comput. Sci. **164**(1&2), 287–298 (1996)
16. Hajiaghayi, M., Nishimura, N.: Subgraph isomorphism, log-bounded fragmentation, and graphs of (locally) bounded treewidth. J. Comput. Syst. Sci. **73**(5), 755–768 (2007)
17. Jansen, B.M.P., Marx, D.: Characterizing the easy-to-find subgraphs from the viewpoint of polynomial-time algorithms, kernels, and Turing kernels. In: SODA 2015, pp. 616–629 (2015)
18. Kannan, R.: Minkowski's convex body theorem and integer programming. Math. Oper. Res. **12**(3), 415–440 (1987)
19. Kijima, S., Otachi, Y., Saitoh, T., Uno, T.: Subgraph isomorphism in graph classes. Discrete Math. **312**(21), 3164–3173 (2012)
20. Kiyomi, M., Otachi, Y.: Finding a chain graph in a bipartite permutation graph. Inf. Process. Lett. **116**(9), 569–573 (2016)
21. Konagaya, M., Otachi, Y., Uehara, R.: Polynomial-time algorithms for subgraph isomorphism in small graph classes of perfect graphs. Discrete Appl. Math. **199**, 37–45 (2016)
22. Lampis, M.: Algorithmic meta-theorems for restrictions of treewidth. Algorithmica **64**(1), 19–37 (2012)
23. Lenstra Jr., H.W.: Integer programming with a fixed number of variables. Math. Oper. Res. **8**(4), 538–548 (1983)
24. Lingas, A.: Subgraph isomorphism for biconnected outerplanar graphs in cubic time. Theor. Comput. Sci. **63**(3), 295–302 (1989)
25. Marx, D., Pilipczuk, M.: Everything you always wanted to know about the parameterized complexity of subgraph isomorphism (but were afraid to ask). In: STACS 2014. LIPIcs, vol. 25, pp. 542–553 (2014)
26. Matoušek, J., Thomas, R.: On the complexity of finding iso- and other morphisms for partial k-trees. Discrete Math. **108**(1–3), 343–364 (1992)
27. Matula, D.W.: Subtree isomorphism in $O(n^{5/2})$. In: Alspach, B., Hell, P., Miller, D. (eds.) Algorithmic Aspects of Combinatorics. Annals of Discrete Mathematics, vol. 2, pp. 91–106. Elsevier (1978)
28. Nesetril, J., de Mendez, P.O.: Sparsity - Graphs, Structures, and Algorithms. Algorithms and Combinatorics, vol. 28. Springer, Heidelberg (2012). https://doi.org/10.1007/978-3-642-27875-4

Your Rugby Mates Don't Need
to Know Your Colleagues:
Triadic Closure with Edge Colors

Laurent Bulteau[1], Niels Grüttemeier[2(✉)], Christian Komusiewicz[2],
and Manuel Sorge[3]

[1] CNRS, Université Paris-Est Marne-la-Vallée, Paris, France
`laurent.bulteau@u-pem.fr`
[2] Fachbereich Mathematik und Informatik,
Philipps-Universität Marburg, Marburg, Germany
`{niegru,komusiewicz}@informatik.uni-marburg.de`
[3] Faculty of Mathematics, Informatics and Mechanics,
University of Warsaw, Warsaw, Poland
`manuel.sorge@mimuw.edu.pl`

Abstract. Given an undirected graph $G = (V, E)$ the NP-hard STRONG TRIADIC CLOSURE (STC) problem asks for a labeling of the edges as *weak* and *strong* such that at most k edges are weak and for each induced P_3 in G at least one edge is weak. In this work, we study the following generalizations of STC with c different strong edge colors. In MULTI-STC an induced P_3 may receive two strong labels as long as they are different. In EDGE-LIST MULTI-STC and VERTEX-LIST MULTI-STC we may additionally restrict the set of permitted colors for each edge of G. We show that, under the ETH, EDGE-LIST MULTI-STC and VERTEX-LIST MULTI-STC cannot be solved in time $2^{o(|V|^2)}$, and that MULTI-STC is NP-hard for every fixed c. We then extend previous fixed-parameter tractability results and kernelizations for STC to the three variants with multiple edge colors or outline the limits of such an extension.

1 Introduction

Social networks represent relationships between humans such as acquaintance and friendship in online social networks. One task in social network analysis is to determine the strength [15,16] and type [3,17] of the relationship signified by each edge of the network. One approach to infer strong ties goes back to the notion of *strong triadic closure* [6] which postulates that, if an agent has strong

Supported by the People Programme (Marie Curie Actions) of the European Union's Seventh Framework Programme (FP7/2007-2013) under REA grant agreement number 631163.11, the Israel Science Foundation (grant no. 551145/14), and by the European Research Council (ERC) under the European Union's Horizon 2020 research and innovation programme under grant agreement number 714704.

© Springer Nature Switzerland AG 2019
P. Heggernes (Ed.): CIAC 2019, LNCS 11485, pp. 99–111, 2019.
https://doi.org/10.1007/978-3-030-17402-6_9

relations to two other agents, then these two should have at least a weak relation. Following this assertion, Sintos and Tsaparas [16] proposed to find strong ties in social networks by labeling the edges as weak or strong such that the strong triadic closure property is fulfilled and the number of strong edges is maximized.

Sintos and Tsaparas [16] also formulated an extension where agents may have c different types of strong relationships. In this model, the strong triadic closure property only applies to edges of the same strong type. This is motivated by the following observation: agents may very well have close relations to agents that do not know each other if these relations arise in segregated contexts. For example, it is quite likely that one's rugby teammates do not know all of one's close colleagues. The edge labelings that model this variant of strong triadic closure and the corresponding problem are defined as follows.

Definition 1. *A c-labeling $L = (S_L^1, \ldots, S_L^c, W_L)$ of an undirected graph $G = (V, E)$ is a partition of the edge set E into $c+1$ color classes. The edges in S_L^i, $i \in [c]$, are* strong *and the edges in W_L are* weak; *L is an STC-labeling if there exists no pair of edges $\{u, v\} \in S_L^i$ and $\{v, w\} \in S_L^i$ such that $\{u, w\} \notin E$.*

MULTI STRONG TRIADIC CLOSURE (MULTI-STC)
Input: An undirected graph $G = (V, E)$ and integers $c \in \mathbb{N}$ and $k \in \mathbb{N}$.
Question: Is there a c-colored STC-labeling L with $|W_L| \le k$?

We refer to the special case $c = 1$ as STRONG TRIADIC CLOSURE (STC). STC, and thus Multi-STC, is NP-hard [16]. We study the complexity of MULTI-STC and two generalizations of MULTI-STC which are defined as follows.

The first generalization deals with the case when one restricts the set of possible relations for some agents. Assume, for example, that strong edges correspond to family relations or professional relations. If one knows the profession of some agents, then this knowledge can be modeled by introducing different strong colors for each profession and constraining the sought edge labeling in such a way that each agent may receive only a strong edge corresponding to a familial relation or to his profession. In other words, for each agent we are given a list Λ of strong colors that may be assigned to incident relationships.

Definition 2. *Let $G = (V, E)$ be a graph, $\Lambda : V \to 2^{\{1,2,\ldots,c\}}$ a mapping for some $c \in \mathbb{N}$, and $L = (S_L^1, \ldots, S_L^c, W_L)$ a c-colored STC-labeling. We say that an edge $\{v, w\} \in E$ satisfies the Λ-list property under L if $\{v, w\} \in W_L$ or $\{v, w\} \in S_L^\alpha$ for some $\alpha \in \Lambda(v) \cap \Lambda(w)$. We call a c-colored STC-labeling Λ-satisfying if every edge $e \in E$ satisfies the Λ-list property under L.*

VERTEX-LIST MULTI STRONG TRIADIC CLOSURE (VL-MULTI-STC)
Input: An undirected graph $G = (V, E)$, integers $c \in \mathbb{N}$ and $k \in \mathbb{N}$, and vertex lists $\Lambda : V \to 2^{\{1,2,\ldots,c\}}$.
Question: Is there a Λ-satisfying STC-labeling L with $|W_L| \le k$?

MULTI-STC is the special case where $\Lambda(v) = \{1, \ldots, c\}$ for all $v \in V$. One might also specify a set of possible strong colors for each edge. This can be useful if

Table 1. An overview of the parameterized complexity results.

Parameter	Multi-STC	VL-Multi-STC	EL-Multi-STC
k		FPT if $c \leq 2$, NP-hard for $k = 0$ for all $c \geq 3$	
k_1	$4k_1$-vertex kernel	W[1]-hard	
(c, k_1)	$4k_1$-vertex kernel	$\mathcal{O}((c+1)^{k_1} \cdot (cm + nm))$ time no polynomial kernel $2^{c+1}k_1$-vertex kernel	

certain relations are not possible. For example, if two rugby players live far apart, it is unlikely that they play together. This more general constraint is formalized as for vertex lists with two differences: we are given edge lists $\Psi : E \rightarrow 2^{\{1,2,\dots,c\}}$ and for each edge e we have $e \in W_L$ or $e \in S_L^\alpha$ for some $\alpha \in \Psi(e)$.

EDGE-LIST MULTI STRONG TRIADIC CLOSURE (EL-MULTI-STC)
Input: An undirected graph $G = (V, E)$, integers $c \in \mathbb{N}$ and $k \in \mathbb{N}$ and edge lists $\Psi : E \rightarrow 2^{\{1,2,\dots,c\}}$.
Question: Is there a Ψ-satisfying STC-labeling L with $|W_L| \leq k$?

From a more abstract point of view, in STC we are to cover all induced P_3s, the paths on three vertices, in a graph by selecting at most k edges. Moreover, all STC-problems studied here have close ties to finding proper vertex colorings in a related graph, the Gallai graph [4] of the input graph G. Hence we are motivated to study these problems from a pure combinatorial and computational complexity point of view in addition to the known applications of MULTI-STC in social network analysis [16]. So far, algorithmic work has focused on STC [5,7,11,16]. Motivated by the NP-hardness of STC [11,16], the parameterized complexity of STC was studied. The two main parameters so far are the number k of weak edges and the number $\ell := |E| - k$ of strong edges in an STC-labeling with a minimal number of weak edges. For k, STC is fixed-parameter tractable [5,7,16] and admits a $4k$-vertex kernel [7]. For ℓ, STC is fixed-parameter tractable but does not admit a polynomial problem kernel [5,7].

Our Results. We show that for all $c \geq 1$ MULTI-STC, VL-MULTI-STC, and EL-MULTI-STC are NP-hard. In particular, for all $c \geq 3$, we obtain NP-hardness even if $k = 0$. We then show that, assuming the ETH, there is no $2^{o(|V|^2)}$-time algorithm for VL-MULTI-STC and EL-MULTI-STC even if $k = 0$ and $c \in \mathcal{O}(|V|)$.

We then proceed to a parameterized complexity analysis; see Table 1 for an overview. Since all variants are NP-hard even if $k = 0$, we consider a structural parameter related to k. This parameter, denoted by k_1, is the minimum number of weak edges needed in an STC-labeling for $c = 1$. Thus, if k_1 is known, then we may immediately accept all instances with $k \geq k_1$; in this sense one may assume $k < k_1$ for MULTI-STC. For VL-MULTI-STC and EL-MULTI-STC this is not necessarily true due to some border cases of the definition.

The parameter k_1 is relevant for two reasons: First, it allows us to determine to which extent the FPT algorithms for STC carry over to MULTI-STC, VL-MULTI-STC, and EL-MULTI-STC. Second, k_1 has a structural interpretation: it is the vertex cover number of the Gallai graph of the input graph G. We believe that this parameterization might be useful for other problems. The specific results are as follows. We extend the $4k_1$-vertex kernelization [7] from STC to MULTI-STC. This yields a $2^{c+1} \cdot k_1$-vertex kernel for VL-MULTI-STC and EL-MULTI-STC. We show that VL-MULTI-STC and EL-MULTI-STC are more difficult than MULTI-STC: parameterization by k_1 alone leads to W[1]-hardness and both are unlikely to admit a kernel that is polynomial in $c + k_1$. We complement these results by a providing an $\mathcal{O}((c + 1)^{k_1} \cdot (c \cdot |E| + |V| \cdot |E|))$-time algorithm for the most general EL-MULTI-STC. Due to lack of space several proofs are deferred to a full version.

Notation. We consider undirected graphs $G = (V, E)$ where $n := |V|$ denotes the number of vertices and $m := |E|$ denotes the number of edges in G. For a vertex $v \in V$ we denote by $N_G(v) := \{u \in V \mid \{u, v\} \in E\}$ the *open neighborhood of v* and by $N_G[v] := N(v) \cup \{v\}$ the *closed neighborhood of v*. For any two vertex sets $V_1, V_2 \subseteq V$, we let $E_G(V_1, V_2) := \{\{v_1, v_2\} \in E \mid v_1 \in V_1, v_2 \in V_2\}$ and $E_G(V') := E_G(V', V')$. We may omit the subscript G if the graph is clear from the context. The *subgraph induced by a vertexset S* is denoted by $G[S] := (S, E_G(S))$. A *proper vertex coloring* with c strong colors for some $c \in \mathbb{N}$ is a mapping $a \colon V \to \{1, \dots, c\}$ such that there is no edge $\{u, v\} \in E$ with $a(u) = a(v)$. For the relevant definitions of parameterized complexity refer to [2].

Gallai Graphs, c-Colorable Subgraphs, and Their Relation to STC. MULTI-STC can be formulated in terms of so-called *Gallai graphs* [4].

Definition 3. *Given a graph $G = (V, E)$, the* Gallai graph $\tilde{G} = (\tilde{V}, \tilde{E})$ *of G is defined by $\tilde{V} := E$ and $\tilde{E} := \{\{e_1, e_2\} \mid e_1$ and e_2 form an induced P_3 in $G\}$.*

The Gallai graph of an n-vertex and m-edge graph has $\mathcal{O}(m)$ vertices and $\mathcal{O}(mn)$ edges. For $c = 1$, in other words, for STC, a graph $G = (V, E)$ has an STC-labeling with at most k weak edges if and only if its Gallai graph has a vertex cover of size at most k [16]. This gives an $\mathcal{O}(1.28^k + nm)$-time algorithm by using the current fastest algorithm for VERTEX COVER [1]. More generally, a graph $G = (V, E)$ has a c-colored STC-labeling with at most k weak edges if and only if the Gallai graph of G has a c-colorable subgraph on $m - k$ vertices [16].

In the following we extend the relation to EL-MULTI-STC by considering list-colorings of the Gallai graph. The special cases VL-MULTI-STC, MULTI-STC, and STC nicely embed into the construction. First, let us formally define the problem that we need to solve in the Gallai graph. Given a graph $G = (V, E)$, we call a mapping $\chi : V \to \{0, 1, \dots, c\}$ a *subgraph-c-coloring* if there is no edge $\{u, v\} \in E$ with $\chi(u) = \chi(v) \neq 0$. Vertices v with $\chi(v) = 0$ correspond to deleted vertices. The LIST-COLORABLE SUBGRAPH problem is, given a graph $G = (V, E)$, integers $c, k \in \mathbb{N}$, and lists $\Gamma : V \to 2^{\{1, \dots, c\}}$, to decide whether there is a subgraph-c-coloring $\chi : V \to \{0, 1, \ldots, c\}$ with $|\{v \in V \mid \chi(v) = 0\}| \leq k$ and $\chi(w) \in \Gamma(w) \cup \{0\}$ for every $w \in V$.

Proposition 1. *An instance* (G, c, k, Ψ) *of* EL-MULTI-STC *is a Yes-instance if and only if* $(\tilde{G}, c, k, \Psi)$ *is a Yes-instance of* LIST-COLORABLE SUBGRAPH, *where* $\tilde{G}$ *is the Gallai graph of* G.

We make use of this correspondence in one FPT algorithm and below.

Proposition 2. LIST-COLORABLE SUBGRAPH *can be solved in* $\mathcal{O}(3^n \cdot c^2(n + m))$ *time.* EL-MULTI-STC *can be solved in* $\mathcal{O}(3^m \cdot c^2 mn)$ *time.*

2 Classical and Fine-Grained Complexity

We first observe that MULTI-STC is NP-hard for all c. For $c = 2$ it was claimed that MULTI-STC is NP-hard since in the Gallai graph this is exactly the NP-hard ODD CYCLE TRANSVERSAL problem [16]. It is not known, however, whether ODD CYCLE TRANSVERSAL is NP-hard on Gallai graphs. Hence, we provide a proof of NP-hardness for $c = 2$ and further hardness results for all $c \geq 3$.

Theorem 1. MULTI-STC *is NP-hard (a) for* $c = 2$ *even on graphs with maximum degree four, and (b) for every* $c \geq 3$, *even if* $k = 0$.

We now provide a stronger hardness result for VL-MULTI-STC and EL-MULTI-STC: we show that they are unlikely to admit a single-exponential-time algorithm with respect to the number n of vertices. Thus, the algorithm behind Proposition 2 is optimal in the sense that m cannot be replaced by n. The reduction behind this hardness result is inspired by a reduction used to show that RAINBOW COLORING cannot be solved in $2^{o(n^{3/2})}$ time under the ETH [12]. We remark that for LIST-EDGE COLORING an ETH-based lower bound of $2^{o(n^2)}$ has been shown recently [13]. While LIST-EDGE COLORING is related to EL-MULTI-STC, the reduction does not work directly for EL-MULTI-STC because the instances created in this reduction contain triangles. Moreover, we consider the more restricted VL-MULTI-STC problem.

Theorem 2. *If the ETH is true, then* VL-MULTI-STC *cannot be solved in* $2^{o(|V|^2)}$ *time even if restricted to instances with* $k = 0$.

Proof. We give a reduction from 3-SAT to VL-MULTI-STC such that the resulting graph has $\mathcal{O}(\sqrt{|\phi|})$ vertices, where ϕ is the input formula and $|\phi|$ is the number of variables plus the number of clauses. The ETH and the Sparsification Lemma [9] then imply the claimed lower bound. Let ϕ be a 3-CNF formula with a set $X = \{x_1, x_2, \ldots, x_n\}$ of n variables and a set $\mathcal{C} := \{C_1, C_2, \ldots, C_m\}$ of $m \leq \frac{4}{3}n$ clauses. We can furthermore assume that each variable occurs in at most four clauses in ϕ [18]. Observe that, then, ϕ has at most $\frac{4}{3}n$ clauses. Let C_j be a clause and x_i a variable occurring in C_j. We define the *occurrence number* $\Omega(C_j, x_i)$ as the number of clauses in $\{C_1, C_2, \ldots, C_j\}$ that contain x_i. Since

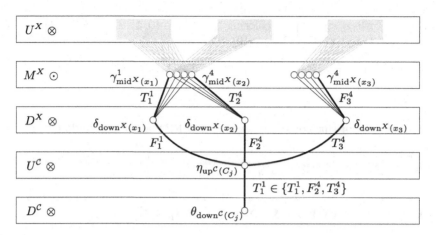

Fig. 1. A sketch of the construction. Gray rectangles represent the variable-soundness gadget, $\otimes$ a clique, and $\odot$ an independent set. Edge $\{\eta_{\mathrm{up}^{\mathcal{C}}(C_j)}, \theta_{\mathrm{down}^{\mathcal{C}}(C_j)}\}$ represents a clause $C_j = (x_1 \vee \overline{x_2} \vee x_3)$ with $\Omega(C_j, x_1) = 1$ and $\Omega(C_j, x_2) = \Omega(C_j, x_3) = 4$. The edge $\{\gamma^1_{\mathrm{mid}^X(x_1)}, \delta_{\mathrm{down}^X(x_1)}\}$ has strong color T_1^1 which models an assignment where x_1 is true, which satisfies C_j. Due to the compression, we may have $\mathrm{mid}(x_1) = \mathrm{mid}(x_2)$ and therefore x_1 and x_2 may share the four middle vertices.

each variable occurs in at most four clauses, we have $\Omega(C_j, x_i) \in \{1, 2, 3, 4\}$. We now construct an equivalent instance $(G = (V, E), c = 9n + 4, k = 0, \Lambda)$ for VL-MULTI-STC such that $|V| \in \mathcal{O}(\sqrt{n})$. First, we give some intuition.

The strong colors $1, \ldots, 8n$ represent the true and false assignments of the occurrences of the variables. Throughout this proof we refer to these strong colors as T_i^r, F_i^r with $i \in \{1, \ldots, n\}$ and $r \in \{1, 2, 3, 4\}$. The idea is that T_i^r represents assigning 'true' and F_i^r represents assigning 'false' to the rth occurrence of a variable $x_i \in X$. The strong colors $8n + 1, \ldots, 9n + 4$ are auxiliary colors which we need for the correctness of our construction. We refer to these strong colors as $R_1, \ldots, R_n$ and Z_1, Z_2, Z_3, Z_4. Due to lack of space, we give only a sketch of the construction. Herein, we only mention the auxiliary colors Z_2, Z_3 and Z_4. In the variable gadget, there are four distinct edges e_1, e_2, e_3, e_4 for each variable x_i representing the (at most) four occurrences of the variable x_i. Every such edge e_r can only be labeled with the strong colors T_i^r and F_i^r. The coloring of these edges represents a truth assignment to the variable x_i. In the clause gadget, there are m distinct edges such that the coloring of these edges represents a choice of literals that satisfies ϕ. The edges between the two gadgets make the values of the literals from the clause gadget consistent with the assignment of the variable gadget. The construction consists of five layers. In the variable gadget we have an upper- a middle- and a down layer (U^X, M^X and D^X). In the clause gadget we have an upper and a down layer ($U^{\mathcal{C}}$ and $D^{\mathcal{C}}$). Figure 1 shows a sketch of the construction.

The Variable Gadget. The vertices in the middle layer and the down layer form a *variable-representation gadget*, where each edge between the two layers represents one occurrence of a variable. The vertices in the upper layer form a *variable-soundness gadget*, which we need to ensure that for each variable either all occurrences are assigned 'true' or all occurrences are assigned 'false'. We start by describing the variable-representation gadget. Let

$$M^X := \{\gamma_t^r \mid t \in \{1, \ldots, \lceil \sqrt{n} \rceil\}, r \in \{1, 2, 3, 4\}\} \quad \text{be the set of middle vertices,}$$
$$D^X := \{\delta_t \mid t \in \{1, \ldots, \lceil \sqrt{n} \rceil + 9\} \quad\quad\quad \text{be the set of down vertices.}$$

We add edges such that D^X becomes a clique in G. To specify the correspondence between the variables in X and the edges in the variable-representation gadget, we define below two mappings $\text{mid}^X : X \to \{1, \ldots, \lceil \sqrt{n} \rceil\}$ and $\text{down}^X : X \to \{1, \ldots, \lceil \sqrt{n} \rceil + 9\}$. Then, for each variable $x_i \in X$ we add four edges $\{\gamma^r_{\text{mid}^X(x_i)}, \delta_{\text{down}^X(x_i)}\}$ for $r \in \{1, 2, 3, 4\}$. The truth assignment for each variable will be transmitted to a clause by edges between the variable and clause gadgets. To ensure that each such transmitter edge is used for exactly one occurrence of one variable, we first define the *variable-conflict graph* $H_\phi^X := (X, \text{Confl}^X)$ by $\text{Confl}^X := \{\{x_i, x_j\} \mid x_i \text{ and } x_j \text{ occur in the same clause } C \in \mathcal{C}\}$. We use H_ϕ^X to define mid^X and down^X. Since every variable of ϕ occurs in at most four clauses, the maximum degree of H_ϕ^X is at most 8. Hence, a proper vertex 9-coloring $\chi : X \to \{1, 2, \ldots, 9\}$ for H_ϕ^X can be computed in polynomial time by a greedy algorithm giving 9 color classes $\chi^{-1}(1), \ldots, \chi^{-1}(9)$. Then, we partition each color class $\chi^{-1}(i)$ into $\frac{|\chi^{-1}(i)|}{\lceil \sqrt{n} \rceil}$ groups arbitrarily such that each group has size at most $\lceil \sqrt{n} \rceil$. Let s be the overall number of such groups and let $\mathcal{S} := \{S_1, S_2, \ldots, S_s\}$ be the family of all such groups of vertices in H_ϕ^X. The definition of $\mathcal{S}$ implies the following.

Claim 1. *We have $|S_i| \le \lceil \sqrt{n} \rceil$ for each $i \in \{1, \ldots, s\}$, and $s \le \lceil \sqrt{n} \rceil + 9$.*

For any $x_i \in X$ let $\text{down}^X(x_i) := j$ be the index of the group S_j containing x_i. By Claim 1, $\text{down}^X(x_i) \le \lceil \sqrt{n} \rceil + 9 = |D_X|$. The mapping down^X is well-defined since $\mathcal{S}$ is a partition of the variable set X. The following is needed to ensure, for example, that no transmitter edge is used twice.

Claim 2. *If $x_i, x_j \in X$ occur in the same clause, then $\text{down}^X(x_i) \ne \text{down}^X(x_j)$.*

Next, we define the mapping $\text{mid}^X : X \to \{1, \ldots, \lceil \sqrt{n} \rceil\}$. To this end, consider the finite sequence $\text{Seq}_1^n := (\text{down}^X(x_1), \text{down}^X(x_2), \ldots, \text{down}^X(x_n)) \in \{1, \ldots, \lceil \sqrt{n} \rceil + 9\}^n$. Define $\text{mid}^X(x_i)$ as the number of occurrences of $\text{down}^X(x_i)$ in the partial sequence $\text{Seq}_1^i := (\text{down}^X(x_1), \ldots, \text{down}^X(x_i))$. From Claim 1 we conclude $\text{mid}^X(x_i) \in \{1, 2, \ldots, \lceil \sqrt{n} \rceil\}$ for every $x_i \in X$.

Claim 3. *Let $x_i, x_j \in X$ and let $r \in \{1, 2, 3, 4\}$. If $x_i \ne x_j$, then it follows that $\{\gamma^r_{\text{mid}^X(x_i)}, \delta_{\text{down}^X(x_i)}\} \ne \{\gamma^r_{\text{mid}^X(x_j)}, \delta_{\text{down}^X(x_j)}\}$.*

Thus we assigned a unique edge in $E(M^X, D^X)$ to each occurrence of a variable in X. Furthermore, the assigned edges of variables that occur in the same clause do not share an endpoint in D^X (Claim 2). We complete the variable-representation gadget by defining the vertex list $\Lambda(v)$ for every $v \in M^X \cup D^X$:

$$\Lambda(\gamma_t^r) := \bigcup_{\substack{x_i \in X \\ \mathrm{mid}^X(x_i)=t}} \{T_i^r, F_i^r, R_j\} \text{ for every } \gamma_t^r \in M^X, \text{ and}$$

$$\Lambda(\delta_t) := \bigcup_{\substack{x_i \in X \\ \mathrm{down}^X(x_i)=t}} \{T_i^1, T_i^2, T_i^3, T_i^4, F_i^1, F_i^2, F_i^3, F_i^4, Z_2\} \text{ for every } \delta_t \in D^X.$$

With these vertex lists, every edge that represents an occurrence of some variable can only be labeled with colors that match the truth assignment:

Claim 4. *Let $x_i \in X$ and $r \in \{1, 2, 3, 4\}$. Then, $\Lambda(\gamma_{\mathrm{mid}^X(x_i)}^r) \cap \Lambda(\delta_{\mathrm{down}^X(x_i)}) = \{T_i^r, F_i^r\}$.*

Note that for each variable x_i there are four edges $\{\gamma_{\mathrm{mid}^X(x_i)}^r, \delta_{\mathrm{down}^X(x_i)} \mid r \in \{1, 2, 3, 4\}\}$ that can only be colored with the strong colors T_i^r and F_i^r representing the truth assignments of the four occurrences of variable x_i. We need to ensure that there is no variable x_i, where, for example, the first occurrence is set to 'true' (T_i^1) and the second occurrence is set to 'false' (F_i^2) in a Λ-satisfying STC-labeling with no weak edges. To this end, we construct a variable-soundness gadget whose description is deferred to a full version.

The Clause Gadget. The clause gadget consists of an upper layer and a down layer. Let $U^{\mathcal{C}} := \{\eta_i \mid i \in \{1, \ldots, 12\lceil\sqrt{n}\rceil + 1\}\}$ be the set of upper vertices and $D^{\mathcal{C}} := \{\theta_i \mid i \in \{1, \ldots, \lceil\sqrt{n}\rceil\}\}$ be the set of lower vertices. We add edges such that $U^{\mathcal{C}}$ and $D^{\mathcal{C}}$ each form cliques in G. Below we define two mappings $\mathrm{up}^{\mathcal{C}} : \mathcal{C} \to \{1, 2, \ldots, 12\lceil\sqrt{n}\rceil + 1\}$, $\mathrm{down}^{\mathcal{C}} : \mathcal{C} \to \{1, 2, \ldots, \lceil\sqrt{n}\rceil\}$, and vertex lists $\Lambda : V \to 2^{\{1, \ldots, c\}}$. Then, for each clause $C_j \in \mathcal{C}$ we add an edge $\{\eta_{\mathrm{up}^{\mathcal{C}}(C_j)}, \theta_{\mathrm{down}^{\mathcal{C}}(C_j)}\}$. Next, we ensure that this edge can only be labeled with the strong colors that match the literals in C_i. This means, for example, if $C_i = (x_1 \vee \overline{x_2} \vee x_3)$ we have $\Lambda(\eta_{\mathrm{up}^{\mathcal{C}}(C_j)}) \cap \Lambda(\theta_{\mathrm{down}^{\mathcal{C}}(C_j)}) = \{T_1^{\Omega(C_j, x_1)}, F_2^{\Omega(C_j, x_2)}, T_3^{\Omega(C_j, x_3)}\}$.

As above, we need to ensure that each variable occurring in a clause has a unique edge between the clause and variable gadgets which transmits the variable's truth assignment to the clause. To achieve this, we define the *clause-conflict graph* $H_\phi^{\mathcal{C}} := (\mathcal{C}, \mathrm{Confl}^{\mathcal{C}})$ by

$$\mathrm{Confl}^{\mathcal{C}} := \{\{C_i, C_j\} \mid C_i \text{ contains a variable } x_i \text{ and } C_j \text{ contains a variable } x_j,$$
$$\text{such that } \mathrm{down}^X(x_i) = \mathrm{down}^X(x_j)\}.$$

Since each variable occurs in at most four clauses and by Claim 1, it follows that the maximum degree of $H_\phi^{\mathcal{C}}$ is at most $12 \cdot \lceil\sqrt{n}\rceil$. Thus, a proper vertex coloring $\chi : \mathcal{C} \to \{1, 2, \ldots, 12 \cdot \lceil\sqrt{n}\rceil + 1\}$ such that each color class $\chi^{-1}(i)$ contains at most $\lceil\frac{m}{12 \cdot \lceil\sqrt{n}\rceil + 1}\rceil + 1 \leq \lceil\sqrt{n}\rceil$ can be computed in polynomial time [10].

For a clause $C_i \in C$ we define $\mathrm{up}^C(C_i) := j$ as the index of the color class $\chi^{-1}(j)$ that contains C_i. Together with Claim 2 (which argues about the endpoint in the variable gadget), the following claim (which argues about the endpoint in the clause gadget) ensures that no transmitter edge is used twice.

Claim 5. *If clause $C_{j_1} \in C$ contains x_{i_1} and clause $C_{j_2} \in C$ contains x_{i_2} such that* $\mathrm{down}^X(x_{i_1}) = \mathrm{down}^X(x_{i_2})$, *then* $\mathrm{up}^C(C_{j_1}) \neq \mathrm{up}^C(C_{j_2})$.

Next, we define down^C analogously to up^X. Consider the sequence $\mathrm{Seq}_1^m = (\mathrm{up}^C(C_1), \mathrm{up}^C(C_2), \dots, \mathrm{up}^C(C_n))$ and define $\mathrm{down}^C(C_j)$ as the number of occurrences of $\mathrm{up}^C(C_j)$ in the sequence $\mathrm{Seq}_1^j := (\mathrm{up}^C(C_1), \dots, \mathrm{up}^C(C_j))$. Since each color class contains at most $\lceil \sqrt{n} \rceil$ elements, we have $\mathrm{down}^C(C_j) \leq \lceil \sqrt{n} \rceil$.

Claim 6. *If $C_i \neq C_j$, then* $\{\eta_{\mathrm{up}^C(C_i)}, \theta_{\mathrm{down}^C(C_i)}\} \neq \{\eta_{\mathrm{up}^C(C_j)}, \theta_{\mathrm{down}^C(C_j)}\}$.

Thus we assigned a unique edge in $E(U^C, D^C)$ to each clause. We complete the description of the clause gadget by defining the vertex lists $\Lambda(v)$ for every $v \in U^C \cup D^C$. For a given clause $C_j \in C$ we define the *color set* $\mathfrak{X}(C_j)$ and the *literal color set* $\mathfrak{L}(C_j)$ of C_j by $\mathfrak{X}(C_j) := \{T_i^{\Omega(C_j, x_i)}, F_i^{\Omega(C_j, x_i)} \mid x_i \text{ occurs in } C_j\}$ and by $\mathfrak{L}(C_j) := \{T_i^{\Omega(C_j, x_i)} \mid x_i \text{ occurs as a positive literal in } C_j\} \cup \{F_i^{\Omega(C_j, x_i)} \mid x_i \text{ occurs as a negative literal in } C_j\}$.

Note that $\mathfrak{L}(C_j) \subseteq \mathfrak{X}(C_j)$. The vertex lists for the vertices in $U^C \cup D^C$ are

$$\Lambda(\eta_t) := \bigcup_{\substack{C_j \in C \\ \mathrm{up}^C(C_j) = t}} \mathfrak{X}(C_j) \cup \{Z_3\} \qquad \text{for every } \eta_t \in U^C, \text{ and}$$

$$\Lambda(\theta_t) := \bigcup_{\substack{C_j \in C \\ \mathrm{down}^C(C_j) = t}} \mathfrak{L}(C_j) \cup \{Z_4\} \qquad \text{for every } \theta_t \in D^C.$$

Claim 7. *Let $C_j \in C$. Then,* $\Lambda(\eta_{\mathrm{up}^C(C_j)}) \cap \Lambda(\theta_{\mathrm{down}^C(C_j)}) = \mathfrak{L}(C_j)$.

By Claim 7, for every clause the assigned edge can only be labeled with strong colors that match its literals.

Connecting the Gadgets. To complete the construction of G, we add edges between D^X and U^C that model the occurrences of variables in clauses. For each clause $C_j \in C$, we add edges $\{\delta_{\mathrm{down}^X(x_{i_1})}, \eta_{\mathrm{up}^C(C_j)}\}$, $\{\delta_{\mathrm{down}^X(x_{i_2})}, \eta_{\mathrm{up}^C(C_j)}\}$, and $\{\delta_{\mathrm{down}^X(x_{i_3})}, \eta_{\mathrm{up}^C(C_j)}\}$ where x_{i_1}, x_{i_2}, and x_{i_3} are the variables that occur in C_j. Intuitively, an edge $\{\delta_{\mathrm{down}^X(x_i)}, \eta_{\mathrm{up}^C(C_j)}\}$ transmits the truth value of x_i to C_j, where x_i occurs as a positive or negative literal. The following claim states that the possible strong colors for such an edge are only $T_i^{\Omega(C_j, x_i)}$ and $F_i^{\Omega(C_j, x_i)}$, which correspond to the truth assignment of the $\Omega(C_j, x_i)$-th occurrence of x_i.

Claim 8. *Let $C_j \in C$ be a clause and let $x_i \in X$ be some variable that occurs in C_j. Then* $\Lambda(\delta_{\mathrm{down}^X(x_i)}) \cap \Lambda(\eta_{\mathrm{up}^C(C_j)}) = \{T_i^{\Omega(C_j, x_i)}, F_i^{\Omega(C_j, x_i)}\}$.

This completes the description of the construction and basic properties of the VL-MULTI-STC instance $(G, 9n + 4, 0, \Lambda)$. Note that G has $\mathcal{O}(\sqrt{n})$ vertices. The correctness proof is deferred to a full version. $\square$

Note that in the instance constructed in the proof of Theorem 2, every edge has at most three possible strong colors and $c \in \mathcal{O}(n)$. This implies the following.

Corollary 1. *If the ETH is true, then*

(a) EL-MULTI-STC *cannot be solved in* $2^{o(|V|^2)}$ *time even if restricted to instances* (G, c, k, Ψ) *where* $k = 0$ *and* $\max_{e \in E} |\Psi(e)| = 3$.

(b) VL-MULTI-STC *cannot be solved in* $c^{o(|V|^2 / \log |V|)}$ *time even if* $k = 0$.

3   Parameterized Complexity

The most natural parameter is the number k of weak edges. The case $c = 1$ (STC) is fixed-parameter tractable [16]. For $c = 2$, we also obtain an FPT algorithm: one may solve ODD CYCLE TRANSVERSAL in the Gallai graph $\tilde{G}$ which is fixed-parameter tractable with respect to k [2]. This extends to EL-MULTI-STC with $c = 2$ by applying standard techniques. In contrast, for every fixed $c \geq 3$, MULTI-STC is NP-hard even if $k = 0$. Hence, FPT algorithms for (c, k) are unlikely. We thus define the parameter k_1 and analyze the parameterized complexity of (VL-/EL-)MULTI-STC regarding the parameters k_1 and (c, k_1).

Definition 4. *Let* $G = (V, E)$ *be a graph with a 1-colored STC-labeling* $L = (S_L, W_L)$ *with a minimal number of weak edges. Then* $k_1 = k_1(G) := |W_L|$.

For a given graph G, the value k_1 equals the size of a minimal vertex cover of the Gallai graph $\tilde{G}$ due to Proposition 1. We now provide an FPT result for EL-MULTI-STC parameterized by (c, k_1). The main idea of the algorithm is to solve LIST-COLORABLE SUBGRAPH on the Gallai graph of G.

Theorem 3. EL-MULTI-STC *can be solved in* $\mathcal{O}((c + 1)^{k_1} \cdot (cm + nm))$ *time.*

We conclude that MULTI-STC parameterized by k_1 is fixed-parameter tractable.

Theorem 4. MULTI-STC *can be solved in* $\mathcal{O}((k_1 + 1)^{k_1} \cdot (k_1 m + nm))$ *time.*

Theorem 4 follows from a relationship between c and k_1, which leads to an FPT result for MULTI-STC parameterized only by k_1. Instances with $c > k_1$ are trivial yes-instances, otherwise Theorem 3 provides FPT running time. However, there is little hope that something similar holds for VL-MULTI-STC.

Theorem 5. VL-MULTI-STC *parameterized by* k_1 *is W[1]-hard, even if* $k = 0$. VL-MULTI-STC *parameterized by* (c, k_1) *does not admit a polynomial kernel unless* $NP \subseteq coNP/poly$.

Algorithm 1. EL-Multi-STC kernel reduction

1: **Input:** $G = (V, E)$ graph, $K \subseteq V$ closed critical clique in G
2: **for each** $v \in \mathcal{N}(K)$ **do**
3:     **for each** $\psi \in \{\Psi(e) \neq \emptyset \mid e \in E(\{v\}, K)\}$ **do**
4:         $i := 0$
5:         **for each** $w \in N(v) \cap K$ **do**
6:             **if** $\Psi(\{v, w\}) = \psi$ **then**
7:                 Mark w as *important* and set $i := i + 1$
8:             **if** $i = |E(\{v\}, \mathcal{N}^2(K))|$ **then break**
9: Delete all vertices $u \in K$ which are not marked as important from G
10: Decrease the value of k by the number of edges e that are incident with a deleted vertex u and $\Psi(e) = \emptyset$.

On Problem Kernelization. Since EL-MULTI-STC is a generalization of VL-MULTI-STC, we conclude from Theorem 5 that there is no polynomial kernel for EL-MULTI-STC parameterized by (c, k_1) unless NP $\subseteq$ coNP/poly and thus we give a $2^{c+1} \cdot k_1$-vertex kernel for EL-MULTI-STC. To this end we define a new parameter τ as follows. Let $I := (G, c, k, \Psi)$ be an instance of EL-MULTI-STC. Then $\tau := |\Psi(E) \setminus \{\emptyset\}|$ is defined as the number of different non-empty edge lists occurring in the instance I. It clearly holds that $\tau \leq 2^c - 1$.

For this kernelization we use *critical cliques* and *critical clique graphs* [14]. The kernelization described here generalizes the linear-vertex kernel for STC [7].

Definition 5. *A* critical clique *of a graph G is a clique K where the vertices of K all have the same neighbors in $V \setminus K$, and K is maximal under this property. Given a graph $G = (V, E)$, let $\mathcal{K}$ be the collection of its critical cliques. The* critical clique graph $\mathcal{C}$ *of G is the graph $(\mathcal{K}, E_{\mathcal{C}})$ with $\{K_i, K_j\} \in E_{\mathcal{C}} \Leftrightarrow \forall u \in K_i, v \in K_j : \{u, v\} \in E$.*

For a critical clique K we let $\mathcal{N}(K) := \bigcup_{K' \in N_{\mathcal{C}}(K)} K'$ denote the union of its neighbor cliques in the critical clique graph and $\mathcal{N}^2(K) := \bigcup_{K' \in N_{\mathcal{C}}^2(K)} K'$ denote the union of the critical cliques at distance exactly two from K. The critical clique graph can be constructed in $\mathcal{O}(n + m)$ time [8].

Critical cliques are an important tool for EL-MULTI-STC because every edge between the vertices of some critical clique is not part of any induced P_3 in G. Hence, each such edge e is strong under any STC-Labeling unless $\Psi(e) = \emptyset$. In the following, we distinguish between two types of critical cliques. We say that K is *closed* if $\mathcal{N}(K)$ forms a clique in G and that K is *open* otherwise. We will see that the number of vertices in open critical cliques is at most $2k_1$. The following reduction rule describes how to deal with large closed critical cliques.

Rule 1. *If G has a closed critical clique K with $|K| > \tau \cdot |E(\mathcal{N}(K), \mathcal{N}^2(K))|$, then apply Algorithm 1 on G and K.*

Proposition 3. *Rule 1 is safe and can be applied in polynomial time.*

Rule 1 leads to the following kernel result.

Theorem 6. EL-MULTI-STC *admits a problem kernel with at most* $(\tau+1) \cdot 2k_1$ *vertices,* EL-MULTI-STC *admits a kernel with at most* $2^{c+1}k_1$ *vertices, and* MULTI-STC *admits a problem kernel with at most* $4k_1$ *vertices.*

For the last two statements of Theorem 6, recall that for any EL-MULTI-STC instance (G, c, k, Ψ) we have $\tau \leq 2^c - 1$. Also, MULTI-STC is the special case of EL-MULTI-STC where every edge has the list $\{1, 2, \ldots, c\}$, and thus $\tau = 1$.

References

1. Chen, J., Kanj, I.A., Xia, G.: Improved upper bounds for vertex cover. Theor. Comput. Sci. **411**(40–42), 3736–3756 (2010)
2. Cygan, M., et al.: Parameterized Algorithms. Springer, Cham (2015). https://doi.org/10.1007/978-3-319-21275-3
3. Diehl, C.P., Namata, G., Getoor, L.: Relationship identification for social network discovery. In: Proceedings of the 22nd AAAI, pp. 546–552. AAAI Press (2007)
4. Gallai, T.: Transitiv orientierbare Graphen. Acta Math. Hung. **18**(1–2), 25–66 (1967)
5. Golovach, P.A., Heggernes, P., Konstantinidis, A.L., Lima, P.T., Papadopoulos, C.: Parameterized aspects of strong subgraph closure. In: Proceedings of the 16th SWAT. LIPIcs, vol. 101, pp. 23:1–23:13. Schloss Dagstuhl - Leibniz-Zentrum fuer Informatik (2018)
6. Granovetter, M.: The strength of weak ties. Am. J. Sociol. **78**, 1360–1380 (1973)
7. Grüttemeier, N., Komusiewicz, C.: On the relation of strong triadic closure and cluster deletion. In: Brandstädt, A., Köhler, E., Meer, K. (eds.) WG 2018. LNCS, vol. 11159, pp. 239–251. Springer, Cham (2018). https://doi.org/10.1007/978-3-030-00256-5_20. https://arxiv.org/abs/1803.00807
8. Hsu, W.L., Ma, T.H.: Substitution decomposition on chordal graphs and applications. In: Hsu, W.L., Lee, R.C.T. (eds.) ISA 1991. LNCS, vol. 557, pp. 52–60. Springer, Heidelberg (1991). https://doi.org/10.1007/3-540-54945-5_49
9. Impagliazzo, R., Paturi, R., Zane, F.: Which problems have strongly exponential complexity? J. Comput. Syst. Sci. **63**(4), 512–530 (2001)
10. Kierstead, H.A., Kostochka, A.V., Mydlarz, M., Szemerédi, E.: A fast algorithm for equitable coloring. Combinatorica **30**(2), 217–224 (2010)
11. Konstantinidis, A.L., Nikolopoulos, S.D., Papadopoulos, C.: Strong triadic closure in cographs and graphs of low maximum degree. Theor. Comput. Sci. **740**, 76–84 (2018)
12. Kowalik, L., Lauri, J., Socala, A.: On the fine-grained complexity of rainbow coloring. SIAM J. Discrete Math. **32**, 1672–1705 (2018)
13. Kowalik, L., Socala, A.: Tight lower bounds for list edge coloring. In: Proceedings of the 16th SWAT. LIPIcs, vol. 101, pp. 28:1–28:12. Schloss Dagstuhl - Leibniz-Zentrum fuer Informatik (2018)
14. Protti, F., da Silva, M.D., Szwarcfiter, J.L.: Applying modular decomposition to parameterized cluster editing problems. Theory Comput. Syst. **44**(1), 91–104 (2009)
15. Rozenshtein, P., Tatti, N., Gionis, A.: Inferring the strength of social ties: a community-driven approach. In: Proceedings of the 23rd KDD, pp. 1017–1025. ACM (2017)

16. Sintos, S., Tsaparas, P.: Using strong triadic closure to characterize ties in social networks. In: Proceedings of the 20th KDD, pp. 1466–1475. ACM (2014)
17. Tang, J., Lou, T., Kleinberg, J.M.: Inferring social ties across heterogenous networks. In: Proceedings of the 5th WSDM, pp. 743–752. ACM (2012)
18. Tovey, C.A.: A simplified NP-complete satisfiability problem. Discrete Appl. Math. **8**(1), 85–89 (1984)

k-cuts on a Path

Xing Shi Cai[1]([⊠]) [ID], Luc Devroye[2], Cecilia Holmgren[1] [ID], and Fiona Skerman[1] [ID]

[1] Mathematics Department, Uppsala University, 75237 Uppsala, Sweden
{xingshi.cai,cecilia.holmgren,fiona.skerman}@math.uu.se
[2] School of Computer Science, McGill University, Montréal, QC H3A 2A7, Canada
lucdevroye@gmail.com

Abstract. We define the (random) k-cut number of a rooted graph to model the difficulty of the destruction of a resilient network. The process is as the cut model of Meir and Moon [14] except now a node must be cut k times before it is destroyed. The first order terms of the expectation and variance of $\mathcal{X}_n$, the k-cut number of a path of length n, are proved. We also show that $\mathcal{X}_n$, after rescaling, converges in distribution to a limit $\mathcal{B}_k$, which has a complicated representation. The paper then briefly discusses the k-cut number of general graphs. We conclude by some analytic results which may be of interest.

Keywords: Cutting · k-cut · Network · Record · Permutation

1 Introduction and Main Results

1.1 The k-cut Number of a Graph

Consider $\mathbb{G}_n$, a connected graph consisting of n nodes with exactly one node labeled as the *root*, which we call a *rooted* graph. Let k be a positive integer. We remove nodes from the graph as follows:

1. Choose a node uniformly at random from the component that contains the root. Cut the selected node once.
2. If this node has been cut k times, remove the node together with edges attached to it from the graph.
3. If the root has been removed, then stop. Otherwise, go to step 1.

We call the (random) total number of cuts needed to end this procedure the k-cut number and denote it by $\mathcal{K}(\mathbb{G}_n)$. (Note that in traditional cutting models, nodes are removed as soon as they are cut once, i.e., $k = 1$. But in our model, a node is only removed after being cut k times.)

One can also define an edge version of this process. Instead of cutting nodes, each time we choose an edge uniformly at random from the component that contains the root and cut it once. If the edge has been cut k-times then we

This work is supported by the Knut and Alice Wallenberg Foundation, the Swedish Research Council, and the Ragnar Söderbergs foundation.

ⓒ Springer Nature Switzerland AG 2019
P. Heggernes (Ed.): CIAC 2019, LNCS 11485, pp. 112–123, 2019.
https://doi.org/10.1007/978-3-030-17402-6_10

remove it. The process stops when the root is isolated. We let $\mathcal{K}_e(\mathbb{G}_n)$ denote the number of cuts needed for the process to end.

Our model can also be applied to botnets, i.e., malicious computer networks consisting of compromised machines which are often used in spamming or attacks. The nodes in $\mathbb{G}_n$ represent the computers in a botnet, and the root represents the bot-master. The effectiveness of a botnet can be measured using the size of the component containing the root, which indicates the resources available to the bot-master [6]. To take down a botnet means to reduce the size of this root component as much as possible. If we assume that we target infected computers uniformly at random and it takes at least k attempts to fix a computer, then the k-cut number measures how difficult it is to completely isolate the bot-master.

The case $k = 1$ and $\mathbb{G}_n$ being a rooted tree has aroused great interests among mathematicians in the past few decades. The edge version of one-cut was first introduced by Meir and Moon [14] for the uniform random Cayley tree. Janson [12,13] noticed the equivalence between one-cuts and records in trees and studied them in binary trees and conditional Galton-Watson trees. Later Addario-Berry, Broutin, and Holmgren [1] gave a simpler proof for the limit distribution of one-cuts in conditional Galton-Watson trees. For one-cuts in random recursive trees, see [7,11,15]. For binary search trees and split trees, see [9,10].

1.2   The *k*-cut Number of a Tree

One of the most interesting cases is when $\mathbb{G}_n = \mathbb{T}_n$, where $\mathbb{T}_n$ is a rooted tree with n nodes.

There is an equivalent way to define $\mathcal{K}(\mathbb{T}_n)$. Imagine that each node is given an alarm clock. At time zero, the alarm clock of node v is set to ring at time $T_{1,v}$, where $(T_{i,v})_{i\geq 1, v\in\mathbb{T}_n}$ are i.i.d. (independent and identically distributed) $\mathrm{Exp}(1)$ random variables. After the alarm clock of node v rings the i-th time, we set it to ring again at time $T_{i+1,v}$. Due to the memoryless property of exponential random variables (see [8, pp. 134]), at any moment, which alarm clock rings next is always uniformly distributed. Thus, if we cut a node that is still in the tree when its alarm clock rings, and remove the node with its descendants if it has already been cut k-times, then we get exactly the k-cut model. (The random variables $(T_{i,v})_{i\geq 1}$ can be seen as the holding times in a Poisson process $N(t)_v$ of parameter 1, where $N(t)_v$ is the number of cuts in v during the time $[0,t]$ and has a Poisson distribution with parameter t.)

How can we tell if a node is still in the tree? When node v's alarm clock rings for the r-th time for some $r \leq k$, and no node above v has already rung k times, we say v has become an r-*record*. And when a node becomes an r-record, it must still be in the tree. Thus, summing the number of r-records over $r \in \{1,\dots,k\}$, we again get the k-cut number $\mathcal{K}(\mathbb{T}_n)$. One node can be a 1-record, a 2-record, etc., at the same time, so it can be counted multiple times. Note that if a node is an r-record, then it must also be a i-record for $i \in \{1,\dots,r-1\}$.

To be more precise, we define $\mathcal{K}(\mathbb{T}_n)$ as a function of $(T_{i,v})_{i\geq 1, v\geq 1}$. Let

$$G_{r,v} \stackrel{\text{def}}{=} \sum_{i=1}^{r} T_{i,v},$$

i.e., $G_{r,v}$ is the moment when the alarm clock of node v rings for the r-th time. Then $G_{r,v}$ has a gamma distribution with parameters $(r, 1)$ (see [8, Theorem 2.1.12]), which we denote by $\text{Gamma}(r)$. Let

$$I_{r,v} \stackrel{\text{def}}{=} [\![G_{r,v} < \min\{G_{k,u} : u \in \mathbb{T}_n, u \text{ is an ancestor of } v\}]\!], \qquad (1.1)$$

where $[\![\cdot]\!]$ denotes the Iverson bracket, i.e., $[\![S]\!] = 1$ if the statement S is true and $[\![S]\!] = 0$ otherwise. In other words, $I_{r,v}$ is the indicator random variable for node v being an r-record. Let

$$\mathcal{K}_r(\mathbb{T}_n) \stackrel{\text{def}}{=} \sum_{v \in \mathbb{T}_n} I_{r,v}, \qquad \mathcal{K}(\mathbb{T}_n) \stackrel{\text{def}}{=} \sum_{r=1}^{k} \mathcal{K}_r(\mathbb{T}_n).$$

Then $\mathcal{K}_r(\mathbb{T}_n)$ is the number of r-records and $\mathcal{K}(\mathbb{T}_n)$ is the total number of records.

1.3   The k-cut Number of a Path

Let $\mathbb{P}_n$ be a one-ary tree (a path) consisting of n nodes labeled $1, \ldots, n$ from the root to the leaf. To simplify notations, from now on we use $I_{r,i}, G_{r,i}$, and $T_{r,i}$ to represent $I_{r,v}, G_{r,v}$ and $T_{r,v}$ respectively for a node v at depth i.

Let $\mathcal{X}_n \stackrel{\text{def}}{=} \mathcal{K}(\mathbb{P}_n)$ and $\mathcal{X}_{n,r} = \mathcal{K}_r(\mathbb{P}_n)$. In this paper, we mainly consider $\mathcal{X}_n$ and we let $k \geq 2$ be a fixed integer.

The first motivation of this choice is that, as shown in Sect. 4, $\mathbb{P}_n$ is the fastest to cut among all graphs. (We make this statement precise in Lemma 4.) Thus $\mathcal{X}_n$ provides a universal stochastic lower bound for $\mathcal{K}(\mathbb{G}_n)$. Moreover, our results on $\mathcal{X}_n$ can immediately be extended to some trees of simple structures: see Sect. 4. Finally, as shown below, $\mathcal{X}_n$ generalizes the well-known record number in permutations and has very different behavior when $k = 1$, the usual cut-model, and $k \geq 2$, our extended model.

The name record comes from the classic definition of *records* in random permutations. Let $\sigma_1, \ldots, \sigma_n$ be a uniform random permutation of $\{1, \ldots, n\}$. If $\sigma_i < \min_{1 \leq j < i} \sigma_j$, then i is called a *(strictly lower) record*. Let $\mathcal{R}_n$ denote the number of records in $\sigma_1, \ldots, \sigma_n$. Let $W_1, \ldots, W_n$ be i.i.d. random variables with a common continuous distribution. Since the relative order of $W_1, \ldots, W_n$ also gives a uniform random permutation, we can equivalently define σ_i as the rank of W_i. As gamma distributions are continuous, we can in fact let $W_i = G_{k,i}$. Thus, being a record in a uniform permutation is equivalent to being a k-record and $\mathcal{R}_n \stackrel{\mathcal{L}}{=} \mathcal{X}_{n,k}$. Moreover, when $k = 1$, $\mathcal{R}_n \stackrel{\mathcal{L}}{=} \mathcal{X}_n$.

Starting from Chandler's article [5] in 1952, the theory of records has been widely studied due to its applications in statistics, computer science, and physics. For more recent surveys on this topic, see [2].

A well-known result of $\mathcal{R}_n$ (and thus also $\mathcal{X}_{n,k}$) [16] is that $(I_{k,j})_{1 \le j \le n}$ are independent. It follows from the Lindeberg–Lévy–Feller Theorem that

$$\frac{E\left[\mathcal{R}_n\right]}{\log n} \to 1, \qquad \frac{\mathcal{R}_n}{\log n} \overset{a.s.}{\to} 1, \qquad \mathcal{L}\left(\frac{\mathcal{R}_n - \log n}{\sqrt{\log n}}\right) \overset{d}{\to} \mathcal{N}(0,1),$$

where $\mathcal{N}(0,1)$ denotes the standard normal distribution.

In the following, Theorem 1 gives the expectation of $\mathcal{X}_{n,r}$ which implies that the number of one-records dominates the number of other records. Subsequently Theorems 2 and 3 estimate the variance and higher moments of $\mathcal{X}_{n,1}$.

Theorem 1. *For all fixed $k \in \mathbb{N}$,*

$$E\left[\mathcal{X}_{n,r}\right] \sim \begin{cases} \eta_{k,r} n^{1-\frac{r}{k}} & (1 \le r < k), \\ \log n & (r = k), \end{cases}$$

where the constants $\eta_{k,r}$ are defined by

$$\eta_{k,r} \overset{\text{def}}{=} \frac{(k!)^{\frac{r}{k}}}{k-r} \frac{\Gamma\left(\frac{r}{k}\right)}{\Gamma(r)},$$

where $\Gamma(z)$ denotes the gamma function. Therefore $E\left[\mathcal{X}_n\right] \sim E\left[\mathcal{X}_{n,1}\right]$. Also, for $k = 2$,

$$E\left[\mathcal{X}_n\right] \sim E\left[\mathcal{X}_{n,1}\right] \sim \sqrt{2\pi n}.$$

Theorem 2. *For all fixed $k \in \{2, 3, \dots\}$,*

$$E\left[\mathcal{X}_{n,1}(\mathcal{X}_{n,1} - 1)\right] \sim E\left[(\mathcal{X}_{n,1})^2\right] \sim \gamma_k n^{2-\frac{2}{k}},$$

where

$$\gamma_k = \frac{\Gamma\left(\frac{2}{k}\right)(k!)^{\frac{2}{k}}}{k-1} + 2\lambda_k,$$

and

$$\lambda_k = \begin{cases} \dfrac{\pi \cot\left(\frac{\pi}{k}\right) \Gamma\left(\frac{2}{k}\right)(k!)^{\frac{2}{k}}}{2(k-2)(k-1)} & k > 2, \\ \dfrac{\pi^2}{4} & k = 2. \end{cases}$$

Therefore

$$\mathrm{Var}\left(\mathcal{X}_{n,1}\right) \sim \left(\gamma_k - \eta_{k,1}^2\right) n^{2-\frac{2}{k}}.$$

In particular, when $k = 2$

$$\mathrm{Var}\left(\mathcal{X}_{n,1}\right) \sim \left(\frac{\pi^2}{2} + 2 - 2\pi\right) n.$$

Theorem 3. *For all fixed* $k \in \{2, 3, \dots\}$ *and* $\ell \in \mathbb{N}$

$$\limsup_{n \to \infty} \boldsymbol{E}\left[\left(\frac{\mathcal{X}_{n,1}}{n^{1-\frac{1}{k}}}\right)^{\ell}\right] \leq \rho_{k,\ell} \stackrel{\text{def}}{=} \ell! \Gamma\left(\ell + 1 - \frac{\ell}{k}\right)^{-1}\left(\frac{\pi}{k}(k!)^{1/k} \sin\left(\frac{\pi}{k}\right)^{-1}\right)^{\ell}.$$

The upper bound is tight for $\ell = 1$ *since* $\rho_{k,1} = \eta_{k,1}$.

The above theorems imply that the correct rescaling parameter should be $n^{1-\frac{1}{k}}$. However, unlike the case $k = 1$, when $k \geq 2$ the limit distribution of $\mathcal{X}_n/n^{1-\frac{1}{k}}$ has a rather complicated representation $\mathcal{B}_k$ defined as follows: Let $U_1, E_1, U_2, E_2, \dots$ be mutually independent random variables with $E_j \stackrel{\mathcal{L}}{=} \mathrm{Exp}(1)$ and $U_j \stackrel{\mathcal{L}}{=} \mathrm{Unif}[0, 1]$. Let

$$S_p \stackrel{\text{def}}{=} \left(k! \sum_{1 \leq s \leq p} \left(\prod_{s \leq j < p} U_j\right) E_s\right)^{\frac{1}{k}},$$

$$B_p \stackrel{\text{def}}{=} (1 - U_p)\left(\prod_{1 \leq j < p} U_j\right)^{1-\frac{1}{k}} S_p,$$

$$\mathcal{B}_k \stackrel{\text{def}}{=} \sum_{1 \leq p} B_p,$$

where we use the convention that an empty product equals one.

Remark 1. An equivalent recursive definition of S_p is

$$S_p = \begin{cases} k! E_1 & (p = 1), \\ (U_{p-1} S_{p-1}^k + k! E_p)^{\frac{1}{k}} & (p \geq 2). \end{cases}$$

Theorem 4. *Let* $k \in \{2, 3, \dots\}$. *Let* $\mathcal{L}(\mathcal{B}_k)$ *denote the distribution of* $\mathcal{B}_k$. *Then*

$$\mathcal{L}\left(\frac{\mathcal{X}_n}{n^{1-\frac{1}{k}}}\right) \stackrel{d}{\to} \mathcal{L}(\mathcal{B}_k).$$

Thus, by Theorems 1, 2 and 3, the convergence also holds in L^p *for all* $p > 0$ *and*

$$\boldsymbol{E}[\mathcal{B}_k] = \eta_{k,1}, \quad \boldsymbol{E}[\mathcal{B}_k^2] = \gamma_k, \quad \boldsymbol{E}[\mathcal{B}_k^p] \in [\eta_{k,1}^p, \rho_{k,p}] \quad (p \in \mathbb{N}).$$

Remark 2. It is easy to see that $\mathcal{X}_{n+1}^e \stackrel{\text{def}}{=} \mathcal{K}_e(P_{n+1}) \stackrel{\mathcal{L}}{=} \mathcal{X}_n$ by treating each edge on a length $n + 1$ path as a node on a length n path.

The rest of the paper is organized as follows: Sect. 2 sketches the proofs for the moment results Theorems 1, 2, and 3. Section 3 deals with the distributional result Theorem 4. Section 4 discusses some easy results for general graphs. Finally, Sect. 5 collects analytic results used in the proofs, which may themselves be of interest. For detailed proofs, see the full version of this paper [3]. For k-cuts in complete binary trees, see our follow-up paper [4].

2 The Moments

2.1 The Expectation

Lemma 1. *Uniformly for all $i \geq 1$ and $r \in \{1, \ldots, k\}$,*

$$E\left[I_{r,i+1}\right] = \left(1 + O\left(i^{-\frac{1}{2k}}\right)\right) \frac{(k!)^{\frac{r}{k}}}{k} \frac{\Gamma\left(\frac{r}{k}\right)}{\Gamma(r)} i^{-\frac{r}{k}}.$$

Proof. By (1.1), $E\left[I_{r,i+1}\right] = P\left\{G_{k,1} > G_{r,i+1}, \ldots, G_{k,i} > G_{r,i+1}\right\}$. Conditioning on $G_{r,i+1} = x$ yields $E\left[I_{r,i+1}\right] = \int_0^\infty x^{r-1}e^{-x}/\Gamma(r)P\left\{G_{k,1} > x\right\}^i \, dx$. Lemma 1 thus follows from Lemma 7.

Proof (Proof of Theorem 1). A simply computation shows that for $a \in (0,1)$

$$\sum_{1 \leq i \leq n} \frac{1}{i^a} = \frac{1}{1-a} n^{1-a} + O(1).$$

It then follows from Lemma 1 that for $r \in \{1, \ldots, k-1\}$.

$$E\left[\mathcal{X}_{n,r}\right] = \sum_{0 \leq i < n} E\left[I_{r,i+1}\right] = \frac{(k!)^{\frac{r}{k}}}{k} \frac{\Gamma\left(\frac{r}{k}\right)}{\Gamma(r)} \frac{1}{1-\frac{r}{k}} n^{1-\frac{r}{k}} + O\left(n^{1-\frac{r}{k}-\frac{1}{2k}}\right) + O(1).$$

When $r = k$, $E\left[\mathcal{X}_{n,k}\right] = E\left[\mathcal{R}_n\right] \sim \log(n)$ is already well-known.

2.2 The Variance

In this section we prove Theorem 2.

Let $E_{i,j}$ denote the event that $[I_{1,i+1}I_{1,j+1} = 1]$. Let $A_{x,y}$ denote the event that $[G_{1,i+1} = x \cap G_{1,j+1} = y]$. Then conditioning on $A_{x,y}$

$$E_{i,j} = \left[\bigcap_{1 \leq s \leq i} G_{k,s} > x \vee y\right] \cap [G_{k,i+1} > y] \cap \left[\bigcap_{i+2 \leq s \leq j} G_{k,s} > y\right],$$

where $x \vee y \overset{\text{def}}{=} \max\{x, y\}$. Since conditioning on $A_{x,y}$, $G_{k,i+1} \overset{\mathcal{L}}{=}$ Gamma$(k-1)+x$, $G_{k,s} \overset{\mathcal{L}}{=}$ Gamma(k) for $s \notin \{i+1, j+1\}$, and all these random variables are independent, we have

$$P\left\{E_{i,j}|A_{x,y}\right\} = P\left\{G_{k-1,1} + x > y\right\} P\left\{G_{k,1} > x \vee y\right\}^i P\left\{G_{k,1} > y\right\}^{j-i-1}.$$

It follows from $G_{1,i+1} \overset{\mathcal{L}}{=} G_{1,j+1} \overset{\mathcal{L}}{=}$ Exp(1) that

$$P\left\{E_{i,j}\right\} = \int_0^\infty \int_y^\infty e^{-x-y} P\left\{E_{i,j}|A_{x,y}\right\} \, dx \, dy$$

$$+ \int_0^\infty \int_0^y e^{-x-y} P\left\{E_{i,j}|A_{x,y}\right\} \, dx \, dy$$

$$\overset{\text{def}}{=} A_{1,i,j} + A_{2,i,j}.$$

Thus Theorem 2 follows from $\mathcal{X}_{n,1}(\mathcal{X}_{n,1} - 1) = 2\sum_{1 \leq i < j \leq n} I_{1,i}I_{1,j}$ and the following two lemmas whose proofs rely on Lemmas 8, 9, 10.

Lemma 2. *Let $k \in \{2, 3, \ldots\}$. We have*

$$A_{2,i,j} = \left(1 + O\left(j^{-\frac{1}{2k}}\right)\right) \frac{(k!)^{\frac{2}{k}}}{k} \Gamma\left(\frac{2}{k}\right) j^{-\frac{2}{k}}.$$

Lemma 3. *Let $k \in \{2, 3, \ldots\}$. Let $a = i$ and $b = j - i - 1$. Then for all $a \geq 1$ and $b \geq 1$,*

$$A_{1,i,j} = \xi_k(a, b) + O\left(\left(a^{-\frac{1}{2k}} + b^{-\frac{1}{2k}}\right)\left(a^{-\frac{2}{k}} + b^{-\frac{2}{k}}\right)\right),$$

where

$$\xi_k(a, b) \stackrel{\text{def}}{=} \int_0^\infty \int_y^\infty \exp\left(-a\frac{x^k}{k!} - b\frac{y^k}{k!}\right) dx\, dy.$$

2.3 Higher Moments

The computations of higher moments of $\mathcal{X}_{n,1}$ are rather complicated. However, an upper bound is readily available. Let $1 \leq i_1 < i_2 < \cdots < i_\ell \leq n$. Then

$$\boldsymbol{E}\left[I_{1,i_1} I_{1,i_2} \cdots I_{1,i_\ell}\right] \leq \boldsymbol{E}\left[I_{1,i_1}\right] \boldsymbol{E}\left[I_{1,i_2-i_1}\right] \cdots \boldsymbol{E}\left[I_{1,i_\ell-i_{\ell-1}}\right].$$

The above inequality holds since if i_j is a one-record in the whole path, then it must also be a one-record in the segment $(i_{j-1} + 1, \ldots, i_j)$ ignoring everything else, and what happens in each of such segments are independent. Theorem 3 follows easily from this observation.

3   Convergence to the k-cut Distribution

By Theorem 1 and Markov's inequality, $\mathcal{X}_{n,r}/n^{1-\frac{1}{k}} \overset{P}{\to} 0$ for $r \in \{2, \ldots, k\}$. So it suffices to prove Theorem 4 for $\mathcal{X}_{n,1}$ instead of $\mathcal{X}_n$. Throughout Sect. 3, unless otherwise emphasized, we assume that $k \geq 2$.

The idea of the proof is to condition on the positions and values of the k-records, and study the distribution of the number of one-records between two consecutive k-records.

We use $(R_{n,p})_{p \geq 1}$ to denote the k-record values and $(P_{n,p})_{p \geq 1}$ the positions of these k-records. To be precise, let $R_{n,0} \stackrel{\text{def}}{=} 0$, and $P_{n,0} \stackrel{\text{def}}{=} n + 1$; for $p \geq 1$, if $P_{n,p-1} > 1$, then let

$$R_{n,p} \stackrel{\text{def}}{=} \min\{G_{k,j} : 1 \leq j < P_{n,p-1}\},$$

$$P_{n,p} \stackrel{\text{def}}{=} \operatorname{argmin}\{G_{k,j} : 1 \leq j < P_{n,p-1}\},$$

i.e., $P_{n,p}$ is the unique positive integer which satisfies that $G_{k,P_{n,p}} \leq G_{k,i}$ for all $1 \leq i < P_{n,p-1}$; otherwise let $P_{n,p} = 1$ and $R_{n,p} = \infty$. Note that $R_{n,1}$ is simply the minimum of n i.i.d. Gamma(k) random variables.

According to $(P_{n,p})_{p\geq 1}$, we split $\mathcal{X}_{n,1}$ into the following sum

$$\mathcal{X}_{n,1} = \sum_{1\leq j\leq n} I_{1,j} = \mathcal{X}_{n,k} + \sum_{1\leq p}\sum_{1\leq j} [\![P_{n,p-1} > j > P_{n,p}]\!]\, I_{1,j} \stackrel{\text{def}}{=} \mathcal{X}_{n,k} + \sum_{1\leq p} B_{n,p}.$$
(3.1)

Figure 1 gives an example of $(B_{n,p})_{p\geq 1}$ for $n = 12$. It depicts the positions of the k-records and the one-records. It also shows the values and the summation ranges for $(B_{n,p})_{p\geq 1}$.

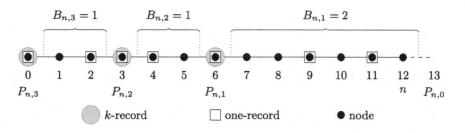

Fig. 1. An example of $(B_{n,p})_{p\geq 1}$ for $n = 12$.

Recall that $T_{r,j} \stackrel{\mathcal{L}}{=} \operatorname{Exp}(1)$, is the lapse of time between the alarm clock of j rings for the $(r-1)$-st time and the r-th time. Conditioning on $(R_{n,p}, P_{n,p})_{n\geq 1, p\geq 1}$, for $j \in (P_{n,p}, P_{n,p-1})$, we have

$$\boldsymbol{E}\,[I_{1,j}] = \boldsymbol{P}\,\{T_{1,j} < R_{n,p}\,|G_{k,j} > R_{n,p-1}\}.$$

Then the distribution of $B_{n,p}$ conditioning on $(R_{n,p}, P_{n,p})_{n\geq 1, p\geq 1}$ is simply that of

$$\operatorname{Bin}\left(P_{n,p-1} - P_{n,p} - 1, \boldsymbol{P}\,\{T_{1,j} < R_{n,p}\,|G_{k,j} > R_{n,p-1}\}\right),$$

where $\operatorname{Bin}(m,p)$ denotes a binomial (m,p) random variable. When $R_{n,p-1}$ is small and $P_{n,p-1} - P_{n,p}$ is large, this is roughly

$$\operatorname{Bin}\left(P_{n,p-1} - P_{n,p}, \boldsymbol{P}\,\{T_{1,j} < R_{n,p}\}\right) \stackrel{\mathcal{L}}{=} \operatorname{Bin}\left(P_{n,p-1} - P_{n,p}, 1 - e^{-R_{n,p}}\right). \quad (3.2)$$

Therefore, we first study a slightly simplified model. Let $(T^*_{r,j})_{r\geq 1, j\geq 1}$ be i.i.d. $\operatorname{Exp}(1)$ which are also independent from $(T_{r,j})_{r\geq 1, j\geq 1}$. Let

$$I^*_j \stackrel{\text{def}}{=} [\![T^*_{1,j} < \min\{G_{k,i} : 1 \leq i \leq j\}]\!], \qquad \mathcal{X}^*_n \stackrel{\text{def}}{=} \sum_{1\leq j\leq n} I^*_j.$$

We say a node j is an *alt-one-record* if $I^*_j = 1$. As in (3.1), we can write

$$\mathcal{X}^*_n = \sum_{1\leq j\leq n} I^*_j = \sum_{1\leq p}\sum_{1\leq j} [\![P_{n,p-1} > j \geq P_{n,p}]\!]\, I^*_j \stackrel{\text{def}}{=} \sum_{1\leq p} B^*_{n,p}.$$

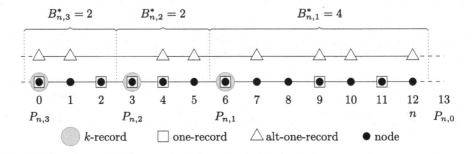

Fig. 2. An example of $(B^*_{n,p})_{p\geq 1}$ for $n = 12$.

Then conditioning on $(R_{n,p}, P_{n,p})_{n\geq 1, p\geq 1}$, $B^*_{n,p}$ has exactly the distribution as (3.2). Figure 2 gives an example of $(B^*_{n,p})_{p\geq 1}$ for $n = 12$. It shows the positions of alt-one-records, as well as the values and the summation ranges of $(B^*_{n,p})_{p\geq 1}$.

The main part of the proof for Theorem 4 consist of showing the following

Proposition 1. *For all fixed $p \in \mathbb{N}$ and $k \geq 2$,*

$$\mathcal{L}\left(\left(\frac{B^*_{n,1}}{n^{1-\frac{1}{k}}}, \ldots, \frac{B^*_{n,p}}{n^{1-\frac{1}{k}}}\right)\right) \xrightarrow{d} \mathcal{L}\left((B_1, \ldots B_p)\right),$$

which implies by the Cramér–Wold device that

$$\mathcal{L}\left(\sum_{1\leq j\leq p} \frac{B^*_{n,j}}{n^{1-\frac{1}{k}}}\right) \xrightarrow{d} \mathcal{L}\left(\sum_{1\leq j\leq p} B_j\right),$$

Then we can prove that p can be chosen large enough so that $\sum_{p<j} B^*_{n,j}/n^{1-\frac{1}{k}}$ is negligible. Thus,

$$\mathcal{L}\left(\frac{\mathcal{X}^*_n}{n^{1-\frac{1}{k}}}\right) \overset{\text{def}}{=} \mathcal{L}\left(\frac{\sum_{1\leq j} B^*_{n,j}}{n^{1-\frac{1}{k}}}\right) \xrightarrow{d} \mathcal{L}\left(\sum_{1\leq j} B_j\right) \overset{\text{def}}{=} \mathcal{L}\left(\mathcal{B}_k\right).$$

Following this, we can use a coupling argument to show that $\mathcal{X}_{n,1}/n^{1-\frac{1}{k}}$ and $\mathcal{X}^*_n/n^{1-\frac{1}{k}}$ converge to the same limit, which finishes the proof of Theorem 4.

4   Some Extensions

4.1   A Lower Bound and an Upper Bound for General Graphs

Let $\mathcal{G}_n$ be the set of rooted graphs with n nodes. It is obvious that $\mathbb{P}_n$ is the easiest to cut among all graphs in $\mathcal{G}_n$. We formalize this by the following lemma:

Lemma 4. *Let* $k \in \mathbb{N}$. *For all* $\mathbb{G}_n \in \mathcal{G}_n$, $\mathcal{X}_n \overset{\text{def}}{=} \mathcal{K}(\mathbb{P}_n) \preceq \mathcal{K}(\mathbb{G}_n)$. *Therefore,*

$$\min_{\mathbb{G}_n \in \mathcal{G}_n} E\mathcal{K}(\mathbb{G}_n) \geq E\mathcal{X}_n \sim \begin{cases} \dfrac{(k!)^{\frac{1}{k}}}{k-1} \Gamma\left(\dfrac{1}{k}\right) n^{1-\frac{1}{k}} & (k \geq 2), \\ \log n & (k = 1), \end{cases}$$

by Theorem 1.

The most resilient graph is obviously $\mathbb{K}_n$, the complete graph with n vertices. Thus, we have the following upper bound:

Lemma 5. *Let* $k \in \mathbb{N}$.

(i) *Let* $Y \overset{\mathcal{L}}{=} \text{Gamma}(k)$, $Z \overset{\mathcal{L}}{=} \text{Poi}(Y)$, *and* $W \overset{\mathcal{L}}{=} Z \wedge k$, *i.e.,* $W \overset{\mathcal{L}}{=} \min\{Z, k\}$. *Then*

$$\mathcal{L}\left(\frac{\mathcal{K}(\mathbb{K}_n)}{n}\right) \overset{d}{\to} \mathcal{L}\left(E\left[W|Y\right]\right) = \mathcal{L}\left(\frac{\Gamma(k+1,Y) - e^{-Y}Y^{k+1}}{k!} + k\right),$$

where $\Gamma(\ell, z)$ *denotes the upper incomplete gamma function. Note that when* $k = 1$, *the right-hand-side is simply* $\text{Unif}[0, 1]$.

(ii) *For all* $\mathbb{G}_n \in \mathcal{G}_n$, $\mathcal{K}(\mathbb{G}_n) \preceq \mathcal{K}(\mathbb{K}_n)$. *Therefore,*

$$\max_{\mathbb{G}_n \in \mathcal{G}_n} E\mathcal{K}(\mathbb{G}_n) \leq E\mathcal{K}(\mathbb{K}_n) \sim k\left(1 - \frac{1}{2^{2k}}\binom{2k}{k}\right) n.$$

4.2 Path-Like Graphs

If a graph $\mathbb{G}_n$ consists of only long paths, then the limit distribution $\mathcal{K}(\mathbb{G}_n)$ should be related to $\mathcal{B}_k$, the limit distribution of $\mathcal{K}(\mathbb{P}_n)/n^{1-\frac{1}{k}}$ (see Theorem 4). We give two simple examples with $k \in \{2, 3, \dots\}$.

Example 1 (Long path). Let $(\mathbb{G}_n)_{n \geq 1}$ be a sequence of rooted graphs such that $\mathbb{G}_n$ contains a path of length $m(n)$ starting from the root with $n - m(n) = o(n^{1-\frac{1}{k}})$. Since it takes at most $k(n - m(n))$ cuts to remove all the nodes outside the long path,

$$\mathcal{K}(P_{m(n)}) \preceq \mathcal{K}(\mathbb{G}_n) \preceq \mathcal{K}(P_{m(n)}) + ko\left(n^{1-1/k}\right).$$

Thus, by Lemma 4, this implies that $\mathcal{K}(\mathbb{G}_n)/n^{1-\frac{1}{k}}$ converges in distribution to $\mathcal{B}_k$.

5 Some Auxiliary Results

Lemma 6. *Let* $G_k \overset{\mathcal{L}}{=} \text{Gamma}(k)$. *Let* $\alpha \overset{\text{def}}{=} \frac{1}{2}\left(\frac{1}{k} + \frac{1}{k+1}\right)$ *and* $x_0 \overset{\text{def}}{=} m^{-\alpha}$. *Then uniformly for all* $x \in [0, x_0]$,

$$P\{G_k > x\}^m = \left(\frac{\Gamma(k, x)}{\Gamma(k)}\right)^m = \left(1 + O\left(m^{-\frac{1}{2k}}\right)\right) \exp\left(-\frac{mx^k}{k!}\right),$$

where $\Gamma(\ell, z)$ *denotes the upper incomplete gamma function.*

Lemma 7. *Let $G_k \overset{\mathcal{L}}{=} \mathrm{Gamma}(k)$. Let $a \geq 0$ and $b \geq 1$ be fixed. Then uniformly for $m \geq 1$,*

$$\int_0^\infty x^{b-1} e^{-ax} P\{G_k > x\}^m \, \mathrm{d}x = \left(1 + O\left(m^{-\frac{1}{2k}}\right)\right) \frac{(k!)^{\frac{b}{k}}}{k} \Gamma\left(\frac{b}{k}\right) m^{-\frac{b}{k}}.$$

Lemma 8. *For $a > 0$, $b > 0$ and $k \geq 2$,*

$$\xi_k(a,b) \overset{def}{=} \int_0^\infty \int_y^\infty e^{-ax^k/k! - by^k/k!} \, \mathrm{d}x \, \mathrm{d}y$$

$$= \frac{\Gamma\left(\frac{2}{k}\right)}{k} \left(\frac{k!}{a}\right)^{\frac{2}{k}} F\left(\frac{2}{k}, \frac{1}{k}; 1 + \frac{1}{k}; -\frac{b}{a}\right),$$

where F denotes the hypergeometric function. In particular,

$$\xi_2(a,b) = \arctan\left(\sqrt{\frac{b}{a}}\right) (ab)^{-\frac{1}{2}}.$$

Lemma 9. *For $a > 0$, $b > 0$ and $k \geq 2$,*

$$(a+b)^{-\frac{2}{k}} \leq \frac{k}{\Gamma\left(\frac{2}{k}\right) (k!)^{\frac{2}{k}}} \xi_k(a,b) \leq a^{-\frac{2}{k}} + b^{-\frac{2}{k}}.$$

Moreover, $\xi_k(a,b)$ is monotonically decreasing in both a and b.

Lemma 10. *For $k \geq 2$, let*

$$\lambda_k \overset{def}{=} \int_0^1 \int_0^{1-s} \xi_k(s,t) \, \mathrm{d}t \, \mathrm{d}s.$$

Then

$$\lambda_k = \begin{cases} \dfrac{\pi \cot\left(\frac{\pi}{k}\right) \Gamma\left(\frac{2}{k}\right) (k!)^{\frac{2}{k}}}{2(k-2)(k-1)} & k > 2, \\[2mm] \dfrac{\pi^2}{4} & k = 2. \end{cases}$$

References

1. Addario-Berry, L., Broutin, N., Holmgren, C.: Cutting down trees with a Markov chainsaw. Ann. Appl. Probab. **24**(6), 2297–2339 (2014)
2. Ahsanullah, M.: Record Values-Theory and Applications. University Press of America Inc., Lanham (2004)
3. Cai, X.S., Devroye, L., Holmgren, C., Skerman, F.: k-cut on paths and some trees. ArXiv e-prints, January 2019

4. Cai, X.S., Holmgren, C.: Cutting resilient networks - complete binary trees. arXiv e-prints, November 2018

5. Chandler, K.N.: The distribution and frequency of record values. J. R. Stat. Soc. Ser. B. **14**, 220–228 (1952)

6. Dagon, D., Gu, G., Lee, C.P., Lee, W.: A taxonomy of botnet structures. In: Twenty-Third Annual Computer Security Applications Conference (ACSAC 2007), pp. 325–339 (2007)

7. Drmota, M., Iksanov, A., Moehle, M., Roesler, U.: A limiting distribution for the number of cuts needed to isolate the root of a random recursive tree. Random Struct. Algorithms **34**(3), 319–336 (2009)

8. Durrett, R.: Probability: Theory and Examples, Cambridge Series in Statistical and Probabilistic Mathematics, vol. 31, 4th edn. Cambridge University Press, Cambridge (2010)

9. Holmgren, C.: Random records and cuttings in binary search trees. Combin. Probab. Comput. **19**(3), 391–424 (2010)

10. Holmgren, C.: A weakly 1-stable distribution for the number of random records and cuttings in split trees. Adv. Appl. Probab. **43**(1), 151–177 (2011)

11. Iksanov, A., Möhle, M.: A probabilistic proof of a weak limit law for the number of cuts needed to isolate the root of a random recursive tree. Electron. Comm. Probab. **12**, 28–35 (2007)

12. Janson, S.: Random records and cuttings in complete binary trees. In: Mathematics and Computer Science III, Trends Math, pp. 241–253. Birkhäuser, Basel (2004)

13. Janson, S.: Random cutting and records in deterministic and random trees. Random Struct. Algorithms **29**(2), 139–179 (2006)

14. Meir, A., Moon, J.W.: Cutting down random trees. J. Austral. Math. Soc. **11**, 313–324 (1970)

15. Meir, A., Moon, J.: Cutting down recursive trees. Math. Biosci. **21**(3), 173–181 (1974)

16. Rényi, A.: Théorie des éléments saillants d'une suite d'observations. Ann. Fac. Sci. Univ. Clermont-Ferrand No. **8**, 7–13 (1962)

Extension of Vertex Cover
and Independent Set in Some
Classes of Graphs

Katrin Casel[1,2]([⊠]), Henning Fernau[2], Mehdi Khosravian Ghadikoalei[3],
Jérôme Monnot[3], and Florian Sikora[3]

[1] Hasso Plattner Institute, University of Potsdam, 14482 Potsdam, Germany
[2] Universität Trier, Fachbereich 4, Informatikwissenschaften, 54286 Trier, Germany
{casel,fernau}@informatik.uni-trier.de
[3] Université Paris-Dauphine, PSL University, CNRS, LAMSADE,
75016 Paris, France
{mehdi.khosravian-ghadikolaei,jerome.monnot,
florian.sikora}@lamsade.dauphine.fr

Abstract. We study extension variants of the classical problems VER-
TEX COVER and INDEPENDENT SET. Given a graph $G = (V, E)$ and a
vertex set $U \subseteq V$, it is asked if there exists a *minimal* vertex cover
(resp. *maximal* independent set) S with $U \subseteq S$ (resp. $U \supseteq S$). Possibly
contradicting intuition, these problems tend to be NP-complete, even
in graph classes where the classical problem can be solved efficiently.
Yet, we exhibit some graph classes where the extension variant remains
polynomial-time solvable. We also study the parameterized complexity
of theses problems, with parameter $|U|$, as well as the optimality of sim-
ple exact algorithms under ETH. All these complexity considerations are
also carried out in very restricted scenarios, be it degree or topological
restrictions (bipartite, planar or chordal graphs). This also motivates presenting some explicit branching algorithms for degree-bounded instances.
e further discuss the *price of extension*, measuring the distance of U to
the closest set that can be extended, which results in natural optimization
problems related to extension problems for which we discuss polynomial-
time approximability.

Keywords: Extension problems · Special graph classes ·
Approximation algorithms · NP-completeness

1 Introduction

We will consider *extension problems* related to the classical graph problems
VERTEX COVER and INDEPENDENT SET. Informally in the extension version
of VERTEX COVER, the input consists of both a graph G and a subset U of
vertices, and the task is to extend U to an inclusion-wise minimal vertex cover
of G (if possible). With INDEPENDENT SET, given a graph G and a subset U

© Springer Nature Switzerland AG 2019
P. Heggernes (Ed.): CIAC 2019, LNCS 11485, pp. 124–136, 2019.
https://doi.org/10.1007/978-3-030-17402-6_11

of vertices, we are looking for an inclusion-wise maximal independent set of G contained in U.

Studying such version is interesting when one wants to develop efficient enumeration algorithms or also for branching algorithms, to name two examples of a list of applications given in [6].

Related Work. In [5], it is shown that extension of partial solutions is NP-hard for computing prime implicants of the dual of a Boolean function; a problem which can also be seen as trying to find a minimal hitting set for the prime implicants of the input function. Interpreted in this way, the proof from [5] yields NP-hardness for the minimal extension problem for 3-HITTING SET (but polynomial-time solvable if $|U|$ is constant). This result was extended in [2] to prove NP-hardness for computing the extensions of vertex sets to minimal dominating sets (EXT DS), even restricted to planar cubic graphs. Similarly, it was shown in [1] that extensions to minimal vertex covers restricted to planar cubic graphs is NP-hard. The first *systematic* study of this type of problems was exhibited in [6] providing quite a number of different examples of this type of problem.

An *independent system* is a set system $(V, \mathcal{E})$, $\mathcal{E} \subseteq 2^V$, that is hereditary under inclusion. The extension problem EXT IND SYS (also called FLASHLIGHT) for independent system was proposed in [17]. In this problem, given as input $X, Y \subseteq V$, one asks for the existence of a maximal independent set including X and that does not intersect with Y. Lawler et al. proved that EXT IND SYS is NP-complete, even when $X = \emptyset$ [17]. In order to enumerate all (inclusion-wise) minimal dominating sets of a given graph, Kanté et al. studied a restriction of EXT IND SYS: finding a minimal dominating set containing X but excluding Y. They proved that EXT DS is NP-complete, even in special graph classes like split graphs, chordal graphs and line graphs [14,15]. Moreover, they proposed a linear algorithm for split graphs when X, Y is a partition of the clique part [13].

Organization of the Paper. After some definitions and first results in Sect. 2, we focus on bipartite graphs in Sect. 3 and give hardness results holding with strong degree or planarity constraints. We also study parameterized complexity at the end of this section and comment on lower bound results based on ETH. In Sect. 4, we give positive algorithmic results on chordal graphs, with a combinatorial characterization for the subclass of trees. We introduce the novel concept of *price of extension* in Sect. 5 and discuss (non-)approximability for the according optimization problems. In Sect. 6, we prove several algorithmic results for bounded-degree graphs, based on a list of reduction rules and simple branching. Finally, in Sect. 7, we give some prospects of future research.

2   Definitions and Preliminary Results

Throughout this paper, we consider simple undirected graphs only, to which we refer as *graphs*. A graph can be specified by the set V of vertices and the set E of edges; every edge has two endpoints, and if v is an endpoint of e, we also say that e and v are *incident*. Let $G = (V, E)$ be a graph and $U \subseteq V$; $N_G(U) = \{v \in V : \exists u \in U (vu \in E)\}$ denotes *the neighborhood* of U in G and $N_G[U] = U \cup N_G(U)$ denotes *the closed neighborhood* of U. For singleton sets $U = \{u\}$, we simply write $N_G(u)$ or $N_G[u]$, even omitting G if clear from the context. The cardinality of $N_G(u)$ is called *degree* of u, denoted $d_G(u)$. A graph where all vertices have degree k is called *k-regular*; 3-regular graphs are called *cubic*. If 3 upper-bounds the degree of all vertices, we speak of *subcubic graphs*.

A vertex set U induces the graph $G[U]$ with vertex set U and $e \in E$ being an edge in $G[U]$ iff both endpoints of e are in U. A vertex set U is called *independent* if $U \cap N_G(U) = \emptyset$; U is called *dominating* if $N_G[U] = V$; U is a *vertex cover* if each edge e is incident to at least one vertex from U. A graph is called *bipartite* if its vertex set decomposes into two independent sets. A vertex cover S is *minimal* if any proper subset $S' \subset S$ of S is not a vertex cover. Clearly, a vertex cover S is minimal iff each vertex v in S possesses a *private edge*, i.e., an edge vu with $u \notin S$. An independent set S is *maximal* if any proper superset $S' \supset S$ of S is not an independent set. The two main problems discussed in this paper are:

Ext VC
Input: A graph $G = (V, E)$, a set of vertices $U \subseteq V$.
Question: Does G have a minimal vertex cover S with $U \subseteq S$?

Ext IS
Input: A graph $G = (V, E)$, a set of vertices $U \subseteq V$.
Question: Does G have a maximal independent set S with $S \subseteq U$?

For Ext VC, the set U is also referred to as the set of *required* vertices.

Remark 1. (G, U) is a yes-instance of Ext VC iff $(G, V \setminus U)$ is a yes-instance of Ext IS, as complements of maximal independent sets are minimal vertex covers.

Since adding or deleting edges between vertices of U does not change the minimality of feasible solutions of Ext VC, we can first state the following.

Remark 2. For Ext VC (and for Ext IS) one can always assume the required vertex set (the set $V \setminus U$) is either a clique or an independent set.

The following theorem gives a combinatorial characterization of yes-instances of Ext VC that is quite important in our subsequent discussions.

Theorem 3. *Let $G = (V, E)$ be a graph and $U \subseteq V$ be a set of vertices. The three following conditions are equivalent:*

(i) *(G, U) is a yes-instance of Ext VC.*
(ii) *$(G[N_G[U]], N_G[U] \setminus U)$ is a yes-instance of Ext IS.*
(iii) *There exists an independent dominating set $S' \subseteq N_G[U] \setminus U$ of $G[N_G[U]]$.*

3   Bipartite Graphs

In this section, we focus on bipartite graphs. We prove that EXT VC is NP-complete, even if restricted to cubic, or planar subcubic graphs. Due to Remark 1, this immediately yields the same type of results for EXT IS. We add some algorithmic notes on planar graphs that are also valid for the non-bipartite case. Also, we discuss results based on ETH. We conclude the section by studying the parameterized complexity of EXT VC in bipartite graphs when parameterized by the size of U.

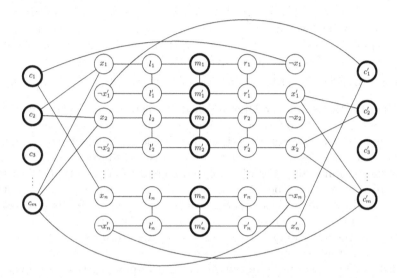

Fig. 1. Graph $G = (V, E)$ for EXT VC built from I. Vertices of U have a bold border.

Theorem 4. EXT VC *(and* EXT IS*) is* NP-*complete in cubic bipartite graphs.*

Proof. We reduce from 2-BALANCED 3-SAT, denoted $(3, B2)$-SAT ,which is NP-hard by [3, Theorem 1], where an instance I is given by a set C of CNF clauses over a set X of Boolean variables such that each clause has exactly 3 literals and each variable appears exactly 4 times, twice negative and twice positive. The bipartite graph associated to I is $BP = (C \cup X, E(BP))$ with $C = \{c_1, \ldots, c_m\}$, $X = \{x_1, \ldots, x_n\}$ and $E(BP) = \{c_j x_i \colon x_i$ or $\neg x_i$ is literal of $c_j\}$.

For an instance $I = (C, X)$ of $(3, B2)$-SAT, we build a cubic bipartite graph $G = (V, E)$ by duplicating instance I (here, vertices $C' = \{c'_1, \ldots, c'_m\}$ and $X' = \{x'_1, \ldots, x'_n\}$ are the duplicate variants of vertices $C = \{c_1, \ldots, c_m\}$ and $X = \{x_1, \ldots, x_n\}$) and by connecting gadgets as done in Fig. 1. We also add the following edges between the two copies: $l_i l'_i$, $m_i m'_i$ and $r_i r'_i$ for $i = 1, \ldots, n$. The construction is illustrated in Fig. 1 and clearly, G is a cubic bipartite graph. Finally we set $U = \{c_i, c'_i \colon i = 1, \ldots, m\} \cup \{m_j, m'_j \colon j = 1, \ldots, n\}$.

We claim that I is satisfiable iff G admits a minimal vertex cover containing U. Assume I is satisfiable and let T be a truth assignment which satisfies all clauses. We set $S = \{\neg x_i, l_i, \neg x_i', r_i' \colon T(x_i) = true\} \cup \{x_i, r_i, x_i', l_i' \colon T(x_i) = false\} \cup U$. We can easily check that S is a minimal vertex cover containing U.

Conversely, assume that G possesses a minimal vertex cover S containing U. For a fixed i, we know that $|\{l_i, l_i', r_i, r_i'\} \cap S| \geq 2$ to cover the edges $l_i l_i'$ and $r_i r_i'$. If $\{l_i, r_i\} \subseteq S$ (resp. $\{l_i', r_i'\} \subseteq S$), then S is not a minimal vertex cover because m_i (resp. m_i') can be deleted, a contradiction. If $\{l_i, l_i'\} \subseteq S$ (resp. $\{r_i, r_i'\} \subseteq S$), then S must contain another vertex to cover $r_i r_i'$ (resp. $l_i l_i'$), leading to the previous case, a contradiction. Hence, if $\{l_i, r_i'\} \subseteq S$ (resp., $\{r_i, l_i'\} \subseteq S$), then $\{\neg x_i, \neg x_i'\} \subseteq S$ (resp., $\{x_i, x_i'\} \subseteq S$), since the edges $l_i' \neg x_i'$ and $r_i \neg x_i$ (resp., $l_i x_i$ and $r_i' x_i$) must be covered. In conclusion, by setting $T(x_i) = true$ if $\neg x_i \in S$ and $T(x_i) = false$ if $x_i \in S$ we obtain a truth assignment T which satisfies all clauses, because $\{C_i, C_i' \colon i = 1, \ldots, m\} \subseteq U \subseteq S$.   □

Theorem 5. EXT IS *is* NP-*complete on planar bipartite subcubic graphs.*

Algorithmic Notes for the Planar Case. By distinguishing between whether a vertex belongs to the cover or not and further, when it belongs to the cover, if it already has a private edge or not, it is not hard to design a dynamic programming algorithm that decides in time $\mathcal{O}^*(c^t)$ if (G, U) is a yes-instance of EXT VC or not, given a graph G together with a tree decomposition of width t. With some more care, even $c = 2$ can be achieved, but this is not so important here. Rather, below we will make explicit another algorithm for trees that is based on several combinatorial properties and hence differs from the DP approach sketched here for the more general notion of treewidth-bounded graphs.

Moreover, it is well-known that planar graphs of order n have treewidth bounded by $\mathcal{O}(\sqrt{n})$. In fact, we can obtain a corresponding tree decomposition in polynomial time, given a planar graph G. Piecing things together, we obtain:

Theorem 6. EXT VC *can be solved in time* $\mathcal{O}^*(2^{\mathcal{O}(\sqrt{n})})$ *on planar graphs.*

Remarks on the Exponential Time Hypothesis. Assuming ETH, there is no $2^{o(n+m)}$-algorithm for solving n-variable, m-clause instances of $(3, B2)$-SAT. As our reduction from $(3, B2)$-SAT increases the size of the instances only in a linear fashion, we can immediately conclude:

Theorem 7. *There is no* $2^{o(n+m)}$-*algorithm for* n-*vertex,* m-*edge bipartite subcubic instances of* EXT VC, *unless ETH fails.*

This also motivates us to further study exact exponential-time algorithms. We can also deduce optimality of our algorithms for planar graphs based on the following auxiliary result.

Proposition 8. *There is no algorithm that solves* 4-BOUNDED PLANAR 3-CONNECTED SAT *(see [16]) on instances with* n *variables and* m *clauses in time* $2^{o(\sqrt{n+m})}$, *unless ETH fails.*

Corollary 9. *There is no* $2^{o(\sqrt{n})}$ *algorithm for solving* EXT VC *on planar instances of order* n, *unless ETH fails.*

Remarks on Parameterized Complexity. We now study our problems in the framework of parameterized complexity where we consider the size of the set of fixed vertices as *standard parameter* for our extension problems.

Theorem 10. EXT VC *with standard parameter is* W[1]-*complete, even when restricted to bipartite instances.*

Theorem 11. EXT VC *with standard parameter is in* FPT *on planar graphs.*

4   Chordal and Circular-Arc Graphs

An undirected graph $G = (V, E)$ is *chordal* iff each cycle of G with a length at least four has a chord (an edge linking two non-consecutive vertices of the cycle) and G is *circular-arc* if it is the intersection graph of a collection of n arcs around a circle. We will need the following problem definition.

MINIMUM INDEPENDENT DOMINATING SET (MINISDS for short)
Input: A graph $G = (V, E)$.
Solution: Subset of vertices $S \subseteq V$ which is independent and dominating.
Output: Solution S that minimizes $|S|$.

WEIGHTED MINIMUM INDEPENDENT DOMINATING SET (or WMINISDS for short) corresponds to the vertex-weighted variant of MINISDS, where each vertex $v \in V$ has a non-negative weight $w(v) \geq 0$ associated to it and the goal consists in minimizing $w(S) = \sum_{v \in S} w(v)$. If $w(v) \in \{a, b\}$ with $0 \leq a < b$, the weights are called *bivaluate*, and $a = 0$ and $b = 1$ corresponds to *binary weights*.

Remark 12. MINISDS for chordal graphs has been studied in [10], where it is shown that the restriction to binary weights is solvable in polynomial-time. Bivalued MINISDS with $a > 0$ however is already NP-hard on chordal graphs, see [7]. WMINISDS (without any restriction on the number of distinct weights) is also polynomial-time solvable in circular-arc graphs [8].

Corollary 13. EXT VC *is polynomial-time decidable in chordal and in circular-arc graphs.*

Farber's algorithm [10] (used in Corollary 13) runs in linear-time and is based on the resolution of a linear programming using primal and dual programs. Yet, it would be nice to find a (direct) combinatorial linear-time algorithm for chordal and circular-arc graphs, as this is quite common in that area. We give a first step in this direction by presenting a characterization of *yes*-instances of EXT VC on trees. Consider a tree $T = (V, E)$ and a set of vertices U. A subtree $T' = (V', E')$ (i.e., a connected induced subgraph) of a tree T is called *edge full with respect to* (T, U) if $U \subseteq V'$, $d_{T'}(u) = d_T(u)$ for all $u \in U$. A subtree $T' = (V', E')$ is *induced edge full with respect to* (T, U) if it is edge full with respect to $(T, U \cap V')$.

For our characterization, we use a coloring of vertices with colors black and white. If $T = (V, E)$ is a tree and $X \subseteq V$, we use $T[X \to \text{black}]$ to denote the colored tree where exactly the vertices from X are colored black. Further define the following class of black and white colored trees $\mathcal{T}$, inductively as follows.

Base case: A tree with a single vertex x belongs to $\mathcal{T}$ if x is black.

Inductive step: If $T \in \mathcal{T}$, the tree resulting from the addition of a P_3 (3 new vertices that form a path p), one endpoint of p being black, the two other vertices being white and the white endpoint of p linked to a black vertex of T, is in $\mathcal{T}$.

The following theorem can be viewed as an algorithm for EXT VC on trees.

Theorem 14. *Let $T = (V, E)$ be a tree and $U \subseteq V$ be an independent set. Then, (T, U) is a yes-instance of EXT VC iff there is no subtree $T' = (V', E')$ of T that is induced edge full with respect to (T, U) such that $T'[U \to \text{black}] \in \mathcal{T}$.*

5   Price of Extension

Considering the possibility that some set U might not be extendible to any minimal solution, one might ask how wrong U is as a choice for an extension problem. One idea to evaluate this, is to ask how much U has to be altered when aiming for a minimal solution. Described differently for our extension problems at hand, we want to discuss how many vertices of U have to be deleted for EXT VC (added for EXT IS) in order to arrive at a yes-instance of the extension problem. The magnitude of how much U has to be altered can be seen as the price that has to be paid to ensure extendibility. To formally discuss this concept, we consider according optimization problems. From an instance $I = (G, U)$ of EXT VC or EXT IS, we define the two NPO problems:

MAX EXT VC
Input: A graph $G = (V, E)$, a set of vertices $U \subseteq V$.
Solutions: Minimal vertex cover S of G.
Output: Solution S that maximizes $|S \cap U|$.

MIN EXT IS
Input: A graph $G = (V, E)$, a set of vertices $U \subseteq V$.
Solutions: Maximal independent set S of G.
Output: Solution S that minimizes $|U| + |S \cap (V \setminus U)|$.

For $\Pi =$ MAX EXT VC or MIN EXT IS, we denote by $opt_{\Pi}(I, U)$ the value of an optimal solution of MAX EXT VC or MIN EXT IS, respectively. Since for both of them, $opt_{\Pi}(I, U) = |U|$ iff (G, U) is a yes-instance of EXT VC or EXT IS, respectively, we deduce that MAX EXT VC and MIN EXT IS are NP-hard as soon as EXT VC and EXT IS are NP-complete. Alternatively, we could write $opt_{\text{MAX EXT VC}}(G, U) = \arg\max\{U' \subseteq U : (G, U')$ is a yes-instance of EXT VC$\}$, $opt_{\text{MIN EXT IS}}(G, U) = \arg\min\{U' \supseteq U : (G, U')$ is a yes-instance of EXT IS$\}$.

Similarly to Remark 1, one observes that the decision variants of MAX EXT VC and MIN EXT IS are equivalent, more precisely:

$$opt_{\text{MAX EXT VC}}(G, U) + opt_{\text{MIN EXT IS}}(G, V \setminus U) = |V| . \tag{1}$$

We want to discuss polynomial-time approximability of MAX EXT VC and MIN EXT IS. Considering MAX EXT VC on $G = (V, E)$ and the particular subset $U = V$ (resp., MIN EXT IS with $U = \emptyset$), we obtain two well known optimization problems called UPPER VERTEX COVER (UVC for short, also called MAXIMUM MINIMAL VERTEX COVER) and MINIMUM MAXIMAL INDEPENDENT SET (ISDS for short). In [18], the computational complexity of these problems are studied (among 12 problems), and (in)approximability results are given in [4,19] for UVC and in [11] for ISDS where lower bounds of $O(n^{\varepsilon - 1/2})$ and $O(n^{1-\varepsilon})$, respectively, for graphs on n vertices are given for every $\varepsilon > 0$. Analogous bounds can be derived depending on the maximum degree Δ. In particular, we deduce:

Corollary 15. *For any constant $\varepsilon > 0$, any $\rho \in \mathcal{O}\left(n^{1-\varepsilon}\right)$ and $\rho \in \mathcal{O}\left(\Delta^{1-\varepsilon}\right)$, there is no polynomial-time ρ-approximation for MIN EXT IS on graphs of n vertices and maximum degree Δ, even when $U = \emptyset$, unless $\mathsf{P} = \mathsf{NP}$.*

Theorem 16. MAX EXT VC *is as hard as* MAXIS *to approximate even if the set U of required vertices forms an independent set.*

Sketch. Let $G = (V, E)$ be an instance of MAXIS. Construct $H = (V_H, E_H)$ from G, where vertex set V_H contains two copies of V, V and $V' = \{v' : v \in V\}$. Let $E_H = E \cup \{vv' : v \in V\}$. Consider $I = (H, U)$ as instance of MAX EXT VC, where the required vertex subset is given by $U = V'$.
We claim: H has a minimal vertex cover containing k vertices from U iff G has a maximal independent set of size k.

Using the strong inapproximability results for MAXIS given in [20,21], observing $\Delta(H) = \Delta(G) + 1$ and $|V_H| = 2|V|$, we deduce the following result.

Corollary 17. *For any constant $\varepsilon > 0$, any $\rho \in \mathcal{O}\left(\Delta^{1-\varepsilon}\right)$ and $\rho \in \mathcal{O}\left(n^{1-\varepsilon}\right)$, there is no polynomial-time ρ-approximation for MAX EXT VC on graphs of n vertices and maximum degree Δ, unless $\mathsf{P} = \mathsf{NP}$.*

In contrast to the hardness results on these restricted graph classes from the previous sections, we find that restriction to bipartite graphs or graphs of bounded degree improve approximability of MAX EXT VC. For the following results, we assume, w.l.o.g., that the input graph is connected, non-trivial and therefore without isolated vertices, as we can solve our problems separately on each connected component and then combine the results. By simply selecting the side containing the largest number of vertices from U, we can show the following.

Theorem 18. *A 2-approximation for MAX EXT VC on bipartite graphs can be computed in polynomial time.*

Theorem 19. *A Δ-approximation for MAX EXT VC on graphs of maximum degree Δ can be computed in polynomial time.*

Proof. Let $G = (V, E)$ be connected of maximum degree Δ, and $U \subseteq V$ be an instance of MAX EXT VC. If $\Delta \leq 2$, or if $G = K_{\Delta+1}$ (the complete graph on $\Delta+1$ vertices), it is easy to check MAX EXT VC is polynomial-time solvable; actually in these two cases, G is either chordal or circular-arc and Theorem 20 gives the conclusion. Hence, assume $\Delta \geq 3$ and $G \neq K_{\Delta+1}$. By Brooks's Theorem, we can color G properly with at most Δ colors in polynomial-time (even linear). Let $(S_1, \ldots, S_\ell)$ be such coloring of G with $\ell \leq \Delta$. For $i \leq \ell$, set $U_i = U \cap N_G(S_i)$ where we recall $N_G(S_i)$ is the open neighborhood of S_i. By construction, S_i is an independent set which dominates U_i in G so it can be extended to satisfy (*iii*) of Theorem 3, so (G, U_i) is a *yes*-instance of EXT VC. Choosing $U' = \arg\max |U_i|$ yields a Δ-approximation, since on the one hand $\sum_{i=1}^{\ell} |U_i| \geq |U \cap (\cup_{i=1}^{\ell} N_G(S_i))| = |U \cap V|$ and on the other hand $\Delta \times |U'| \geq \sum_{i=1}^{\ell} |U_i| \geq |U| \geq opt_{\text{MAX EXT VC}}(G, U)$. $\qquad\square$

Along the lines of Corollary 13 with more careful arguments, we can prove:

Theorem 20. MAX EXT VC *can be solved optimally for chordal graphs and circular-arc graphs in polynomial time.*

Proof. Let (G, U) be an instance of MAX EXT VC where $G = (V, E)$ is a chordal graph (resp., a circular-arc graph) and U is an independent set. We build a weighted graph G' for WMINISDS such that G' is the subgraph of G induced by $N_G[U]$ and the weights on vertices are given by $w(v) = 1$ if $v \in U$ and $w(v) = 0$ for $v \in N_G[U] \setminus U$. Thus, we get: $opt_{\text{WMINISDS}}(G', w) = |U| - opt_{\text{MAX EXT VC}}(G, U)$. $\qquad\square$

6   Bounded Degree Graphs

Our NP-hardness results also work for the case of graphs of bounded degree, hence it is also interesting to consider EXT VC with standard parameter with an additional degree parameter Δ.

Theorem 21. EXT VC *is in* FPT *when parameterized both by the standard parameter and by the maximum degree Δ of the graph.*

Sketch. Recursively, the algorithm picks some $u \in U$ and branches on every neighbor $x \in N(u) \setminus U$ to be excluded from the vertex cover to ensure a private edge xu for u. This is a limited choice of at most Δ neighbors, and considering the new instance $(G - N[x], U \setminus N[x])$, this yields a running time in $\mathcal{O}^*(\Delta^k)$.

Let us look at this algorithm more carefully in the case of $\Delta = 3$ analyzing it from the standpoint of exact algorithms, i.e., dependent on the number of vertices n of the graph. Our algorithm has a branching vector of $(2, 2, 2)$ (in each branch, u and a neighbor of u is removed, so n reduces by 2), resulting in a branching number upper-bounded by 1.733. However, the worst case is a vertex in U that has three neighbors of degree one. Clearly, this can be improved. We propose the following reduction rules for EXT VC on an instance (G, U), $G = (V, E)$, which have to be applied exhaustively and in order:

0. If $U = \emptyset$, then answer *yes*.
1. If some $u \in U$ is of degree zero, then (G, U) is a *no*-instance.
2. If some $x \notin U$ is of degree zero, then delete x from V.
3. If $u, u' \in U$ with $uu' \in E$, then delete uu' from E.
4. If $u \in U$ is of degree one, then the only incident edge $e = ux$ must be private, hence we can delete $N[x]$ from V and all u' from U that are neighbors of x.
5. If $u \in U$ has a neighbor x that is of degree one, then assume $e = ux$ is the private edge of u, so that we can delete u and x from V and u from U.

After executing the reduction rules exhaustively, the resulting graph has only vertices of degree two and three (in the closed neighborhood of U) if we start with a graph of maximum degree three. This improves the branching vector to $(3, 3, 3)$, resulting in a branching number upper-bounded by 1.443. However, the rules are also valid for arbitrary graphs, as we show in the following.

Lemma 22. *The reduction rules are sound for general graphs when applied exhaustively and in order.*

Theorem 23. EXT VC *can be solved in time* $\mathcal{O}^*((\sqrt[3]{\Delta})^n)$ *on graphs of order n with maximum degree Δ.*

This gives interesting branching numbers for $\Delta = 3$: 1.443, $\Delta = 4$: 1.588, $\Delta = 5$: 1.710, etc., but from $\Delta = 8$ on this is no better than the trivial $\mathcal{O}^*(2^n)$-algorithm.

Let us remark that the same reasoning that resulted in Rule 5 is valid for:

5'. If $x \notin U$ satisfies $N(x) \subseteq U$, then delete $N[x]$ from V and from U.
6. Delete $V \setminus N_G[U]$. (inspired by Theorem 3)

We now run the following branching algorithm:

1. Apply all reduction rules exhaustively in the order given by the numbering.
2. On each connected component, do:
   - Pick a vertex v of lowest degree.
   - If $v \in U$: Branch on all possible private neighbors.
   - If $v \notin U$: Branch on if v is not in the cover or one of its neighbors.

A detailed analysis of the suggested algorithm gives the following result.

Theorem 24. EXT VC *on subcubic graphs can be solved in time* $\mathcal{O}^*(1.26^n)$ *on graphs of order n.*

Corollary 25. EXT VC *on subcubic graphs can be solved in time* $\mathcal{O}^*(2^{|U|})$ *with fixed vertex set U.*

Our reduction rules guarantee that each vertex not in U (and hence in $N_G(U)$) has one or two neighbors in U, and each vertex in U has two or three neighbors in $N_G(U)$. Hence, $|N_G(U)| \leq 3|U|$. In general, due to Rule 6:

Theorem 26. EXT VC *on graphs of maximum degree Δ allows for a vertex kernel of size $(\Delta + 1)|U|$, parameterized by the size of the given vertex set U.*

Looking at the dual parameterization (i.e., EXT IS with standard parameter), we can state due to all reduction rules:

Theorem 27. EXT VC *on graphs of maximum degree Δ allows for a vertex kernel of size $\frac{\Delta-1}{2}|V \setminus U|$, parameterized by $|V \setminus U|$.*

For $\Delta = 3$, we obtain vertex kernel bounds of $4|U|$ and $2|V \setminus U|$, respectively. With the computations of [9, Cor. 3.3 & Cor. 3.4], we can state the following.

Corollary 28. *Unless $P = NP$, for any $\varepsilon > 0$, there is no size $(2-\varepsilon)|U|$ and no size $(\frac{4}{3} - \varepsilon)|V \setminus U|$ vertex kernel for EXT VC on subcubic graphs, parameterized by $|U|$ or $|V \setminus U|$, respectively.*

This shows that our (relatively simple) kernels are quite hard to improve on.

Remark 29. Note that the arguments that led to the FPT-result for EXT VC on graphs of bounded degree (by providing a branching algorithm) also apply to graph classes that are closed under taking induced subgraphs and that guarantee the existence of vertices of small degree. This idea leads to a branching algorithm with running time $\mathcal{O}^*(5^{|U|})$ or $\mathcal{O}^*(1.32^{|V|})$.

Remark 30. Let us mention that we also derived several linear-time algorithms for solving EXT VC (and hence EXT IS) on trees in this paper. (1) A simple restriction of the mentioned DP algorithm on graphs of bounded treewidth solves this problem. (2) Apply our reduction rules exhaustively. (3) Check the characterization given in Theorem 14. Also, Theorem 20 provides another polynomial-time algorithm on trees.

7   Conclusions

We have found many graph classes where EXT VC (and hence also EXT IS) remains NP-complete, but also many classes where these problems are solvable in poly-time. The latter findings could motivate looking into parameterized algorithms that consider the distance from favorable graph classes in some way.

It would be also interesting to study further optimization problems that could be related to our extension problems, for instance the following ones, here formulated as decision problems. (a) Given G, U, k, is it possible to delete at most k vertices from the graph such that (G, U) becomes a yes-instance of EXT VC? Clearly, this problem is related to the idea of the price of extension discussed in this paper, in particular, if one restricts the possibly deleted vertices to be vertices from U. (b) Given G, U, k, is it possible to add at most k edges from the graph such that (G, U) becomes a yes-instance of EXT VC? Recall that adding edges among vertices from U does not change our problem, as they can never be private edges, but adding edges elsewhere might create private edges for certain vertices. Such problems would be defined according to the general idea of graph editing problems studied quite extensively in recent years. These problems are particularly interesting in graph classes where EXT VC is solvable in poly-time.

Considering the underlying classical optimization problems, it is also a rather intriguing question to decide for a given set U if it can be extended not just to any inclusion minimal vertex cover but to a globally smallest one, as a kind of *optimum-extension* problem. However, it has been shown in [12, Cor. 4.13] that the VERTEX COVER MEMBER problem (given a graph G and a vertex v, does there exist a vertex cover of minimum size that has v as a member, or, in other words, that extends $\{v\}$ is complete for the complexity class $P_\parallel^{NP}$, which is above NP and co-NP.

Acknowledgements. The first author was partially supported by the Deutsche Forschungsgemeinschaft DFG (FE 560/6-1).

References

1. Bazgan, C., Brankovic, L., Casel, K., Fernau, H.: On the complexity landscape of the domination chain. In: Govindarajan, S., Maheshwari, A. (eds.) CALDAM 2016. LNCS, vol. 9602, pp. 61–72. Springer, Cham (2016). https://doi.org/10.1007/978-3-319-29221-2_6
2. Bazgan, C., et al.: The many facets of upper domination. Theor. Comput. Sci. **717**, 2–25 (2018)
3. Berman, P., Karpinski, M., Scott, A.D.: Approximation hardness of short symmetric instances of MAX-3SAT. In: ECCC, no. 049 (2003)
4. Boria, N., Croce, F.D., Paschos, V.T.: On the max min vertex cover problem. Disc. Appl. Math. **196**, 62–71 (2015)
5. Boros, E., Gurvich, V., Hammer, P.L.: Dual subimplicants of positive Boolean functions. Optim. Meth. Softw. **10**(2), 147–156 (1998)
6. Casel, K., Fernau, H., Ghadikolaei, M.K., Monnot, J., Sikora, F.: On the complexity of solution extension of optimization problems. CoRR, abs/1810.04553 (2018)
7. Chang, G.J.: The weighted independent domination problem is NP-complete for chordal graphs. Disc. Appl. Math. **143**(1–3), 351–352 (2004)
8. Chang, M.: Efficient algorithms for the domination problems on interval and circular-arc graphs. SIAM J. Comput. **27**(6), 1671–1694 (1998)
9. Chen, J., Fernau, H., Kanj, I.A., Xia, G.: Parametric duality and kernelization: lower bounds and upper bounds on kernel size. SIAM J. Comput. **37**(4), 1077–1106 (2007)
10. Farber, M.: Independent domination in chordal graphs. Oper. Res. Lett. **4**(1), 134–138 (1982)
11. Halldórsson, M.M.: Approximating the minimum maximal independence number. Inf. Proc. Lett. **46**(4), 169–172 (1993)
12. Hemaspaandra, E., Spakowski, H., Vogel, J.: The complexity of kemeny elections. Theor. Comput. Sci. **349**(3), 382–391 (2005)
13. Kanté, M.M., Limouzy, V., Mary, A., Nourine, L.: On the enumeration of minimal dominating sets and related notions. SIAM J. Disc. Math. **28**(4), 1916–1929 (2014)
14. Kanté, M.M., Limouzy, V., Mary, A., Nourine, L., Uno, T.: Polynomial delay algorithm for listing minimal edge dominating sets in graphs. In: Dehne, F., Sack, J.-R., Stege, U. (eds.) WADS 2015. LNCS, vol. 9214, pp. 446–457. Springer, Cham (2015). https://doi.org/10.1007/978-3-319-21840-3_37

15. Kanté, M.M., Limouzy, V., Mary, A., Nourine, L., Uno, T.: A polynomial delay algorithm for enumerating minimal dominating sets in chordal graphs. In: Mayr, E.W. (ed.) WG 2015. LNCS, vol. 9224, pp. 138–153. Springer, Heidelberg (2016). https://doi.org/10.1007/978-3-662-53174-7_11
16. Kratochvíl, J.: A special planar satisfiability problem and a consequence of its NP-completeness. Disc. Appl. Math. **52**, 233–252 (1994)
17. Lawler, E.L., Lenstra, J.K., Kan, A.H.G.R.: Generating all maximal independent sets: NP-hardness and polynomial-time algorithms. SIAM J. Comp. **9**, 558–565 (1980)
18. Manlove, D.F.: On the algorithmic complexity of twelve covering and independence parameters of graphs. Disc. Appl. Math. **91**(1–3), 155–175 (1999)
19. Mishra, S., Sikdar, K.: On the hardness of approximating some NP-optimization problems related to minimum linear ordering problem. RAIRO Inf. Théor. Appl. **35**(3), 287–309 (2001)
20. Trevisan, L.: Non-approximability results for optimization problems on bounded degree instances. In: Vitter, J.S., Spirakis, P.G., Yannakakis, M. (eds.) Proceedings on 33rd Annual ACM Symposium on Theory of Computing, STOC, pp. 453–461. ACM (2001)
21. Zuckerman, D.: Linear degree extractors and the inapproximability of max clique and chromatic number. Theory Comp. Syst. **3**(1), 103–128 (2007)

On Hedonic Games with Common Ranking Property

Bugra Caskurlu[(✉)] and Fatih Erdem Kizilkaya

Computer Engineering Department, TOBB University of Economics and Technology,
06560 Ankara, Turkey
{bcaskurlu,f.kizilkaya}@etu.edu.tr

Abstract. Hedonic games are a prominent model of coalition formation, in which each agent's utility only depends on the coalition she resides. The subclass of hedonic games that models the formation of general partnerships [21], where output is shared equally among affiliates, is called hedonic games with common ranking property (HGCRP). Aside from their economic motivation, HGCRP came into prominence since they are guaranteed to have core stable solutions that can be found efficiently [2]. Nonetheless, a core stable solution is not necessarily a socially desirable (Pareto optimal) outcome. We improve upon existing results by proving that every instance of HGCRP has a solution that is both Pareto optimal and core stable. We establish that finding such a solution is, however, NP-HARD, by proving the stronger statement that finding any Pareto optimal solution is NP-HARD. We show that the gap between the total utility of a core stable solution and that of the socially optimal solution (OPT) is bounded by $|N|$, where N is the set of agents, and that this bound is tight. Our investigations reveal that finding a solution, whose total utility is within a constant factor of that of OPT, is intractable.

Keywords: Algorithmic game theory · Computational complexity · Hedonic games · Pareto optimality · Core stability

1 Introduction

The class of games, where a finite set of agents are to be partitioned into groups (coalitions) is known as coalition formation games [7]. A coalition formation game is said to be a hedonic coalition formation game (or simply hedonic game) if the utilities of the agents exclusively depend on the group they belong to [1], i.e., the agents do not worry about how the remaining agents are partitioned. Hedonic games subsume a wide range of problems, which includes well-known matching problems, such as the stable marriage [5], the stable roommates [6], and the hospital/residents problems [5].

This work is supported by The Scientific and Technological Research Council of Turkey (TÜBİTAK) through grant 118E126.

ⓒ Springer Nature Switzerland AG 2019
P. Heggernes (Ed.): CIAC 2019, LNCS 11485, pp. 137–148, 2019.
https://doi.org/10.1007/978-3-030-17402-6_12

An important application of hedonic games is the formation of partnerships, which is a formal agreement made by the founders of for-profit start-up ventures when the company is founded [21]. Though there are several different types of partnerships, the most common form is referred to as general partnership, where all parties share the legal and financial liabilities as well as the profit of the partnership equally. A hedonic game, where all agents of a coalition receive the same utility (as is the case in general partnerships) is said to possess *common ranking property* [2].

In game theory, the main goal is to determine the viable outcomes of a game under the assumption that the agents are rational. This is typically done by means of solution concepts. Each solution concept, designed with the intent of modeling the behavior of rational agents, comes with a set of axioms to be satisfied by the outcome of the game. Existence and tractable computability of Nash-stable [8], individually stable [9], contractually individually stable [10], core stable [11] outcomes of various subclasses of hedonic games are extensively studied in the literature.

The most prominent solution concept used in hedonic games is the core stability, where a partition of agents is defined as stable if no subset of agents can form a new coalition together so that the utilities of all the agents in the subset are increased [11]. Notice that core stability is the stability notion used in the definition of classical matching problems such as the stable marriage [5], and the stable roommates [6]. Some subclasses of hedonic games, such as the stable roommates, may not have any core stable outcome. However, every hedonic game that possesses common ranking property (HGCRP) is guaranteed to have a nonempty set of core stable partitions and such a partition can be computed via a simple greedy algorithm [2].

Pareto optimality is the most widely adopted concept of efficiency in game theory [3]. In hedonic games, a partition π of agents is called a Pareto optimal coalition structure, if no other partition π' of agents make some agents better off without making some agents worse off. Pareto optimality is a desirable property for the outcome of a game, and every game is guaranteed to have a Pareto optimal outcome. However, Pareto optimal outcomes do not necessarily coincide with stable outcomes of a game. As an extreme example, in the classical Prisoner's Dilemma game [23], all unstable outcomes are Pareto optimal, whereas the unique stable outcome is not. Similarly, in a HGCRP instance, a core stable coalition structure is not necessarily Pareto optimal (see Example 1), and a Pareto optimal coalition structure is not necessarily core stable (see Example 2). Thus, a natural research question (RQ) is as follows:

RQ 1: *Is it always possible to find a partition π of agents of a given HGCRP instance, such that π is both core stable and Pareto optimal?*

Theorem 1 answers to this question affirmatively via a constructive proof using a potential function [13].

Since the existence problem is resolved by Theorem 1, the next immediate problem is establishing the computational complexity of finding a core stable and Pareto optimal partition of a given HGCRP instance. In order to address this problem, we first need to state how a HGCRP instance is represented.

In a hedonic game, the utility function of each agent is defined over all subsets of agents containing her. Thus, if N is the set of agents, the input size is $O(|N| \times 2^{|N|-1})$. For HGCRP instances, we do not need to define separate utility functions for each agent, rather, we only need to have a joint utility value for each nonempty coalition (since all agents in the coalition have the same utility), and thus the input size is $O(2^{|N|} - 1)$.

Since the representation of a hedonic game instance requires exponential space in the number of agents, a lot of effort is spent in defining subclasses of hedonic games with concise (polynomial in the number of agents) representations [12]. The succinct representations in the literature are, either incapable of representing all HGCRP instances, or require space that is exponential in the number of agents. In this paper, we assume that a HGCRP instance is represented as *individually rational coalition lists* [4], i.e., the utility value of a subset $S \subseteq N$ is omitted, if the utility value of S is smaller than that of one of its singleton subsets. This is because no stable coalition structure can contain such a coalition. Since we have a model for input representation, we are ready to state the next immediate RQ as follows:

RQ 2: *What is the computational complexity of finding a Pareto optimal and core stable partition of agents of a given HGCRP instance?*

The problem is NP-HARD as stated by Corollary 1, since finding any Pareto optimal partition is NP-HARD due to Theorem 4.

In algorithmic game theory, a hot topic is quantifying the inefficiency due to selfish behavior of the agents. The outcome of the game maximizing the sum of the utilities of the agents, i.e., the total utility that could be achieved if the agents were not selfish, is referred to as the socially optimal solution (OPT).

There are several metrics in the literature to quantify the loss of total utility due to selfish acts, the most popular of which are the *price of anarchy* [14], and the *price of stability* [15]. The price of anarchy, and the price of stability of a given game are defined as the supremum of the ratio of the total utility of OPT to that of the socially worst and best stable solutions, respectively, over all instances of the game. In this paper, by a stable solution, we mean a core stable partition of agents. Immediate research questions along this line and our answers to them are stated in the following paragraphs.

RQ 3: *Given a HGCRP instance, is it tractable to compute OPT? If not, is it tractable to find a solution, whose total utility is within a constant factor of that of OPT? Is the problem of computing OPT fixed-parameter tractable, where the parameter is the maximum number of agents in a coalition?*

Finding a solution to a given HGCRP instance, whose total utility is within a constant factor of that of OPT, is intractable due to Theorem 5. However, finding a solution with a total utility of at least $1/|N|$ times that of OPT, is tractable due to Corollary 5, where N is the set of agents. The problem of computing OPT of a given HGCRP instance is not fixed-parameter tractable with respect to the maximum number of agents in a coalition, since the problem is APX-COMPLETE by Corollary 4, even under the restricted setting, where the size of a coalition is bounded above by 3.

RQ 4: *What are the price of anarchy, and the price of stability of HGCRP?*

Both the price of anarchy and the price of stability of HGCRP are $|N|$, where N is the set of agents, due to Theorem 6.

The rest of the paper is organized as follows: In Sect. 2, we introduce the notation used in this paper. In Sect. 3, we prove existence of a Pareto optimal and core stable partition of a given HGCRP instance, and show that finding such a partition is NP-HARD. In Sect. 4, we present our hardness of approximation and fixed-parameter intractability results for the problem of computing OPT. In Sect. 5, we give tight bounds for the price of anarchy, and the price of stability of HGCRP. In Sect. 6, we conclude and point out some future research directions.

2   Notation and Preliminaries

We define an instance of HGCRP as a binary pair $\mathcal{G} = (N, U)$, where N is a finite set of n agents, and $U : 2^N \smallsetminus \emptyset \to \mathbb{R}_+$ is a non-negative real-valued function defined over the nonempty subsets of N. We assume that an instance is represented as *individually rational coalition lists* (IRCL) [4], i.e., if $U(S) < U(\{i\})$ for some agent $i \in S$ for a subset of agents $S \subseteq N$, then $U(S)$ is omitted from the list, because coalition S is not a viable coalition in any stable coalition structure.

The solution (outcome) of a game is a partition (coalition structure) π over the set of agents N. The coalition containing an agent $i \in N$ in partition π is denoted by $\pi(i)$. In a partition π, the utilities of all the agents in the same coalition $S \in \pi$ are the same, and equal to the *joint utility* $U(S)$. We use $u_i(\pi)$ to denote the *utility* of some agent i in partition π. Notice that $u_i(\pi) = U(\pi(i))$.

A nonempty subset $S \subseteq N$ of agents is said to be a *blocking coalition with respect to partition* π, if $U(S) > u_i(\pi)$ for all agents $i \in S$, i.e., any agent $i \in S$ is *strictly* better off in S than she is in $\pi(i)$. A coalition structure π is *core stable* if there is no blocking coalition with respect to it.

For a partition π that is not core stable, and a blocking coalition S with respect to π, we define π_S as the partition induced on π by S. π_S is the partition that would arise if the agents in S collectively deviated from π to form coalition S, i.e., $\pi_S(i) = S$ for all $i \in S$, and $\pi_S(j) = \pi(j) \smallsetminus S$ for all $j \in N \smallsetminus S$. Notice that π_S may or may not be core stable.

Pareto optimality is a measure to assess the social quality of the solutions of a game with respect to alternative solutions of the same game. For two coalition structures π and π' over the set of agents N, we say that π' *Pareto dominates* π if, $u_i(\pi') \geq u_i(\pi)$ for all agents $i \in N$, and there exists an agent i for which the inequality is strict. In other words, if π' Pareto dominates π, then all agents are at least as good in π' than in π, and there is an agent that is strictly better off in π' than in π. A coalition structure π is said to be *Pareto optimal* if no coalition structure Pareto dominates it. Notice that if π is a Pareto optimal coalition structure, and if there is an agent i that is strictly better off in some other coalition structure π' than in π, then there is necessarily an agent j that is strictly worse off in π' than in π. Every finite game is known to possess a Pareto optimal solution.

We next present two examples to illustrate the notions of core stability and Pareto optimality, as well as the notation given above. In Example 1 we present a core stable partition that is not Pareto optimal, and in Example 2 we present a Pareto optimal partition that is not core stable.

Example 1. Let $\mathcal{G} = (N, U)$ be a HGCRP instance, where $N = \{1, 2\}$, and U is defined as $U(\{1\}) = U(\{1, 2\}) = 1$, and $U(\{2\}) = 0$. Notice that coalition structure $\pi = \{\{1\}, \{2\}\}$ is core stable since there is no blocking deviation with respect to it. But π is not Pareto optimal since the partition $\pi' = \{\{1, 2\}\}$ Pareto dominates π.

Example 2. Let $\mathcal{G} = (N, U)$ be a HGCRP instance, where $N = \{1, 2, 3\}$, and U is defined as $U(\{1\}) = 0$, $U(\{2, 3\}) = 2$ and $U(S) = 1$ for all other subsets $S \subseteq N$ of agents. Notice that partition $\pi = \{\{1, 2\}, \{3\}\}$ is Pareto optimal since no partition Pareto dominates it. However, π is not core stable since $S = \{2, 3\}$ is a blocking coalition with respect to π. Notice that $\pi_S = \{\{1\}, \{2, 3\}\}$.

3   Pareto Optimal and Core Stable Partitions

We devote this section to proving the existence of Pareto optimal and core stable partitions of any given HGCRP instance, and establishing the computational complexity of finding one such partition. We first prove the existence result, given by Theorem 1, by presenting a potential function [13] defined over the set of partitions of a given HGCRP instance, which is maximized at a Pareto optimal and core stable partition.

Theorem 1. *Every HGCRP instance* $\mathcal{G} = (N, U)$ *has a coalition structure that is both Pareto optimal and core stable.*

Proof. Let $\mathcal{G} = (N, U)$ be a given HGCRP instance with n agents, and let π and π' be two partitions over N. We define $\psi(\pi)$ as the sequence of the utilities of the agents in partition π in a *non-increasing order*. We denote the ith element in the sequence $\psi(\pi)$ by $\psi_i(\pi)$. We use the symbols $\rhd$ and $\unrhd$, respectively, to denote the binary relations "lexicographically greater than" and "lexicographically greater than or equal to" over the set of sequences of utilities of agents.

We next show, by Lemma 1, that if π' Pareto dominates π, then $\psi(\pi')$ is lexicographically greater than $\psi(\pi)$.

Lemma 1. *Let π and π' be partitions of N such that π' Pareto dominates π. Then, $\psi(\pi') \rhd \psi(\pi)$.*

Proof. We rename the agents so that $\psi_i(\pi)$ is the utility of agent i in partition π. Notice that we have $u_1(\pi) \geq u_2(\pi) \geq \ldots \geq u_n(\pi)$. Let $\psi'(\pi')$ be the permutation of $\psi(\pi')$, where $\psi_i'(\pi')$ and $\psi_i(\pi)$ are the respective utilities of the same agent for all $i \in N$. Notice that $\psi_i'(\pi') = u_i(\pi')$ and $\psi_i(\pi) = u_i(\pi)$. Since π' Pareto dominates π, we have $u_i(\pi') \geq u_i(\pi)$ for all $i \in N$, and $u_j(\pi') > u_j(\pi)$ for some agent j. Hence, $\psi_i'(\pi') \geq \psi_i(\pi)$ for all $i \in N$, and $\psi_j'(\pi') > \psi_j(\pi)$ for some agent $j \in N$. Therefore, $\psi'(\pi') \rhd \psi(\pi)$.

Notice that $\psi(\pi') \unrhd \psi'(\pi')$ since $\psi(\pi')$ is the same sequence as $\psi'(\pi')$ but sorted in descending order. Hence, $\psi(\pi') \unrhd \psi'(\pi') \rhd \psi(\pi)$, which completes the proof. □

We next show, by Lemma 2, that for a partition π that is not core stable, and a blocking coalition S with respect to π, we have $\psi(\pi_S)$ lexicographically greater than $\psi(\pi)$.

Lemma 2. *Let π be a partition of N that is not core stable, and let S be a blocking coalition with respect to π. Then, $\psi(\pi_S) \rhd \psi(\pi)$.*

Proof. Due to common ranking property, we can assume without loss of generality, that the utilities of agents that are in the same group in partition π are listed consecutively in $\psi(\pi)$. Moreover, we can assume without loss of generality, that for a group G in partition π, the utilities of agents in $G \cap S$ precede in the ordering of $\psi(\pi)$ those in $G \setminus S$. We also rename agents such that $\psi_i(\pi)$ correspond to the utility of agent i in partition π. Notice that we have $u_1(\pi) \geq u_2(\pi) \geq \ldots \geq u_n(\pi)$.

Let $\psi'(\pi_S)$ be a permutation of $\psi(\pi_S)$, where $\psi_i'(\pi_S)$ and $\psi_i(\pi)$ are the respective utilities of the same agent for all $i \in N$. Then, $\psi_i'(\pi_S) = u_i(\pi_S)$ and $\psi_i(\pi) = u_i(\pi)$. Let i be the agent with the smallest index such that $\psi_i(\pi) > \psi_i'(\pi_S)$. That is, $u_i(\pi) > u_i(\pi_S)$. This implies $i \notin S$, since otherwise we would have $u_i(\pi) < u_i(\pi_S)$. Additionally, there must be an agent $j \in \pi(i)$ such that $j \in S$, since otherwise we would have $\pi(i) = \pi_S(i)$, and it would mean $u_i(\pi) = u_i(\pi_S)$. Note that $u_j(\pi_S) > u_j(\pi)$ since $j \in S$, which means $\psi_j'(\pi_S) > \psi_j(\pi)$. Also recall that $u_j(\pi)$ precedes $u_i(\pi)$ in the ordering of $\psi(\pi)$. Therefore, $\psi'(\pi_S) \rhd \psi(\pi)$ since i is the agent with the smallest index such that $\psi_i(\pi) > \psi_i'(\pi_S)$, and there exists an agent $j < i$ such that $\psi_j(\pi) < \psi_j'(\pi_S)$.

Notice that $\psi(\pi_S) \unrhd \psi'(\pi_S)$ since $\psi(\pi_S)$ is the same sequence as $\psi'(\pi_S)$ but sorted in descending order. Therefore, $\psi(\pi_S) \unrhd \psi'(\pi_S) \rhd \psi(\pi)$, which completes the proof. □

Let π^* be a coalition structure such that $\psi(\pi^*) \unrhd \psi(\pi)$ for all partitions π over N. Notice that such a partition π^* exists, since the set of partitions over N is finite. π^* is Pareto optimal, since otherwise there is a partition π such that π Pareto dominates π^*. But then $\psi(\pi) \rhd \psi(\pi^*)$ by Lemma 1, which contradicts the fact that $\psi(\pi^*) \unrhd \psi(\pi)$ for all partitions π over N. π^* is also core stable, since otherwise there is a subset of agents S, which is a blocking coalition with respect to π^*. But then $\psi(\pi_S^*) \rhd \psi(\pi^*)$ by Lemma 2, which again contradicts the fact that $\psi(\pi^*) \unrhd \psi(\pi)$ for all partitions π over N. □

The rest of the section is devoted to proving that finding a Pareto optimal and core stable partition of a given HGCRP instance is NP-HARD. To do that, we first establish intractability of finding any Pareto optimal partition of a given HGCRP, which trivially reduces to finding a Pareto optimal and core stable partition. In our proof, we make use of an interesting result in the literature [3] that relates the computational complexity of finding a Pareto optimal partition of a subclass of hedonic games to that of finding a perfect partition in the same. A partition of agents in a hedonic game is said to be a *perfect partition*, if every agent is in her most preferred group, i.e., receives the maximum utility she can attain. Notice that a hedonic game (and also an instance of HGCRP) does not necessarily have a perfect partition. Therefore, the problem of finding a perfect partition is formally specified as follows.

PERFECT-PARTITION = *"Given a hedonic game, return a perfect partition if exists, return $\emptyset$ otherwise."*

The aforementioned relation between the computational complexities of finding a Pareto optimal partition, and PERFECT-PARTITION is established via the following result in [3].

Theorem 2 (Aziz et al. [3]). *For every class of hedonic games, where it can be checked whether a given partition is perfect in polynomial time, NP-hardness of PERFECT-PARTITION implies NP-hardness of computing a Pareto optimal coalition structure.*

Since it can be efficiently checked whether a given partition π of a given HGCRP instance is perfect or not in IRCL representation, Theorem 2 implies that all we need to complete the proof is to show that PERFECT-PARTITION is NP-HARD for the subclass HGCRP, which is stated as Theorem 3.

Theorem 3. PERFECT-PARTITION $\in$ NP-HARD *for HGCRP.*

Proof. Since search version of an NP-COMPLETE decision problem is NP-HARD [17], all we need is to show that deciding existence of a perfect partition of a given HGCRP instance is NP-COMPLETE. We do that by giving a polynomial time mapping reduction from EXACT-COVER [16].

In the EXACT-COVER problem, we are given a universe $U = \{1, \ldots, n\}$ and a family $S = \{S_1, \ldots, S_k\}$ of subsets of U. We are asked to decide whether there exists an exact cover $C \subseteq S$, i.e., each element in U is contained in exactly one subset in C. Given an instance (U, S) of EXACT-COVER, we construct a corresponding instance (N, U) of HGCRP as follows:

- For every element of $i \in U$ of (U, S), there is a corresponding agent $i \in N$ of (N, U),
- Each subset $S_i \in S$ of (U, S) corresponds to a subset of agents $S_i \subseteq N$ of (N, U) with $U(S_i) = 2$,
- For each subset $G \subseteq N$ of agents of (N, U), for which there is no corresponding subset in (U, S), the joint utility $U(G) = 1$ if $|G| = 1$, and $U(G) = 0$ otherwise.

Notice that in the IRCL representation, the constructed HGCRP instance (N, U) have joint utilities given only for the singleton groups, and the groups with a corresponding subset in (U, S). Hence, the size of the constructed HGCRP instance (N, U) is polynomial in the size of (U, S). We next show that $(U, S) \in$ EXACT-COVER if and only if there exists a perfect partition in (N, U).

(If) Suppose $(U, S) \in$ EXACT-COVER, i.e., there exists an exact cover C of (U, S). Let π be the coalition structure of (N, U), where each coalition $G \in \pi$ is the correspondent of a subset $S_i \in C$ of (U, S). Notice that π is a partition over the set of agents of (N, U). This is because for any distinct pair of subsets $S_i \in C$ and $S_j \in C$, we have $S_i \cap S_j = \emptyset$ since C is an exact cover. But then, $u_i(\pi) = 2$ for all agents $i \in N$. Since no coalition of (N, U) has a joint utility greater than 2, all agents of (N, U) are in their most preferred group, and thus, π is a perfect partition of (N, U).

(Only If) Suppose that there exists a perfect partition π of (N, U), i.e., $u_i(\pi) = 2$ for every agent $i \in N$. But then, for each coalition in π, there is a corresponding set S_i of (U, S). Notice that these sets not only cover all elements of U but also do not overlap since π is a partition, and hence, form an exact cover.

Since it is trivial in IRLC representation to check efficiently whether a partition of a given HGCRP instance is perfect, the decision version of PERFECT-PARTITION is in NP, and thus NP-COMPLETE by the above reduction.   □

Since it can be efficiently checked, whether a given partition π of a given HGCRP instance is perfect in IRCL representation, Theorem 4 is a direct consequence of Theorems 2 and 3.

Theorem 4. *Finding a Pareto optimal coalition structure of a given HGCRP instance is* NP-HARD.

Even though a core stable partition can be computed in polynomial time by a simple greedy algorithm [2], Theorem 4 implies that we cannot find a Pareto optimal core stable partition of a given HGCRP instance in polynomial time, as stated by Corollary 1.

Corollary 1. *Finding a coalition structure that is both core stable and Pareto optimal of a given HGCRP instance is* NP-HARD.

Since the EXACT-COVER is NP-COMPLETE, even under the restriction that $|S_i| = 3$ for all $S_i \in S$ [18], the mapping reduction used in the proof of Theorem 3 establishes the fixed parameter intractability result that PERFECT-PARTITION $\in$ NP-HARD for HGCRP, even when the sizes of coalitions are bounded above by 3. As a consequence of the reduction given in [3] from finding a perfect partition to finding a Pareto optimal partition, and the trivial reduction from finding a Pareto optimal partition to finding a Pareto optimal and core stable partition, we obtain the following result stated as Corollary 2.

Corollary 2. *The three problems, (i) finding a perfect partition, (ii) finding a Pareto optimal partition, and (iii) finding a Pareto optimal and core stable* *partition, are all* NP-HARD *for HGCRP, even under the restriction that the sizes of coalitions are bounded above by* 3.

4   Computing the Socially Optimal Solution

This section is devoted to establishing the computational complexity of finding a socially optimal solution π^* of a given HGCRP instance $\mathcal{G} = (N, U)$. The metric we use to evaluate the social welfare of a given solution π is the utilitarian objective function [23], i.e., the sum of the utilities of all agents. The social welfare $W(\pi)$ of a coalition structure π is then defined as $W(\pi) = \sum_{i \in N} u_i(\pi)$. A socially optimal solution π^* is a partition for which the social welfare is maximized.

Every socially optimal solution is a perfect partition, provided it exists, since in a perfect partition all agents achieve their respective maximum attainable utilities. Notice that PERFECT-PARTITION polynomially reduces to the problem of finding a socially optimal solution. This is because all it takes to decide whether a given socially optimal solution is also a perfect partition is to verify that each agent is in her most preferred coalition, and that can trivially be done efficiently in IRCL representation. Hence, all the hardness results presented for the PERFECT-PARTITION applies to the problem of computing the socially optimal solution as stated by Corollary 3.

Corollary 3. *Finding a socially optimal solution of a given HGCRP instance is* NP-HARD*, even under the restriction that the sizes of coalitions are bounded above by* 3.

In this section, we improve upon this immediate result by proving that finding a socially optimal solution of a given HGCRP instance is APX-COMPLETE, even under the restriction that the coalition sizes are bounded above by 3. For the general case, we prove that finding a socially optimal solution is constant factor inapproximable, as stated by Theorem 5.

Theorem 5. *Finding a socially optimal solution of a given HGCRP instance cannot be constant factor approximated.*

Proof Sketch. The proof is via an approximation preserving A-reduction [22] from MAXIMUM-INDEPENDENT-SET, which is known to be constant factor inapproximable [19]. For a given undirected graph $G = (V, E)$, we construct a corresponding instance (N, U) of HGCRP as follows:

- For each edge $e \in E$, there is a corresponding agent in N,
- Each $v \in V$ corresponds to a group $C_v \subseteq N$, that consists of agents corresponding to the incident edges of v, with joint utility $U(C_v) = \frac{1}{|C_v|}$,
- For each subset $C \subseteq N$ of agents of (N, U), for which there is no corresponding vertex in G, if $|C| = 1$ then $U(C) = \varepsilon$ where $0 < \varepsilon \leq \frac{1}{n^2}$, else $U(C) = 0$.

Notice that the size of the constructed HGCRP instance (N, U) is polynomial in the size of G.

Let I^* be a maximum independent set of G, and let π^* be a socially optimal solution of (N, U). Then, it is possible to show that any partition π of (N, U) induces an independent set I of G such that if $\frac{W(\pi^*)}{W(\pi)} \leq r$ then $\frac{|I^*|}{|I|} \leq r + 1$. Therefore, if we could have found a r-approximation π of a socially optimal solution π^* of (N, U), then we could have also found a $(r + 1)$-approximation I of a maximum independent set I^* of G.   □

Since MAXIMUM-INDEPENDENT-SET problem on cubic graphs (i.e., graphs in which all the vertices have degree 3) is APX-COMPLETE [20], we also have the following corollary.

Corollary 4. *Finding a socially optimal solution of a given instance of HGCRP is* APX-COMPLETE, *even when the sizes of coalitions are bounded above by 3.*

5   Quantification of Inefficiency

We now give tight bounds for the price of anarchy and the price of stability of HGCRP. The price of anarchy and the price of stability of a given game are defined as the supremum of the ratio between the total utility of OPT and that of the socially worst and best core stable solutions over all instances of the game, respectively [23].

Theorem 6. *The price of anarchy and the price of stability of HGCRP are* n.

Proof. For a given HGCRP instance, let π and π^* be a core stable solution, and a socially optimal solution, respectively. That is, π^* is a partition maximizing $W(\pi)$. Assume for the sake of contradiction $W(\pi^*) > nW(\pi)$. Then, there is an agent $i \in N$ such that $u_i(\pi^*) > W(\pi)$ by pigeonhole principle, i.e., $U(\pi^*(i)) > W(\pi)$. This means that $\pi^*(i)$ is a blocking coalition with respect to π. This contradicts the fact that π is a core stable partition. Therefore, $W(\pi^*) \leq nW(\pi)$ for any core stable partition π, specifically the socially worst one. Thus we have the following upper bound for the price of anarchy:

$$Price\ of\ Anarchy \leq \frac{W(\pi^*)}{W(\pi)} \leq n$$

We now give a lower bound for the price of stability of HGCRP by giving an example. Let $\mathcal{G} = (N, U)$ be an HGCRP instance, where U is defined as $U(N) = 1$, $U(\{1\}) = 1 + \varepsilon$ for some $\varepsilon > 0$, and $U(G) = 0$ for all other subsets $G \subset N$ of agents. In a core stable partition π of this game, $\{1\} \in \pi$, since otherwise $\{1\}$ is a blocking coalition with respect to π. Then, $N \notin \pi$ since $1 \in N$. Therefore, we have $W(\pi) = 1 + \varepsilon$. On the other hand, the socially optimal solution π^* is the one that N is formed, i.e., $W(\pi^*) = n$. Therefore, we have the following lower bound for the price of stability:

$$\frac{W(\pi^*)}{W(\pi)} = \frac{n}{1 + \varepsilon} \leq Price\ of\ Stability$$

Since the price of stability is always less than or equal to the price of anarchy, both of them are equal to n. □

Since the price of anarchy of HGCRP is n, by Theorem 6, the social welfare of the solution returned by the greedy algorithm [2] for finding a core stable partition is within a factor of n of that of the socially optimal solution. Thus, it is an n-approximation algorithm for finding the socially optimal solution, which proves Corollary 5. Notice that this approximation bound is tight for the greedy algorithm since the price of stability of HGCRP is also n.

Corollary 5. *Finding a social optimal solution of a given HGCRP instance is n-approximable.*

6 Conclusion and Future Research Direction

We presented a comprehensive study of hedonic games possessing common ranking property, which is a natural model for formation of general partnerships [21]. We strengthened the landmark result that every instance of HGCRP has a core stable partition, by proving that every instance of HGCRP has a partition that is both core stable and Pareto optimal. The economic significance of our result is that efficiency is not to be totally sacrificed for the sake of stability in HGCRP.

We established the computational complexity of several problems related to HGCRP both for the general case, and for the restricted case, where the size of the coalitions are bounded above by 3. Our investigations revealed that all the computational problems we considered are intractable. The restricted case, where the size of the coalitions are bounded above by 2 remains as a future research direction.

We quantified the loss of efficiency in HGCRP due to selfish behavior of agents by proving tight bounds on the price of anarchy, and the price of stability. In this way, we determined that finding a socially optimal solution can be approximated within a factor of n; however, it cannot be approximated within a constant factor. Finding a tighter bound remains as a future research direction.

References

1. Dreze, J.H., Greenberg, J.: Hedonic coalitions: optimality and stability. Econometrica **48**(4), 987 (1980)
2. Farrell, J., Scotchmer, S.: Partnerships. Q. J. Econ. **103**(2), 279 (1988)
3. Aziz, H., Brandt, F., Harrenstein, P.: Pareto optimality in coalition formation. Games Econ. Behav. **82**, 562–581 (2013)
4. Ballester, C.: NP-completeness in hedonic games. Games Econ. Behav. **49**(1), 1–30 (2004)
5. Gale, D., Shapley, L.S.: College admissions and the stability of marriage. Am. Math. Mon. **69**(1), 9 (1962)
6. Irving, R.W.: An efficient algorithm for the stable roommates problem. J. Algorithms **6**(4), 577–595 (1985)

7. Shenoy, P.P.: On coalition formation: a game-theoretical approach. Int. J. Game Theory **8**(3), 133–164 (1979)
8. Olsen, M.: Nash stability in additively separable hedonic games and community structures. Theory Comput. Syst. **45**(4), 917–925 (2009)
9. Bogomolnaia, A., Jackson, M.O.: The stability of hedonic coalition structures. Games Econ. Behav. **38**(2), 201–230 (2002)
10. Aziz, H., Brandt, F., Seedig, H.G.: Computing desirable partitions in additively separable hedonic games. Artif. Intell. **195**, 316–334 (2013)
11. Banerjee, S., Konishi, H., Sönmez, T.: Core in a simple coalition formation game. Soc. Choice Welf. **18**(1), 135–153 (2001)
12. Elkind, E., Wooldridge, M.: Hedonic coalition nets. In: The Proceedings of AAMAS 2009, vol. 1, pp. 417–424 (2009)
13. Monderer, D., Shapley, L.S.: Potential games. Games Econ. Behav. **14**(1), 124–143 (1996)
14. Koutsoupias, E., Papadimitriou, C.: Worst-case equilibria. Comput. Sci. Rev. **3**(2), 65–69 (2009)
15. Anshelevich, E., Dasgupta, A., Kleinberg, J., Tardos, É., Wexler, T., Roughgarden, T.: The price of stability for network design with fair cost allocation. SIAM J. Comput. **38**(4), 1602–1623 (2008)
16. Karp, R.M.: Reducibility among combinatorial problems. In: Miller, R.E., Thatcher, J.W., Bohlinger, J.D. (eds.) Complexity of Computer Computations. The IBM Research Symposia Series, pp. 85–103. Springer, Heidelberg (1972). https://doi.org/10.1007/978-1-4684-2001-2_9
17. Bellare, M., Goldwasser, S.: The complexity of decision versus search. SIAM J. Comput. **23**(1), 97–119 (1994)
18. Dyer, M., Frieze, A.: Planar 3DM is NP-complete. J. Algorithms **7**(2), 174–184 (1986)
19. Bazgan, C., Escoffier, B., Paschos, V.T.: Completeness in standard and differential approximation classes: Poly-(D)APX- and (D)PTAS-completeness. Theor. Comput. Sci. **339**(2–3), 272–292 (2005)
20. Alimonti, P., Kann, V.: Some APX-completeness results for cubic graphs. Theor. Comput. Sci. **237**(1–2), 123–134 (2000)
21. Larson, A.: What Is a Partnership. Expertlaw.com (2018). https://www.expertlaw.com/library/business/partnership.html. Accessed 28 Oct 2018
22. Crescenzi, P.: A short guide to approximation preserving reductions. In: Proceedings of Computational Complexity. Twelfth Annual IEEE Conference (1997)
23. Nisan, N., Roughgarden, T., Tardos, E., Vazirani, V.V. (eds.): Algorithmic Game Theory (2007)

Complexity of Scheduling for DARP
with Soft Ride Times

Janka Chlebíková[1], Clément Dallard[1(✉)], and Niklas Paulsen[2]

[1] School of Computing, University of Portsmouth, Portsmouth, UK
{janka.chlebikova,clement.dallard}@port.ac.uk
[2] Institut für Informatik, Christian-Albrechts Universität zu Kiel, Kiel, Germany
npau@informatik.uni-kiel.de

Abstract. The Dial-a-Ride problem may contain various constraints for pickup-delivery requests, such as time windows and ride time constraints. For a tour, given as a sequence of pickup and delivery stops, there exist polynomial time algorithms to find a schedule respecting these constraints, provided that there exists one. However, if no feasible schedule exists, the natural question is to find a schedule minimising constraint violations. We model a generic fixed-sequence scheduling problem, allowing lateness and ride time violations with linear penalty functions and prove its APX-hardness. We also present an approach leading to a polynomial time algorithm if only the time window constraints can be violated (by late visits). Then, we show that the problem can be solved in polynomial time if all the ride time constraints are bounded by a constant. Lastly, we give a polynomial time algorithm for the instances where all the pickups precede all the deliveries in the sequence of stops.

Keywords: Dial-A-Ride · Scheduling · Vehicle Routing Problem · NP-hardness · Ride times · Time windows

1 Introduction

The Dial-A-Ride Problem (DARP) is a well studied variant of the Vehicle Routing Problem. The DARP, with its various restrictions, serves as a model for many real-world problems from logistics, *e.g.* passenger transportation, or pickup-delivery of perishable goods. For a review of DARPs, we refer the readers to [2].

The study of the Dial-A-Ride Problem can be split into three main subproblems: the *clustering* of the requests into tours, the *routing* of the stops within each tour into a sequence, and the *scheduling* of the stops inside the tours [4]. These problems are the source of major research topics in operation research, each of them intensively studied. To get a better understanding of the inherent complexity of the problems, and eventually obtain faster algorithms, many restricted models have been studied.

In this paper, we focus on the scheduling subproblem, where the input of the problem is a tour with fixed sequence of stops, a set of pickup-delivery requests,

© Springer Nature Switzerland AG 2019
P. Heggernes (Ed.): CIAC 2019, LNCS 11485, pp. 149–160, 2019.
https://doi.org/10.1007/978-3-030-17402-6_13

and time constraints with their corresponding penalty functions. Each pickup-delivery request is represented by two stops, the first one as a pickup, and the second one as a delivery. The visit of each stop has to be performed within a given *time window*. Furthermore, the time between the scheduled pickup and delivery of a same request is bounded by a given *ride time*.

Time windows and ride times constraints are naturally arising when scheduling pickups and deliveries. When both constraints must be respected in the solution, there exist efficient algorithms [6,9,11]. However, in all these approaches, ride time and time window constraints are *hard* in the sense that a solution must respect all the constraints.

In case there is no feasible schedule, one may look for a schedule "close" to a feasible one with minimal penalties. Therefore, variants of the problem with *soft* constraints, in which the violation of constraints is allowed but penalised, have been introduced. Depending on the type of constraints and their corresponding penalty functions, various results can be obtained. For instance, when the only constraints are time windows for the stops, Dumas et al. [5] proposed a linear programming approach for convex penalty functions with a linear time complexity, but their algorithm does not incorporate ride time constraints. To the best of our knowledge, the complexity of the problem with soft ride times constraints was previously unknown.

In this paper, we propose a systematic study of the complexity of the problem when allowing lateness at stops (scheduling after the time windows) and ride time violation.

2   Problem Statement

In the following, we assume that $0 \in \mathbb{N}$. For $\triangle \in \{\leq, <, \geq, >\}$, and X a well ordered set, let $X_{\triangle x} := \{y \in X : y \triangle x\}$ and $X[i]$ be the i-th smallest element of X.

We are given a sequence $\mathsf{S} = (1, 2, \ldots, 2n)$, $n \in \mathbb{N}$, of $2n$ stops in the order in which their visits must be scheduled (in case of no ambiguity $s \in \mathsf{S}$ also represents an integer). Each stop $s \in \mathsf{S}$ is associated with a time interval (a_s, b_s), $0 \leq a_s \leq b_s$, representing a time window in which a visit of the stop s should take place (without loss of generality we suppose that $a_1 = 0$). Furthermore, we have a set $\mathcal{P}$ of n requests representing the pairs (p, d) of stops from S, $p < d$, where p is a pickup and d is a delivery stop. Each request (p, d) has a time constraint $r_{p,d}$ ($r_{p,d} \geq 0$) on the ride time: a visit at stop d should be scheduled at most $r_{p,d}$ time units after the visit at stop p. Each stop s serves exactly one request (either as a pickup or as a delivery stop) and all times are represented as non-negative integers.

As it has been mentioned in Sect. 1, it is not always possible to schedule the visits for all stops (in a given order) with respect to their time windows and ride time constraints. Therefore we introduce the model in which the time window and ride time constraints can be violated for penalties (soft constraints).

In order to model soft constraints: (i) each stop $s \in S$ is associated with a penalty function $\sigma_s^L : \mathbb{N} \to \mathbb{Q}$, mapping visit times which are later than the time window bounds to a non-negative penalty, and (ii) each request $(p, d) \in \mathcal{P}$ is associated with a penalty function $\sigma_{p,d}^{RT} : \mathbb{N} \to \mathbb{Q}$, mapping ride times exceeding ride time constraints to a non-negative penalty.

In this paper we suppose that all penalty functions are linear non-decreasing functions. We consider a restricted model where earliness at stops is not allowed, hence a stop must be either scheduled within or after its time window. Therefore, for each stop $s \in S$ and the visit time x at stop s, then $x \geq a_s$. Moreover, we have the function $\sigma_s^L(x)$ such that $\sigma_s^L(x) = 0$ for $x \leq b_s$ and otherwise $\sigma_s^L(x) = \alpha_s \cdot (x - b_s) + \beta_s$ for given $\alpha_s, \beta_s \in \mathbb{Q}_{\geq 0}$. Analogously, for each request $(p, d) \in \mathcal{P}$ we have the function $\sigma_{p,d}^{RT}$ with $\sigma_{p,d}^{RT}(x) = 0$ for $x \leq r_{p,d}$ and otherwise $\sigma_{p,d}^{RT}(x) = \alpha_{p,d} \cdot (x - r_{p,d}) + \beta_{p,d}$ for given $\alpha_{p,d}, \beta_{p,d} \in \mathbb{Q}_{\geq 0}$.

A *schedule* $\mathsf{t} = (\mathsf{t}_1, \ldots, \mathsf{t}_{2n})$ is a sequence of visit times for all the stops in S (in the given order), where we say that t schedules a stop $s \in S$ at time t_s. We say that a schedule t is *feasible* if and only if for all stops $s \in S$, $\mathsf{t}_s \geq a_s$ and for any $s \in S_{<2n}$, $\mathsf{t}_s \leq \mathsf{t}_{s+1}$. Obviously, each time window and ride time constraint is either *violated* or *satisfied* by a schedule t. The *cost* $c(\mathsf{t})$ of a schedule t is the sum of the penalties of violated constraints:

$$c(\mathsf{t}) = \sum_{s \in S} \sigma_s^L(\mathsf{t}_s) + \sum_{(p,d) \in \mathcal{P}} \sigma_{p,d}^{RT}(\mathsf{t}_d - \mathsf{t}_p). \tag{1}$$

When we solve the Min Pickup-Delivery Scheduling problem, we look for a feasible schedule t with a minimum cost.

Min Pickup-Delivery Scheduling (Min PDS)
Input An instance I of Min PDS.
Task Find a feasible schedule t of I such that $c(\mathsf{t})$ is minimum.

We also consider two special cases of the main problem:

Min Pickup-Delivery Scheduling with Hard Ride Time Constraints (Min PDS-HRT)
Input An instance I of Min PDS.
Task Find a feasible schedule t of I respecting all ride time constraints and minimising $c(\mathsf{t})$.

Min Pickup-Delivery Scheduling with Hard Time Windows (Min PDS-HTW)
Input An instance I of Min PDS.
Task Find a feasible schedule t of I respecting all time windows constraints and minimising $c(\mathsf{t})$.

2.1 Our Contribution

The paper investigates how penalisation of the time window and ride time constraints contributes to the computational complexity of the problem. We show that an essential factor for the complexity are soft maximum ride time constraints. In Subsect. 2.2, we give some remarks on our model. An overview of complexity results is shown in Table 1. We prove the NP-hardness of the main problem MIN PICKUP-DELIVERY SCHEDULING and its special case with hard time window constraints, MIN PDS-HTW (Sect. 3.1). Nevertheless, we show that the problem can be solved in polynomial time in case of hard ride time constraints, MIN PDS-HRT (Sect. 3.2). Further underlining the role of ride time constraints, we give a parameterised algorithm that solves MIN PICKUP-DELIVERY SCHEDULING in polynomial time if all ride time constraints are bounded by a constant (Sect. 4). In Sect. 5 we show that some structural properties in the sequence of the stops can be exploited to find a polynomial time algorithm. Namely, we present an $\mathcal{O}(n^4)$ time algorithm when all pickups precede all the deliveries in the sequence.

Table 1. Overview of complexity results classified by constraints. Arrows mean results are inferred.

Ride time constraints	Time window constraints	
	Hard	Soft
Hard	$\mathcal{O}(n)$ [6]	$\mathcal{O}(n)$ [Sect. 3.2]
0, $\beta_{p,d} = 0$	$\mathcal{O}(n)$	$\leftarrow \mathcal{O}(n)$ [Sect. 4, from [5]]
Soft, bounded values	P	$\leftarrow$ P [Sect. 4]
Soft, unbounded values	NP-hard, APX-hard [Sect. 3.1]	$\rightarrow$ NP-hard, APX-hard

2.2 Remarks on the Model

Driving times, (un)loading times. In favour of simplicity, our model neglects times needed to travel between stops as well as loading or unloading times. We emphasise that this is not restrictive, since we focus on the scheduling of fixed sequences. An instance with given driving and (un)loading times can be transformed to an equivalent instance of our form with a simple preprocessing.

Waiting times. Constraints w_s on the time to wait between two consecutive stops s and $s + 1$ (as in [6]) are omitted in our model since they can be expressed by ride time constraints: assume for $s \in \mathsf{S}_{<2n}$ the constraint $t_s + w_s \geq t_{s+1}$ is given for schedules t, $w_s \geq 0$. Simply insert two additional stops: the stop p immediately before s and the stop d immediately after $s+1$ into S and add a request (p, d) to $\mathcal{P}$ with $r_{p,d} = w_s$. Replacing all waiting time constraints leads to an equivalent instance with at most $6n - 2 \in \mathcal{O}(n)$ requests.

Increasing time windows opening times. Since earliness is not allowed in our model, we expect that any instance of $2n$ stops has $a_s \leq a_{s+1}$ for all

$s \in S_{<2n}$. If for a stop s, $s \in S_{<2n}$, we have $a_s > a_{s+1}$, for any feasible schedule t it holds $t_s \geq a_s$ and $t_{s+1} \geq t_s$ and therefore $t_{s+1} < a_s$ cannot hold for any feasible scheduling. We can therefore preprocess the instance in such a way that for all $s \in S_{<2n}$, $a_{s+1} := \max\{a_s, a_{s+1}\}$. Notice that due to this property and the fact that the last stop $2n$ is a delivery stop, it always exists an optimal schedule t^* such that $t_{2n}^* = a_{2n}$.

As soon as possible deliveries. All deliveries can be scheduled at a time as soon as possible without increasing costs. Let d be a delivery of a request $(p, d) \in \mathcal{P}$ and t a feasible schedule. We define a schedule t' with $t_s' := t_s$ for all stops $s \in S \setminus \{d\}$ and $t_d' := \max\{a_d, t_{d-1}\}$. Clearly, t' is feasible. Obviously, t_d' can only decrease the lateness at d as well as the ride time for $(p, d) \in \mathcal{P}$ with no changes in scheduling of the other stops.

3 Complexity Study

3.1 Min Pickup-Delivery Scheduling with Hard Time Windows

In this subsection we study the variant of MIN PICKUP-DELIVERY SCHEDULING in which ride time constraints may be violated in return for a penalty (soft constraints), but the time windows must be respected (hard constraints). We show that such a problem, called MIN PICKUP-DELIVERY SCHEDULING WITH HARD TIME WINDOWS (MIN PDS-HTW), is NP-hard and APX-hard even in case of restricted time windows and very circumscribed penalty functions. The proof is based on a reduction from the MAXIMUM DICUT problem which is known to be NP-hard and APX-hard when restricted to directed acyclic graphs (DAG) [7,10].

A directed cut (A, B) of a directed graph $G = (V, E)$ is a partition of V into two subsets A, B. Its size $s(A, B) := |\{(u, v) \in E : u \in A, v \in B\}|$ is the number of outgoing arcs from A to B. The MAXIMUM DICUT problem is defined as follows:

MAXIMUM DICUT
Input A directed graph $G = (V, E)$.
Task Find a directed cut (A, B) of maximum size in G.

Theorem 1. MIN PICKUP-DELIVERY SCHEDULING WITH HARD TIME WINDOWS *is NP-hard.*

Proof. Firstly, the decision version of MIN PDS-HTW is clearly in NP. Let $G = (V, E)$ be a connected DAG such that $|V| = n$, $|E| = m$. Since G is a DAG, the vertices of G can be labelled by $1, 2, \ldots, n$ in a topological ordering in such a way that for any arc $(u, v) \in E$ it holds $lab(u) < lab(v)$, where $lab(z)$ represents the number used for labelling the vertex z [12].

In the following we show how the graph G can be transformed into an instance I of MIN PDS-HTW. The sequence of stops for I is defined as the concatenation

$\mathsf{S} := \mathsf{S}^1 \mathsf{S}^2 \ldots \mathsf{S}^n$, where each S^v represents a gadget of stops for each vertex $v \in V$. Let $v \in V$ be fixed, then the gadget S^v contains the stop s_e^v for each arc e of G incident to v, i.e. $e = (v, u)$ or $e = (u, v)$ for $u \in V$. An example of such a gadget is depicted in Fig. 1.

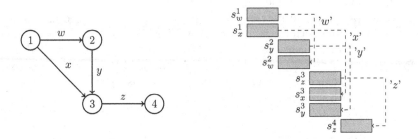

Fig. 1. A DAG (left) transformed into an instance of PDS (right). Gray boxes are time windows of length 1. Note there are two stops for every arc of G and the stops are grouped in gadgets for each vertex of G.

The stops within the gadget S^v are ordered in such a way that all stops belonging to outgoing arcs precede all stops belonging to ingoing arcs. Note that S has $2m$ stops. For each vertex $v \in V$ the time windows of all stops $s \in \mathsf{S}^v$ are set to $a_s := lab(v) - 1$ and $b_s := lab(v)$. The requests correspond to the arcs in G, hence $\mathcal{P} = \{(s_e^u, s_e^v) : e = (u, v) \in E\}$ and for each $(p, d) \in \mathcal{P}$ the ride time is set to be $r_{p,d} = a_d - b_p$. Due to the specific numbering of vertices and sizes of windows, the stop p always precedes the stop d and $r_{p,d} \geq 0$ for all $(p, d) \in \mathcal{P}$. Setting the penalty coefficients $\alpha_{p,d} = 0$ and $\beta_{p,d} = 1$, the cost of a schedule corresponds to the number of violated ride time constraints. Obviously, the transformation from G to the instance I can be done in polynomial time.

Now we show that G has a directed cut of size at least $(m - k)$ if and only if there exists a schedule t violating at most k ride time constraints, for any $k \in \mathbb{N}$.

$\Rightarrow$ Suppose there exists a directed cut (A, B) in G of size at least $(m - k)$. Define a schedule t such that for every vertex $v \in A$ and every stop $s \in \mathsf{S}^v$ we set $\mathsf{t}_s := b_s$ and for all other stops $\mathsf{t}_s := a_s$. Clearly, t is a feasible schedule. Each arc $(u, v) \in E$ corresponds to the unique $(p, d) \in \mathcal{P}$ with $p \in \mathsf{S}^u$ and $d \in \mathsf{S}^v$. If $u \in A$ and $v \in B$, then $\mathsf{t}_d - \mathsf{t}_p = b_d - a_p \leq r_{p,d}$, hence the ride time constraint is respected. As we suppose $s(A, B) \geq m - k$, the previous holds for at least $(m - k)$ requests. With $|\mathcal{P}| = m$, it implies t violates at most k ride time constraints.

$\Leftarrow$ Now suppose there exists a schedule t for I violating at most k ride time constraints. For each vertex $v \in V$ let $s[v]$ be the first delivery stop in S^v and if there is no such stop, then $s[v]$ be the last pickup stop in S^v. This allows us to define a partition of V in the following way: for each vertex $v \in V$, if $\mathsf{t}_{s[v]} = b_{s[v]}$ then $v \in A$, otherwise $v \in B$. Fix a $(p, d) \in \mathcal{P}$ and let $u, v \in V$ be such vertices that p (resp. d) is from the gadget S^u (resp. S^v). If t satisfies the ride time constraint of (p, d), then by the definition of I it must hold $\mathsf{t}_p = b_p$ and

$t_d = a_d$. Since p is from the gadget S^u and t is feasible, $\mathsf{t}_{s[v]} \geq b_p$ and necessarily $b_{s[u]} = b_p$, hence $\mathsf{t}_{s[u]} = b_{s[u]}$. Analogously, $\mathsf{t}_{s[v]} = a_{s[v]}$. Therefore, $u \in A$ and $v \in B$. As we suppose that in the schedule t at most k ride time constraints are violated, then at least $|\mathcal{P}| - k = m - k$ are satisfied. Since each satisfied ride time leads to a distinct arc going from A to B, $s(A, B) \geq m - k$.

The reduction defined in Theorem 1 is in fact a Strict-reduction [3] between the optimisation problems. As Lampis et al. [10] proved the MAXIMUM DICUT problem is APX-hard even when restricted on DAGs, the following result follows:

Corollary 1. MIN PICKUP-DELIVERY SCHEDULING WITH HARD TIME WINDOWS *is APX-hard.*

Now we argue that the main problem MIN PICKUP-DELIVERY SCHEDULING is NP-hard and APX-hard too. The idea is to set the penalties for lateness at each stop to such values that any optimal schedule must respect the time windows. Let I be an instance of MIN PDS-HTW with n requests. As mentioned in Subsect. 2.2, there exists an optimal schedule t of I such that $\mathsf{t}_{2n} = a_{2n}$. Therefore the actual ride time of each request is bounded by the value a_{2n}. Let $k = \max_{(p,d)\in\mathcal{P}} \sigma_{p,d}^{RT}(a_{2n})$. Then the instance I can be transformed into an instance I' of MIN PDS by setting $\alpha_s = 0$ and $\beta_s = nk + 1$ for all $s \in \mathsf{S}$ (hence $\sigma_s^L(x) = kn + 1$ for $x > b_s$). The cost of any schedule respecting time windows is at most nk, hence there exists an optimal schedule of I' with cost strictly less than $nk + 1$. Such a schedule must respect time windows and therefore is also valid for the instance I of MIN PDS-HTW, hence MIN PDS-HTW can be seen as a special case of MIN PDS. Therefore we can conclude

Corollary 2. MIN PICKUP-DELIVERY SCHEDULING *is NP-hard and APX-hard.*

3.2 Min Pickup-Delivery Scheduling with Hard Ride Time Constraints

In this subsection we study the variant of MIN PICKUP-DELIVERY SCHEDULING in which the ride time constraints must be respected (hard constraints), while time windows may be violated in return for penalty (soft constraints). As it was mentioned in Sect. 1, we consider a model in which *lateness* is the only possible way to violate a time window restriction. We prove that this variant of the problem, called MIN PICKUP-DELIVERY SCHEDULING WITH HARD RIDE TIME CONSTRAINTS (MIN PDS-HRT), can be solved in linear time, compared to NP-hardness of MIN PICKUP-DELIVERY SCHEDULING WITH HARD TIME WINDOWS shown in Sect. 3.1.

When both ride times and time window constraints are hard, a linear time algorithm was proposed by Firat and Woeginger in [6]. It has also been adapted to handle additional *minimum* ride time constraints in [8]. We show how the same approach can be used to minimise lateness penalties in MIN PICKUP-DELIVERY SCHEDULING WITH HARD RIDE TIME CONSTRAINTS.

Our idea, similarly to the one used in [6,8], is to formulate a difference constraint system (DCS) with variables of the schedule and interpret it as a graph in which the existence of negative weight cycles is equivalent to infeasibility of the DCS. In these papers it is shown how to apply the single-source shortest path algorithm for interval graphs presented in [1] to test the existence of negative weight cycles in linear time. In case of feasible instances, a solution can be extracted in linear time as well. We point out that this approach will lead to a schedule visiting every stop as late as possible: the scheduled time of each stop is chosen by the length of a shortest path from the start vertex. This path corresponds to a chain of difference equations and can be seen as the tightest upper bound on the timing value. Since the shortest path lengths are upper bounds this implies that no feasible schedule can visit any of the stops later.

Theorem 2. MIN PICKUP-DELIVERY SCHEDULING WITH HARD RIDE TIME CONSTRAINTS *can be solved in linear time.*

4   Bounded Ride Time Constraints

The MIN PICKUP-DELIVERY SCHEDULING problem is NP-hard as it follows from Sect. 3.1. In this section we show that some restrictions on the parameters of the problem improve the complexity of the problem.

We suppose that $\mu \in \mathbb{N}$ is a fixed constant. Let μ-MIN PICKUP-DELIVERY SCHEDULING (μ-MIN PDS) be the restriction of the MIN PICKUP-DELIVERY SCHEDULING problem to the instances with the ride time constraints bounded by μ, *i.e.* $r_{p,d} \leq \mu$ for all $(p,d) \in \mathcal{P}$. In the following we propose a polynomial-time algorithm for μ-MIN PDS.

Given an instance of MIN PDS, let $\mathcal{W}$ be the set of all time window bounds for all stops, *i.e.* $\mathcal{W} = \bigcup_{s \in S}\{a_s, b_s\}$, and $\mathcal{J}_s := \{(p,d) \in \mathcal{P} : d > s \text{ and } p \leq s\}$ be the set of the *loaded requests* after the stop $s \in S$, and its size $\mathsf{load}(s) := |\mathcal{J}_s|$. Firstly we observe that the visit times of an optimal solution can be chosen from a restricted set of time values. We define the set

$$\widetilde{\mathcal{W}} := (\bigcup_{w \in \mathcal{W}} [w - n\mu, w + n\mu])_{\geq a_1, \leq a_{2n}} \, .$$

Note that $\widetilde{\mathcal{W}} = \mathcal{W}$ in case of 0−MIN PDS.

We say that a schedule t is *defined in* $\widetilde{\mathcal{W}}$ if and only if $\mathsf{t}_s \in \widetilde{\mathcal{W}}$ for all $s \in S$. The following lemma states that in fact there is an optimal schedule defined in $\widetilde{\mathcal{W}}$.

Theorem 3. *For a given instance of* μ-MIN PDS *there is an optimal schedule* t *defined in* $\widetilde{\mathcal{W}}$.

Definition 1. *For a given schedule* t *of an instance of* μ-MIN PDS *and a stop* $\ell \in S$ *we define the* partial cost $\tilde{c}(t, \ell)$ *of* t *up to the stop* $\ell \in S$ *as*

$$\tilde{c}(t, \ell) := \sum_{s \in S_{\leq \ell}} \sigma_s^L(t_s) + \sum_{(p,d) \in \mathcal{J}_\ell} \sigma_{p,d}^{RT}(t_\ell - t_p) + \sum_{\substack{(p,d) \in \mathcal{P} \\ d \leq \ell}} \sigma_{p,d}^{RT}(t_d - t_p) \, .$$

In the following lemma we prove some observations regarding the partial cost function.

Lemma 1. *For a given schedule* t *of the instance* μ-MIN PDS *and the stop* $\ell \in S$ *the following hold*

(i) $\tilde{c}(t, 1) = \sigma_1^L(t_1)$;

(ii) $\tilde{c}(t, \ell + 1) = \tilde{c}(t, \ell) + \sigma_{\ell+1}^L(t_{\ell+1}) + \sum_{(p,d) \in \mathcal{J}_\ell} f_{p,d}^\ell(t)$

$\quad$ *with* $f_{p,d}^\ell(t) := \begin{cases} \alpha_{p,d}(t_{\ell+1} - t_\ell), & \text{if } t_\ell - t_p > r_{p,d} \\ \sigma_{p,d}^{RT}(t_{\ell+1} - t_p), & \text{if } t_\ell - t_p \le r_{p,d} \end{cases}$,

(iii) $\tilde{c}(t, 2n) = c(t)$,

Let I be an instance of μ-MIN PDS. For each stop l, $l = 1, 2, \ldots, 2n$ we define so called l-*labels* to capture the structure of 'similar' schedules for I. The labels enable to restrict the number of schedules for I in each step and therefore to use the ideas of dynamic programming.

As it follows from Theorem 3, we can focus on schedules defined in $\widetilde{\mathcal{W}}$ only. For a schedule t defined in $\widetilde{\mathcal{W}}$ and $\ell \in S$, the ℓ-*label of* t is defined as

$$Label_\ell(t) = (t_\ell, s^0, s^1, \ldots, s^\mu, \tilde{c}(t, \ell)),$$

where $s^m := \min\{s \in S \text{ such that } t_s \ge t_\ell - m\}$, i.e. s^m is the first stop of the schedule t visited at or after time $(t_\ell - m)$ for any m, $0 \le m \le \mu$.

Note that every schedule has one such label for each $\ell \in S$, but a label may describe more (different) schedules. We say that a label $L \in \widetilde{\mathcal{W}} \times S^{\mu+1} \times \mathbb{Q}^+$ is a feasible ℓ-label if there exists a feasible schedule t for I with $Label_\ell(t) = L$.

In the following lemma we prove that in each stop there is a restriction on the number of labels to consider to find an optimal schedule.

Lemma 2 (Domination rule). *Let* I *be an instance of* μ-MIN PDS *and the stop* $\ell \in S$ *be fixed. Let* $L^1 = (\tau, s^0, s^1, \ldots, s^\mu, \tilde{c}_1)$ *and* $L^2 = (\tau, s^0, s^1, \ldots, s^\mu, \tilde{c}_2)$ *be feasible* ℓ-*labels with* $\tilde{c}_1 \le \tilde{c}_2$. *Then there is a feasible schedule* t *with* $Label_\ell(t) = L^1$ *such that* $c(t) \le c(t')$ *for any feasible schedule* t' *with* $Label_\ell(t') = L^2$. *We say that the label* L_1 *dominates the label* L_2.

Now, according to Lemma 2, we can give an upper bound on the number of non-dominated labels for any fixed stop $l \in S$. There are at most $|\widetilde{\mathcal{W}}|$ possibilities for the first item of the label, hence $\mathcal{O}(\mu \cdot n^2)$ if $\mu > 0$ (in case $\mu = 0$ only $\mathcal{O}(n)$), and $\mathcal{O}(n)$ choices for each of the next $(\mu + 1)$ items of the label.

Remark 1. For each instance of μ-MIN PDS and a stop $\ell \in S$ the number of non-dominated ℓ-labels is bounded by $\mathcal{O}(\mu \cdot n^{\mu+3})$ if $\mu > 0$ and by $\mathcal{O}(n^2)$ if $\mu = 0$.

This leads to the following results:

Theorem 4. *An instance of* μ-MIN PDS *with* $\mu > 0$ *can be solved in time* $\mathcal{O}(\mu^2 \cdot n^\mu \cdot \text{poly}(n))$.

Proof. Starting with the initial labels $(\tau, 1, \ldots, 1, \sigma_1^L(\tau))$ for each $\tau \in \widetilde{\mathcal{W}}_{\geq a_1}$ we have all the labels for the first stop for any feasible schedule defined on $\widetilde{\mathcal{W}}$, by Lemma 1 (i). The labels for the stop $\ell \in S_{>1}$ can be calculated from the labels of the stop $(\ell - 1)$. For a non-dominated label $(\tau, s^0, \ldots, s^\mu, \tilde{c})$ of the stop $(\ell - 1)$ (there are $\mathcal{O}(\mu \cdot n^{\mu+3})$ such labels) do the following: for every possible visit time $\tau' \in \widetilde{\mathcal{W}}_{\geq \tau}$ at the stop ℓ (there are $\mathcal{O}(\mu \cdot n^2)$ such possible time visits), generate a new label:

- the first item of the label is τ';
- the s^* items are defined in the following way: $(\tau' - \tau)$ items have the value ℓ (truncate to at most $\mu + 1$ items), if $(\tau' - \tau) < \mu + 1$, then start to add the items $s^0, \ldots, s^\mu$ from the previous $(l - 1)$-label until there are $(\mu + 1)$ s^*-items,
- the new cost can be calculated in a linear time from the given label using Lemma 1 (ii).

Overall each new label is generated in time $\mathcal{O}(n)$. Any label at the stop $2n$ minimising the last item of the label (cost) represents only optimal schedules by Lemma 1 (iii).

Corollary 3. *An instance of* $0-$MIN PDS *can be solved in time* $\mathcal{O}(n^5)$.

Proof. The result follows from the proof of Theorem 4 considering $\widetilde{\mathcal{W}} = \mathcal{W}$ and the fact that the number of non-dominated labels per stop is bounded by $\mathcal{O}(n^2)$.

We consider another specific case, when the goal is to minimise the sum of the lateness penalties and the sum of the ride times. This implies that $\mu = 0$. Since driving times are excluded from instances of our model (see Subsect. 2.2), all ride times can be seen as excess ride times (excess ride times are defined as the actual ride time minus the driving time). The problem can be solved with the algorithm of Dumas et al. [5] in linear time (*). The ride times can also be minimised in a weighted manner, using $\alpha_{p,d} \geq 0$ for $(p, d) \in \mathcal{P}$. The waiting time before a stop $s \in S_{>1}$ is then simply weighted by $\sum_{(p,d)\in\mathcal{J}_{s-1}} \alpha_{p,d}$.

5   Special Patterns in the Sequence of Stops

In this section we study a class of polynomial time solvable instance of MIN PICKUP-DELIVERY SCHEDULING. We introduce the *First Pickup Then Deliveries (FPTD)* instances in which all the stops $1, \ldots, n$ are pickup stops, and the stops $n + 1, \ldots, 2n$ are delivery stops. We show that MIN PICKUP-DELIVERY SCHEDULING can be solved in polynomial time in the class of *FPTD* instances despite the NP-hardness of the problem (Sect. 3.1).

Firstly, we prove that for each stop we can reduce the set of potential scheduling times to a subset polynomial in size. Each time in this subset is calculated from the time windows and maximal ride time values of the instance.

Lemma 3. *Let I be an FPTD instance with $2n$ stops. Then there exists an optimal schedule t of I such that for each $s \in S$:*

- *if $t_s < t_n$, then $t_s \in \mathcal{B}^s(t_n) := \Big(\{b_{s'} : s' \in S, \ s \leq s' \leq n\}_{<t_n} \cup$*

$$\bigcup\nolimits_{(p,d) \in \mathcal{P}, \ p \leq s} \{\max\{t_n, a_d\} - r_{p,d}\} \Big)_{\geq a_s};$$

- *if $t_s = t_n$, then*

$$t_s \in \mathcal{C} := \{a_n\} \cup \bigcup_{(p,d) \in \mathcal{P}} \Big\{ \{b_p, a_d - r_{p,d}\} \cup \bigcup_{(p',d') \in \mathcal{P}} \{b_p + r_{p',d'}, a_d - r_{p,d} + r_{p',d'}\} \Big\};$$

- *if $t_s > t_n$, then $t_s = a_s$.*

According to Lemma 3, there is an optimal schedule t with $t_n \in \mathcal{C}$, and $|\mathcal{C}|$ is quadratic in the instance size. Moreover, when t_n is fixed, each stop $s \in S_{<n}$ with $t_s < t_n$ belongs to $\mathcal{B}^s(t_n)$, linear in size. Obviously, when t_n is fixed, one can schedule all deliveries $d \in S_{>n}$ at $t_d := \max\{t_n, a_d\}$. In the following, we show how one can efficiently calculate an optimal schedule for a fixed t_n.

Let $p \in S_{\leq n}$ be a pickup stop and $(p, d) \in \mathcal{P}$ its corresponding request. We define the function σ_p^k as the partial cost of scheduling p when $t_n = k$ such that if p is scheduled at l then $\sigma_p^k(l) = \sigma_p^L(l) + \sigma_{p,d}^{RT}(\max\{k, a_d\} - l)$. In order to define a recursive equation for calculating the cost of the optimal schedule, we define for each pickup stop $s \in S_{\leq n}$ the function $T_s^k : \mathbb{N} \to \mathbb{N}$:

$$T_s^k(j) := \begin{cases} \max \left(\mathcal{B}^s(k) \cup \{k\} \right)_{\leq j} & \text{if } \left(\mathcal{B}^s(k) \cup \{k\} \right)_{\leq j} \neq \emptyset, \\ -1 & \text{otherwise.} \end{cases}$$

The call of $T_s^k(j)$ yields the largest time of the set $\mathcal{B}^s(k) \cup \{k\}$ which is smaller than j, or returns -1 if there is no such time. Thus, given $t_n = k$ and a bound j on the visit time for the pickup s, we are able to iterate over the candidate times for s in $\mathcal{B}^s(k) \cup \{k\}$.

Thereby, we can define a recursion table calculating the minimum cost of a schedule t when the value of t_n is fixed.

Lemma 4. *Let I be an FPTD instance, $k \in \mathcal{C}$, and t a schedule of I such that $t_n = k$ and t has minimum cost. Then, $c(t) = C[n, k] + \sum_{d \in S_{>n}} \sigma_d^L(\max\{k, a_d\})$*

$$C[i, j] = \begin{cases} \min \begin{Bmatrix} C[i - 1, T_{i-1}^k(j)] + \sigma_i^k(j), \\ C[i, T_i^k(j - 1)] \end{Bmatrix} & \text{if } i \geq 1, \ j \geq a_i, \\ 0 & \text{if } i = 0, \\ \infty & \text{otherwise.} \end{cases} \qquad (2)$$

Finally, to find the cost of an optimal schedule, we have to compute the value $C[n, k]$ for each $k \in \mathcal{C}$. A dynamic programming algorithm $(*)$ can solve it in $\mathcal{O}(n^4)$ time.

Theorem 5. *An FPTD instance I with $2n$ stops can be solved in $\mathcal{O}(n^4)$.*

6 Conclusion

We study a new model of the Dial-A-Ride Problem for the scheduling of fixed sequences with several time constraints typical in the Pickup-and-Delivery scenario. We highlight the key role of soft maximal ride time constraints in the combinatorial complexity of the problem, as they induce the NP-hardness of the problem. We also prove that if the maximal ride times are bounded by a constant, we can obtain a polynomial-time algorithm. Finally, we show that instances of the problem with a special structure can be solved efficiently, independently of the timing constraints. We believe that this result can be generalised whenever the number of times a pickup is followed by a delivery in the sequence is bounded.

To get a better understanding of the overall complexity of the problem further research may consider other constraints, *e.g.* allow earliness at stops, and more complex penalty functions.

References

1. Atallah, M.J., Chen, D.Z., Lee, D.T.: An optimal algorithm for shortest paths on weighted interval and circular-arc graphs, with applications. Algorithmica **14**(5), 429–441 (1995). https://doi.org/10.1007/BF01192049
2. Cordeau, J.F., Laporte, G.: The dial-a-ride problem (DARP): variants, modeling issues and algorithms. Q. J. Belg. Fr. Ital. Oper. Res. Soc. **1**(2), 89–101 (2003)
3. Crescenzi, P.: A short guide to approximation preserving reductions. In: Proceedings of the 12th Annual IEEE Conference on Computational Complexity. CCC '97, pp. 262–273. IEEE Computer Society, Washington (1997)
4. Desrosiers, J., Dumas, Y., Solomon, M.M., Soumis, F.: Time constrained routing and scheduling. Handb. Oper. Res. Manag. Sci. **8**, 35–139 (1995)
5. Dumas, Y., Soumis, F., Desrosiers, J.: Optimizing the schedule for a fixed vehicle path with convex inconvenience costs. Transp. Sci. **24**(2), 145–152 (1990)
6. Firat, M., Woeginger, G.J.: Analysis of the dial-a-ride problem of Hunsaker and Savelsbergh. Oper. Res. Lett. **39**(1), 32–35 (2011)
7. Gatto, M., Jacob, R., Peeters, L., Schöbel, A.: The computational complexity of delay management. In: Kratsch, D. (ed.) WG 2005. LNCS, vol. 3787, pp. 227–238. Springer, Heidelberg (2005). https://doi.org/10.1007/11604686_20
8. Gschwind, T.: Route feasibility testing and forward time slack for the synchronized pickup and delivery problem. Technical report, Citeseer (2015)
9. Hunsaker, B., Savelsbergh, M.: Efficient feasibility testing for dial-a-ride problems. Oper. Res. Lett. **30**(3), 169–173 (2002)
10. Lampis, M., Kaouri, G., Mitsou, V.: On the algorithmic effectiveness of digraph decompositions and complexity measures. Discrete Optim. **8**(1), 129–138 (2011). https://doi.org/10.1016/j.disopt.2010.03.010
11. Tang, J., Kong, Y., Lau, H., Ip, A.W.: A note on efficient feasibility testing for dial-a-ride problems. Oper. Res. Lett. **38**(5), 405–407 (2010)
12. Tarjan, R.E.: Edge-disjoint spanning trees and depth-first search. Acta Informatica **6**(2), 171–185 (1976)

Vertex Deletion on Split Graphs: Beyond 4-Hitting Set

Pratibha Choudhary[1]([✉]), Pallavi Jain[2], R. Krithika[3], and Vibha Sahlot[2]

[1] Indian Institute of Technology Jodhpur, Jodhpur, India
pratibhac247@gmail.com
[2] Institute of Mathematical Sciences, HBNI, Chennai, India
pallavij@imsc.res.in, sahlotvibha@gmail.com
[3] Indian Institute of Technology Palakkad, Palakkad, India
krithika@iitpkd.ac.in

Abstract. In vertex deletion problems on graphs, the task is to find a set of minimum number of vertices whose deletion results in a graph with some specific property. The class of vertex deletion problems contains several classical optimization problems, and has been studied extensively in algorithm design. Recently, there was a study on vertex deletion problems on split graphs. One of the results shown was that transforming a split graph into a block graph and a threshold graph using minimum number of vertex deletions is NP-hard. We call the decision version of these problems as SPLIT TO BLOCK VERTEX DELETION (SBVD) and SPLIT TO THRESHOLD VERTEX DELETION (STVD), respectively. In this paper, we study these problems in the realm of parameterized complexity with respect to the number of vertex deletions k as parameter. These problems are "implicit" 4-HITTING SET, and thus admit an algorithm with running time $\mathcal{O}^\star(3.0755^k)$, a kernel with $\mathcal{O}(k^3)$ vertices, and a 4-approximation algorithm. In this paper, we exploit the structure of the input graph to obtain a kernel for SBVD with $\mathcal{O}(k^2)$ vertices and FPT algorithms for SBVD and STVD with running times $\mathcal{O}^\star(2.3028^k)$ and $\mathcal{O}^\star(2.7913^k)$.

Keywords: Vertex deletion problems · Split graphs ·
Parameterized algorithms · Kernelization · Approximation algorithms

1 Introduction

Graphs are one of the most versatile mathematical objects endowed with immense modelling power. Many problems of practical interest can be represented as problems on graphs and therefore the study of graph problems and algorithms have become an integral part of computer science. In vertex deletion problems on graphs, the task is to find a minimum number of vertices whose

We thank Saket Saurabh for his invaluable advice and several helpful suggestions.
P. Jain— Supported by SERB-NPDF fellowship (PDF/2016/003508) of DST, India.

P. Heggernes (Ed.): CIAC 2019, LNCS 11485, pp. 161–173, 2019.
https://doi.org/10.1007/978-3-030-17402-6_14

deletion results into a graph that belongs to a graph class with some specific property. The family of vertex deletion problems contains several classical optimization problems. For example, VERTEX COVER, FEEDBACK VERTEX SET and ODD CYCLE TRANSVERSAL are vertex deletion problems with the desired target class being edgeless graphs, forests, and bipartite graphs, respectively. A general result by Lewis and Yannakakis shows a complete dichotomy of the complexity of these problems [18]. In particular, the problem is NP-hard for most choices of the desired target class. As vertex deletion problems contain several problems of both practical and theoretical interest, they have been studied extensively in various algorithmic paradigms like parameterized complexity, approximation algorithms, and exact algorithms.

Another natural direction of research is to study vertex deletion problems on special graph classes. In this paper, we focus our study on split graphs – a natural family of graphs. A *split graph* is a graph whose vertex set can be partitioned into a clique and an independent set. Every split graph is a chordal graph[1] as well. Split graphs can be recognised in polynomial time, and they admit elegant polynomial-time algorithms for several problems that are NP-hard in general. Notable examples include MAXIMUM INDEPENDENT SET, MAXIMUM CLIQUE, MINIMUM COLORING, and CLUSTER DELETION [3,12]. However, there are many problems which remain NP-hard even when restricted to split graphs. For instance, STEINER TREE, DISJOINT PATHS, CUTWIDTH, and RAINBOW COLOURING are NP-hard on split graphs [5,13,15]. This contrast in the complexity of classical problems makes the class of split graphs an important and well-studied graph class.

A recent study on vertex deletion problems on split graphs showed interesting NP-hardness results [4]. In particular, the following problems of transforming a split graph into a block graph, and a threshold graph, using the minimum number of vertex deletions were shown to be NP-hard.

SPLIT TO BLOCK VERTEX DELETION (SBVD)
Input: A split graph G and an integer k.
Question: Does there exist a set $S \subseteq V(G)$ of at most k vertices such that $G - S$ is a block graph?

SPLIT TO THRESHOLD VERTEX DELETION (STVD)
Input: A split graph G and an integer k.
Question: Does there exist a set $S \subseteq V(G)$ of at most k vertices such that $G - S$ is a threshold graph?

A graph is a *block graph* if every biconnected component is a clique. Equivalently, block graphs are those chordal graphs that do not contain a diamond (a complete graph on four vertices with exactly one edge deleted) as an induced subgraph. A graph G is a *threshold graph* if there exists a real number t and a function $f : V(G) \to \mathbb{R}$ such that two vertices u and v are adjacent in G

---

[1] A *chordal graph* is a graph in which every induced (or chordless) cycle is a triangle.

if and only if $f(u) + f(v) \geq t$. It is easy to verify that threshold graphs are split graphs as well. Specifically, threshold graphs are those split graphs that do not contain an induced path on four vertices [12]. Vertex deletion to split graphs, block graphs and threshold graphs are well-studied in the framework of parameterized algorithms [1,6,8,11,16]. In this paper, we study SBVD and STVD with respect to the number of vertex deletions k as parameter. We also design approximation algorithms for the minimization version of these problems, referred to as MinSBVD and MinSTVD, respectively. We begin our study by observing that these problems can be cast as restricted cases of the popular d-HITTING SET problem. For a set system $(U, \mathcal{F})$ comprising of a finite universe U, and a collection $\mathcal{F}$ of subsets of U, a *hitting set* is a set $T \subseteq U$ that has a non-empty intersection with each set in $\mathcal{F}$. For a fixed integer $d > 0$, given a set system $(U, \mathcal{F})$ with each set in $\mathcal{F}$ consisting of at most d elements, the d-HITTING SET problem requires finding a minimum hitting set. As SBVD and STVD can be seen as "implicit" 4-HITTING SET, they admit an algorithm with running time $\mathcal{O}^\star(3.0755^k)^2$, a kernel with $\mathcal{O}(k^3)$ vertices and a 4-approximation algorithm [9,14,19]. In this paper, we exploit the structure of the input (a split graph) and obtain improved polynomial kernel, parameterized algorithms, and approximation algorithms. Due to space constraint, approximation algorithms have been omitted, and will appear in the full version of the paper. Our results comprise of the following:

- SBVD can be solved in $\mathcal{O}^\star(2.3028^k)$ time and admits a kernel with $\mathcal{O}(k^2)$ vertices. MinSBVD can be solved in $\mathcal{O}^\star(1.5658^{n+o(n)})$ time, where n is the number of vertices in the input graph.
- MinSBVD admits a factor 2-approximation algorithm but is APX-hard. Further, no $(2-\epsilon)$-approximation algorithm is possible under the Unique Games Conjecture.
- STVD can be solved in $\mathcal{O}^\star(2.7913^k)$ time and MinSTVD can be solved in $\mathcal{O}^\star(1.6418^{n+o(n)})$ time, where n is the number of vertices in the input graph.
- MinSTVD admits a factor 2-approximation algorithm but is APX-hard.

The parameterized algorithms are based on the branching technique and the exact exponential-time algorithms are obtained using the results of [10] in conjunction with our FPT algorithms. The quadratic vertex kernel for SBVD is based on the recently introduced *new expansion lemma* [17].

Preliminaries. The set $\{1, \cdots, n\}$ of consecutive integers from 1 to n is denoted by $[n]$. For a set S, $\binom{S}{2}$ denotes the set $\{\{u, v\} : u, v \in S, u \neq v\}$. We use standard terminology from the book of Diestel [7] for the graph related terminologies that are not explicitly defined here. All graphs considered in this paper are simple and undirected. The vertex set and the edge set of a graph G are denoted by $V(G)$ and $E(G)$, respectively. For an edge $e = uv$, vertices u and v are called *endpoints* of e. For a vertex $v \in V(G)$, its *neighbourhood* $N(v)$ is the set of all vertices

---

^{2} $\mathcal{O}^\star$ notation suppresses polynomial factors. That is, $\mathcal{O}^\star(f(k)) = \mathcal{O}(f(k)n^{\mathcal{O}(1)})$.

adjacent to it, and its *non-neighbourhood* $\overline{N}(v)$ is the set $V(G)\backslash(N(v) \cup \{v\})$. For $S \subseteq V(G)$, $N(S) = \bigcup_{v \in S} N_G(v)$. The *degree* of a vertex $v \in V(G)$, denoted by $deg_G(v)$, is the size of $N(v)$. For a set $S \subseteq V(G)$, $G[S]$ and $G - S$ denote the subgraphs of G induced by the set S and $V(G)\backslash S$ respectively. For a singleton set $S = \{u\}$, we denote $G - \{u\}$ as $G - u$. A complete graph on q vertices is denoted by K_q. A *diamond* is denoted by D_4. A *block vertex deletion set* of G is a set $S \subseteq V(G)$ of vertices such that $G - S$ is a block graph. The partition (C, I) of the vertex set of a split graph G into clique C and independent set I is called a *split partition* of G. A split graph with split partition (C, I) is called *complete split* if every vertex of C is adjacent to every vertex of I. In a graph G, an induced path on four vertices is denoted by $P_4 = (a, b, c, d)$, where $ab, bc, cd \in E(G)$, and $ac, ad, bd \notin E(G)$. A split graph is a block graph if and only if it does not have a D_4 as an induced subgraph [2]. A split graph is a threshold graph if and only if it does not have a P_4 as an induced subgraph [12]. All reduction rules mentioned in the paper are applied in the sequence stated and each rule is applied as long as it is applicable on the instance. Due to space constraint, safeness of all reduction rules and branching rules will appear in the full version of the paper.

2   A Quadratic Vertex Kernel for SBVD

In this section, we show that SBVD parameterized by k admits a kernel with $\mathcal{O}(k^2)$ vertices.

2.1   Tools Used: Expansion Lemma, 3-Hitting Set and Approximation Algorithm

First, we list the tools and techniques that we crucially use in our kernelization algorithm. We begin with the notions of *ℓ-expansion* and *new expansion lemma* from [17].

Definition 1. *(ℓ-expansion) [17] Let ℓ be a positive integer and H be a bipartite graph with bipartition (A, B). Let $\hat{A} \subseteq A$ and $\hat{B} \subseteq B$. We say that $\hat{A}$ has an ℓ-expansion into $\hat{B}$ in H if $|N(Y) \cap \hat{B}| \geq \ell|Y|$, for all $Y \subseteq \hat{A}$.*

An *ℓ-star* at a vertex u is a set of ℓ distinct edges incident on u. As discussed in [17], the existence of an ℓ-expansion from $\hat{A}$ to $\hat{B}$ is equivalent to having an ℓ-star at each $u \in \hat{A}$ with the other endpoints in $\hat{B}$, and the ℓ-stars being pairwise disjoint with respect to the vertices they use from $\hat{B}$. For an ℓ-expansion from $\hat{A}$ to $\hat{B}$, the vertices of $\hat{B}$ that are endpoints of ℓ-star are termed as *saturated*, while the remaining vertices of $\hat{B}$ are termed as *unsaturated*.

Lemma 1. *(New Expansion Lemma) [17] Let ℓ be a positive integer and H be a bipartite graph with bipartition (A, B). Then there exists $\hat{A} \subseteq A$ and $\hat{B} \subseteq B$ such that $\hat{A}$ has an ℓ-expansion into $\hat{B}$ in H, and, $N(\hat{B}) \subseteq \hat{A}$ and $|B \setminus \hat{B}| \leq \ell|A \setminus \hat{A}|$. Moreover, the sets $\hat{A}$ and $\hat{B}$ can be computed in polynomial time.*

Note that $\hat{A}$ and $\hat{B}$ may be empty. In that case, since $|B \setminus \hat{B}| \leq \ell |A \setminus \hat{A}|$, we have $|B| \leq \ell |A|$. Therefore, if $|B| > \ell |A|$, then $\hat{B} \neq \emptyset$.

The next tool that we use is based on the 3-HITTING SET problem. Consider a 3-HITTING SET instance $(U, \mathcal{F}, k)$. An element $u \in U$ is called *essential* for $(U, \mathcal{F}, k)$ if every solution to $(U, \mathcal{F}, k)$ contains u. Observe that if u is essential, then $(U, \mathcal{F}, k)$ is a yes-instance if and only if $(U \setminus u, \mathcal{F}', k - 1)$ is a yes-instance where $\mathcal{F}' = \{F : F \in \mathcal{F}, u \notin F\}$. A family $\widehat{\mathcal{F}} \subseteq \binom{U}{2}$ is called a family of *essential pairs* for $(U, \mathcal{F}, k)$ if for every $F \in \mathcal{F}$, there is a set $F' \in \widehat{\mathcal{F}}$ such that $F' \subseteq F$.

Lemma 2. [20]⊛[3] *Given an instance $(U, \mathcal{F}, k)$ of 3-HITTING SET, there is a polynomial-time algorithm that either declares that $(U, \mathcal{F}, k)$ is a no-instance, or returns an essential element $u \in U$, or returns a set $\widehat{\mathcal{F}}$ of essential pairs with $|\widehat{\mathcal{F}}| = \mathcal{O}(k^2)$.*

Theorem 1. ⊛ MinSBVD *admits a factor 2-approximation algorithm.*

2.2   The Kernelization Algorithm

Let (G, k) be an instance of SBVD and S be a block vertex deletion set for G obtained using the factor 2-approximation algorithm for SBVD due to Theorem 1. Observe that if $|S| > 2k$, then (G, k) is a no-instance and the required kernel is a trivial no-instance of constant size. Otherwise, let (C, I) and $(C^\star, I^\star)$ be split partitions of $G[S]$ and $G - S$, such that $C \cup C^\star$ is a clique and $I \cup I^\star$ is an independent set. If $|C^\star| \leq 2$, then we add the vertices of $C^\star$ to S and delete $C^\star$ from $G - S$. Now, S is a block vertex deletion set of G of size $\mathcal{O}(k)$.

To obtain the required kernel, it suffices to bound $|C^\star|$ and $|I^\star|$. We will define a sequence of reduction rules for the same. After the application of each rule, we reuse the notations G to denote the resultant graph, and (C, I) and $(C^\star, I^\star)$ to denote the split partitions of the (new) $G[S]$ and $G - S$, respectively.

Before proceeding to the kernelization algorithm, we state a crucial observation which is used subsequently.

Observation 1. ⊛ *Suppose $|C^\star| > 2$. Then, there is at most one vertex in $I^\star$ that is adjacent to every vertex in $C^\star$. Further, every vertex in $I^\star$ that has a non-neighbour in $C^\star$ is adjacent to at most one vertex in $C^\star$.*

Preprocessing Rule 1. *If there is a vertex $v \in I^\star$ that is adjacent to every vertex in $C^\star$, add v to S, and delete v from $G - S$.*

Note that $|S|$ remains $\mathcal{O}(k)$ as a result of this preprocessing rule. Subsequently, every vertex $v \in I^\star$ has at most one neighbour in $C^\star$.

Reduction Rule 1.   – *If $k \leq 0$ and G is not a block graph, then a no-instance of constant size is the required kernel.*
– *If $k \geq 0$ and G is a block graph, then a yes-instance of constant size is the required kernel.*

---

[3] Proofs of results marked with ⊛ will be given in full version of the paper.

– If there exists a vertex $v \in V(G)$ that is not a part of any D_4, then delete v from G and the reduced instance is $(G - v, k)$.

Next, we define the notion of an *auxiliary bipartite graph* that is used in several reduction rules subsequently.

Definition 2. (Auxiliary Bipartite Graph) *Given an ordered pair (Γ, Λ) with $\Gamma \subseteq \binom{V(G)}{2}$ and $\Lambda \subseteq V(G)$, the auxiliary bipartite graph $\hat{G}(\Gamma, \Lambda)$ with bipartition (A, B) is defined as $A = \Gamma$, $B = \Lambda$, and a vertex $\{a, b\} \in A$ is adjacent to $u \in B$ if and only if $ua, ub \in E(G)$.*

Now, we describe reduction rules to independently bound $|I^\star|$ and $|C^\star|$.

Bounding $|I^\star|$ when $|C^\star| \le k + 1$. We show that if $|C^\star|$ is $\mathcal{O}(k)$, then $|I^\star|$ is $\mathcal{O}(k^2)$ (and hence $|V(G)|$ is $\mathcal{O}(k^2)$) using Reduction Rule 2.

Reduction Rule 2. *Suppose $|C^\star| \le k + 1$. Let (A, B) be the bipartition of the auxiliary bipartite graph $\hat{G}(\binom{C \cup C^\star}{2}, I^\star)$ given by Definition 2. Let $X \subseteq A$ and $Y \subseteq B$ be the sets obtained from Lemma 1 such that, X has a 2-expansion into Y with $N(Y) \subseteq X$, and $|B \setminus Y| \le 2|A \setminus X|$. If there is an unsaturated vertex u in Y, then delete u from G and the reduced instance is $(G - u, k)$.*

Lemma 3. ⊛ *Suppose $|C^\star| \le k + 1$, and neither of the Reduction Rules 1 and 2 are applicable. Then $|I^\star| = \mathcal{O}(k^2)$ and hence $|V(G)| = \mathcal{O}(k^2)$.*

Bounding $|I^\star|$ when $|C^\star| \ge k + 2$. Here, for every vertex $v \in I^\star$, we have $|\overline{N}(v) \cap C^\star| \ge k + 1$ using Observation 1. Before presenting more reduction rules, we show the following result.

Lemma 4. ⊛ *Suppose $|C^\star| \ge k + 2$, and Reduction Rule 1 is not applicable. Then any solution Z of (G, k) satisfies $Z \cap \{v, a, b\} \ne \emptyset$, where $v \in I^\star$ and $\{a, b\}$ is a pair of distinct neighbours of v in $C \cup C^\star$.*

Reduction Rule 3. *Suppose $|C^\star| \ge k + 2$. For a vertex $v \in I^\star$, let $\mathcal{F}_v$ denote the family $\{\{v, a, b\} : a, b \in N(v) \cap (C \cup C^\star)$ and $a \ne b\}$. Let $\mathcal{F} = \bigcup_{v \in I^\star} \mathcal{F}_v$ and $U = \bigcup_{F \in \mathcal{F}} F$.*
Case 1. If Lemma 2 declares that $(U, \mathcal{F}, k)$ is a no-instance, then return a no-instance of constant size.
Case 2. If Lemma 2 returns an essential element u for $(U, \mathcal{F}, k)$, then delete u from G, and the reduced instance is $(G - u, k - 1)$.
Case 3. If Lemma 2 returns an essential family $\widehat{\mathcal{F}}$ for $(U, \mathcal{F}, k)$, then let $\mathcal{D}$ be the set $\{\{x, y\} \in \widehat{\mathcal{F}} : x, y \in C \cup C^\star\}$, and R be the set of vertices that appear in some pair in $\widehat{\mathcal{F}}$. If $\mathcal{D} \ne \emptyset$, let (A, B) be the bipartition of the auxiliary bipartite graph $\hat{G}(\mathcal{D}, I^\star \setminus R)$ given by Definition 2. Let $X \subseteq A$ and $Y \subseteq B$ be the sets obtained from Lemma 1 such that X has a 1-expansion into Y, $N(Y) \subseteq X$ and $|B \setminus Y| \le |A \setminus X|$. If there is an unsaturated vertex u in Y, then delete u from G and the reduced instance is $(G - u, k)$.

Lemma 5. ⊛ *Suppose $|C^\star| \geq k+2$ and neither of the Reduction Rules 1 and 3 is applicable. Then $|I^\star|$ is $\mathcal{O}(k^2)$.*

Bounding $|C^\star|$. Assume that $|C^\star| \geq k+2$, otherwise we have the required kernel. For every vertex $v \in I^\star$, we have $|\overline{N}(v) \cap C^\star| \geq k+1$ using Observation 1. Let $C_0^\star = \{v \in C^\star : N(v) \cap I^\star = \emptyset\}$ and $C_1^\star = \{v \in C^\star : N(v) \cap I^\star \neq \emptyset\}$. We define a marking scheme to identify vertices of C that may be present in any solution. We first bound the size of $C_1^\star$.

Marking Scheme 1.

For every vertex $v \in C$, we mark the vertices in $C^\star$ and $I^\star$ using the following procedure.

– Initialize $M_1 = \emptyset$, and $mark(v) = \emptyset$ for each $v \in C$.
– For each $v \in C$, if there exists $a \in C^\star \setminus M_1$ and $b \in I^\star \setminus M_1$ such that $ab, vb \in E(G)$, then set $M_1 = M_1 \cup \{a,b\}$, and $mark(v) = mark(v) \cup \{ab\}$.

Lemma 6. ⊛ *Let Z be a solution to (G,k) and $v \in C \setminus Z$. Then, $Z \cap \{a,b\} \neq \emptyset$, for every $ab \in mark(v)$.*

Reduction Rule 4. *If there exists a vertex $v \in C$ such that $|mark(v)| \geq k+1$, then delete v from C and the reduced instance is $(G - v, k - 1)$.*

Lemma 7. ⊛ *If Reduction Rules 1 and 4 are not applicable, then $|C_1^\star|$ is $\mathcal{O}(k^2)$.*

Now, it remains to bound the size of $C_0^\star$. We define another marking scheme.

Marking Scheme 2.

– Initialize $M_2 = \emptyset$.
– For each $v \in I$
  • Add $\min\{k + 2, |N(v) \cap C_0^\star|\}$ neighbours of v in $C_0^\star$ to M_2.
  • Add $\min\{k + 2, |\overline{N}(v) \cap C_0^\star|\}$ non-neighbours of v in $C_0^\star$ to M_2.

Let $W = C_0^\star \setminus M_2$. Clearly, $|M_2|$ is $\mathcal{O}(k^2)$.

Reduction Rule 5. *Let (A, B) be the bipartition of the auxiliary bipartite graph $\hat{G}(\binom{I}{2}, W)$ given by Definition 2. Let $X \subseteq A$ and $Y \subseteq B$ be the sets obtained from Lemma 1 such that X has a 2-expansion into Y, $N(Y) \subseteq X$, and $|B \setminus Y| \leq 2|A \setminus X|$. If there is an unsaturated vertex u in Y, then delete u from G and the reduced instance is $(G - u, k)$.*

Lemma 8. ⊛ *If neither of the Reduction Rules 1 and 5 are applicable, then $|C_0^\star|$ is $\mathcal{O}(k^2)$.*

Theorem 2. ⊛ *SBVD admits a kernel with $\mathcal{O}(k^2)$ vertices.*

3   An FPT Algorithm for SBVD

In this section, we describe an FPT algorithm for SBVD which runs in $\mathcal{O}^\star(2.3028^k)$ time. Before proceeding to the algorithm, we analyze certain special cases where SBVD can be solved in polynomial time.

Lemma 9. ⊛ *If G is a complete split graph, then the instance* (G, k) *of* SBVD *can be solved in polynomial time.*

Observation 2. ⊛ *If* $|C| < 40$, *then the instance* (G, k) *of* SBVD *can be solved in polynomial time, where* (C, I) *is a split partition of G.*

Let (G, k) be an instance of SBVD and (C, I) be a split partition of G. First, we apply the following reduction rules exhaustively.

Reduction Rule 6. – *If* $k \leq 0$ *and G is not a block graph, then declare that* (G, k) *is a no-instance.*
– *If* $k \geq 0$ *and G is a block graph, then declare that* (G, k) *is a yes-instance.*
– *If there is a vertex* $v \in V(G)$ *that does not participate in any* D_4, *then delete* v *from G and recurse on the instance* $(G - v, k)$.

Subsequently, we will assume that Reduction Rule 6 is not applicable on (G, k). Further, G is not a complete split graph and $|C| \geq 40$. Next, we apply the following branching rules.

Branching Rule 1. *Suppose there is a vertex* $v \in I$ *that has exactly two neighbours* a, b *in C. Let* $y_1, \cdots, y_{38}$ *be distinct non-neighbours of v in C. Branch into adding a or b or* $\{y_1, \cdots, y_{38}\}$ *to the solution. Recurse on the instances* $(G - a, k - 1)$, $(G - b, k - 1)$, *and* $(G - \{y_1, \cdots, y_{38}\}, k - 38)$.

Branching Rule 2. *Suppose there is a vertex* $v \in I$ *that has at least three neighbours* a, b, c *in C. Let u be a non-neighbour of v in C.*
Case 1: Suppose v has at least 20 non-neighbours in C, say $x_1, \cdots, x_{20}$. Branch into adding v or $\{a, b\}$ or $\{b, c\}$ or $\{a, c\}$ or $\{x_1, \cdots, x_{20}\}$ to the solution. Recurse on the instances $(G - v, k - 1)$, $(G - \{a, b\}, k - 2)$, $(G - \{b, c\}, k - 2)$, $(G - \{a, c\}, k - 2)$, and $(G - \{x_1, \cdots, x_{20}\}, k - 20)$.
Case 2: Suppose v has at least 20 neighbours in C, say $x_1, \cdots, x_{20}$. Let $X = \{x_1, \cdots, x_{20}\}$ and $\mathcal{A} = \binom{X}{19}$. Let $\mathcal{A} = \{A_1, \cdots, A_{20}\}$. Branch into adding v or u or $A_i \in \mathcal{A}$ for each $i \in [20]$ to the solution. Recurse on the instances $(G - v, k - 1)$, $(G - u, k - 1)$, $(G - A_1, k - 19)$, $\cdots$, and $(G - A_{20}, k - 19)$.

Theorem 3. ⊛ SBVD *can be solved in* $\mathcal{O}^\star(2.3028^k)$ *time.*

Theorem 3 along with the machinery of designing exact exponential-time algorithms using FPT algorithms given in [10] leads to the following result.

Theorem 4. MinSBVD *can be solved in* $\mathcal{O}^\star(1.5658^{n+o(n)})$ *time, where n is the number of vertices in the input graph.*

4   An FPT Algorithm for STVD

In this section, we give an FPT algorithm for STVD which runs in $\mathcal{O}^\star(2.7913^k)$ time.

4.1   Description of the Algorithm

Let (G, k) be a STVD instance and (C, I) be a split partition of G. For $v \in I$, let $P^I(v)$ denote the set $\{u \in I : \exists P_4(v, a, b, u) \text{ in } G\}$, $P^C(v)$ denote the set $\{u \in C : uv \notin E(G) \text{ and } \exists P_4(v, a, u, b) \text{ in } G\}$, and $L^C(v)$ denote the set $\{u \in C : uv \in E(G) \text{ and } \exists P_4(v, u, a, b) \text{ in } G\}$.

Reduction Rule 7. – *If $k \leq 0$ and G is not a threshold graph, then declare that (G, k) is a no-instance of STVD.*

– *If $k \geq 0$ and G is a threshold graph, then declare that (G, k) is a yes-instance of STVD.*

– *If there exists a vertex $v \in V(G)$ that does not participate in any P_4, then delete v from G and recurse on the instance $(G - v, k)$.*

Henceforth, we will assume that Reduction Rule 7 is not applicable on (G, k).

Branching Rule 3. *Suppose there exist two vertices u, v in I such that $a, b \in N(u) \setminus N(v)$, and $c, d \in N(v) \setminus N(u)$. Then branch into adding either u or v or $\{a, b\}$ or $\{c, d\}$ to the solution. Recurse on the instances $(G - u, k - 1)$, $(G - v, k - 1)$, $(G - \{a, b\}, k - 2)$, and $(G - \{c, d\}, k - 2)$.*

Branching Rule 4. *Suppose there exist three vertices u, v, w in I such that $c, d \in N(u) \setminus (N(v) \cup N(w))$, $a \in N(v) \setminus N(u)$, and $b \in N(w) \setminus N(u)$. Branch into adding either $\{u\}$, $\{c, d\}$, $\{v, w\}$, $\{a, b\}$, $\{a, w\}$, or $\{b, v\}$ to the solution. Recurse on the instances $(G - u, k - 1)$, $(G - \{c, d\}, k - 2)$, $(G - \{v, w\}, k - 2)$, $(G - \{a, b\}, k - 2)$, $(G - \{a, w\}, k - 2)$, and $(G - \{b, v\}, k - 2)$.*

Branching Rule 5. *Suppose there exist three vertices u, v, w in I such that $c, d \in N(u) \setminus (N(v) \cup N(w))$, $a \in (N(v) \cap N(w)) \setminus N(u)$. Branch into adding either $\{u\}$, $\{a\}$, $\{v, w\}$, or $\{c, d\}$ to the solution. Recurse on the instances $(G - u, k - 1)$, $(G - a, k - 1)$, $(G - \{v, w\}, k - 2)$, and $(G - \{c, d\}, k - 2)$.*

Branching Rule 6. *Suppose there exist three vertices u, v, w in I such that $a \in (N(v) \cap N(w)) \setminus N(u)$, $c \in N(u) \setminus N(v)$, $d \in N(u) \setminus (N(v) \cup N(w))$, and $e \in N(u) \setminus N(w)$. Branch into adding $\{u\}$, $\{a\}$, $\{v, w\}$, $\{v, d, e\}$, $\{c, d, e\}$, or $\{c, d, w\}$ to the solution. Recurse on the instances $(G - u, k - 1)$, $(G - a, k - 1)$, $(G - \{v, w\}, k - 2)$, $(G - \{v, d, e\}, k - 3)$, $(G - \{c, d, e\}, k - 3)$, and $(G - \{c, d, w\}, k - 3)$.*

Branching Rule 7. *Suppose $u, v \in I$ such that $v \in P^I(u)$, and $|P^C(u)| = 1$. Let $a \in P^C(u)$, and $c, d \in N(u) \setminus N(v)$. Branch into adding a, v, or $\{c, d\}$ to the solution. Recurse on the instances $(G - a, k - 1)$, $(G - v, k - 1)$, and $(G - \{c, d\}, k - 2)$.*

Branching Rule 8. *Suppose there exist three vertices u, v, w in I such that $a \in N(v) \setminus (N(u) \cup N(w))$, $b \in N(w) \setminus N(v)$, $c \in N(u) \setminus N(v)$, and $d \in (N(u) \cap N(w)) \setminus N(v)$. Branch into adding v, a, $\{u, w\}$, $\{u, b, d\}$, $\{c, d, w\}$, or $\{c, d, b\}$ to the solution. Recurse on the instances $(G - v, k - 1)$, $(G - a, k - 1)$, $(G - \{u, w\}, k - 2)$, $(G - \{u, b, d\}, k - 3)$, $(G - \{c, d, w\}, k - 3)$, and $(G - \{c, d, b\}, k - 3)$.*

Branching Rule 9. *Suppose $u \in I$ that forms a P_4 with a vertex $v \in I$.*
Case 1: *Suppose $|L^C(u)| = |L^C(v)| = 1$. Let $a \in L^C(u)$, and $b \in L^C(v)$. Branch into adding either $\{a\}$, or $\{b\}$ to the solution. Recurse on the instances $(G - \{a\}, k - 1)$, or $(G - \{b\}, k - 1)$.*
Case 2: *Suppose $|P^C(u)| = |P^C(v)| = 1$. Let $a \in P^C(u)$, and $b \in P^C(v)$. Branch into adding either $\{a\}$, or $\{b\}$ to the solution. Recurse on the instances $(G - \{a\}, k - 1)$ and $(G - \{b\}, k - 1)$.*

4.2   Correctness and Running Time

Suppose $u, v \in V(G)$ such that $N(u) = N(v)$. Then observe that, $P^I(u) = P^I(v)$, $P^C(u) = P^C(v)$, and $L^C(u) = L^C(v)$.

Observation 3. ✽ *Suppose (G, k) is an instance of STVD. If $u, v \in I$ such that $deg(v) < deg(u)$, and $|N(v) \backslash N(u)| \geq 1$, then $|N(u) \backslash N(v)| \geq 2$.*

Observation 4. ✽ *Suppose (G, k) is an instance of STVD, where Branching Rule 3 is not applicable. If $u, v \in I$ and $deg(v) \leq deg(u)$, then $|N(v) \setminus N(u)| \leq 1$.*

Lemma 10. ✽ *Suppose (G, k) is an instance of STVD, where Reduction Rule 7 and Branching Rule 3 are not applicable. Suppose for any $u \in I$, for each $v \in P^I(u)$, we have $deg(v) = deg(u)$. For $u \in I$, let $X_u = \{v \in I : deg(v) = deg(u)\}$. Then for all $x \in X_u$ either $|P^C(x)| = 1$, or $|L^C(x)| = 1$.*

Theorem 5. *There exists an algorithm for STVD which runs in $\mathcal{O}^\star(2.7913^k)$ time.*

Proof. Given an instance (G, k) of STVD, let (C, I) be a split partition of G. We first apply Reduction Rule 7 exhaustively. Suppose there exists two vertices u and v in I such that there exist at least two vertices in $N(u) \cap \overline{N}(v)$, and at least two vertices in $N(v) \cap \overline{N}(u)$, then we apply Branching Rule 3. Next we consider the following cases.

(A) Suppose u is a highest degree vertex in I such that $P^I(u)$ has at least one vertex of degree lesser than that of u.

   (1) Suppose there exists at least two vertices v, w in $P^I(u)$ such that $deg(v) < deg(u)$, and $deg(w) < deg(u)$. Clearly, $|N(v) \setminus N(u)| \geq 1$, otherwise v and u cannot form a P_4 together. Similarly, $|N(w) \setminus N(u)| \geq 1$. Using Observation 4, we have $|N(v) \backslash N(u)| \leq 1$ and $|N(w) \backslash N(u)| \leq 1$. Hence, there exists a unique vertex, say a in $N(v) \setminus N(u)$, and a unique vertex, say b in $N(w) \backslash N(u)$. By Observation 3, there exist at least two vertices in $N(u) \backslash N(v)$, say c and d. Without loss of generality, let $deg(w) \leq deg(v)$.
   Case (i) Suppose $a \neq b$. Since $N(v) \backslash \{a\} \subseteq N(u)$, it follows that $b \notin N(v)$. Moreover, since $deg(w) \leq deg(v)$, and $b \notin N(v)$, by Observation 4, $N(w) \backslash \{b\} \subseteq N(v)$. Note that since b is a non-neighbour of u, $b \notin \{c, d\}$. Hence, $c, d \notin N(w)$. Therefore, Branching Rule 4 is applicable.

Case (ii) Suppose $a = b$. By Observation 4, we have $|N(w) \setminus N(v)| \leq 1$. Suppose $|N(w) \setminus N(v)| = 0$. Since $N(w) \subseteq N(v)$, $c, d \notin N(w)$. Hence Branching Rule 5 is applicable. Now, suppose $|N(w) \setminus N(v)| = 1$. Let $x \in N(w) \setminus N(v)$. Since $N(w) \setminus \{a\} \subseteq N(u)$, $x \in N(u)$. Since $|N(w) \setminus N(v)| = 1$, by Observation 4, $N(w) \setminus \{x\} \subseteq N(v)$. If $x \notin \{c, d\}$, then $c, d \notin N(w)$. In this case Branching Rule 5 is applicable. Otherwise, without loss of generality let $x = c$. Then $d \notin N(w)$. Since $deg(w) < deg(u)$, by Observation 3, there exists a vertex, say $e \in N(u) \setminus N(w)$, where $e \neq d$. In this case Branching Rule 6 is applicable.

(2) Suppose that there exists only one vertex v in $P^I(u)$ such that $deg(v) < deg(u)$. Note that $|N(v) \setminus N(u)| \geq 1$ as v forms a P_4 with u. Due to Observation 3, there exist at least two vertices in the neighbourhood of u that are non-neighbours of v. Let $c, d \in N(u) \setminus N(v)$. Suppose $|P^C(u)| = 1$, then Branching Rule 7 is applicable. Now, suppose $|P^C(u)| > 1$. Since u and v form a P_4, v has at least one neighbour which is a non-neighbour of u. However, by Observation 4, since $deg(v) < deg(u)$, it follows that v has exactly one neighbour a which is a non-neighbour of u. Let $b \in P^C(u) \setminus \{a\}$. Note that $vb \notin E(G)$. Therefore, there exists a vertex, say w in $P^I(u)$ such that $wb \in E(G)$. Since $P^I(u)$ has only one vertex of degree lesser than $deg(u)$, it follows that $deg(w) \geq deg(u)$. If $deg(w) > deg(u)$, then it contradicts the fact that u is a highest degree vertex which forms a P_4 with a vertex of lesser degree. Hence, $deg(w) = deg(u)$. Since u forms a P_4 with w, using Observation 4, we have $|N(u) \setminus N(w)| = 1$. Therefore, either c or d is in the neighbourhood of w. Without loss of generality, let $d \in N(w)$. By Observation 4, there exists at most one neighbour of w which is a non-neighbour of u. Therefore, $N(w) \setminus \{b\} \subseteq N(u)$. Thus $a \notin N(w)$. Since there exists vertices $a \in N(v) \setminus (N(u) \cup N(w))$, $b \in N(w) \setminus N(v)$, $c \in N(u) \setminus N(v)$, and $d \in (N(u) \cap N(w)) \setminus N(v)$, Branching Rule 8 is applicable.

(B) Suppose there exists a vertex $u \in I$, such that for all $v \in P^I(u)$, we have $deg(v) = deg(u)$. By Lemma 10, if $v \in P^I(u)$, then either $|P^C(u)| = |P^C(v)| = 1$, or $|L^C(u)| = |L^C(v)| = 1$. Hence, Branching Rule 9 is applicable.

The correctness of the algorithm follows from the correctness of reduction rules and branching rules. No reduction rule increases k, and k always reduces in each branch of every branching rule. Moreover, all rules can be applied in polynomial time. As the depth of recursion tree is at most k, the running time of the algorithm is $\mathcal{O}^\star(2.7913^k)$ which follows from solving the recurrence corresponding to the most expensive branching vector (Branching Rule 4), i.e, $(1, 2, 2, 2, 2, 2)$.   □

Using Theorem 5 along with the machinery of designing exact exponential-time algorithms using FPT algorithms given in [10], we have the following result.

Theorem 6. MinSTVD *can be solved in* $\mathcal{O}^\star(1.6418^{n+o(n)})$ *time where n is the number of vertices in the input graph.*

5   Conclusion

In this paper, we study the parameterized complexity of SBVD and STVD with respect to the solution size as parameter. We also give factor 2-approximation algorithms for MinSBVD and MinSTVD(both proven APX hard in full version). Further, if there is an α-approximation algorithm for MinSBVD with $\alpha < 1.36$, then P = NP. Also, assuming the Unique Games Conjecture, if there is an α-approximation algorithm for MinSBVD with $\alpha < 2$, then P = NP.

References

1. Agrawal, A., Kolay, S., Lokshtanov, D., Saurabh, S.: A faster FPT algorithm and a smaller kernel for BLOCK GRAPH VERTEX DELETION. In: Kranakis, E., Navarro, G., Chávez, E. (eds.) LATIN 2016. LNCS, vol. 9644, pp. 1–13. Springer, Heidelberg (2016). https://doi.org/10.1007/978-3-662-49529-2_1
2. Bandelt, H., Mulder, H.M.: Distance-hereditary graphs. J. Comb. Theory Ser. B **41**(2), 182–208 (1986)
3. Bonomo, F., Durán, G., Valencia-Pabon, M.: Complexity of the cluster deletion problem on subclasses of chordal graphs. Theor. Comput. Sci **600**, 59–69 (2015)
4. Cao, Y., Ke, Y., Otachi, Y., You, J.: Vertex deletion problems on chordal graphs. In: Proceedings of FSTTCS, pp. 22:1–22:14 (2017)
5. Chandran, L.S., Rajendraprasad, D., Tesar, M.: Rainbow colouring of split graphs. Discrete Appl. Math. **216**, 98–113 (2017)
6. Cygan, M., Pilipczuk, M.: Split vertex deletion meets vertex cover: new fixed-parameter and exact exponential-time algorithms. Inf. Process. Lett. **113**(5), 179–182 (2013)
7. Diestel, R.: Graph Theory. Springer, Berlin (2005)
8. Drange, P.G., Dregi, M.S., Lokshtanov, D., Sullivan, B.D.: On the threshold of intractability. In: Bansal, N., Finocchi, I. (eds.) ESA 2015. LNCS, vol. 9294, pp. 411–423. Springer, Heidelberg (2015). https://doi.org/10.1007/978-3-662-48350-3_35
9. Fomin, F.V., Gaspers, S., Kratsch, D., Liedloff, M., Saurabh, S.: Iterative compression and exact algorithms. Theor. Comput. Sci. **411**(7), 1045–1053 (2010)
10. Fomin, F.V., Gaspers, S., Lokshtanov, D., Saurabh, S.: Exact algorithms via monotone local search. In: Proceedings of STOC, pp. 764–775. ACM (2016)
11. Ghosh, E., et al.: Faster parameterized algorithms for deletion to split graphs. Algorithmica **71**(4), 989–1006 (2015)
12. Golumbic, M.C.: Algorithmic graph theory and perfect graphs, 2nd edn. Elsevier Science B.V., Amsterdam (2004)
13. Heggernes, P., Lokshtanov, D., Mihai, R., Papadopoulos, C.: Cutwidth of split graphs and threshold graphs. SIAM J. Discrete Math. **25**(3), 1418–1437 (2011)
14. Hochbaum, D.S.: Approximation Algorithms for NP-hard Problems. PWS Publishing Co., Boston (1997)
15. Illuri, M., Renjith, P., Sadagopan, N.: Complexity of steiner tree in split graphs - dichotomy results. In: Govindarajan, S., Maheshwari, A. (eds.) CALDAM 2016. LNCS, vol. 9602, pp. 308–325. Springer, Cham (2016). https://doi.org/10.1007/978-3-319-29221-2_27
16. Kim, E.J., Kwon, O.: A polynomial kernel for block graph deletion. Algorithmica **79**(1), 251–270 (2017)

17. Le, T., Lokshtanov, D., Saurabh, S., Thomassé, S., Zehavi, M.: Subquadratic kernels for implicit 3-hitting set and 3-set packing problems. In: Proceedings of SODA, pp. 331–342 (2018)
18. Lewis, J.M., Yannakakis, M.: The node-deletion problem for hereditary properties is NP-complete. J. Comput. Syst. Sci. **20**(2), 219–230 (1980)
19. Moser, H.: Finding optimal solutions for covering and matching problems. Ph.D. thesis, Institut für Informatik, Friedrich-Schiller Universitüt Jena (2009)
20. Saurabh, S.: Personal Communication (2018)

Fair Hitting Sequence Problem: Scheduling Activities with Varied Frequency Requirements

Serafino Cicerone[1], Gabriele Di Stefano[1], Leszek Gasieniec[2],
Tomasz Jurdzinski[3], Alfredo Navarra[4], Tomasz Radzik[5(✉)],
and Grzegorz Stachowiak[3]

[1] University of L'Aquila, L'Aquila, Italy
{serafino.cicerone,gabriele.distefano}@univaq.it
[2] University of Liverpool, Liverpool, UK
l.a.gasieniec@liverpool.ac.uk
[3] University of Wroclaw, Wrocław, Poland
{tju,gst}@cs.uni.wroc.pl
[4] University of Perugia, Perugia, Italy
alfredo.navarra@unipg.it
[5] King's College London, London, UK
tomasz.radzik@kcl.ac.uk

Abstract. Given a set $V = \{v_1, \ldots, v_n\}$ of n elements and a family $\mathcal{S} = \{S_1, S_2, \ldots, S_m\}$ of (possibly intersecting) subsets of V, we consider a scheduling problem of perpetual monitoring (attending) these subsets. In each time step one element of V is visited, and all sets in $\mathcal{S}$ containing v are considered to be attended during this step. That is, we assume that it is enough to visit an arbitrary element in S_j to attend to this whole set. Each set S_j has an urgency factor h_j, which indicates how frequently this set should be attended relatively to other sets. Let $t_i^{(j)}$ denote the time slot when set S_j is attended for the i-th time. The objective is to find a perpetual schedule of visiting the elements of V, so that the maximum value $h_j \left(t_{i+1}^{(j)} - t_i^{(j)} \right)$ is minimized. The value $h_j \left(t_{i+1}^{(j)} - t_i^{(j)} \right)$ indicates how urgent it was to attend to set S_j at the time slot $t_{i+1}^{(j)}$. We call this problem the *Fair Hitting Sequence* (FHS) problem, as it is related to the minimum hitting set problem. In fact, the uniform FHS, when all urgency factors are equal, is equivalent to the minimum hitting set problem, implying that there is a constant $c_0 > 0$ such that it is NP-hard to compute $(c_0 \log m)$-approximation schedules for FHS.

We demonstrate that scheduling based on one hitting set can give poor approximation ratios, even if an optimal hitting set is used. To counter

The work has been supported in part by the European project "Geospatial based Environment for Optimisation Systems Addressing Fire Emergencies" (GEO-SAFE), contract no. H2020-691161, by the Italian National Group for Scientific Computation GNCS-INdAM, by Networks Sciences and Technologies (NeST) initiative at University of Liverpool, and by the Polish National Science Center (NCN) grant 2017/25/B/ST6/02010.

© Springer Nature Switzerland AG 2019
P. Heggernes (Ed.): CIAC 2019, LNCS 11485, pp. 174–186, 2019.
https://doi.org/10.1007/978-3-030-17402-6_15

this, we design a deterministic algorithm which partitions the family $\mathcal{S}$ into sub-families and combines hitting sets of those sub-families, giving $O(\log^2 m)$-approximate schedules. Finally, we show an LP-based lower bound on the optimal objective value of FHS and use this bound to derive a randomized algorithm which with high probability computes $O(\log m)$-approximate schedules.

Keywords: Scheduling · Periodic maintenance · Hitting set · Approximation algorithms

1 Introduction

The combinatorial problem studied in this paper is a natural extension of the following perpetual scheduling proposed in [11]. Nodes $v_1, v_2, \ldots, v_n$ of a network need to be indefinitely monitored (visited) by a mobile agent according to their *urgency factors* $h_1, h_2, \ldots, h_n$, which indicate how often each node should be visited relatively to other nodes in the network. The (current) *urgency indicator* of node v_i is defined as $t \cdot h_i$, where t is the time which has elapsed since the last visit to this node. The objective of scheduling visits to nodes is to minimise the maximum value ever observed on the urgency indicators. Two variants of this problem were considered in [11]. In the *discrete variant* the time needed to visit each node is assumed to be uniform, corresponding to a single round of the monitoring process. The *continuous variant* assumes that the nodes are distributed in a geometric space and the time required to move to, and attend, the next node depends on the current location of the mobile agent.

In this paper we consider a generalization of the discrete variant of the perpetual scheduling, where the emphasis is on monitoring a given family of sets $\mathcal{S} = \{S_1, S_2, \ldots, S_m\}$, which are (possibly intersecting) subsets of the set of n network nodes. We assume that it is enough to visit an arbitrary node in a set S_j to attend to this whole set. Moreover, by visiting a node we assume that *all* sets in $\mathcal{S}$ containing this node are attended. As in the perpetual scheduling problem studied in [11], we schedule visits to nodes, but these visits are now only means of attending to sets S_j and the urgency factors $h_1, h_2, \ldots, h_m$ are associated with these sets, not with the nodes.

This generalization of the perpetual scheduling problem is motivated by dissemination (or collection) of information across different, possibly overlapping, communities in social (media) networks. Participants of such a network can provide access to all communities to which they belong. While a lot of work has been done on recognition/detection of communities, starting with the seminal studies presented in [12,18], much less is known about efficient ways of informing or monitoring such communities, especially when they are highly overlapping and dynamic and have their own frequency requirements. One way of modeling such problems is to decide whom and when to contact to ensure regular, but proportionate to the requirements, access to all communities.

Other scenarios motivating our scheduling problem arise in the context of overlapping sensor or data networks. Consider overlapping networks

$S_1, S_2, \ldots, S_m$ and *access nodes* $v_1, v_2, \ldots, v_n$. Each node v_i is an access node of one or more networks $S_{i_1}, S_{i_2}, \ldots, S_{i_k}$, $k \geq 1$. In the context of our abstract scheduling problem, these overlapping networks correspond to overlapping communities of the previous scenario. Each network S_j has a specified *access rate* $h_j > 0$, which indicates how often this network should be accessed relative to other networks. If an access node v_i is used at the current time slot, then all networks $S_{i_1}, S_{i_2}, \ldots, S_{i_k}$ containing v_i are accessed during this time slot. Accessing a network can be thought of, for example, as gathering data from that network, or providing some other service, maintenance or update for that network. We want to find an infinite schedule $\mathcal{A} = (v_{q_1}, \ldots, v_{q_t}, \ldots)$, where v_{q_t} is the access node used in the time slot $t \geq 1$, so that each network is accessed as often as possible and in a fair way according to the specified access rates.

Fair Hitting Sequence Problem. We formalize the objective of the regular and fair access to networks $S_1, S_2, \ldots, S_m$ in the following way. When progressing through a schedule $\mathcal{A}$, if a network S_j was accessed for the last time at a time slot t', then the value $h_j (t - t')$ indicates the urgency of accessing this network at the current time slot $t > t'$. We refer to this value as the *urgency indicator* of network S_j, or simply as the (current) urgency or the *height* of S_j. The urgency indicator of S_j grows with the rate h_j over the time when S_j is not accessed and is reset to 0 when S_j is accessed. Hence we will refer to numbers h_j also as *growth rates* (of urgency indicators). We want to find a schedule which minimizes the maximum $h_j \left(t_{i+1}^{(j)} - t_i^{(j)} \right)$, over all networks S_j, $j = 1, 2, \ldots, m$ and all $i \geq 0$, where $t_i^{(j)}$ is the i-th time slot when network S_j is accessed (setting $t_0^{(j)} \equiv 0$). That is, $h_j \left(t_{i+1}^{(j)} - t_i^{(j)} \right)$ is the height of S_j at the time when this network is (about to be) accessed for the $(i + 1)$-st time.

For a given schedule $\mathcal{A} = (v_{q_1}, v_{q_2}, \ldots)$ and $1 \leq j \leq m$, the number

$$Height(\mathcal{A}, j) = \sup \left\{ h_j \left(t_{i+1}^{(j)} - t_i^{(j)} \right) : i \geq 0 \right\} \tag{1}$$

is the maximum value, or the maximum height, of the urgency indicator of S_j, when schedule $\mathcal{A}$ is followed, and the number

$$Height(\mathcal{A}) = \max \left\{ Height(\mathcal{A}, j) : 1 \leq j \leq m \right\} \tag{2}$$

is the maximum height of any urgency indicator and is called the *height of schedule* $\mathcal{A}$. We want to find an *optimal schedule* $\mathcal{A}_{opt}$ which minimizes (2). We refer to this problem as the *Fair Hitting Sequence* (FHS) problem, and show below that it includes the *hitting set* problem as a special case. We say that a schedule $\mathcal{A}$ is ρ-approximate, if $Height(\mathcal{A}) \leq \rho \cdot Height(\mathcal{A}_{opt})$.

We denote by $V = \{v_1, v_2, \ldots, v_n\}$ the set of all access nodes (or all participants in a social network), which from now on will be simply referred to as nodes, and we identify each network (or a community) S_j with the set $\{v_{j_1}, v_{j_2}, \ldots, v_{j_q}\} \subseteq V$ of all (access) nodes of this network (or all members of this community). The simplest, and trivial, instance of the FHS problem is when $m = n$, $S_j = \{v_j\}$ and $h_j = 1$, for all $1 \leq j \leq n$. For this instance a schedule is optimal if, and only if, it is a repetition of the same permutation of V. The height of such a schedule is equal to n.

A still special case, but more interesting and non-trivial, is when sets S_j are arbitrary, with possibly $m \neq n$, but all h_j remain equal to 1. It is not difficult to see that for such instances of FHS a schedule is optimal if, and only if, it is a repetition of the same permutation of the same minimum-size *hitting set* $W \subseteq V$. That is, $|S_j \cap W| \geq 1$, for each $1 \leq j \leq m$, and W has the minimum size among all subsets of V with this property. The height of such optimal schedule is equal to $|W|$. NP-hardness of the *minimum hitting set* problem, which is equivalent to the *minimum cover set* problem, implies NP-hardness of the more general FHS problem. The natural greedy algorithm for the minimum hitting-set problem, which selects in each iteration a node *hitting* (belonging to) the maximum number of the remaining sets S_j, gives an $O(\log m)$-approximate hitting set. On the other hand, it is known that there is a constant $c_0 > 0$ such that finding a $(c_0 \log m)$-approximate hitting set is NP-hard [21]. This implies NP-hardness of $(c_0 \log m)$-approximation for the more general FHS problem.

Continuing with the case of uniform growth rates, if all sets S_j have size 2, then such an instance is represented by the graph $G = (V, E)$, where $E = \{S_1, S_2, \ldots, S_m\}$. In this case the FHS problem becomes a problem of efficient monitoring of the edges of graph G (by visiting veritces of G), which is equivalent to the *vertex cover* problem.

Another non-trivial special case of the FHS problem is when $S_j = \{v_j\}$, for each $1 \leq j \leq n$, but the access rates h_j are non-uniform. This is the perpetual scheduling problem considered in [4,8,11]. If we further assume that all input parameters h_j are inverses of positive integer numbers, then the question whether there exists a schedule of height not greater than 1 is known as the *Pinwheel scheduling* problem [14].

We are interested in deriving good approximation algorithms for the FHS problem. While schedules are defined as infinite sequences, it can be shown that there is always an optimal schedule which has a periodic form $\mathcal{B}_{init}(\mathcal{B}_{period})^*$, where $\mathcal{B}_{init}$ and $\mathcal{B}_{period}$ are finite schedules (see e.g. [1]). The period of a periodic optimal schedule can have exponential length, but our approximate algorithms compute in polynomial time schedules with periods polynomial in m.

Our Results. If we denote by $\mathcal{A}(W)$ the schedule obtained by repeating the same hitting set W, then the height of $\mathcal{A}(W)$ is at most $h_{\max}|W|$, where $h_{\max} = \max_{1 \leq j \leq m}\{h_j\}$. Actually it is possible to show instances for which the $\mathcal{A}(W)$ schedule is only $\Theta(m/\log m)$ approximate. To get better schedules, we have to handle the variations in the growth rates h_j. In Sect. 2, we present simple $O(\log^2 m)$-approximate schedules. Such schedules are obtained by partitioning the whole family of sets S_j into $O(\log m)$ sub-families of sets which have similar growth rates, and by combining $O(\log m)$-approximate hitting sets of these sub-families. To improve further the approximation ratio of computed schedules, we first derive in Sect. 3 a lower bound on the height of any schedule. This lower bound can be viewed as the optimal solution to a fractional version of the FHS problem. Then we show in Sect. 4 a randomized algorithm which uses the optimal fractional solution to compute schedules which are $O(\log m)$-approximate with probability at least $1 - 1/m$.

Previous Related Results. Several constant approximation algorithms for the discrete variant and $O(\log n)$ approximation for the continuous variant of this perpetual scheduling problem are discussed in [11] and further work on this problem is presented in [4,8]. In [4], the authors consider monitoring by two agents of n nodes located on a line and requiring different frequencies of visits. The authors provide several approximation algorithms concluding with the best currently known $\sqrt{3}$-approximation.

The perpetual scheduling problem considered in [4,8,11] is closely related to periodic scheduling [22], general Pinwheel scheduling [2,3], periodic Pinwheel scheduling [14,15], and to other problems motivated by Pinwheel scheduling [20]. This problem is also related to several classical algorithmic problems which focus on monitoring and mobility. These include the Art Gallery Problem [19] and its dynamic alternative called the k-Watchmen Problem [17,23]. In further work on fence patrolling [5,6] the authors focus on monitoring vital (possibly disconnected) parts of a linear environment where each point is expected to be visited with the same frequency. The authors of [7] study monitoring linear environments by agents prone to faults.

2   Deterministic $O(\log^2 m)$-Approximate Schedules

In this section, we show a deterministic approximation algorithm for the FHS problem. The algorithm exploits the properties of schedules which are based on hitting sets.

2.1   Algorithm Based on Hitting Sets

We first formalize an observation that if there is not much variation among the growth rates of the sets, then the minimum hitting set gives a good approximate solution. Consider an input instance with $h_{\max} \leq C \cdot h_{\min}$, where $h_{\min} = \min_{1 \leq j \leq m}\{h_j\}$ and $C \geq 1$ is a parameter. Let W_{opt} be a minimum hitting set and compare the heights of the schedule $\mathcal{A}(W_{\mathrm{opt}})$ and an optimal schedule $\mathcal{A}_{\mathrm{opt}}$. We note that an optimal schedule exists since the schedule $\mathcal{A}(V)$ (the round-robin schedule $(v_1, v_2, \ldots, v_n)^*$) has height $nh_{\max}$ and all (infinitely many) schedules with heights at most $nh_{\max}$ have heights in the finite set $\{ih_j : j = 1, 2, \ldots, m, \ i - \text{ positive integer}, \ ih_j \leq h_{\max} \cdot n\}$.

Let $[1, t]$ be the shortest initial time interval in schedule $\mathcal{A}_{\mathrm{opt}}$ when each set is accessed at least once. We have $t \geq |W_{\mathrm{opt}}|$, since the set of nodes used in the first t time slots in schedule $\mathcal{A}_{\mathrm{opt}}$ is a hitting set. Let S_j be any set accessed for the first time in schedule $\mathcal{A}_{\mathrm{opt}}$ at time t. We have

$$Height(\mathcal{A}(W_{\mathrm{opt}})) \leq h_{\max}|W_{\mathrm{opt}}| \leq C\,h_j\,|W_{\mathrm{opt}}| \leq C\,h_j\,t \leq C \cdot Height(\mathcal{A}_{\mathrm{opt}}),$$

where the last inequality follows from the fact that in schedule $\mathcal{A}_{\mathrm{opt}}$, the height of set S_j (that is, the height of its urgency indicator) at time t is equal to $h_j t$. Thus $\mathcal{A}(W_{\mathrm{opt}})$ is a C-approximate schedule. If W_{apx} is a D-approximate hitting set ($|W_{\mathrm{apx}}| \leq D \cdot |W_{\mathrm{opt}}|$), then a similar argument shows that $\mathcal{A}(W_{\mathrm{apx}})$ is a (CD)-approximate schedule. This and the $O(\log m)$ approximation of the greedy algorithm for the hitting set problem give the following lemma.

Lemma 1. *If W_{apx} is a D-approximate hitting set, then the schedule $\mathcal{A}(W_{apx})$ is $(Dh_{\max}/h_{\min})$-approximate. There is a polynomial-time algorithm which computes $O((\log m)h_{\max}/h_{\min})$-approximate schedules for the FHS problem.*

If there is considerable variation in the growth rates h_j, then the schedule $\mathcal{A}(W_{\mathrm{opt}})$, which relies on one common minimum hitting set, can be far from optimal. To get a better approximation, we consider separately sets with similar growth rates. More precisely, we partition the whole family of sets $\mathcal{S} = \{S_1, S_2, \ldots, S_m\}$ into the following $k_{\max} = \lfloor \log m \rfloor + 1$ families.

$$\mathcal{F}_k = \{S_j : h_{\max}/2^k < h_j \leq h_{\max}/2^{k-1}\}, \quad \text{for } k = 1, 2, \ldots, k_{\max} - 1,$$
$$\mathcal{F}_{k_{\max}} = \{S_j : h_j \leq h_{\max}/2^{k_{\max}-1}\}.$$

Let W_k be a D-approximate hitting set for the family $\mathcal{F}_k$, $1 \leq k \leq k_{\max} - 1$, and let $W_{k_{\max}}$ be any hitting set for the family $\mathcal{F}_{k_{\max}}$ such that $|W_{k_{\max}}| \leq |\mathcal{F}_{k_{\max}}|$. For $1 \leq k \leq k_{\max} - 1$, the schedule $\mathcal{A}(W_k)$, which repeats the same permutation of W_k, is a $(2D)$-approximate schedule for the family $\mathcal{F}_k$ (from Lemma 1). The schedule $\mathcal{A}(W_{k_{\max}})$, which repeats the same permutation of $W_{k_{\max}}$, is a schedule for the family $\mathcal{F}_{k_{\max}}$ with height at most $|W_{k_{\max}}| \left(h_{\max}/2^{k_{\max}-1} \right) \leq m \left(h_{\max}/2^{k_{\max}-1} \right) \leq 2h_{\max} \leq 2 \cdot H(\mathcal{A}_{\mathrm{opt}})$. Therefore the schedule $\mathcal{A}$ which interleaves the $k_{\max}$ schedules $\mathcal{A}(W_1), \mathcal{A}(W_2), \ldots, \mathcal{A}(W_{k_{\max}-1}), \mathcal{A}(W_{k_{\max}})$ is a $(2Dk_{\max})$-approximate schedule for the whole family $\mathcal{S}$. This is because for each $1 \leq k \leq k_{\max}$, the lengths of the periods in the schedule $\mathcal{A}(W_k)$ between the consecutive accesses to the same set $S_j \in \mathcal{F}_k$ increase $k_{\max}$ times in the schedule $\mathcal{A}$. (Set S_j may have some additional accesses in $\mathcal{A}$, which come from other schedules $\mathcal{A}(W_{k'})$, $k' \neq k$).

Theorem 1. *The schedule $\mathcal{A}$ constructed above using D-approximate hitting sets is $O(D \log m)$ approximate.*

Corollary 1. *There is a polynomial-time algorithm which computes $O(\log^2 m)$-approximate schedules for the FHS problem.*

2.2 A Tight Example for Using $\log m$ Hitting Sets

We showed in Sect. 2.1 that the schedule $\mathcal{A}$ which is based on $\log m$ hitting sets computed separately for the groups of sets with similar growth rates is $O(D \log m)$-approximate, where D is an upper bound on the approximation ratio of the used hitting sets (Theorem 1). We provide now an instance of FHS such that even if optimal hitting sets are used, the schedule $\mathcal{A}$ is only $\Theta(\log m)$-approximate.

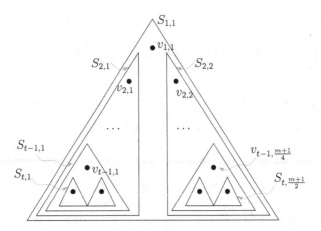

Fig. 1. An instance of the FHS problem for algorithm $\mathcal{A}$ defined in Theorem 1.

Consider the following instance for the FHS problem, illustrated in Fig. 1. Given an integer $t > 0$, let $m = 2^t - 1$ be the number of sets. The sets are defined as follows:

- $S_{t,i} = \{v_{t,i}\}$, for each $i = 1, 2, \ldots, \frac{m+1}{2} = 2^{t-1}$;
- $S_{\ell,i} = S_{\ell+1,2i-1} \cup S_{\ell+1,2i} \cup \{v_{\ell,i}\}$, for each $\ell = t-1, t-2, \ldots, 1$ and for each $i = 1, 2, \ldots, \frac{m+1}{2^{t-\ell+1}} = 2^{\ell-1}$.

For the growth rates, we take $h(S_{\ell,i}) = \frac{1}{2^t}$ for each $\ell = t, t-1, \ldots, 1$ and $i = 1, 2, \ldots, \frac{m+1}{2^{t-\ell+1}} = 2^{\ell-1}$.

On this instance, we now compare the performance of an optimum schedule with the schedule $\mathcal{A}$ defined in Theorem 1.

Any schedule must cover separately the sets $S_{t,i}$, $i = 1, \ldots, \frac{m+1}{2} = 2^{t-1}$, since each of these sets is a singleton containing a different element $v_{t,i}$. Thus the height of any schedule is at least $\frac{1}{2}$. A schedule given by an interleaved round-robin on elements $v_{t,i}$, $i = 1, \ldots, \frac{m+1}{2}$ achieves this lower bound. In fact, these elements form a hitting set for the whole family $\mathcal{S}$. By interleaved schedule we mean that the elements are picked according to a permutation which ensures that each set $S_{\ell,i}$ at level ℓ is served every $2^{\ell-1}$ time slots, so it grows to the maximum height of $\frac{1}{2}$. For instance, for $t = 4$, the schedule can be $v_{4,1}$, $v_{4,5}$, $v_{4,3}$, $v_{4,7}$, $v_{4,2}$, $v_{4,6}$, $v_{4,4}$, $v_{4,8}$.

On the other hand, the schedule $\mathcal{A}$ constructed as in Sect. 2.1 has height $\frac{\log(m+1)}{2}$. Indeed, the $\frac{m+1}{2}$ sets at level t have growth rates $\frac{1}{2^t}$, and each of these sets is served every $\frac{m+1}{2}t$ time slots, giving the height

$$\frac{m+1}{2} \cdot t \cdot \frac{1}{2^t} = \frac{t}{2} = \frac{\log(m+1)}{2}.$$

The heights of the sets at other levels are never greater than $\frac{t}{2}$, so the approximation ratio of schedule $\mathcal{A}$ is $\Theta(\log m)$.

3   A Lower Bound via the Fractional Solution

We derive a lower bound on the height of any schedule $\mathcal{A}$ of the FHS problem. Consider a schedule $\mathcal{A} = (v_{q_1}, \ldots, v_{q_t}, \ldots)$ in which each S_j, $1 \le j \le m$, is accessed infinitely many times (otherwise the schedule has infinite height) and take a large time slot T. We look at the first T slots of schedule $\mathcal{A}$, that is, at the schedule $\mathcal{A}[T] = (v_{q_1}, v_{q_2}, \ldots, v_{q_T})$. For $i = 1, 2, \ldots, n$, let z_i denote the fraction of the time slots $1, 2, \ldots, T$ when the node v_i is used, that is, $z_i = |\{1 \le t \le T : v_{q_t} = v_i\}|/T$. For $j = 1, 2, \ldots, m$, let $1 \le t_1^{(j)} < t_2^{(j)} < \cdots < t_{I(j,T)}^{(j)} \le T$ be the time slots in the period $[1, T]$ when S_j is accessed. We assume that T is large enough so that for each $1 \le j \le m$, $I(j, T) \ge 1$, that is, each S_j is accessed at least once in the period $[1, T-1]$. Defining $t_0^{(j)} = 0$ and $t_{I(j,T)+1}^{(j)} = T$, the maximum height of $S_j = \{v_{j_1}, v_{j_2}, \ldots, v_{j_{q(j)}}\}$ in the period $[1, T]$ is

$$Height(\mathcal{A}[T], j) = \max \left\{ h_j \left(t_i^{(j)} - t_{i-1}^{(j)} \right) : 1 \le i \le I(j, T) + 1 \right\} \tag{3}$$

$$\ge \frac{h_j}{I(j, T) + 1} \sum_{i=1}^{i = I(j,T)+1} \left(t_i^{(j)} - t_{i-1}^{(j)} \right) = \frac{h_j T}{I(j, T) + 1} \tag{4}$$

$$= \frac{h_j}{z_{j_1} + z_{j_2} + \cdots + z_{j_{q(j)}}} \frac{I(j, T)}{I(j, T) + 1}. \tag{5}$$

Inequality (4) simply says that the maximum of $I(j, T) + 1$ numbers is at least their mean value. The equality on the last line above holds because z_{j_r} is the fraction of the time slots $1, 2, \ldots, T$ when node v_{j_r} is used, so $z_{j_1} + z_{j_2} + \cdots + z_{j_q}$ is the fraction of the time slots $1, 2, \ldots, T$ when S_j is accessed, which is equal to $I(j, T)/T > 0$. For the height of schedule $\mathcal{A}$, we have

$$Height(\mathcal{A}) \tag{6}$$
$$\ge Height(\mathcal{A}[T]) \equiv \max\{Height(\mathcal{A}[T], j) : j = 1, 2, \ldots, m\}$$
$$\ge \left(1 - \frac{1}{I(T) + 1}\right) \max \left\{ \frac{h_j}{z_{j_1} + z_{j_2} + \cdots + z_{j_{q(j)}}} : j = 1, 2, \ldots, m \right\}, \tag{7}$$

where $I(T) = \min_{1 \le j \le m}\{I(j, T)\}$ is the minimum number of times any S_j is accessed in the period $[1, T]$.

Consider the following linear program. (To get an equivalent proper linear program, substitute X with $1/Z$ and maximize Z.)

$(\mathcal{P})$ minimize X;

subject to:

$$x_1 + x_2 + \cdots + x_n = 1,$$
$$x_{j_1} + x_{j_2} + \cdots + x_{j_{q(j)}} \ge h_j / X, \quad \text{for each } j = 1, 2, \ldots, m, \tag{8}$$
$$x_i \ge 0, \quad \text{for } i = 1, 2, \ldots, n,$$
$$X > 0. \tag{9}$$

Comparing Inequalities (7) with Inequalities (8), we see that by setting x_1, x_2, ..., x_n to numbers $z_1, z_2, \ldots, z_n$ and X to $Height(\mathcal{A})/\left(1 - \frac{1}{I(T)+1}\right)$, we satisfy all constraints of this linear program. Thus denoting by X_{opt} the minimum feasible value of X in this linear program, we have $Height(\mathcal{A}) \geq X_{opt}\left(1 - \frac{1}{I(T)+1}\right)$, and by increasing T to infinity (so $I(T)$ increases to infinity) we conclude that

$$Height(\mathcal{A}) \geq X_{opt}. \tag{10}$$

The linear program $(\mathcal{P})$ can be viewed as giving the optimal solution for the following fractional variant of the FHS problem. For the discrete FHS problem, a schedule $\mathcal{A}$ can be represented by binary values $y_{i,t} \in \{0,1\}$, $1 \leq i \leq n$, $t \geq 1$, with $y_{i,t} = 1$ indicating that node v_i is used in the time slot t. For the fractional variant of FHS, a schedule is represented by numbers $0 \leq y_{i,t} \leq 1$ indicating the fraction of commitment during the time slot t to node v_i. (Think about the nodes being dealt with during the time period $(t - 1, t]$ concurrently, with the fraction $y_{i,t}$ of the total effort spent on node v_i.) In both discrete and fractional cases we require that $\sum_{i=1}^{n} y_{i,t} = 1$, for each time slot $t \geq 1$. For the discrete variant, the time slot $t_i^{(j)}$ when S_j is accessed for the i-th time is the time slot τ such that

$$\sum_{t=1}^{\tau} \left(y_{j_1,t} + y_{j_2,t} + \cdots + y_{j_{q(j)},t}\right) = i.$$

For the fractional variant, the time $t_i^{(j)}$ when the i-th "cycle" of access to S_j is completed (and the urgency indicator of S_j is reset to 0) is the fractional time $\tau + \delta$, where τ is a positive integer and $0 \leq \delta < 1$, such that

$$\sum_{t=1}^{\tau} \left(y_{j_1,t} + y_{j_2,t} + \cdots + y_{j_{q(j)},t}\right) + \delta \left(y_{j_1,\tau+1} + y_{j_2,\tau+1} + \cdots + y_{j_{q(j)},\tau+1}\right) = i.$$

In both cases, the fraction of the period $(0, T]$ when a node v_i is used is equal to $z_i = \left(\sum_{t=1}^{T} y_{i,t}\right)/T$ and (3)–(7) and (10) apply. For the fractional variant, the schedule $y_{i,t} = x_i^*$, for $1 \leq i \leq n$ and $t \geq 1$, where $(x_1^*, x_2^*, \ldots, x_n^*, X_{opt})$ is an optimal solution of $(\mathcal{P})$, has the optimal (minimum) height X_{opt}.

4   Randomized $O(\log m)$-Approximate Algorithm

We use an optimal solution $(x_1^*, x_2^*, \ldots, x_n^*, X_{opt})$ of linear program $(\mathcal{P})$ to randomly select nodes for the first $T = \Theta(m)$ slots of a schedule $\mathcal{A}$, so that with high probability each set S_j is accessed at least once during each period $[t + 1, t + \tau_j] \subseteq [1, T]$, where $\tau_j = \Theta((X_{opt}/h_j) \log n)$. Thus during the first T slots of the schedule, the heights of the urgency indicators remain $O(X_{opt} \log n)$. The full (infinite) schedule keeps repeating the schedule from the first T slots. In our calculations we assume that $m \geq m_0$, for a sufficiently large constant m_0.

We take $T = 2m$ and construct a random schedule $\mathcal{A}_R = (v_{q_1}, v_{q_2}, \ldots, v_{q_T})$ for T time slots in the following way. We put aside the even time slots for some deterministic assignment of nodes. Specifically, for each time slot $t = 2j$, $j = 1, 2, \ldots, m$, we (deterministically) take for the node v_{q_t} for this time slot an arbitrary node in S_j. This way we guarantee that each set S_j is accessed at least once when the schedule $\mathcal{A}_R$ is followed. For each odd time slot t, $1 \leq t \leq T$, node v_{q_t} is a random node selected according to the distribution $(x_1^*, x_2^*, \ldots, x_n^*)$ and independently of the selection of other nodes. Thus for each odd time slot $t \in [1, T]$ and for each node $v_i \in V$, $\mathbf{Pr}(v_{q_t} = v_i) = x_i^*$.

Lemma 2. *The random schedule $\mathcal{A}_R$ has the properties that each set S_j, $j = 1, 2, \ldots, m$, is accessed at least once and with probability at least $1 - 1/m$, $Height(\mathcal{A}_R) \leq (5 \ln m) X_{opt}$.*

Proof. The first property is obvious from the construction. We show that with probability at least $1 - 1/m$, no urgency indicator grows above $(5 \ln m) X_{opt}$. A set S_j with the rate growth $h_j < (2.5 X_{opt} \ln m)/m$ cannot grow above the height $h_j T < 5 X_{opt} \ln m$, so it suffices to look at the growth of the sets S_j with $h_j \geq (2.5 \cdot X_{opt} \ln m)/m$. Observe that $X_{opt} \geq h_{\max} = \max\{h_1, h_2, \ldots, h_m\}$, from (8).

Let $J \subseteq \{1, 2, \ldots, m\}$ be the set of indices of the sets S_j for which $h_j \geq (2.5 \cdot X_{opt} \ln m)/m$. For each $j \in J$ and for each odd time slot $t \in [1, T]$, the probability that set S_j is accessed during this time slot is equal to $x_{j_1}^* + x_{j_2}^* + \cdots + x_{j_q}^* \geq h_j/X_{opt}$. In each period $[t, t + \tau - 1] \subseteq [1, T]$ of τ consecutive time slots, there are at least $\lfloor \tau/2 \rfloor$ odd time slots, so the probability that S_j is not accessed during this period is at most $(1 - h_j/X_{opt})^{\lfloor \tau/2 \rfloor}$. We take $\tau_j = 5(X_{opt}/h_j) \ln m$ (observe that $\ln m \leq \tau_j \leq T$) and use the union bound over all $j \in J$ and all $[t, t + \tau_j - 1] \subseteq [1, T]$ to conclude that the probability that there is a set S_j, $j \in J$, which is not accessed during consecutive τ_j time slots (and its urgency indicator goes above $(5 \ln m) X_{opt}$) is at most

$$T \cdot \sum_{j \in J} \left(1 - \frac{h_j}{X_{opt}}\right)^{(\tau_j - 1)/2} \leq T \cdot \sum_{j \in J} \left(1 - \frac{h_j}{X_{opt}}\right)^{2.4(X_{opt}/h_j) \ln m}$$

$$\leq 2m \cdot e^{-2.4 \ln m} \leq \frac{1}{m}.$$

$\square$

Theorem 2. *For the infinite schedule $\mathcal{A}_R^*$ which keeps repeating the same random schedule $\mathcal{A}_R$ (all copies are the same), $Height(\mathcal{A}_R^*) \leq (10 \ln m) X_{opt}$ with probability at least $1 - 1/m$.*

Proof. With probability at least $1 - 1/m$, $Height(\mathcal{A}_R) \leq (5 \ln m) X_{opt}$ (Lemma 2). Assuming that $Height(\mathcal{A}_R) \leq (5 \ln m) X_{opt}$, we show that $Height(\mathcal{A}_R^*) \leq (10 \ln m) X_{opt}$.

Let $T = 2m$ be the length of the schedule $\mathcal{A}_R$. We consider an arbitrary set S_j and show that its height is never greater than $(10 \ln m) X_{\text{opt}}$ when the schedule $\mathcal{A}_R^*$ is followed. Since S_j is accessed in $\mathcal{A}_R$ at least once, the height of S_j under the schedule $\mathcal{A}_R^*$ is the same at the end of the time slots kT, for all positive integers k (and is equal to $h_j \left(T - t_{\text{last}}^{(j)} \right)$, where $t_{\text{last}}^{(j)}$ is the last time slot in $\mathcal{A}_R$ when S_j is accessed). The maximum height of S_j during the period $[1, T]$ is at most $(5 \log m) X_{\text{opt}}$. For each integer $k \geq 1$, the maximum height of set S_j during the period $[kT + 1, (k+1)T]$ is at most the height of S_j at the end of time slot kT, which is at most $(5 \ln m) X_{\text{opt}}$, plus the maximum growth of S_j under the schedule $\mathcal{A}_R$, which is again at most $(5 \ln m) X_{\text{opt}}$. Thus the height of S_j is never greater than $(10 \ln m) X_{\text{opt}}$.   □

5   Concluding Remarks

We studied the Fair Hitting Sequence problem, showing its wide range of applications. We provide both deterministic and randomized approximation algorithms, with approximation ratios of $O(\log^2 m)$ and $O(\log m)$, respectively. These upper bounds should be compared with the lower bound of $\Omega(\log m)$ on the approximation ratio of polynomial-time algorithms, which is inherited from the well-known minimum hitting set problem. As a natural question one may ask whether it is possible to provide a deterministic algorithm with approximation ratio guarantee of $O(\log m)$. Due to the deep relation shown for FHS with the hitting set problem, one may be interested in understanding whether introducing some restriction on the sets might result in better approximation ratios. For instance, interesting cases might be when the size of each set S_j is bounded, when each element is contained in a bounded number of sets, or when the intersection of each pair of sets is bounded. In particular, when the size of each set is two, then the sets can be seen as edges of a graph, as mentioned in Sect. 1, and one may consider special graph topologies.

When we consider more than two elements per set, then instead of graphs we actually deal with hypergraphs. In the finite hypergraph setting, a (minimal) hitting set of the edges is called a (minimal) transversal of the hypergraph [9]. Fixed-parameter tractability results have been obtained for the related *transversal hypergraph recognition* problem with a wide variety of parameters, including vertex degree parameters, hyperedge size or number parameters, and hyperedge intersection or union size parameters [13]. Concerning special classes of hypergraph, it is known that the transversal recognition is solvable in polynomial time for special cases of acyclic hypergraphs [9,10]. These results for transversal of hypergraphs may be useful in further study of the FHS problem.

Furthermore, some variants of the FHS problem may be interesting from the theoretical or practical point of view. For instance, one may consider the elements embedded in the plane and the time required by a visiting agent to move from one element to another defined by the distance between those elements. In such setting, it may be useful to consider the following geometric version of the hitting set problem given in [16]. Given a set of geometric objects and a set of points, the goal is to compute the smallest subset of points that hit all geometric objects. The authors of [16] provide $(1 + \epsilon)$-approximation schemes for the minimum geometric hitting set problem for a wide class of geometric range spaces. It would be interesting to investigate how these results could be applied in the wider context of the FHS problem. Finally, further investigations can come from the variant where sets dynamically evolve, as it would be expected in the context of evolving communities in a social network.

References

1. Anily, S., Glass, C.A., Hassin, R.: The scheduling of maintenance service. Discret. Appl. Math. **82**(1–3), 27–42 (1998)
2. Chan, M.Y., Chin, F.Y.L.: General schedulers for the pinwheel problem based on double-integer reduction. IEEE Trans. Comput. **41**(6), 755–768 (1992)
3. Chan, M.Y., Chin, F.: Schedulers for larger classes of pinwheel instances. Algorithmica **9**(5), 425–462 (1993)
4. Chuangpishit, H., Czyzowicz, J., Gąsieniec, L., Georgiou, K., Jurdziński, T., Kranakis, E.: Patrolling a path connecting a set of points with unbalanced frequencies of visits. In: Tjoa, A.M., Bellatreche, L., Biffl, S., van Leeuwen, J., Wiedermann, J. (eds.) SOFSEM 2018. LNCS, vol. 10706, pp. 367–380. Springer, Cham (2018). https://doi.org/10.1007/978-3-319-73117-9_26
5. Collins, A., et al.: Optimal patrolling of fragmented boundaries. In: SPAA, pp. 241–250 (2013)
6. Czyzowicz, J., Gąsieniec, L., Kosowski, A., Kranakis, E.: Boundary patrolling by mobile agents with distinct maximal speeds. In: Demetrescu, C., Halldórsson, M.M. (eds.) ESA 2011. LNCS, vol. 6942, pp. 701–712. Springer, Heidelberg (2011). https://doi.org/10.1007/978-3-642-23719-5_59
7. Czyzowicz, J., Gasieniec, L., Kosowski, A., Kranakis, E., Krizanc, D., Taleb, N.: When patrolmen become corrupted: monitoring a graph using faulty mobile robots. In: ISAAC, pp. 343–354 (2015)
8. D'Emidio, M., Di Stefano, G., Navarra, A.: Priority scheduling in the Bamboo Garden Trimming Problem. In: Catania, B., Královič, R., Nawrocki, J., Pighizzini, G. (eds.) SOFSEM 2019. LNCS, vol. 11376, pp. 136–149. Springer, Cham (2019). https://doi.org/10.1007/978-3-030-10801-4_12
9. Eiter, T., Gottlob, G.: Identifying the minimal transversals of a hypergraph and related problems. SIAM J. Comput. **24**(6), 1278–1304 (1995)
10. Eiter, T., Gottlob, G., Makino, K.: New results on monotone dualization and generating hypergraph transversals. SIAM J. Comput. **32**(2), 514–537 (2003)
11. Gąsieniec, L., Klasing, R., Levcopoulos, C., Lingas, A., Min, J., Radzik, T.: Bamboo Garden Trimming Problem (perpetual maintenance of machines with different attendance urgency factors). In: Steffen, B., Baier, C., van den Brand, M., Eder, J., Hinchey, M., Margaria, T. (eds.) SOFSEM 2017. LNCS, vol. 10139, pp. 229–240. Springer, Cham (2017). https://doi.org/10.1007/978-3-319-51963-0_18
12. Girvan, M., Newman, M.E.J.: Community structure in social and biological networks. Proc. Natl. Acad. Sci. **99**(12), 7821–7826 (2002)
13. Hagen, M.: Algorithmic and Computational Complexity Issues of MONET, Dr. rer. nat., Friedrich-Schiller-Universit at Jena (2008)
14. Holte, R., Rosier, L., Tulchinsky, I., Varvel, D.: Pinwheel scheduling with two distinct numbers. Theor. Comput. Sci. **100**(1), 105–135 (1992)

15. Lin, S.-S., Lin, K.-J.: A pinwheel scheduler for three distinct numbers with a tight schedulability bound. Algorithmica **19**(4), 411–426 (1997)
16. Mustafa, N.H., Ray, S.: Improved results on geometric hitting set problems. Discret. Comput. Geom. **44**(4), 883–895 (2010)
17. Nilsson, B.: Guarding art galleries - methods for mobile guards. Ph.D. thesis, Department of Computer Science, Lund University, Sweden (1995)
18. Newman, M.E.J., Girvan, M.: Finding and evaluating community structure in networks. Phys. Rev. E **69**(2), 026113 (2004)
19. Ntafos, S.: On gallery watchmen in grids. Inf. Process. Lett. **23**(2), 99–102 (1986)
20. Romer, T.H., Rosier, L.E.: An algorithm reminiscent of euclidean-gcd for computing a function related to pinwheel scheduling. Algorithmica **17**(1), 1–10 (1997)
21. Raz, R., Safra, M.: A sub-constant error-probability low-degree test, and a sub-constant error-probability PCP characterization of NP. In: Proceedings of STOC, pp. 475–484 (1997)
22. Serafini, P., Ukovich, W.: A mathematical model for periodic scheduling problems. SIAM J. Discret. Math. **2**(4), 550–581 (1989)
23. Urrutia, J.: Art gallery and illumination problems. In: Handbook of Computational Geometry, vol. 1, no. 1, pp. 973–1027 (2000)

Towards a Theory of Mixing Graphs:
A Characterization of Perfect Mixability
(Extended Abstract)

Miguel Coviello Gonzalez and Marek Chrobak[(✉)]

University of California at Riverside,
Riverside, CA 92521, USA
marek@cs.ucr.edu

Abstract. We study the problem of fluid mixing in microfluidic chips represented by mixing graphs, that model a network of micro-mixers (vertices) connected by micro-pipes (edges). We address the following *perfect mixability* problem: given a collection C of droplets, is there a mixing graph that mixes C perfectly, producing only droplets whose concentration is the average concentration of C? We provide a complete characterization of such perfectly mixable sets and an efficient algorithm for testing perfect mixability. Further, we prove that any perfectly mixable set has a perfect-mixing graph of polynomial size, and that this graph can be computed in polynomial time.

Keywords: Algorithms · Graph theory · Lab-on-chip · Fluid mixing

1 Introduction

Research advances in microfluidics led to the development of *lab-on-chip* (LoC) devices that integrate on a tiny chip various functions of bulky and costly biochemical systems. LoCs play an important role in applications that include environmental monitoring, protein analysis, and physiological sample analysis.

One of the most fundamental functions of LoC devices is *mixing* of different fluids, where the objective is to produce desired volumes of pre-specified mixtures of fluids. In typical applications only two fluids are involved, in which case the process of mixing is often referred to as *dilution*. The fluid to be diluted is called *reactant* and the diluting fluid is called *buffer*. For example, in clinical diagnostics common reactants include blood, serum, plasma and urine, and phosphate buffered saline is often used as buffer.

In this work we consider LoCs that involve a collection of tiny components called *micro-mixers* connected by *micro-channels*. In such chips, input fluids are injected into the chip using fluid dispensers, then they travel, following appropriate micro-channels, through a sequence of micro-mixers in which they are

Research supported by NSF grant CCF-1536026.

P. Heggernes (Ed.): CIAC 2019, LNCS 11485, pp. 187–198, 2019.
https://doi.org/10.1007/978-3-030-17402-6_16

subjected to mixing operations, and are eventually discharged into output reservoirs. We focus on *droplet-based* LoCs, where fluids are manipulated in discrete units called *droplets*. In such chips, a micro-mixer has exactly two input and two output channels. It receives one droplet of fluid from each input, and mixes them perfectly producing two identical droplets on its outputs. Specifically, if the input droplets have (reactant) concentrations a, b, then the produced droplets have concentration $\frac{1}{2}(a+b)$. It follows that all droplets flowing through the chip have concentrations of the form $c/2^d$, where c and $d \geq 0$ are integers. This simply means that their binary representations are finite, and we refer to such numbers as *binary numbers*. The number d, called *precision*, is the number of fractional bits (assuming c is odd when $d \geq 1$).

Processing of droplets on such chips can be represented by a directed acyclic graph G that we call a *mixing graph*. The edges of G represent micro-channels. Source vertices represent dispensers, internal vertices represent micro-mixers, and sink vertices represent output reservoirs. Given a set I of input droplets injected into the source nodes, G converts it into a set T of droplets in its sink nodes. We refer to this set T as a *target set*. An example of a mixing graph is shown in Fig. 1. Here, and elsewhere, we represent each droplet by its reactant concentration (which uniquely determines the buffer concentration).

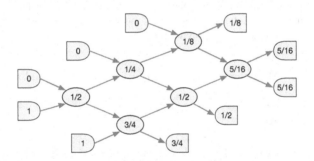

Fig. 1. Mixing graph for input set $I = \{0, 0, 0, 1, 1\}$ and target set $T = \left\{ \frac{1}{8}, \frac{5}{16}, \frac{5}{16}, \frac{1}{2}, \frac{3}{4} \right\}$. The numbers on the nodes represent the reactant concentration of the output droplets.

There is growing literature in the embedded systems and bioengineering communities on designing microfluidic chips represented by such mixing graphs. The most fundamental algorithmic problem emerging in this area is the following:

MIXREACHABILITY: Given an input set I and a target set T of droplets, design a mixing graph that converts I into T (if at all possible).

If there is a mixing graph that converts I into T then we say that T is *mix-reachable*, or just *reachable*, from I. For T to be reachable from I, clearly, I and T must have the same cardinality and equal reactant volumes. However, these conditions are not sufficient. For example, $T = \left\{ \frac{1}{4}, \frac{3}{4} \right\}$ is not reachable from

$I = \{0, 1\}$, because producing $\frac{1}{4}$ from I requires at least two buffer droplets and one reactant droplet, but T itself contains only two droplets.

In typical applications the input set I consists of pure buffer and reactant droplets (that is, only 0's and 1's). We denote this variant by MIXPRODUCIBIL-ITY, and target sets reachable from such input sets are caled *mix-producible*, or just *producible*. MIXPRODUCIBILITY is not likely to be computationally easier than MIXREACHABILITY. For example, via a simple linear mapping, any algorithm that solves MIXPRODUCIBILITY can also solve the variant of MIXREACH-ABILITY where the input set has droplets of *any* two given concentrations.

Related work. The previous work in the literature focuses on the MIXPRO-DUCIBILITY problem. To generate target sets that are not producible, one can consider mixing graphs that besides a target set T also produce some amount of superfluous fluid called *waste*. If we allow waste then, naturally, MIXPRO-DUCIBILITY can be extended to an optimization problem where the objective is to design a mixing graph that generates T while minimizing waste. Alternative objective functions have been studied, for example minimizing the reactant waste, minimizing the number of micro-mixers, and other.

Most of the previous papers study designing mixing graphs using heuristic approaches. Earlier studies focused on producing *single-concentration targets*, where only one droplet of some desired concentration is needed. This line of research was pioneered by Thies *et al.* [1], who proposed an algorithm called Min-Mix that constructs a mixing graph for a single target droplet. Roy *et al.* [2] developed a single-droplet algorithm called DMRW that considered waste reduction and the number of mixing operations. Huang *et al.* [3] and Chiang *et al.* [4] proposed single-droplet algorithms designed to minimize reactant usage.

Many applications, however, require target sets with multiple concentrations (see [5–9]). Huang *et al.* [10] proposed an algorithm called WARA, an extension of [3] for multiple-concentration targets. Mitra *et al.* [11] model the problem as an instance of the Asymmetric TSP on a de Brujin graph.

The papers cited above describe heuristic algorithms with no formal performance guarantees. Dinh *et al.* [12] took a more rigorous approach. Assuming that the depth of a mixing graph does not exceed the maximum target precision, they showed how to formulate the problem as an integer linear program. This leads to an algorithm with doubly exponential running time. Unfortunately, contrary to the claim in [12], their algorithm does not necessarily produce mixing graphs with minimum waste. (A counter example is provided in the full version [13].)

Our results. To our knowledge, the computational complexity of MIXREACH-ABILITY is open; in fact, (given the flaw in [12] mentioned above) it is not even known whether the MIXPRODUCIBILITY variant is decidable. This paper reports partial progress towards resolving this problem. We consider the following sub-problem of MIXREACHABILITY:

PERFECTMIXABILITY: Given a set C of n droplets with average concentration
   $\mu = (\sum_{c \in C} c)/n$, is there a mixing graph that mixes C perfectly, converting
   C into the set of n droplets with concentration μ?

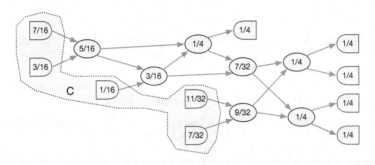

Fig. 2. A mixing graph that perfectly mixes set $C = \left\{ \frac{1}{16}, \frac{3}{16}, \frac{7}{32}, \frac{11}{32}, \frac{7}{16} \right\}$.

Figure 2 shows an example of a perfect-mixing graph. As an example of a set that is not perfectly mixable, consider $D = \left\{ 0, \frac{3}{16}, \frac{9}{16} \right\}$. After any (non-zero) number of mixing operations the resulting set will have form $D' = \{a, a, b\}$ for $a \neq b$, so no finite mixing graph will convert D into its perfect mixture $\left\{ \frac{1}{4}, \frac{1}{4}, \frac{1}{4} \right\}$.

In this paper, addressing the PERFECTMIXABILITY problem, we give a complete characterization of perfectly mixable sets and a polynomial-time algorithm that tests whether a given set is perfectly mixable. We also show that this algorithm can be extended to construct a polynomial-size perfect-mixing graph for such perfect-mixing sets and in polynomial-time.

We represent droplet sets as multisets of concentration values. First, we observe that without loss of generality we can assume that $C \cup \{\mu\} \sqsubset \mathbb{Z}$, for otherwise we can simply rescale all values by an appropriate power of 2. ($\mathbb{Z}$ is the set of integers; $\mathbb{Z}_{>0}$ and $\mathbb{Z}_{\geq 0}$ are the sets of positive and non-negative integers, respectively. Symbol $\sqsubset$ is used to specify a ground set of a multiset.) For any finite multiset $A \sqsubset \mathbb{Z}$ and $b \in \mathbb{Z}_{>0}$, we define A to be *b-congruent* if $x \equiv y$ (mod b) for all $x, y \in A$. (Otherwise we say that A is *b-incongruent*.)

We say that C satisfies *Condition (MC)* if, for each odd $b \in \mathbb{Z}_{>0}$, if C is b-congruent then $C \cup \{\mu\}$ is b-congruent as well, where $\mu = \mathsf{ave}(C)$. The following theorem summarizes our results.

Theorem 1. *Assume that $n \geq 4$ and $C \cup \{\mu\} \sqsubset \mathbb{Z}$, where $\mu = \mathsf{ave}(C)$. Then:* **(a)** *C is perfectly mixable if and only if C satisfies Condition (MC).* **(b)** *If C satisfies Condition (MC) then it can be perfectly mixed with precision at most 1 and in a polynomial number of steps. (That is, C has a perfect-mixing graph of polynomial size where all intermediate concentration values are half-integral.)* **(c)** *There is a polynomial-time algorithm that tests whether C is perfectly mixable and, if so, computes a polynomial-size perfect-mixing graph for C.*

Part (b) says that, providing that C is perfectly mixable, at most one extra bit of precision is needed in the intermediate nodes of a perfect-mixing graph for C. This extra 1-bit of precision is necessary. For example, $C = \{0, 0, 0, 3, 7\}$ (for which $\mu = 2$) cannot be mixed perfectly with precision 0. If we mix 3 and 7, we will obtain multiset $\{0, 0, 0, 5, 5\}$ which is not perfectly mixable, as it violates

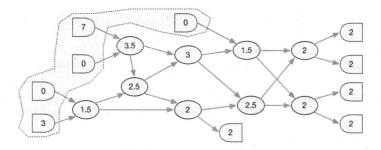

Fig. 3. A perfect mixing graph for $C = \{0, 0, 0, 3, 7\}$ with precision 1.

Condition (MC). Any other mixing creates fractional values. However, C does have a mixing graph where the intermediate precision is at most 1—see Fig. 3.

Due to limited space, most proofs are left out to the full version of the paper [13]. The rest of the paper is organized as follows. The necessity of Condition (MC) in Theorem 1(a) is given in Sect. 3. For the sufficiency of Condition (MC), first, in Sect. 4 (Lemma 1), we claim that for Condition (MC) it is sufficient to consider only the values of b that are odd prime power factors of n. This property is used to show that any set C that satisfies Condition (MC) has a perfect-mixing graph, completing the argument in Theorem 1(a). This property is also used in our polynomial-time algorithm for testing perfect-mixability (part one of Theorem 1(c)), which is given at the end of Sect. 4. Finally, the mixing graph constructed in Sect. 4 has precision at most 1, hence showing the first part of Theorem 1(b).

2   Preliminaries

Let $\mathbb{Q}_0$ be the set of binary numbers. For $c \in \mathbb{Q}_0$, we denote by $\mathsf{prec}(c)$ the precision of c, that is the number of fractional bits in the binary representation of c, assuming there are no trailing 0's. In other words, $\mathsf{prec}(c)$ is the smallest $d \in \mathbb{Z}_{\geq 0}$ such that $c = a/2^d$ for some $a \in \mathbb{Z}$. If $c = a/2^d$ represents actual fluid concentration, then we have $0 \leq a \leq 2^d$. However, it is convenient to relax this restriction and allow "concentration values" that are arbitrary binary numbers, even negative. In fact, it will be convenient to work with integral values.

By a *configuration* we mean a multiset C of n binary numbers, which represent droplet concentrations. We will typically denote a configuration by $C = \{f_1 : c_1, f_2 : c_2, ..., f_m : c_m\} \sqsubset \mathbb{Q}_0$, where each c_i represents a (different) concentration value and f_i denotes the multiplicity of c_i in C, so that $\sum_{i=1}^{m} f_i = n$.

The number of droplets in C is denoted $|C| = n$, while the number of different concentrations is denoted $\|C\| = m$. Occasionally, if it does not lead to confusion, we may say "droplet c_i" or "concentration c_i", referring to some droplet with concentration c_i. If $f_i = 1$, we shorten "$f_i : c_i$" to just "c_i". If $f_i = 1$ we say that droplet c_i is a *singleton*, if $f_i = 2$ we say that droplet c_i is a *doubleton* and if $f_i \geq 2$ we say that droplet c_i is a *non-singleton*. By $\mathsf{ave}(C) = (\sum_{c \in C} c)/n$ we denote the average value of the concentrations in C, which will be also typically denoted by μ.

Mixing graphs were defined in the introduction. As we are not concerned in this paper with the topological properties of mixing graphs, we will often identify a mixing graph G with a corresponding *mixing sequence*, which is a sequence (not necessarily unique) of mixing operations that convert C into its perfect mixture. In other words, a mixing sequence is a sequence of mixing operations in a topological ordering of a mixing graph.

Of course in a perfect-mixing graph (or sequence) G for C, all concentrations in G, including those in $C \cup \{\mu\}$, must have finite precision (that is, belong to $\mathbb{Q}_0$) which is at least $\max\{\text{prec}(C), \text{prec}(\mu)\}$. In addition to the basic question about finding a perfect-mixing graph for C, we are also interested in bounding the precision required to do so.

For $x \in \mathbb{Q}_0$, define multisets $C + x = \{c + x \mid c \in C\}$, $C - x = C + (-x)$, and $C \cdot x = \{c \cdot x \mid c \in C\}$. The next observation says that offsetting all values in C does not affect perfect mixability, as long as the offset value's precision does not exceed that of C or μ.

Observation 1. *Let $\mu = \text{ave}(C)$, $x \in \mathbb{Q}_0$ and $d \geq \max\{\text{prec}(C), \text{prec}(\mu),\}$ $\text{prec}(x)$. Then C is perfectly mixable with precision d if and only if $C' = C + x$ is perfectly mixable with precision d.*

Observation 2. *Let $\mu = \text{ave}(C)$, $\delta = \max\{\text{prec}(C), \text{prec}(\mu)\}$, $C' = C \cdot 2^\delta$ with $\mu' = \text{ave}(C') = 2^\delta \mu$. (Thus $C' \cup \{\mu'\} \sqsubset \mathbb{Z}$.) Then C is perfectly mixable with precision $d \geq \delta$ if and only if C' is perfectly mixable with precision $d' = d - \delta$.*

Integral configurations. Per Observation 2, we can restrict our attention to configurations C with integer values and average, that is $C \cup \{\mu\} \sqsubset \mathbb{Z}$. For $x \in \mathbb{Z}_{>0}$, if each $c \in C$ is a multiple of x, let $C/x = \{c/x \mid c \in C\}$. For integral configurations, we can extend Observation 2 to also multiplying C by an odd integer or dividing it by a common odd factor of all concentrations in C.

Observation 3. *Assume that $C \cup \{\mu\} \sqsubset \mathbb{Z}$ and let $x \in \mathbb{Z}_{>0}$ be odd. (a) Let $C' = C \cdot x$. Then C is perfectly mixable with precision 0 if and only if C' is perfectly mixable with precision 0. (b) Suppose that x is a divisor of all concentrations in $C \cup \{\mu\}$. Then C is perfectly mixable with precision 0 if and only if C/x is perfectly mixable with precision 0.*

See [13] for the proofs for Observations 1, 2 and 3.

3   Necessity of Condition (MC)

We now prove that Condition (MC) in Theorem 1(a) is necessary for perfect mixability. Let $C \cup \{\mu\} \sqsubset \mathbb{Z}$, where $\mu = \text{ave}(C)$, and assume that C is perfectly mixable. We want to prove that C satisfies Condition (MC).

Let G be a graph that mixes C perfectly. Suppose that C is b-congruent for some odd $b \in \mathbb{Z}_{>0}$. Consider an auxiliary configuration $C' = C \cdot 2^\delta$, where δ

is sufficiently large, so that all intermediate concentrations in G when applying G to C' are integral. This C' is b-congruent, and G converts C' into its perfect mixture $\{n : \mu'\}$, for $\mu' = 2^\delta \mu$.

Since C' is b-congruent, there is $\beta \in \{0, ..., b-1\}$ such that for each $x \in C'$ we have $x \equiv \beta \pmod{b}$. We claim that this property is preserved as we apply mixing operations to droplets in C'. Suppose that we mix two droplets with concentrations $x, y \in C'$, producing two droplets with concentration z. Since $x \equiv \beta \pmod{b}$ and $y \equiv \beta \pmod{b}$, we have $x = \alpha b + \beta$ and $y = \alpha' b + \beta$, for some $\alpha, \alpha' \in \mathbb{Z}$, so $z = \frac{1}{2}(x + y) = (\frac{1}{2}(\alpha + \alpha'))b + \beta$. As b is odd, $\alpha + \alpha'$ must be even, and therefore $z \equiv \beta \pmod{b}$, as claimed. Eventually G produces μ', so this must also hold for $z = \mu'$. This implies that $C' \cup \{\mu'\}$ is b-congruent.

Finally, since $C' \cup \{\mu'\}$ is b-congruent, for all $2^\delta x, 2^\delta y \in C' \cup \{\mu'\}$ it holds that $2^\delta x \equiv 2^\delta y \pmod{b}$. But this implies that $x \equiv y \pmod{b}$, because b is odd. So $C \cup \{\mu\}$ is b-congruent, proving that C satisfies Condition (MC).

4   Sufficiency of Condition (MC)

In this section we sketch the proof that Condition (MC) in Theorem 1(a) is sufficient for perfect mixability. A perfect-mixing graph constructed in our argument has precision at most 1, showing also the first part of Theorem 1(b).

Condition (MC) involves all odd $b \in \mathbb{Z}_{>0}$, so it does not directly lead to an efficient test for perfect mixability. Hence, we show in Lemma 1 below (whose proof is given in [13]) that only factors b of n that are odd prime powers need to be considered. This implies that perfect mixability testing can be done in polynomial time. (An algorithm for testing perfect mixability in polynomial time, see [13], is given in Algorithm 1 at the end of this section.)

Lemma 1. *If Condition (MC) holds for all factors $b \in \mathbb{Z}_{>0}$ of n that are a power of an odd prime then it holds for all odd $b \in \mathbb{Z}_{>0}$.*

Note that in Theorem 1 we assume that $n = |C| \geq 4$. Regarding smaller values of n, for $n = 2$, trivially, all configurations C with two droplets are perfectly mixable with precision 0. The case $n = 3$ is exceptional, as in this case Theorem 1 is actually false. (For example, consider configuration $C = \{0, 1, 5\}$, for which $\mu = 2$. This configuration is b-incongruent for all odd $b > 1$, so it satisfies condition (MC), but is not perfectly mixable.) Nevertheless, for $n = 3$, perfectly mixable configurations are easy to characterize: Let $C = \{a, b, c\}$, where $a \leq b \leq c$. Then C is perfectly mixable if and only if $b = \frac{1}{2}(a+c)$. Further, if this condition holds, C is perfectly mixable with precision 0. (That this condition is sufficient is obvious. That it is also necessary can be proven by following the argument given in the introduction for the example configuration D right after the definition of PERFECTMIXABILITY.)

So from now on we assume that $n \geq 4$. Let C be the input configuration and $\mu = \mathsf{ave}(C)$, where $C \cup \{\mu\} \sqsubset \mathbb{Z}$. The outline of our proof is as follows:

– First we show that C is perfectly mixable with precision 0 when n is a power of 2. This easily extends to configurations C called *near-final*, which are disjoint unions of multisets with the same average and cardinalities being powers of 2. (This proves Theorem 1(a) for $n = 4$.)

– Next, we sketch a proof for $n \geq 7$. The basic idea of the proof is to define an invariant (I) and show that any configuration that satisfies (I) has a mixing operation that either preserves invariant (I) or produces a near-final configuration. Condition (I) is stronger than (MC) (it implies (MC), but not vice versa), but we show that any configuration that satisfies Condition (MC) can be modified to satisfy (I).

– Finally, the proofs for $n = 5, 6$ appear in the full version of the paper [13]. These are similar to the case $n \geq 7$, but they require a more subtle invariant.

Algorithm 1. PerfectMixabilityTesting(C)

$n \leftarrow |C|$
$\mu \leftarrow$ ave(C)
$c_{max} \leftarrow$ absolute value of maximum concentration in C
$P \leftarrow$ powers of odd prime factors of n that are at most c_{max}
for all $p \in P$ **do**
    if C is p-congruent but $C \cup \{\mu\}$ is not **then**
        return false
return true

4.1   Perfect Mixability of Near-Final Configurations

Let $C \sqsubset \mathbb{Z}$ with ave(C) $= \mu \in \mathbb{Z}$ be a configuration with $|C| = n = \sigma 2^\tau$, for some odd $\sigma \in \mathbb{Z}_{>0}$ and $\tau \in \mathbb{Z}_{\geq 0}$, and $\|C\| = m$. We say that C is *near-final* if it can be partitioned into multisets $C_1, C_2, ..., C_k$, such that, for each j, ave(C_j) $= \mu$ and $|C_j|$ is a power of 2. In this sub-section we show (Lemma 2 below) that near-final configurations are perfectly mixable with precision 0.

Define $\Psi(C) = \sum_{c \in C}(c - \mu)^2$ as the "variance" of C. Obviously $\Psi(C) \in \mathbb{Z}_{\geq 0}$, $\Psi(C) = 0$ if and only if C is a perfect mixture, and, by a straightforward calculation, mixing any two different same-parity concentrations in C decreases the value of $\Psi(C)$ by at least 1.

Lemma 2. *If C is near-final then C is perfectly mixable with precision 0.*

Note that it is sufficient to prove the lemma for the case when n is a power of 2. (Otherwise, we can apply it separately to each multiset C_j in the partition of C from the definition of near-final configurations.) The idea is to always mix two distinct same-parity concentrations, which strictly decreases $\Psi(C)$ and preserves integrality. Now, the existence of such a pair is trivial when $m \geq 3$. For the case when $m = 2$, showing both the concentrations in C have same-parity is sufficient; this follows by simple number theory after offsetting C by $-c$, for any $c \in C$.

4.2   Proof for Arbitrary $n \geq 7$

In this sub-section we sketch the proof that Condition (MC) in Theorem 1(a) is sufficient for perfect mixability when $n \geq 7$. Let C be a configuration that satisfies Condition (MC), where $C \cup \{\mu\} \sqsubset \mathbb{Z}$ and $|C| = n$. Also, let the factorization of n be $n = 2^{\tau_0} p_1^{\tau_1} p_2^{\tau_2} ... p_s^{\tau_s}$, where $\{p_1, p_2, ..., p_s\} = \bar{p}$ is the set of the odd prime factors of n and $\{\tau_1, \tau_2, ..., \tau_s\}$ are their corresponding multiplicities.

If $A \sqsubset \mathbb{Z}$ is a configuration with $|A| = n$ (where n is as above) and $\mathsf{ave}(A) \in \mathbb{Z}$, then we say that A is $\bar{p}$-incongruent if A is p_r-incongruent for all r. We next show two key properties of $\bar{p}$-incongruent configurations. (See [13] for the proof.)

Observation 4. *Assume that $A \sqsubset \mathbb{Z}$ with $\mathsf{ave}(A) \in \mathbb{Z}$ is a $\bar{p}$-incongruent configuration. Then (a) A satisfies Condition (MC), and (b) if A is not near-final then $\|A\| \geq 3$.*

Proof outline. The outline of the proof is as follows (see Fig. 4): Instead of dealing with C directly, we consider a $\bar{p}$-incongruent configuration $\check{C} \sqsubset \mathbb{Z}$ with $\check{\mu} = \mathsf{ave}(\check{C}) \in \mathbb{Z}$ that is "equivalent" to C in the sense that C is perfectly mixable with precision at most 1 if and only if $\check{C}$ is perfectly mixable with precision 0.

It is thus sufficient to show that $\check{C}$ is perfectly mixable with precision 0. To this end, we first apply some mixing operations to $\check{C}$, producing only integer concentrations, that convert $\check{C}$ into a configuration E such that:

(I.0) $E \sqsubset \mathbb{Z}$ and $\mathsf{ave}(E) = \check{\mu}$,
(I.1) E has at least 2 distinct non-singletons, and
(I.2) E is $\bar{p}$-incongruent.

We refer to the three conditions above as Invariant (I). Then we show that any configuration E that satisfies Invariant (I) has a pair of different concentrations that are "safe" to mix, in the sense that after they are mixed the new configuration is either near-final or satisfies Invariant (I). We can thus repeatedly mix such safe mixing pairs, preserving Invariant (I), until we produce a near-final configuration, that, by Lemma 2, can be perfectly mixed with precision 0.

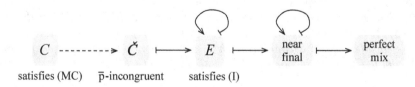

Fig. 4. Proof outline for $n \geq 7$. The first dashed arrow represents replacing C by $\check{C}$. Solid arrows represent sequences of mixing operations.

Replacing C by $\check{C}$. We now explain how to modify C. First, let $C' = C - c_i$, for some arbitrary $c_i \in C$. Note that $\mu' = \mathsf{ave}(C') = \mu - c_i \in \mathbb{Z}$, that $0 \in C'$, and that C' satisfies Condition (MC). By Observation 1, C is perfectly mixable if and only if C' is perfectly mixable (with the same precision), so it is sufficient to show that C' is perfectly mixable.

Let $\theta \in \mathbb{Z}_{>0}$ be the maximum odd integer that divides all concentrations $c \in C'$ (the greatest common odd divisor of C'). Let $C'' = C'/\theta$. By Observation 3(b) and the paragraph above, C is perfectly mixable if and only C'' is perfectly mixable (with the same precision), so from now on we can replace C by C''.

By Condition (MC) applied to C', θ is a divisor of μ', so $\mu'' = \mathsf{ave}(C'') = \mu'/\theta \in \mathbb{Z}$. Next, we claim that C'' is $\bar{p}$-incongruent. To show this, we argue by contradiction. Suppose that C'' is p_r-congruent for some r. This means that there is $\beta \in \{0, 1, ..., p_r - 1\}$ such that $c \equiv \beta \pmod{p_r}$ for all $c \in C''$. Since $0 \in C''$ (because $0 \in C'$), we must have $\beta = 0$. In other words, all $c \in C''$ are multiples of p_r. That would imply, however, that all $c \in C'$ are multiples of θp_r, which contradicts the choice of θ, completing the proof.

Finally, let $\check{C} = 2 \cdot C''$ and $\check{\mu} = 2\mu'' = \mathsf{ave}(\check{C})$. All concentrations in $\check{C}$ are even and, since multiplying all concentrations by 2 does not affect $\bar{p}$-incongruence, $\check{C}$ is $\bar{p}$-incongruent. By Observation 2, and the properties of C'' established above, C is perfectly mixable with precision at most 1 if and only if $\check{C}$ is perfectly mixable with precision 0. Therefore, it is sufficient to show a mixing sequence with integral concentration values that converts $\check{C}$ into its perfect mixture $\{n : \check{\mu}\}$.

Converting $\check{C}$ into E. Let $\check{C}$ be the configuration constructed above. We now show that with at most two mixing operations, producing only integer values, we can convert $\check{C}$ into a configuration E that satisfies Invariant (I).

Let $A \sqsubset \mathbb{Z}$ be a configuration with $\mathsf{ave}(A) \in \mathbb{Z}$ and $|A| = n$. Assume that A is $\bar{p}$-incongruent. For different concentrations $a, a' \in A$ with the same parity, we say that the pair (a, a') is p_r-safe if mixing a and a' converts A into a p_r-incongruent configuration; in other words, there is $a'' \in A$ with $a'' \notin \{a, a'\}$ that satisfies $a'' \not\equiv \frac{1}{2}(a + a') \pmod{p_r}$. (Otherwise, we say that the pair (a, a') is p_r-unsafe, or just unsafe to generalize.) We will also say that (a, a') is $\bar{p}$-safe if it is p_r-safe for all r. For example, let $n = 5$ and $A = \{0, 0, 3, 7, 10\}$, for which $\mathsf{ave}(A) = 4$. Then pair $(0, 10)$ is 5-safe but pair $(3, 7)$ is 5-unsafe. The proof of Lemma 3 below is given in [13].

Lemma 3. *Let A be a $\bar{p}$-incongruent configuration with $\mathsf{ave}(A) \in \mathbb{Z}$ and $|A| = n$. (Recall that $n \geq 7$.) Then **(a)** For each r, there is at most one p_r-unsafe pair in A. **(b)** At most $n - 5$ droplets are involved in same-parity concentration pairs that are unsafe. **(c)** If a concentration $a \in A$ is a non-singleton then for any $b \in A$ with $b \neq a$ and the same parity as a, the pair (a, b) is $\bar{p}$-safe.*

The configuration $\check{C}$ constructed earlier contains only even concentration values, it already satisfies $\check{C} \cup \{\check{\mu}\} \sqsubset \mathbb{Z}$ and is $\bar{p}$-incongruent (that is, it satisfies conditions (I.0) and (I.2) for E). It remains to show that there are mixing operations involving only droplets already present in $\check{C}$ (and thus of even value, to assure that condition (I.0) holds) that preserve condition (I.2), and such that the resulting configuration E is either near-final or it satisfies condition (I.1). If $\check{C}$ already has two or more non-singletons, we can take $E = \check{C}$ and we are done, so assume otherwise, namely that there is either one non-singleton in $\check{C}$ or none. We consider three cases.

Case 1: $\check{C}$ has one non-singleton a with multiplicity $f \geq 3$. Mix a with any singleton $b \in \check{C}$ and let E be the resulting configuration. E has two non-singletons and condition (I.2) is preserved by Lemma 3(c), so E satisfies Invariant (I).

Case 2: $\check{C}$ has one non-singleton a with multiplicity 2. This, and Lemma 3(b) imply that there are at least 3 singletons, say b, c, d, that are not involved in any unsafe pair. Mixing one of pairs (b, c) or (b, d) produces a concentration other than a, which yields to configuration E that satisfies Invariant (I).

Case 3: $\check{C}$ has only singletons. By Lemma 3(b), there are at least 5 singletons, say $a, b, c, d, e \in \check{C}$ with $a < b < c < d < e$, that are not involved in any unsafe pair. Let E be obtained by mixing a with b and c with d. Then, E is $\bar{p}$-incongruent and contains two non-singleton, so E satisfies Invariant (I).

Preserving Invariant (I). We now sketch the last part of the argument, which follows the outline given at the beginning of this section. Let $E \sqsubset \mathbb{Z}$ be the configuration, say $E = \{f_1 : e_1, f_2 : e_2, ..., f_m : e_m\}$, with $\mathsf{ave}(E) = \breve{\mu}$, constructed above from $\check{C}$. If E is near-final then E has a perfect-mixing sequence, by Lemma 2. Otherwise, we show that E has a pair of concentrations whose mixing produces a configuration that is either near-final or satisfies Invariant (I).

Let $e_i, e_j \in E$ with $e_i \neq e_j$. The pair (e_i, e_j) is called *a safe mixing pair* if the configuration E', obtained by mixing e_i and e_j, is either near-final or satisfies Invariant (I). In Lemma 4 below we show that if E satisfies Invariant (I) and is not near-final, then it must contain a safe mixing pair. Thus, we can repeatedly mix E, maintaining Invariant (I), until E becomes near-final. (In particular, it is sufficient to transform E so that $\|E\| = 2$, because each configuration that satisfies both Condition (MC) and $\|E\| = 2$ is near final; see [13] for the proof.) Note that, to preserve conditions (I.0) and (I.2), we always choose a $\bar{p}$-safe pair (e_i, e_j) with same-parity concentration values.

Lemma 4. *Assume that $\|E\| \geq 3$ and that E is not near-final. If E satisfies Invariant (I) then E has a safe mixing pair.*

The proof of this lemma is given in the full version of the paper [13].

Completing the proof. We are now ready to complete the sketch for proof that Condition (MC) in Theorem 1(a) is sufficient for perfect mixability when $n \geq 7$. The argument follows the outline depicted in Fig. 4.

Assume that C satisfies Condition (MC). Replace C by configuration $\check{C} \sqsubset \mathbb{Z}$ such that (i) $\breve{\mu} = \mathsf{ave}(\check{C}) \in \mathbb{Z}$, all values in $\check{C}$ are even, and $\check{C}$ is $\bar{p}$-incongruent, and (ii) C is perfectly mixable with precision at most 1 if and only if $\check{C}$ is perfectly mixable with precision 0.

Then we show that $\check{C}$ has a perfect-mixing sequence (with precision 0), converting $\check{C}$ into its perfect mixture $\{n : \breve{\mu}\}$. To this end, we first perform some mixing operations (at most two) that convert $\check{C}$ into a configuration E that either satisfies Invariant (I) or is near-final. If this E is near-final, we can complete the mixing sequence using Lemma 2. If this E is not near-final then it satisfies Invariant (I) and, by Observation 4(b), $\|E\| \geq 3$. Therefore, we can apply Lemma 4 to show that E has a safe mixing pair, namely a pair of different concentrations whose mixing either preserves Invariant (I) or produces a near-final configuration. We can thus apply the above argument repeatedly to E. As in Sect. 4.1, each mixing decreases the value of $\Psi(E)$. Thus after a finite number of steps we convert E into a near-final configuration, that has a mixing sequence by Lemma 2.

References

1. Thies, W., Urbanski, J.P., Thorsen, T., Amarasinghe, S.: Abstraction layers for scalable microfluidic biocomputing. Nat. Comput. **7**(2), 255–275 (2008)
2. Roy, S., Bhattacharya, B.B., Chakrabarty, K.: Optimization of dilution and mixing of biochemical samples using digital microfluidic biochips. IEEE Trans. Comput. Aided Des. Integr. Circ. Syst. **29**, 1696–1708 (2010)
3. Huang, J.D., Liu, C.H., Chiang, T.W.: Reactant minimization during sample preparation on digital microfluidic biochips using skewed mixing trees. In: Proceedings of the International Conference on Computer-Aided Design, pp. 377–383. ACM (2012)
4. Chiang, T.W., Liu, C.H., Huang, J.D.: Graph-based optimal reactant minimization for sample preparation on digital microfluidic biochips. In: 2013 International Symposium on VLSI Design, Automation and Test (VLSI-DAT), pp. 1–4. IEEE (2013)
5. Xu, T., Chakrabarty, K., Pamula, V.K.: Defect-tolerant design and optimization of a digital microfluidic biochip for protein crystallization. IEEE Trans. Comput. Aided Des. Integr. Circ. Syst. **29**(4), 552–565 (2010)
6. Srinivasan, V., Pamula, V.K., Fair, R.B.: An integrated digital microfluidic lab-on-a-chip for clinical diagnostics on human physiological fluids. Lab Chip **4**(4), 310–315 (2004)
7. Srinivasan, V., Pamula, V.K., Fair, R.B.: Droplet-based microfluidic lab-on-a-chip for glucose detection. Anal. Chim. Acta **507**(1), 145–150 (2004)
8. Hsieh, F., Keshishian, H., Muir, C.: Automated high throughput multiple target screening of molecular libraries by microfluidic MALDI-TOF-MS. J. Biomol. Screen. **3**(3), 189–198 (1998)
9. Hsieh, Y.L., Ho, T.Y., Chakrabarty, K.: A reagent-saving mixing algorithm for preparing multiple-target biochemical samples using digital microfluidics. IEEE Trans. Comput. Aided Des. Integr. Cir. Syst. **31**, 1656–1669 (2012)
10. Huang, J.D., Liu, C.H., Lin, H.S.: Reactant and waste minimization in multitarget sample preparation on digital microfluidic biochips. IEEE Trans. Comput. Aided Des. Integr. Circ. Syst. **32**, 1484–1494 (2013)
11. Mitra, D., Roy, S., Chakrabarty, K., Bhattacharya, B.B.: On-chip sample preparation with multiple dilutions using digital microfluidics. In: IEEE Computer Society Annual Symposium on VLSI (ISVLSI), pp. 314–319. IEEE (2012)
12. Dinh, T.A., Yamashita, S., Ho, T.Y.: A network-flow-based optimal sample preparation algorithm for digital microfluidic biochips. In: 19th Asia and South Pacific Design Automation Conference (ASP-DAC), pp. 225–230. IEEE (2014)
13. Coviello Gonzalez, M., Chrobak, M.: Towards a theory of mixing graphs: a characterization of perfect mixability. CoRR, abs/1806.08875 (2018)

Searching by Heterogeneous Agents

Dariusz Dereniowski[1]($\boxtimes$) , Łukasz Kuszner[2] , and Robert Ostrowski[1]

[1] Faculty of Electronics, Telecommunications and Informatics,
Gdańsk University of Technology,
Gabriela Narutowicza 11/12, 80-233 Gdańsk, Poland
{deren,robostro}@eti.pg.edu.pl
[2] Faculty of Mathematics, Physics and Informatics, University of Gdańsk,
Jana Bażyńskiego 8, 80-309 Gdańsk, Poland
lkuszner@inf.ug.edu.pl

Abstract. In this work we introduce and study a pursuit-evasion game in which the search is performed by heterogeneous entities. We incorporate heterogeneity into the classical edge search problem by considering edge-labeled graphs. In such setting a searcher, once a search strategy initially decides on the label of the searcher, can be present on an edge only if the label of the searcher and the label of the edge are the same. We prove that such searching problem is not monotone even for trees and moreover we give instances in which the number of recontamination events is $\Omega(n^2)$, where n is the size of a tree. Another negative result regards the NP-completeness of the monotone heterogeneous search in trees. The two above properties show that this problem behaves very differently from the classical edge search. On the other hand, if all edges of a particular label form a (connected) subtree of the input tree, then we show that optimal heterogeneous search strategy can be computed efficiently.

Keywords: Graph searching · Mobile agent computing ·
Monotonicity · Pursuit-evasion

1 Introduction

Consider a scenario in which a team of searchers should propose a search strategy, i.e., a sequence of their moves, that results in capturing a fast and invisible fugitive hiding in a graph. This strategy should succeed regardless of the actions of the fugitive and the fugitive is considered captured when at some point it shares the same location with a searcher. In a strategy, the searchers may perform the following moves: a searcher may be placed/removed on/from a vertex of the graph, and a searcher may slide along an edge from currently occupied vertex to its neighbor. The fugitive may represent an entity that does not want to be

Research partially supported by National Science Centre (Poland) grant number 2015/17/B/ST6/01887.

P. Heggernes (Ed.): CIAC 2019, LNCS 11485, pp. 199–211, 2019.
https://doi.org/10.1007/978-3-030-17402-6_17

captured but may as well be an entity that wants to be found but is constantly moving and the searchers cannot make any assumptions on its behavior. There are numerous models that have been studied and these models can be produced by enforcing some properties of the fugitive (e.g., visibility, speed, randomness of its movements), properties of the searchers (e.g., speed, type of knowledge provided as an input or during the search, restricted movements, radius of capture), types of graphs (e.g., simple, directed) or by considering different optimization criteria (e.g., number of searchers, search cost, search time).

One of the central concepts in graph searching theory is *monotonicity*. Informally speaking, if a search strategy has the property that once a searcher traversed an edge (and by this action it has been verified that in this very moment the fugitive is not present on this edge) it is guaranteed (by the future actions of the searchers) that the edge remains inaccessible to the fugitive, then we say that the search strategy is *monotone*. In most graph searching models it is not beneficial to consider search strategies that are not monotone. Such a property is crucial for two main reasons: firstly, knowing that monotone strategies include optimal ones reduces the algorithmic search space when finding good strategies and secondly, monotonicity places the problem in the class NP.

To the best of our knowledge, all searching problems studied to date are considering the searchers to have the same characteristics, and they may only have different 'identities' which allows them to differentiate their actions. However, there exist pursuit-evasion games in which some additional device (like a sensor or a trap) is used by the searchers [6–8,31]. In this work we introduce a searching problem in which searchers are different: each searcher has access only to some part of the graph. More precisely, there are several *types* of searchers, and for each edge e in the graph, only one type of searchers can slide along e. We motivate this type of search twofold. First, referring to some applications of graph searching problems in the field of robotics, one can imagine scenarios in which the robots that should physically move around the environment may not be all the same. Thus some robots, for various reasons, may not have access to the entire search space. Our second motivation is an attempt to understand the concept of monotonicity in graph searching. In general, the graph searching theory lacks of tools for analyzing search strategies that are not monotone, where a famous example is the question whether the connected search problem belongs to NP [2]. In the latter, the simplest examples that show that recontamination may be beneficial for some graphs are quite complicated [32]. The variant of searching that we introduce has an interesting property: it is possible to construct relatively simple examples of graphs in which multiple recontaminations are required to search the graph with the minimum number of searchers. Moreover, it is interesting that this property holds even for trees.

Related Work. In this work we adopt two models of graph searching to our purposes: the classical *edge search* and its connected variant introduced in [3]. As an optimization criterion we consider minimization of the number of searchers. The edge search problem is known to be monotone but the connected search is not [32]. Knowing that the connected search is not monotone, a natural question is what is the 'price of monotonicity', i.e., what is the ratio of the minimum number of searchers required in a monotone strategy and an arbitrary (possibly non-monotone) one? It follows that this ratio is a constant that tends to 2 [13]. We remark that if the searchers do not know the graph in advance and need to learn its structure during execution of their search strategy then this ratio is $\Omega(n/\log n)$ even for trees [22]. A recently introduced model of *exclusive graph searching* shows that internal edge search with additional restriction that at most one searcher can occupy a vertex behaves very differently than edge search. Namely, considerably more searchers are required for trees and exclusive graph searching is not monotone even in trees [5,26]. Few other searching problems are known not to be monotone and we only provide references for further readings [10,20,32]. Also see [19] for a searching problem for which determining whether monotonicity holds turns out to be a challenging open problem.

Since we focus on trees in this work, we briefly survey a few known results for this class of graphs. An edge search strategy that minimizes the number of searchers can be computed in linear time for trees [27]. Connected search is monotone and can be computed efficiently for trees [2] as well. However, if one considers weighted trees (the weight of a vertex or edge indicate how many searchers are required to clean or prevent recontamination), then the problem turns out to be strongly NP-complete, both for edge search [28] and connected search [11]. On the other hand, due to [12,13] both of these weighted problems have constant factor approximations. The class of trees usually turns out to be a very natural subclass to study for many graph searching problems — for some recent algorithmic and complexity examples see e.g. [1,14,16,21].

We conclude by pointing to few works that use heterogeneous agents for solving different computational tasks, mostly in the area of mobile agent computing [15,17,18,25,30]. We also note that heterogeneity can be introduced by providing weights to mobile agents, where the meaning of the weight is specific to a particular problem to be solved [4,9,23].

Our Work. We focus on studying monotonicity and computational complexity of our heterogeneous graph searching problem that we formally define in Sect. 2. We start by proving that the problem is not monotone in the class of trees (Sect. 3). In Sect. 4 we list our two main theorems regarding the complexity: both the monotone and the general heterogeneous search problems are hard for trees. The proofs are postponed to the appendix due to space limitations. Our investigations suggest that the essence of the problem difficulty is hidden in the properties of the availability areas for certain types of agents. For example, the problem becomes hard for trees if such areas are allowed to be disconnected. To formally argue that this is the case we give, in Sect. 5, a polynomial-time algorithm that finds an optimal search strategy for heterogeneous agents in case when the such areas are connected.

2   Preliminaries

In this work we consider simple edge-labeled graphs $G = (V(G), E(G), c)$, i.e., without loops or multiple edges, where $c \colon E(G) \to \{1, \ldots, z\}$ is a function that assigns labels, called *colors*, to the edges of G. Then, if $c(\{u, v\}) = i$, $\{u, v\} \in E(G)$, then we also say that vertices u and v *have color* i. Note that vertices may have multiple colors, so by $c(v) := \{c(\{u, v\}) : \{u, v\} \in E(G)\}$ we will refer to the set of colors of a vertex $v \in V(G)$.

Problem Formulation. We will start by recalling the classical *edge search* problem [29] and then we will formally introduce our adaptation of this problem to the case of heterogeneous agents.

An (edge) *search strategy* $\mathcal{S}$ for a simple graph $G = (V(G), E(G))$ is a sequence of moves $\mathcal{S} = (m_1, \ldots, m_\ell)$. Each move m_i is one of the following actions:

- (M1) placing a searcher on a vertex,
- (M2) removing a searcher from a vertex,
- (M3) sliding a searcher present on a vertex u along an edge $\{u, v\}$ of G, which results in a searcher ending up on v.

Furthermore, we recursively define for each $i \in \{0, \ldots, \ell\}$ a set $\mathcal{C}_i$ such that $\mathcal{C}_i$, $i > 0$, is the set of edges that are *clean* after the move m_i and $\mathcal{C}_0$ is the set of edges that are clean prior to the first move of $\mathcal{S}$. Initially, we set $\mathcal{C}_0 = \emptyset$. For $i > 0$ we compute $\mathcal{C}_i$ in two steps. In the first step, let $\mathcal{C}'_i = \mathcal{C}_{i-1}$ for moves (M1) and (M2), and let $\mathcal{C}'_i = \mathcal{C}_{i-1} \cup \{\{u, v\}\}$ for a move (M3). In the second step compute $\mathcal{R}_i$ to consists of all edges e in $\mathcal{C}'_i$ such that there exists a path P in G such that none of the vertices of P is occupied by a searcher at the end of move m_i, one endpoint of P belongs to e and the other endpoint of P belong to an edge not in $\mathcal{C}_{i-1}$. Then, set $\mathcal{C}_i = \mathcal{C}'_i \setminus \mathcal{R}_i$. If $\mathcal{R}_i \neq \emptyset$, then we say that the edges in $\mathcal{R}_i$ become *recontaminated* (or that *recontamination occurs in* $\mathcal{S}$ if it is not important which edges are involved). If l_e is the number of times the edge e becomes recontaminated during a search strategy, then the value $\sum_{e \in E(G)} l_e$ is referred to as the number of *unit recontaminations*. Finally, we define $\mathcal{D}_i = E(G) \setminus \mathcal{C}_i$ to be the set of edges that are *contaminated* at the end of move m_i, $i > 0$, where again $\mathcal{D}_0$ refers to the state prior to the first move. Note that $\mathcal{D}_0 = E(G)$. We require from a search strategy that $\mathcal{C}_\ell = E(G)$.

Denote by $V(m_i)$ the vertices occupied by searchers at the end of move m_i. We write $|\mathcal{S}|$ to denote the number of searchers used by $\mathcal{S}$ understood as the minimum number k such that at most k searchers are present on the graph in each move. Then, the *search number of* G is

$$\mathsf{s}(G) = \min\left\{ |\mathcal{S}| \;\middle|\; \mathcal{S} \text{ is a search strategy for } G \right\}.$$

If the graph induced by edges in $\mathcal{C}_i$ is connected for each $i \in \{1, \ldots, \ell\}$, then we say that $\mathcal{S}$ is *connected*. We then recall the *connected search number of* G:

$$\mathsf{cs}(G) = \min\left\{ |\mathcal{S}| \;\middle|\; \mathcal{S} \text{ is a connected search strategy for } G \right\}.$$

We now adopt the above classical graph searching definitions to the searching problem we study in this work. For an edge-labeled graph $G = (V(G), E(G), c)$, a search strategy assigns to each of the k searchers used by a search strategy a color: the color of searcher i is denoted by $\tilde{c}(i)$. This is done prior to any move, and the assignment remains fixed for the rest of the strategy. Then again, a search strategy $\mathcal{S}$ is a sequence of moves with the following constraints: in move (M1) that places a searcher i on a vertex v it holds $\tilde{c}(i) \in c(v)$; move (M2) has no additional constraints; in move (M3) that uses a searcher i for sliding along an edge $\{u, v\}$ it holds $\tilde{c}(i) = c(\{u, v\})$. Note that, in other words, the above constraints enforce the strategy to obey the requirement that at any given time a searcher may be present on a vertex of the same color and a searcher may only slide along an edge of the same color. To stress out that a search strategy uses agents with color assignment $\tilde{c}$, we refer to as a *search $\tilde{c}$-strategy*. We write $\tilde{c}_{\mathcal{S}}(i)$ to refer to the number of agents with color i in a search strategy $\mathcal{S}$.

Then we introduce the corresponding graph parameters $\mathbf{hs}(G)$ and $\mathbf{hcs}(G)$ called the *heterogeneous search number* and *heterogeneous connected search number* of G, where $\mathbf{hs}(G)$ (respectively $\mathbf{hcs}(G)$) is the minimum integer k such that there exists a (connected) search $\tilde{c}$-strategy for G that uses k searchers. Whenever we write $\mathbf{s}(G)$ or $\mathbf{cs}(G)$ for an edge-labeled graph $G = (V, E, c)$ we refer to $\mathbf{s}(G')$ and $\mathbf{cs}(G')$, respectively, where $G' = (V, E)$ is isomorphic to G.

We say that a search strategy $\mathcal{S}$ is *monotone* if no recontamination occurs in $\mathcal{S}$. Analogously, for the search numbers given above, we define *monotone, connected monotone, heterogeneous monotone* and *connected heterogeneous monotone* search numbers denoted by $\mathbf{ms}(G)$, $\mathbf{mcs}(G)$, $\mathbf{mhs}(G)$ and $\mathbf{mhcs}(G)$, respectively, to be the minimum number of searchers required by an appropriate monotone search strategy.

The decision versions of the combinatorial problems we study are as follows:

Heterogeneous Graph Searching Problem (HGS)
> Given an edge-labeled graph $G = (V(G), E(G), c)$ and an integer k, does it hold $\mathbf{hs}(G) \leq k$?

Heterogeneous Connected Graph Searching Problem (HCGS)
> Given an edge-labeled graph $G = (V(G), E(G), c)$ and an integer k, does it hold $\mathbf{hcs}(G) \leq k$?

In the optimization versions of both problems an edge-labeled graph G is given as an input and the goal is to find the minimum integer k, a labeling $\tilde{c}$ of k searchers and a (connected) search $\tilde{c}$-strategy for G.

Additional Notation and Remarks. For some nodes v in $V(G)$ we have $|c(v)| > 1$, such connecting nodes we will call *junctions*. Thus, a node v is a junction if there exist two edges with different colors incident to v.

We define an *area* in G to be a maximal subgraph H of G such that for every two edges e, f of H, there exists a path P in H connecting an endpoint of e with and endpoint of f such that P contains no junctions. We further extend our notation to denote by $c(H)$ the color of all edges in area H. Note that two areas of the same color may share a junction. Let $Areas(G)$ denote all areas of G. Two areas are said to be *adjacent* if they include the same junction.

Fact 1. *If T is a tree and v is a junction that belongs to some area H in T, then v is a leaf (its degree is one) in H.* □

Lemma 1. *Given a tree $T = (V(T), E(T), c)$ and any area H in T, any search $\tilde{c}$-strategy for T uses at least $\mathbf{s}(H)$ searchers of color $c(H)$.* □

Lemma 1 leads to the following lower bound:

$$\beta(G) = \sum_{i=1}^{z} \max\{\mathbf{s}(H) \mid H \in Areas(G), c(H) = i\}.$$

Lemma 2. *For each tree T it holds $\mathbf{hs}(T) \geq \beta(T)$.* □

3   Lack of Monotonicity

Restricting available strategies to monotone ones can lead to increase of heterogeneous search number, even in case of trees. We express this statement in form of the following main theorem of this section:

Theorem 1. *There exists a tree T such that $\mathbf{mhs}(T) > \mathbf{hs}(T)$.*

In order to prove this theorem we provide an example of a tree $T_l = (V, E, c)$, where $l \geq 3$ is an integer, which cannot be cleaned with $\beta(T_l)$ searchers using a monotone search strategy, but there exists a non-monotone strategy, provided below, which achieves this goal. Our construction is shown in Fig. 1.

We first define three building blocks needed to obtain T_l, namely subtrees T_1', T_2' and T_l''. We use three colors, i.e., $z = 3$. The construction of T_i', $i \in \{1, 2\}$, starts with a root vertex q_i, which has 3 further children connected by edges of color 1. Each child of q_i has 3 children connected by edges of color 2. For the tree T_l'', $l \geq 3$, take vertices $v_0, \ldots, v_{l+1}$ that form a path with edges $e_x = \{v_x, v_{x+1}\}$, $x \in \{0, \ldots, l\}$. We set $c(e_x) = x \mod 3 + 1$. We attach one pendant edge with color $x \mod 3 + 1$ and one with color $(x - 1) \mod 3 + 1$ to each vertex v_x, $x \in \{1, \ldots, l\}$. Next, we take a path P with four edges in which two internal edges are of color 2 and two remaining edges are of color 3. To finish the construction of T_l'', identify the middle vertex of P, incident to the two edges of color 2, with the vertex v_0 of the previously constructed subgraph. We link two copies of T_i', $i \in \{1, 2\}$, by identifying two endpoints of the path P with the roots q_1 and q_2 of T_1' and T_2', respectively, obtaining the final tree T_l shown in Fig. 1.

Now, we analyze a potential monotone search $\tilde{c}$-strategy $\mathcal{S}$ using $\beta(T_l) = 3$ searchers. By Lemma 1, $\mathcal{S}$ uses one searcher of each color. We define a notion of a *step* for $\mathcal{S} = (m_1, \ldots, m_l)$ to refer to some particular move indices in $\mathcal{S}$. We distinguish the following steps that will be used in the lemmas below:

1. step t_i, $i \in \{1, 2\}$, equals the minimum index j such that at the end of move m_j all searchers are placed on the vertices of T_i' (informally, this is the first move in which all searchers are present in T_i');

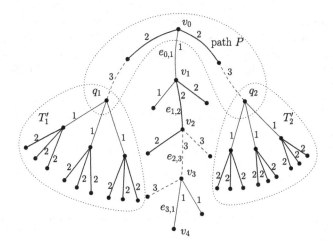

Fig. 1. The construction of T_3 ($l = 3$) from the trees T_1', T_2' and T_3''. Regular, heavy and dashed edges have labels $1, 2$ and 3, respectively.

2. step t_i', $i \in \{1, 2\}$, is the maximum index j such that at the end of move m_j all searchers are placed on the vertices of T_i' (informally, this is the last move in which all searchers are present in T_i');
3. steps t_3, t_3' are, respectively, the minimum and maximum indices j such that at the end of m_j all searchers are placed on the vertices in $V(P) \cup V(T_l'')$.

We skip a simple proof that all above steps are well defined, i.e., for any search strategy using 3 searchers for T each of the steps t_i, t_i', $i \in \{1, 2, 3\}$, must occur (for trees T_1' and T_2' this immediately follows from $\mathsf{s}(T_i') = 3$ for $i \in \{1, 2\}$).

Lemma 3. *For each monotone $\tilde{c}$-search strategy $\mathcal{S}$ for T_3 it holds: $t_1 \leq t_1' < t_3 \leq t_3' < t_2 \leq t_2'$ or $t_2 \leq t_2' < t_3 \leq t_3' < t_1 \leq t_1'$.*

Proof. The arguments used to prove this lemma do not use colors, so atomic statements about search strategies for subgraphs can be analyzed using simple and well known results for edge search model. Furthermore, due to the symmetry of T, it is enough to analyze only the case when $t_1 < t_2$. Note that $t_i \leq t_i'$, $i \in \{1, 2, 3\}$, follows directly from the definition. The vertices q_i, $i \in \{1, 2\}$, have to be guarded at some point between t_i-th move and t_i'-th move because $\mathsf{s}(T_i') = 3$. Because each step t_j, $j \in \{1, 2, 3\}$, uses all searchers, it cannot be performed if a searcher preventing recontamination is required to stay outside of subtree related to the respective step. The subtrees T_1' and T_2' contain no common vertices, so $t_1 < t_2$ implies $t_2 > t_1'$, as stated in the lemma.

Suppose for a contradiction that $t_3 < t_i$ for each $i \in \{1, 2\}$. In t_3-th move, since neither of t_i'-th moves has occurred, both subtrees T_1', T_2' contain contaminated edges. Moreover, some of the contaminated edges are incident to vertices q_i. Thus, any edge of T_l'' that is clean becomes recontaminated in the step $\min\{t_1, t_2\}$. Therefore, $t_1 < t_3$ as required.

Now we prove that $t_1' < t_3$. Suppose for a contradiction that $t_1 < t_3 < t_1'$. Consider the t_1'-th move. By $t_3 < t_1'$, T_l'' contains clean edges. By $t_1' < t_2$, q_2 is incident to contaminated edges in T_2'. Thus, there is a searcher outside of T_1' which prevents recontamination of clean edges in T_l''. Contradiction with the definition of t_1'.

In t_2-th move there are no spare searchers left to guard any contaminated area outside T_2' which bypasses q_2 and could threaten recontamination of T_1'. So all edges, including the ones in T_l'', between those two trees are already clean. Therefore, step t_3' already occurred, which allows us to conclude $t_3' < t_2$.   □

Due to the symmetry of T_l, we consider the case $t_1 \leq t_1' < t_3 \leq t_3' < t_2 \leq t_2'$.

Lemma 4. *During each move of index $t \in [t_1', t_2]$ there is a searcher on a vertex of P.*

Proof. By $t \geq t_1'$, q_1 is incident to some clean edges of T_1'. By $t \leq t_2$, q_2 is incident to some contaminated edges from T_2'. Hence there has to be a searcher on q_1, q_2 or a vertex of the path P between them to prevent recontamination. □

Let $f_i, i \in \{1, \ldots, l-1\}$, be the number of a move such that one of the edges incident to v_i is clean, one of the edges incident to v_i is being cleaned and and all other edges incident to v_i are contaminated.

Notice that $\mathsf{s}(T_l'') = 2$, and therefore an arbitrary search strategy $\mathcal{S}'$ using two searchers to clean a subtree without colors that is isomorphic to T_l'' follows one of these patterns: either the first searcher is placed, in some move of $\mathcal{S}'$, on v_1 and throughout the search strategy it moves from v_1 to v_{l-1} or the first searcher starts at v_{l-1} and moves from v_{l-1} to v_1 while $\mathcal{S}'$ proceeds. If for each $i \in \{1, \ldots, l-1\}$ the edge $\{v_{i-1}, v_i\}$ becomes clean prior to the edge $\{v_i, v_{i+1}\}$— we say that such $\mathcal{S}'$ *cleans T_l'' from v_1 to v_{l-1}* and if the edge $\{v_{i-1}, v_i\}$ becomes clean after $\{v_i, v_{i+1}\}$—we say that such $\mathcal{S}$ *cleans T_l'' from v_{l-1} to v_1*.

Lemma 5. *Each move of index f_i, $i \in \{1, \ldots, l-1\}$, is well defined. Either $f_1 < f_2 < \ldots < f_{l-2} < f_{l-1}$ or $f_{l-1} < f_{l-2} < \ldots < f_2 < f_1$.*

Proof. Consider a f-th move which belongs to $[t_3, t_3']$ in a search strategy $\mathcal{S}$. By Lemmas 3 and 4, a searcher is present on a vertex of P in the f-th move. Hence, only two searchers can be in T_l'' in the f-th move, so $\mathcal{S}$ cleans T_l'' from v_1 to v_{l-1} or cleans T_l'' from v_{l-1} to v_1. Note that during an execution of such a strategy there occur moves which satisfy the definition of f_i, and therefore there exists well defined f_i. When $\mathcal{S}$ cleans T_l'' from v_0 to v_l, then $f_1 < f_2 < \ldots < f_{l-2} < f_{l-1}$ is satisfied and when $\mathcal{S}$ cleans T_l'' from v_l to v_0, then $f_{l-1} < f_{l-2} < \ldots < f_2 < f_1$ is satisfied.   □

Lemma 6. *There exists no monotone search $\tilde{c}$-strategy that uses 3 searchers to clean T_l when $l \geq 7$.*

Proof. The vertex $v_i, i \in \{1, \ldots, l-1\}$, is incident to edges of colors $i \mod 3 + 1$ and $(i-1) \mod 3 + 1$, and therefore each move f_i uses both searchers of colors $i \mod 3 + 1$ and $(i-1) \mod 3 + 1$. By Lemma 4, the third searcher, which is of color $(i-2) \mod 3 + 1$, stays on P.

Consider a sequence $f_6 < f_5 < \ldots < f_2 < f_1$. Note that it implies that T_3'' is cleaned from v_{l-1} to v_1. Let us show that it is impossible to place a searcher on the vertices of P such that no recontamination occurs in each $f_i, i \in \{1, \ldots, 6\}$.

Consider the f_6-th move, where searchers of colors 1 and 3 are in T_l'' and 2 is on P. Before f_6-th move an edge incident to v_6 is clean (by definition of f_6). No edge incident to v_1 is clean and, by Lemma 3, T_j' has a clean edge, $j \in \{1, 2\}$. In order to prevent recontamination of T_j', the searcher is present on P, particularly on a vertex of the path from q_j to v_0. It cannot be the vertex q_j, because $2 \notin c(q_j)$, so the edge of color 3 incident to q_j is clean, and the searcher is on one of the remaining two vertices. Consider the f_5-th move in which the searcher of color 1 is on a vertex v of P. The vertex between q_j and v_0 cannot be occupied, due to its colors, and occupying q_j would cause recontamination—only the vertex v_0 is available, $v = v_0$. Consider the f_4-th move. The vertex v_0 cannot be occupied, due to its colors. The edge e_0 cannot be clean before e_4 is clean, because T_l'' is cleaned from v_{l-1} to v_1. Therefore, the searcher on v_0 cannot be moved towards q_2. Monotone strategy fails.

The argument is analogical for a sequence $f_1 < f_2 < \ldots < f_5 < f_6$. By Lemma 5, T_l'' is cleaned either from v_1 to v_{l-1} or the other way, which implies that considering the two above cases completes the proof. □

The following lemma (whose proof we skip due to space limitation) together with Lemma 6 prove Theorem 1.

Lemma 7. *There exists a non-monotone $\tilde{c}$-strategy S that cleans T_l using three searchers for each $l \geq 3$.*

Theorem 2. *There exist trees such that each search $\tilde{c}$-strategy that uses the minimum number of searchers has $\Omega(n^2)$ unit recontaminations.*

Proof. As a proof we use a tree H_l obtained through a modification of the tree T_l. In order to construct H_l, we replace each edge on the path P with a path P_m containing m vertices, where each edge between them is in the same color as the replaced edge in T_l. Clearly $\text{hs}(H_l) = \text{hs}(T_l)$. Note that we can adjust the number of vertices in T_l'' and P_m of H_l independently of each other. While the total number of vertices is $n = \Theta(m + l)$, we take $m = \Theta(n)$, $l = \Theta(n)$ in H_l.

In order to clean H_l, we employ the strategy provided in Lemma 7 adjusted in such a way, that any sliding moves performed on edges of P are replaced by $O(m)$ sliding moves on the corresponding paths of P_m. As shown previously, the number of times an edge of P in T_l, or path P_m in H_l, which contains $\Theta(m)$ elements, has to be recontaminated depends linearly on size of T_l''. In the later case the $\tilde{c}$-strategy cleaning H_l has $\Omega(ml) = \Omega(n^2)$ unit recontaminations . □

4   Hard Cases

We remark that the decision problem HGS is NP-complete for trees if we restrict available strategies to monotone ones. Formally, we show that the following problem is NP-complete:

Monotone Heterogeneous Graph Searching Problem (MHGS)
Given an edge-labeled graph $G = (V(G), E(G), c)$ and an integer k, does it hold $\mathbf{mhs}(G) \leq k$?

We obtain the following theorem whose proof due to space limitations is omitted.

Theorem 3. *The problem MHGS is NP-complete in the class of trees.*

It turns out that obtaining a similar result for arbitrary, that is, possibly non-monotone strategies, is much more complicated. Our proof of Theorem 4 is a non-trivial refinement of the proof of Theorem 3 and we present it in this way. In particular, we outline the changes and additional arguments that need to be used when dealing with non-monotone strategies. Due to the space limitations the full proof is omitted.

Theorem 4. *The problem HGS is NP-hard in the class of trees.*

5   Polynomially Tractable Instances

If G is a tree then Lemma 2 gives us a lower bound of $\beta(G)$ on the number of searchers. In this section we will look for an upper bound assuming that there is exactly one area for each color. With this assumption we show a constructive, polynomial-time algorithm both for HGS and HCGS.

Let $(E_1, \ldots, E_z)$ be the partition of edges of T so that E_i induces the area of color i in T. This partition induces a tree structure. More formally, consider a graph in which the set of vertices is $P_E = \{E_1, E_2, \ldots, E_z\}$ and $\{E_i, E_j\}$ is an edge if and only if an edge in E_i and and edge in E_j share a common junction in T. Then, let $\tilde{T}$ be the BFS spanning tree with the root E_1 in this graph. We write V_i to denote all vertices of the area with edge set E_i, $i \in \{1, \ldots, z\}$.

Our strategy for clearing T is recursive, starting with the root. The following procedure requires that when it is called, the area that corresponds to the parent of E_i in $\tilde{T}$ has been cleared, and if $i \neq 1$ (i.e., E_i is not the root of $\tilde{T}$), then assuming that E_j is the parent of E_i in $\tilde{T}$, a searcher of color j is present on the junction in $V_i \cap V_j$. With this assumption, the procedure recursively clears the subtree of $\tilde{T}$ rooted in E_i.

procedure CLEAR(labeled tree T, E_i)   ▷ Clear the subtree of T that corresponds to the subtree of $\tilde{T}$ rooted in E_i
  1. For each E_j such that E_j is a child of E_i in $\tilde{T}$ place a searcher of color j on the junction $v \in V_j \cap V_i$.

2. Clear the area of color i using $s(T[V_i])$ searchers. Remove all searchers of color i from vertices in V_i.

3. For each child E_j of E_i in $\tilde{T}$:
   (a) place a searcher of color i on the junction $v \in V_j \cap V_i$,
   (b) remove the searcher of color j from the vertex v,
   (c) call *Clear* recursively with input T and E_j,
   (d) remove the searcher of color i from the vertex v.

end procedure

We skip a simple proof that, for a given tree $G = (V(G), E(G), c)$, procedure Clear(G, E_1) clears G using $\beta(G)$ searchers. We also immediately obtain.

Lemma 8. *If all the strategies used in step 2 of procedure Clear to clear a subtree $T[V_i]$ are monotone, then the resulting $\tilde{c}$-strategy for G is also monotone.* □

It is known that there exists an optimal monotone search strategy for any graph [24] and it can be computed in linear time for a tree [27]. An optimal connected search strategy can be also computed in linear time for a tree [2]. Thus, using Lemma 2 we conclude with the following theorem:

Theorem 5. *Let $G = (V(G), E(G), c)$ be a tree such that the subgraph G_j composed by the edges in E_j is connected for each $j \in \{1, 2, \ldots, z\}$. Then, there exists a polynomial-time algorithm for solving problems HGS and HCGS.*

6 Conclusions and Open Problems

Our main open question, following the same unresolved one for connected searching, is whether problems HGS and HCGS belong to NP?

Our more practical motivation for studying the problems is derived from modeling physical environments to whose parts different robots have different access. More complex scenarios than the one considered in this work are those in which either an edge can have multiple colors (allowing it to be traversed by all agents of those colors), and/or a searcher can have multiple colors, which in turns extends its range of accessible parts of the graph.

References

1. Amini, O., Coudert, D., Nisse, N.: Non-deterministic graph searching in trees. Theoret. Comput. Sci. **580**, 101–121 (2015)
2. Barrière, L., et al.: Connected graph searching. Inf. Comput. **219**, 1–16 (2012)
3. Barrière, L., Flocchini, P., Fraigniaud, P., Santoro, N.: Capture of an intruder by mobile agents. In: Proceedings of the Fourteenth Annual ACM Symposium on Parallelism in Algorithms and Architectures, SPAA 2002, pp. 200–209. ACM (2002)
4. Bärtschi, A., et al.: Energy-efficient delivery by heterogeneous mobile agents. In: 34th Symposium on Theoretical Aspects of Computer Science, STACS, pp. 10:1–10:14 (2017)

5. Blin, L., Burman, J., Nisse, N.: Exclusive graph searching. Algorithmica **77**(3), 942–969 (2017)
6. Clarke, N., Connon, E.: Cops, robber, and alarms. Ars Comb. **81**, 283–296 (2006)
7. Clarke, N., Nowakowski, R.: Cops, robber, and photo radar. Ars Comb. **56**, 97–103 (2000)
8. Clarke, N., Nowakowski, R.: Cops, robber and traps. Utilitas Mathematica **60**, 91–98 (2001)
9. Czyzowicz, J., Gąsieniec, L., Kosowski, A., Kranakis, E.: Boundary patrolling by mobile agents with distinct maximal speeds. In: Demetrescu, C., Halldórsson, M.M. (eds.) ESA 2011. LNCS, vol. 6942, pp. 701–712. Springer, Heidelberg (2011). https://doi.org/10.1007/978-3-642-23719-5_59
10. Dereniowski, D.: Maximum vertex occupation time and inert fugitive: recontamination does help. Inf. Process. Lett. **109**(9), 422–426 (2009)
11. Dereniowski, D.: Connected searching of weighted trees. Theoret. Comput. Sci. **412**, 5700–5713 (2011)
12. Dereniowski, D.: Approximate search strategies for weighted trees. Theoret. Comput. Sci. **463**, 96–113 (2012)
13. Dereniowski, D.: From pathwidth to connected pathwidth. SIAM J. Discrete Math. **26**(4), 1709–1732 (2012)
14. Dereniowski, D., Dyer, D., Tifenbach, R., Yang, B.: The complexity of zero-visibility cops and robber. Theoret. Comput. Sci. **607**, 135–148 (2015)
15. Dereniowski, D., Klasing, R., Kosowski, A., Kuszner, Ł.: Rendezvous of heterogeneous mobile agents in edge-weighted networks. Theoret. Comput. Sci. **608**, 219–230 (2015)
16. Dyer, D., Yang, B., Yasar, Ö.: On the fast searching problem. In: 4th International Conference Algorithmic Aspects in Information and Management, AAIM 2008, pp. 143–154 (2008)
17. Farrugia, A., Gąsieniec, L., Kuszner, Ł., Pacheco, E.: Deterministic rendezvous in restricted graphs. In: Italiano, G.F., Margaria-Steffen, T., Pokorný, J., Quisquater, J.-J., Wattenhofer, R. (eds.) SOFSEM 2015. LNCS, vol. 8939, pp. 189–200. Springer, Heidelberg (2015). https://doi.org/10.1007/978-3-662-46078-8_16
18. Feinerman, O., Korman, A., Kutten, S., Rodeh, Y.: Fast rendezvous on a cycle by agents with different speeds. In: Chatterjee, M., Cao, J., Kothapalli, K., Rajsbaum, S. (eds.) ICDCN 2014. LNCS, vol. 8314, pp. 1–13. Springer, Heidelberg (2014). https://doi.org/10.1007/978-3-642-45249-9_1
19. Fomin, F., Heggernes, P., Telle, J.: Graph searching, elimination trees, and a generalization of bandwidth. Algorithmica **41**(2), 73–87 (2004)
20. Fraigniaud, P., Nisse, N.: Monotony properties of connected visible graph searching. Inf. Comput. **206**(12), 1383–1393 (2008)
21. Gaspers, S., Messinger, M.E., Nowakowski, R., Pralat, P.: Parallel cleaning of a network with brushes. Discrete Appl. Math. **158**(5), 467–478 (2010)
22. Ilcinkas, D., Nisse, N., Soguet, D.: The cost of monotonicity in distributed graph searching. Distrib. Comput. **22**(2), 117–127 (2009)
23. Kawamura, A., Kobayashi, Y.: Fence patrolling by mobile agents with distinct speeds. Distrib. Comput. **28**(2), 147–154 (2015)
24. LaPaugh, A.: Recontamination does not help to search a graph. J. ACM **40**(2), 224–245 (1993)
25. Luna, G.D., Flocchini, P., Santoro, N., Viglietta, G., Yamashita, M.: Self-stabilizing meeting in a polygon by anonymous oblivious robots. CoRR abs/1705.00324 (2017). http://arxiv.org/abs/1705.00324

26. Markou, E., Nisse, N., Pérennes, S.: Exclusive graph searching vs. pathwidth. Inf. Comput. **252**, 243–260 (2017)

27. Megiddo, N., Hakimi, S., Garey, M., Johnson, D., Papadimitriou, C.: The complexity of searching a graph. J. ACM **35**(1), 18–44 (1988)

28. Mihai, R., Todinca, I.: PATHWIDTH is NP-hard for weighted trees. In: Deng, X., Hopcroft, J.E., Xue, J. (eds.) FAW 2009. LNCS, vol. 5598, pp. 181–195. Springer, Heidelberg (2009). https://doi.org/10.1007/978-3-642-02270-8_20

29. Parsons, T.: Pursuit-evasion in a graph. In: Alavi, Y., Lick, D.R. (eds.) Theory and Applications of Graphs. Lecture Notes in Mathematics, vol. 642, pp. 426–441. Springer, Heidelberg (1978). https://doi.org/10.1007/BFb0070400

30. Qian, Z., Li, J., Li, X., Zhang, M., Wang, H.: Modeling heterogeneous traffic flow: a pragmatic approach. Transp. Res. Part B: Methodol. **99**, 183–204 (2017)

31. Sundaram, S., Krishnamoorthy, K., Casbeer, D.: Pursuit on a graph under partial information from sensors. CoRR abs/1609.03664 (2016)

32. Yang, B., Dyer, D., Alspach, B.: Sweeping graphs with large clique number. Discrete Math. **309**(18), 5770–5780 (2009)

Finding a Mediocre Player

Adrian Dumitrescu$^{(\boxtimes)}$

University of Wisconsin–Milwaukee, Milwaukee, WI, USA
dumitres@uwm.edu

Abstract. Consider a totally ordered set S of n elements; as an example, a set of tennis players and their rankings. Further assume that their ranking is a total order and thus satisfies transitivity and anti-symmetry. Following Yao [29], an element (player) is said to be (i, j)-*mediocre* if it is neither among the top i nor among the bottom j elements of S. More than 40 years ago, Yao suggested a stunningly simple algorithm for finding an (i, j)-mediocre element: Pick $i + j + 1$ elements arbitrarily and select the $(i + 1)$-th largest among them. She also asked: "Is this the best algorithm?" No one seems to have found such an algorithm ever since.

We first provide a deterministic algorithm that beats the worst-case comparison bound in Yao's algorithm for a large range of values of i (and corresponding suitable $j = j(i)$). We then repeat the exercise for randomized algorithms; the average number of comparisons of our algorithm beats the average comparison bound in Yao's algorithm for another large range of values of i (and corresponding suitable $j = j(i)$); the improvement is most notable in the symmetric case $i = j$. Moreover, the tight bound obtained in the analysis of Yao's algorithm allows us to give a definite answer for this class of algorithms. In summary, we answer Yao's question as follows: (i) "Presently not" for deterministic algorithms and (ii) "Definitely not" for randomized algorithms. (In fairness, it should be said however that Yao posed the question in the context of deterministic algorithms.)

Keywords: Comparison algorithm · Randomized algorithm ·
Approximate selection · i-th order statistic · Mediocre element ·
Yao's hypothesis · Tournaments · Quantiles

1 Introduction

Given a sequence A of n numbers and an integer (selection) parameter $1 \leq i \leq n$, the selection problem asks to find the i-th smallest element in A. If the n elements are distinct, the i-th smallest is larger than $i - 1$ elements of A and smaller than the other $n - i$ elements of A. By symmetry, the problems of determining the i-th smallest and the i-th largest are equivalent; throughout this paper, we will be mainly concerned with the latter dual problem.

Together with sorting, the selection problem is one of the most fundamental problems in computer science. Sorting trivially solves the selection problem;

© Springer Nature Switzerland AG 2019
P. Heggernes (Ed.): CIAC 2019, LNCS 11485, pp. 212–223, 2019.
https://doi.org/10.1007/978-3-030-17402-6_18

however, a higher level of sophistication is required in order to obtain a linear time algorithm. This was accomplished in the early 1970s, when Blum et al. [6] gave an $O(n)$-time algorithm for the problem. Their algorithm performs at most $5.43n$ comparisons and its running time is linear irrespective of the selection parameter i. Their approach was to use an element in A as a pivot to partition A into two smaller subsequences and recurse on one of them with a (possibly different) selection parameter i. The pivot was set as the (recursively computed) median of medians of small disjoint groups of the input array (of constant size at least 5). More recently, several variants of SELECT with groups of 3 and 4, also running in $O(n)$ time, have been obtained by Chen and Dumitrescu and independently by Zwick; see [7].

The selection problem, and computing the median in particular are in close relation with the problem of finding the quantiles of a set. The k-th *quantiles* of an n-element set are the $k - 1$ order statistics that divide the sorted set in k equal-sized groups (to within 1); see, e.g., [8, p. 223]. The k-th quantiles of a set can be computed by a recursive algorithm running in $O(n \log k)$ time.

In an attempt to drastically reduce the number of comparisons done for selection (down from $5.43n$), Schönhage et al. [28] designed a non-recursive algorithm based on different principles, most notably the technique of mass-production. Their algorithm finds the median (the $\lceil n/2 \rceil$-th largest element) using at most $3n + o(n)$ comparisons; as noted by Dor and Zwick [11], it can be adjusted to find the i-th largest, for any i, within the same comparison count. In a subsequent work, Dor and Zwick [12] managed to reduce the $3n + o(n)$ comparison bound to about $2.95n$; this however required new ideas and took a great deal of effort.

Mediocre Elements (Players). Following Yao, an element is said to be (i, j)-*mediocre* if it is neither among the top (i.e., largest) i nor among the bottom (i.e., smallest) j of a totally ordered set S of n elements. Yao remarked, that historically, finding a mediocre element is closely related to finding the median, with a common motivation being selecting an element that is not too close to either extreme. Observe also that (i, j)-mediocre elements where $i = \lfloor \frac{n-1}{2} \rfloor$, $j = \lfloor \frac{n}{2} \rfloor$ (and symmetrically exchanged), are medians of S.

In her PhD thesis [29], Yao suggested a stunningly simple algorithm for finding an (i, j)-mediocre element: Pick $i + j + 1$ elements arbitrarily and select the $(i + 1)$-th largest among them. It is easy to check that this element satisfies the required condition. Yao asked whether this algorithm is optimal. No improvements over this algorithm were known. An interesting feature of this algorithm is that its complexity does not depend on n (unless i or j do). The author also proved that this algorithm is optimal for $i = 1$. For $i + j + 1 \leq n$, let $S(i, j, n)$ denote the minimum number of comparisons needed in the worst case to find an (i, j)-mediocre element. Yao [29, Sect. 4.3] proved that $S(1, j, n) = V_2(j + 2) = j + \lceil \log(j + 2) \rceil$, and so $S(1, j, n)$ is independent of n. Here $V_2(j + 2)$ denotes the minimum number of comparisons needed in the worst case to find the second largest out of $j + 2$ elements.

The question of whether this algorithm is optimal for all values of i and j has remained open ever since; alternatively, the question is whether $S(i, j, n)$ is independent of n for all other values of i and j. Here we provide two alternative algorithms for finding a mediocre element (one deterministic and one randomized), and thereby confront Yao's algorithm with concrete challenges.

Background and Related Problems. Determining the comparison complexity for computing various order statistics including the median has lead to many exciting questions, some of which are still unanswered today. In this respect, Yao's hypothesis on selection [29, Sect. 4] has stimulated the development of such algorithms [11,27,28]. That includes the seminal algorithm of Schönhage et al. [28], which introduced principles of mass-production for deriving an efficient comparison-based algorithm.

Due to its primary importance, the selection problem has been studied extensively; see for instance [1,4,5,9–12,16–23,27,30]. A comprehensive review of early developments in selection is provided by Knuth [24]. The reader is also referred to dedicated book chapters on selection, such as those in [3,8] and the more recent articles [7,13], including experimental work [2].

In many applications (e.g., sorting), it is not important to find an exact median, or any other precise order statistic, for that matter, and an approximate median suffices [15]. For instance, quick-sort type algorithms aim at finding a balanced partition without much effort; see e.g., [19]. As a concrete example, Battiato et al. [4] gave an algorithm for finding a *weak* approximate median by using few comparisons. While the number of comparisons is at most $3n/2$ in the worst case, their algorithm can only guarantee finding an (i, j)-mediocre element with $i, j = \Omega(n^{\log_3 2})$; however, $n^{\log_3 2} = o(n)$, and so the selection made could be *shallow*.

Our Results. Our main results are summarized in the following. It is worth noting, however, that the list of sample data the theorems provide is not exhaustive.

Theorem 1. *Given a sequence of n elements, an (i, j)-mediocre element, where $i = \alpha n$, $j = (1 - 2\alpha)n - 1$, and $0 < \alpha < 1/3$, can be found by a deterministic algorithm A1 using $c_{A1} \cdot n + o(n)$ comparisons in the worst case, where the constants $c_{A1} = c_{A1}(\alpha)$ for the quantiles 1 through 33 are given in Fig. 2 (column A1 of the second table). In particular, if the number of comparisons done by Yao's algorithm is $c_{Yao} \cdot n + o(n)$, we have $c_{A1} < c_{Yao}$, for each of these quantiles.*

Theorem 2. *Given a sequence of n elements, an (i, j)-mediocre element, where $i = j = n/2 - n^{3/4}$, can be found by a randomized algorithm using $n + O(n^{3/4})$ comparisons on average. If $0 < \alpha < 1/2$ is a fixed constant, an (i, j)-mediocre element, where $i = j = \alpha n$, can be found using $2\alpha n + O(n^{3/4})$ comparisons on average. If $\alpha, \beta > 0$ are fixed constants with $\alpha + \beta < 1$, an (i, j)-mediocre element, where $i = \alpha n, j = \beta n$, can be found using $(\alpha + \beta)n + O(n^{3/4})$ comparisons on average.*

In particular, finding an element near the median requires about $3n/2$ comparisons for any previous algorithm (including Yao's), and finding the precise median requires $3n/2 + o(n)$ comparisons on average, while the main term in this expression cannot be improved [9]. In contrast, our randomized algorithm finds an element near the median using about n comparisons on average, thereby achieving a substantial savings of roughly $n/2$ comparisons.

Preliminaries and Notation. Without affecting the results, the following two standard simplifying assumptions are convenient:(i) the input sequence A contains n distinct numbers; and (ii) the floor and ceiling functions are omitted in the descriptions of the algorithms and their analyses. For example, for simplicity we write the αn-th element instead of the more precise $\lfloor \alpha n \rfloor$-th element. In the same spirit, for convenience we treat $\sqrt{n}$ and $n^{3/4}$ as integers. Unless specified otherwise, all logarithms are in base 2.

Let $E[X]$ and $Var[X]$ denote the expectation and respectively, the variance, of a random variable X. If E is an event in a probability space, $Prob(E)$ denotes its probability.

2 Instances and Algorithms for Deterministic Approximate Selection

We first make a couple of observations on the problem of finding an (i, j)-mediocre element. Without loss of generality (by considering the complementary order), it can be assumed that $i \leq j$; and consequently, $i < n/2$, if convenient. Our algorithm is designed to work for a specific range of values of i, j: $i \leq j \leq n - 2i - 1$; outside this range our algorithm simply proceeds as in Yao's algorithm. With anticipation, we note that our test values for purpose of comparison will belong to the specified range. Note that the conditions $i \leq j$ and $i \leq j \leq n - 2i - 1$ imply that $i < n/3$.

Yao's algorithm is very simple: simply pick $i + j + 1$ elements arbitrarily and select the $(i+1)$-th largest among them. As mentioned earlier, it is easy to check that this element satisfies the required condition.

Our algorithm (for the specified range) is also simple: Group the n elements into $m = n/2$ pairs and perform $n/2$ comparisons; then select the $(i + 1)$-th largest from the m upper elements in the m pairs. Let us first briefly argue about its correctness; denoting the selected element by x, on one hand, observe that x is smaller than i (upper) elements in disjoint pairs; on the other hand, observe that x is larger than $2\left(\frac{n}{2} - i - 1\right) + 1 = n - 2i - 1 \geq j$ (lower) elements in disjoint pairs, by the range assumption. It follows that the algorithm returns an (i, j)-mediocre element, as required.

It should be noted that both algorithms (ours as well as Yao's) make calls to exact selection, however with different input parameters. As such, we use state of the art algorithms and corresponding worst-case bounds for (exact) selection available. In particular, selecting the median can be accomplished with at most $2.95n$ comparisons, by using the algorithm of Dor and Zwick [12]; and if l is any fixed integer, selecting the αn-th largest element can be accomplished with at most $c_{\text{Dor}-\text{Zwick}} \cdot n + o(n)$ comparisons, where

$$c_{\text{Dor}-\text{Zwick}} = c_{\text{Dor}-\text{Zwick}}(\alpha, l) = 1 + (l+2)\left(\alpha + \frac{1-\alpha}{2^l}\right), \tag{1}$$

by using an algorithm tailored for shallow selection by the same authors [11]. In particular, by letting $l = \lfloor \log \frac{1}{\alpha} + \log\log \frac{1}{\alpha} \rfloor$ in Eq. (1), the authors obtain the following upper bound:

$$c_{\text{Dor}-\text{Zwick}}(\alpha) = 1 + (l+2)\left(\alpha + \frac{1-\alpha}{2^l}\right) \tag{2}$$
$$\leq 1 + \left(\log\frac{1}{\alpha} + \log\log\frac{1}{\alpha} + 2\right) \cdot \left(\alpha + \frac{2\alpha(1-\alpha)}{\log\frac{1}{\alpha}}\right).$$

Note that Eqs. (1) and (2) only lead to upper bounds in asymptotic terms.

Here we present an algorithm that outperforms Yao's algorithm for finding an $(\alpha n, \beta n)$-mediocre element for large n and for a broad range of values of α and suitable $\beta = \beta(\alpha)$, using current best comparison bounds for exact selection as described above. A key difference between our algorithm and Yao's lies in the amount of effort put into processing the input. Whereas Yao's algorithm chooses an arbitrary subset of elements of a certain size and ignores the remaining elements, our algorithm looks at all the input elements and gathers initial information based on grouping the elements into disjoint pairs and performing the respective comparisons.

Problem Instances. Consider the instance $(\alpha n, (1 - 2\alpha)n - 1)$ of the problem of selecting a mediocre element, where α is a constant $0 < \alpha < 1/3$.

Algorithms. We next specify our algorithm and Yao's algorithm for our problem instances. We start with our algorithm; and refer to Fig. 1 for an illustration.

Algorithm A1.

STEP 1: Group the n elements into $n/2$ pairs by performing $n/2$ comparisons.
STEP 2: Select and return the $(\alpha n + 1)$-th largest from the $n/2$ upper elements in the $n/2$ pairs. Refer to Fig. 1.

Let x denote the selected element. The general argument given earlier shows that x is $(\alpha n, (1 - 2\alpha)n - 1)$-mediocre: On one hand, there are $(2\alpha)n/2 = \alpha n$ elements larger than x; on the other hand, there are $(1 - 2\alpha)n/2 \cdot 2 - 1 = (1 - 2\alpha)n - 1$ elements smaller than x, as required.

Algorithm Yao.

STEP 1: Choose an arbitrary subset of $k = (1-\alpha)n$ elements from the given n.
STEP 2: Select and return the $(\alpha n + 1)$-th largest element from the k chosen.

Let y denote the selected element. As noted earlier, y is $(\alpha n, (1 - 2\alpha)n - 1)$-mediocre. Observe that the element returned by Yao's algorithm corresponds to a selection problem with a fraction $\alpha' = \frac{\alpha}{1-\alpha}$ from the k available.

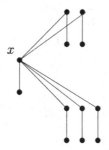

Fig. 1. Illustration of Algorithm A1 for $\alpha = 1/6$ and $n = 12$; large elements are at the top of the respective edges.

Analysis of the Number of Comparisons. For $0 < \alpha < 1$, let $f(\alpha)$ denote the multiplicative constant in the current best upper bound on the number of comparisons in the algorithm of Dor and Zwick for selection of the αn-th largest element out of n elements, according to (1), with one improvement. Instead of considering only one value for l, namely $l = \lfloor \log \frac{1}{\alpha} + \log \log \frac{1}{\alpha} \rfloor$, we also consider the value $l + 1$, and let the algorithm choose the best (i.e., the smallest of the two resulting values in (1) for the number of comparisons in terms of α). This simple change improves the advantage of Algorithm A1 over Yao's algorithm.

Recall that the algorithm of Dor and Zwick [11], which is a refinement of the algorithm of Schönhage et al. [28], is non-recursive, thus the selection target remains the same during its execution, and so choosing the best value for l can be done at the beginning of the algorithm. (Recall that the seminal algorithm of Schönhage et al. [28] is non-recursive as well.)

To be precise, let

$$g(\alpha, l) = \left(1 + (l + 2)\left(\alpha + \frac{1-\alpha}{2^l}\right)\right), \text{ and} \tag{3}$$

$$l = \left\lfloor \log \frac{1}{\alpha} + \log \log \frac{1}{\alpha} \right\rfloor, \tag{4}$$

$$f(\alpha) = \min\left(g(\alpha, l), g(\alpha, l + 1)\right). \tag{5}$$

It follows by inspection that the comparison counts for Algorithm A1 and Algorithm Yao are bounded from above by $c_{A1} \cdot n + o(n)$ and $c_{Yao} \cdot n + o(n)$, respectively, where

$$c_{A1} = \frac{1}{2}\left(1 + f(2\alpha)\right), \tag{6}$$

$$c_{Yao} = (1 - \alpha) \cdot f\left(\frac{\alpha}{1 - \alpha}\right). \tag{7}$$

It is worth noting that Eq. (5) yields values larger than 3 for certain values of α; e.g., for $\alpha = 0.4$, we have $l = 1$, and $g(0.4, 1) = 3.1$, $g(0.4, 2) = 3.2$, and so $f(0.4) = 3.1$. Moreover, a problem instance with $1/6 \leq \alpha < 1/3$ would entail

i	l	$g(\alpha_i, l)$	$g(\alpha_i, l+1)$	$f(\alpha_i)$		i	A1	Yao
1	9	1.1312	1.1316	1.1312		1	1.1191	1.1210
2	8	1.2382	1.2410	1.2382		2	1.2137	1.2175
3	7	1.3382	1.3378	1.3378		3	1.2987	1.3069
4	6	1.4400	1.4275	1.4275		4	1.3775	1.3846
5	6	1.5187	1.5168	1.5168		5	1.4484	1.4625
6	6	1.5975	1.6060	1.5975		6	1.5162	1.5300
7	5	1.6934	1.6762	1.6762		7	1.5812	1.5975
8	5	1.7612	1.7550	1.7550		8	1.6375	1.6637
9	5	1.8290	1.8337	1.8290		9	1.6937	1.7193
10	5	1.8968	1.9125	1.8968		10	1.7500	1.7750
11	4	1.9937	1.9646	1.9646		11	1.7937	1.8306
12	4	2.0500	2.0325	2.0320		12	1.8375	1.8850
13	4	2.1062	2.1003	2.1003		13	1.8812	1.9275
14	4	2.1625	2.1681	2.1625		14	1.9200	1.9700
15	4	2.2187	2.2359	2.2187		15	1.9500	2.0125
16	4	2.2750	2.3037	2.2750		16	1.9800	2.0550
17	3	2.3687	2.3312	2.3312		17	2.0000	2.0925
18	3	2.4125	2.3875	2.3875		18	2.0000	2.1200
19	3	2.4562	2.4437	2.4437		19	2.0000	2.1475
20	3	2.5000	2.5000	2.5000		20	2.0000	2.1750
21	3	2.5437	2.5562	2.5437		21	2.0000	2.2025
22	3	2.5875	2.6125	2.5875		22	2.0000	2.2200
23	3	2.6312	2.6687	2.6312		23	2.0000	2.2300
24	3	2.6750	2.7250	2.6750		24	2.0000	2.2400
25	2	2.7500	2.7187	2.7187		25	2.0000	2.2125
26	2	2.7800	2.7625	2.7625		26	2.0000	2.2200
27	2	2.8100	2.8062	2.8062		27	2.0000	2.1900
28	2	2.8400	2.8500	2.8400		28	2.0000	2.1600
29	2	2.8700	2.8937	2.8700		29	2.0000	2.1300
30	2	2.9000	2.9375	2.9000		30	2.0000	2.1000
31	2	2.9300	2.9812	2.9300		31	2.0000	2.0700
32	2	2.9600	3.0250	2.9600		32	2.0000	2.0400
33	2	2.9900	3.0687	2.9900		33	2.0000	2.0100

Fig. 2. Left: the values of $f(\alpha_i)$, $\alpha_i = i/100$, $i = 1, \ldots, 33$, for the algorithm of Dor and Zwick. Right: the comparison counts per element of A1 versus Yao for the first 33 quantiles. (The tables list the first four digits after the decimal point.)

Algorithm A1 making a call to an exact selection with parameter $1/3 \leq 2\alpha < 2/3$ (see (6) above). However, taking into consideration the possible adaptation of their algorithm pointed out by the authors [11], the expression of $f(\alpha)$ in (5) can be replaced by

$$f(\alpha) = \min\left(g(\alpha, l), g(\alpha, l+1), 3\right), \tag{8}$$

or even by

$$f(\alpha) = \min\left(g(\alpha, l), g(\alpha, l+1), 2.95\right). \tag{9}$$

We next show that (the new) Algorithm A1 outperforms Algorithm Yao with respect to the (worst-case) number of comparisons in selecting a mediocre element for n large enough and for all instances $(\alpha_i n, (1 - 2\alpha_i)n - 1)$, where $\alpha_i = i/100$, and $i = 1, \ldots, 33$; that is, for all quantiles $i = 1, \ldots, 33$ and suitable values of the 2nd parameter. This is proven by the data in the two tables in Fig. 2; the entries are computed using Eqs. (6) and (7), respectively. Moreover, the results remain the same, regardless of whether one uses the expression of $f(\alpha)$ in (8) or (9); to avoid the clutter, we only included the results obtained by using the expression of $f(\alpha)$ in (8).

3   Instances and Algorithms for Randomized Approximate Selection

Problem Instances. Consider the problem of selecting an (i, j)-mediocre element, for the important symmetric case $i = j$. To start with, let $i = j = n/2 - n^{3/4}$ (the first scenario described in Theorem 2); an extended range of values will be given in the end.

Algorithms. We next specify our algorithm[1] and compare it with Yao's algorithm instantiated with these values $(i = j = n/2 - n^{3/4})$.

Algorithm A2.
Input: A set S of n elements over a totally ordered universe.
Output: An (i, j)-mediocre element, where $i = j = n/2 - n^{3/4}$.

STEP 1: Pick a (multi)-set R of $\lceil n^{3/4} \rceil$ elements in S, chosen uniformly and independently at random with replacement.
STEP 2: Let m be median of R (computed by a linear-time deterministic algorithm).
STEP 3: Compare each of the remaining elements of $S \setminus R$ to m.
STEP 4: If there are at least $i = j = n/2 - n^{3/4}$ elements of S on either side of m, return m, otherwise FAIL.

Observe that (i) Algorithm A2 performs at most $n + O(n^{3/4})$ comparisons; and (ii) it either correctly outputs an (i, j)-mediocre element, where $i = j = n/2 - n^{3/4}$, or FAIL.

Analysis of the Number of Comparisons. Our analysis is quite similar to that of the classic randomized algorithm for finding the median; see [16], but also [26, Sect. 3.3] and [25, Sect. 3.4]. In particular, the randomized median finding algorithm and Algorithm A2 both fail for similar reasons.

---

[1] We could formulate a general algorithm for finding an (i, j)-mediocre element, acting differently in a specified range, as we did for the deterministic algorithm in Sect. 2. However, for clarity, we preferred to specify it in this way.

Recall that an execution of Algorithm A2 performs at most $n + O(n^{3/4})$ comparisons. Define a random variable X_i by

$$X_i = \begin{cases} 1 \text{ if the rank of the } i\text{th sample is less than } n/2 - n^{3/4}, \\ 0 \text{ else.} \end{cases}$$

The variables X_i are independent, since the sampling is done with replacement. It is easily seen that

$$p := \text{Prob}(X_i = 1) \le \frac{n/2 - n^{3/4}}{n} = \frac{1}{2} - \frac{1}{n^{1/4}}.$$

Let $X = \sum_{i=1}^{n^{3/4}} X_i$ be the random variable counting the number of samples in R whose rank is less than $n/2 - n^{3/4}$. By the linearity of expectation, we have

$$E[X] = \sum_{i=1}^{n^{3/4}} E[X_i] \le n^{3/4}\left(\frac{1}{2} - \frac{1}{n^{1/4}}\right) = \frac{n^{3/4}}{2} - \sqrt{n}.$$

Observe that the randomized algorithm A2 fails if and only if the rank (in S) of the median m of R is outside the rank interval $[n/2 - n^{3/4}, n/2 + n^{3/4}]$, i.e., the rank of m is smaller than $n/2 - n^{3/4}$ or larger than $n/2 + n^{3/4}$. Note that if algorithm A2 fails then at least $|R|/2 = n^{3/4}/2$ elements of R have rank $\le n/2 - n^{3/4}$ or at least $|R|/2 = n^{3/4}/2$ elements of R have rank $\ge n/2 + n^{3/4}$; denote these two events by E_1 and E_2, respectively. We next bound from above their probability.

Lemma 1.

$$\text{Prob}(E_1) \le \frac{1}{4n^{1/4}}.$$

Proof. Since X_i is a Bernoulli trial, X is a binomial random variable with parameters $n^{3/4}$ and p. Observing that $p(1 - p) \le 1/4$, it follows (see for instance [25, Sect. 3.2.1]) that

$$\text{Var}(X) = n^{3/4}p(1 - p) \le \frac{n^{3/4}}{4}.$$

Applying Chebyshev's inequality yields

$$\text{Prob}(E_1) \le \text{Prob}\left(X \ge \frac{n^{3/4}}{2}\right) \le \text{Prob}\left(|X - E[X]| \ge \sqrt{n}\right)$$

$$\le \frac{\text{Var}(X)}{n} \le \frac{n^{3/4}}{4n} = \frac{1}{4n^{1/4}},$$

as claimed. $\qquad\square$

Similarly, we deduce that $\text{Prob}(E_2) \le 1/(4n^{1/4})$. Consequently, by the union bound it follows that the probability that one execution of Algorithm A2 fails is bounded from above by

$$\text{Prob}(E_1 \cup E_2) \le \text{Prob}(E_1) + \text{Prob}(E_2) \le \frac{1}{2n^{1/4}}.$$

As in [25, Sect. 3.4], Algorithm A2 can be converted (from a Monte Carlo algorithm) to a Las Vegas algorithm by running it repeatedly until it succeeds. By Lemma 1, the FAIL probability is significantly small, and so the expected number of comparisons of the resulting algorithm is still $n + o(n)$). Indeed, the expected number of repetitions until the algorithm succeeds is at most

$$\frac{1}{1 - 1/(2n^{1/4})} \leq 1 + \frac{1}{n^{1/4}}.$$

Since the number of comparisons in each execution of the algorithm is $n + O(n^{3/4})$, the expected number of comparisons until success is at most

$$\left(1 + \frac{1}{n^{1/4}}\right)\left(n + O(n^{3/4})\right) = n + O(n^{3/4}).$$

We now analyze the average number of comparisons done by Yao's algorithm. On one hand, by a classic result of Floyd and Rivest [16], the k-th largest element out of n given, can be found using at most $n + \min(k, n - k) + o(n)$ comparisons on average. On the other hand, by a classic result of Cunto and Munro [9], this task requires $n + \min(k, n - k) + o(n)$ comparisons on average. In particular, the median of $i + j + 1 = n - 2n^{3/4} + 1$ elements can be found using at most $3n/2 + o(n)$ comparisons on average; and the main term in this expression cannot be improved.

Consequently, since $1 < 3/2$, the average number of comparisons done by Algorithm A2 is significantly smaller than the average number of comparisons done by Yao's algorithm for the task of finding an (i, j)-mediocre element, when n is large and $i = j = n/2 - n^{3/4}$.

Generalization. A broad range of symmetric instances for comparison purposes can be obtained as follows. Let $0 < \alpha < 1/2$ be any fixed constant. Consider the problem of selecting an (i, j)-mediocre element in the symmetric case, where $i = j = \alpha n$. Our algorithm first chooses an arbitrary subset of $2\alpha n + 2n^{3/4}$ elements of S to which it applies Algorithm A2; as such, it uses at most $2\alpha n + O(n^{3/4})$ comparisons on average. It is implicitly assumed here that $2\alpha n + 2n^{3/4} \leq n$, i.e., that $n^{1/4}(1 - 2\alpha) \geq 2$, which holds for n large enough. In contrast, Yao's algorithm chooses an arbitrary subset of $2\alpha n + 1$ elements and uses $3\alpha n + o(n)$ comparisons on average. Since $2\alpha < 3\alpha$ for every $\alpha > 0$, the average number of comparisons in Algorithm A2 is significantly smaller than the average number of comparisons in Yao's algorithm for the task of finding an (i, j)-mediocre element, when n is large and $i = j = \alpha n$.

A broad range of asymmetric instances with a gap, as described in Theorem 2, can be constructed using similar principles; in particular, in STEP 2 of Algorithm A2, a different order statistic of R (i.e., a biased partitioning element) is computed rather than the median of R. It is easy to see that the resulting algorithm performs at most $(\alpha + \beta)n + O(n^{3/4})$ comparisons on average. The correctness argument is similar to the one used above in the symmetric case and so we omit further details.

4   Conclusion

We presented two alternative algorithms—one deterministic and one randomized—for finding a mediocre element, i.e., for approximate selection. The deterministic algorithm outperforms Yao's algorithm for large n with respect to the worst-case number of comparisons for about one third of the quantiles (as the first parameter), and suitable values of the 2nd parameter, using state of the art algorithms for exact selection due to Dor and Zwick [11]. Moreover, we suspect that the comparison outcome remains the same for large n and the entire range of $\alpha \in (0, 1/3)$ and suitable $\beta = \beta(\alpha)$ in the problem of selecting an $(\alpha n, \beta n)$-mediocre element. Whether Yao's algorithm can be beaten by a deterministic algorithm in the symmetric case $i = j$ remains an interesting question.

The randomized algorithm outperforms Yao's algorithm for large n with respect to the expected number of comparisons for the entire range of $\alpha \in (0, 1/2)$ in the problem of finding an (i, j)-mediocre element, where $i = j = \alpha n$. These ideas can be also used to generate asymmetric instances with a gap for suitable variants of the randomized algorithm.

Discussions pertaining to lower bounds—for the deterministic and randomized cases, respectively—have been omitted from this version due to space constraints; they can be found in [14].

References

1. Ajtai, M., Komlós, J., Steiger, W.L., Szemerédi, E.: Optimal parallel selection has complexity $O(\log \log n)$. J. Comput. Syst. Sci. **38**(1), 125–133 (1989)
2. Alexandrescu, A.: Fast deterministic selection. In: Proceedings of the 16th International Symposium on Experimental Algorithms (SEA 2017), June 2017, London, pp. 24:1–24:19 (2017)
3. Baase, S.: Computer Algorithms: Introduction to Design and Analysis, 2nd edn. Addison-Wesley, Reading (1988)
4. Battiato, S., Cantone, D., Catalano, D., Cincotti, G., Hofri, M.: An efficient algorithm for the approximate median selection problem. In: Bongiovanni, G., Petreschi, R., Gambosi, G. (eds.) CIAC 2000. LNCS, vol. 1767, pp. 226–238. Springer, Heidelberg (2000). https://doi.org/10.1007/3-540-46521-9_19
5. Bent, S.W., John, J.W.: Finding the median requires $2n$ comparisons. In: Proceedings of the 17th Annual ACM Symposium on Theory of Computing (STOC 1985), pp. 213–216. ACM (1985)
6. Blum, M., Floyd, R.W., Pratt, V., Rivest, R.L., Tarjan, R.E.: Time bounds for selection. J. Comput. Syst. Sci. **7**(4), 448–461 (1973)
7. Chen, K., Dumitrescu, A.: Select with groups of 3 or 4. In: Dehne, F., Sack, J.R., Stege, U. (eds.) WADS 2015. LNCS, vol. 9214, pp. 189–199. Springer, Cham (2015). https://doi.org/10.1007/978-3-319-21840-3_16. arXiv.org/abs/1409.3600
8. Cormen, T.H., Leiserson, C.E., Rivest, R.L., Stein, C.: Introduction to Algorithms, 3rd edn. MIT Press, Cambridge (2009)
9. Cunto, W., Munro, J.I.: Average case selection. J. ACM **36**(2), 270–279 (1989)
10. Dor, D., Håstad, J., Ulfberg, S., Zwick, U.: On lower bounds for selecting the median. SIAM J. Discrete Math. **14**(3), 299–311 (2001)

11. Dor, D., Zwick, U.: Finding the αn-th largest element. Combinatorica **16**(1), 41–58 (1996)
12. Dor, D., Zwick, U.: Selecting the median. SIAM J. Comput. **28**(5), 1722–1758 (1999)
13. Dumitrescu, A.: A selectable sloppy heap. Algorithms **12**(3), 58 (2019). https://doi.org/10.3390/a12030058
14. Dumitrescu, A.: Finding a mediocre player (2019, preprint). arXiv.org/abs/1901.09017
15. Edelkamp, S., Weiß, A.: QuickMergesort: practically efficient constant-factor optimal sorting, preprint available at arXiv.org/abs/1804.10062
16. Floyd, R.W., Rivest, R.L.: Expected time bounds for selection. Commun. ACM **18**(3), 165–172 (1975)
17. Fussenegger, F., Gabow, H.N.: A counting approach to lower bounds for selection problems. J. ACM **26**(2), 227–238 (1979)
18. Hadian, A., Sobel, M.: Selecting the t-th largest using binary errorless comparisons. Comb. Theory Appl. **4**, 585–599 (1969)
19. Hoare, C.A.R.: Algorithm 63 (PARTITION) and algorithm 65 (FIND). Commun. ACM **4**(7), 321–322 (1961)
20. Hyafil, L.: Bounds for selection. SIAM J. Comput. **5**(1), 109–114 (1976)
21. John, J.W.: A new lower bound for the set-partitioning problem. SIAM J. Comput. **17**(4), 640–647 (1988)
22. Kaplan, H., Kozma, L., Zamir, O., Zwick, U.: Selection from heaps, row-sorted matrices and X+Y using soft heaps (2018, preprint). http://arXiv.org/abs/1802.07041
23. Kirkpatrick, D.G.: A unified lower bound for selection and set partitioning problems. J. ACM **28**(1), 150–165 (1981)
24. Knuth, D.E.: The Art of Computer Programming, Volume 3: Sorting and Searching, 2nd edn. Addison-Wesley, Reading (1998)
25. Mitzenmacher, M., Upfal, E.: Probability and Computing: Randomized Algorithms and Probabilistic Analysis. Cambridge University Press, Cambridge (2005)
26. Motwani, R., Raghavan, P.: Randomized Algorithms. Cambridge University Press, Cambridge (1995)
27. Paterson, M.: Progress in selection. In: Karlsson, R., Lingas, A. (eds.) SWAT 1996. LNCS, vol. 1097, pp. 368–379. Springer, Heidelberg (1996). https://doi.org/10.1007/3-540-61422-2_146
28. Schönhage, A., Paterson, M., Pippenger, N.: Finding the median. J. Comput. Syst. Sci. **13**(2), 184–199 (1976)
29. Yao, F.: On lower bounds for selection problems. Technical report MAC TR-121, Massachusetts Institute of Technology, Cambridge (1974)
30. Yap, C.K.: New upper bounds for selection. Commun. ACM **19**(9), 501–508 (1976)

Covering Tours and Cycle Covers
with Turn Costs: Hardness
and Approximation

Sándor P. Fekete[ID] and Dominik Krupke[(✉)][ID]

Department of Computer Science, TU Braunschweig,
38106 Braunschweig, Germany
{s.fekete,d.krupke}@tu-bs.de

Abstract. We investigate a variety of geometric problems of finding tours and cycle covers with minimum turn cost, which have been studied in the past, with complexity and approximation results, and open problems dating back to work by Arkin et al. in 2001. Many new practical applications have spawned variants: For *full coverage*, every point has to be covered, for *subset coverage*, specific points have to be covered, and for *penalty coverage*, points may be left uncovered by incurring a penalty. We make a number of contributions. We first show that finding a minimum-turn (full) cycle cover is NP-hard even in 2-dimensional grid graphs, solving the long-standing open *Problem 53* in *The Open Problems Project* edited by Demaine, Mitchell and O'Rourke. We also prove NP-hardness of finding a *subset* cycle cover of minimum turn cost in *thin* grid graphs, for which Arkin et al. gave a polynomial-time algorithm for full coverage; this shows that their boundary techniques cannot be applied to compute exact solutions for subset and penalty variants.

On the positive side, we establish the first constant-factor approximation algorithms for all considered subset and penalty problem variants for very general classes of instances, making use of LP/IP techniques. For these problems with many possible edge directions (and thus, turn angles, such as in hexagonal grids or higher-dimensional variants), our approximation factors also improve the combinatorial ones of Arkin et al. Our approach can also be extended to other geometric variants, such as scenarios with obstacles and linear combinations of turn and distance costs.

1 Introduction

Finding roundtrips of minimum cost is one of the classic problems of theoretical computer science. In its most basic form, the objective of the *Traveling Salesman Problem (TSP)* is to minimize the total length of a single tour that covers all of a given set of locations. If the tour is not required to be connected, the result may be a *cycle cover*: a set of closed subtours that together cover the whole set.

A full version of this extended abstract can be found at [17].

© Springer Nature Switzerland AG 2019
P. Heggernes (Ed.): CIAC 2019, LNCS 11485, pp. 224–236, 2019.
https://doi.org/10.1007/978-3-030-17402-6_19

This distinction makes a tremendous difference for the computational complexity: while the TSP is NP-hard, computing a cycle cover of minimum total length can be achieved in polynomial time, based on matching techniques.

Evaluating the cost for a tour or a cycle cover by only considering its length may not always be the right measure. Figure 1 shows an example application, in which a drone has to sweep a given region to fight mosquitoes that may transmit dangerous diseases. As can be seen in the right-hand part of the figure, by far the dominant part of the overall travel cost occurs when the drone has to change its direction. (See our related video and abstract [10] for more details, and the resulting tour optimization.) There is an abundance of other related applied work, e.g., mowing lawns or moving huge wind turbines [8].

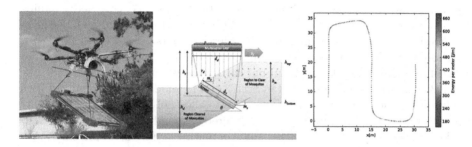

Fig. 1. (Left) A drone equipped with an electrical grid for killing mosquitoes. **(Middle)** Physical aspects of the flying drone. **(Right)** Making turns is expensive. See our related video [10].

For many purposes, two other variants are also practically important: for *subset coverage*, only a prespecified subset of locations needs to be visited, while for *penalty coverage*, locations may be skipped at the expense of an individual penalty. From the theoretical side, Arkin et al. [6] showed that finding minimum-turn tours in grid graphs is NP-hard, even if a minimum-turn cycle cover is given. The question whether a minimum-turn cycle cover can be computed in polynomial time (just like a minimum-length cycle cover) has been open for at least 17 years, dating back to the conference paper [5]; it has been listed for 15 years as *Problem 53* in *The Open Problems Project* edited by Demaine, Mitchell, and O'Rourke [15]. In Sect. 2 we resolve this problem by showing that computing a minimum-turn cycle cover in planar grid graphs is indeed NP-hard.

This raises the need for approximation algorithms. In Sect. 3, we present a technique based on Integer Programming (IP) formulations and their Linear Programming (LP) relaxations. Based on polyhedral results and combinatorial modifications, we prove constant approximation for all problem variants.

1.1 Related Work

Milling with Turn Costs. Arkin et al. [5,6] introduce the problem of milling (i.e., "carving out") with turn costs. They show hardness of finding an optimal tour, even in *thin* 2-dimensional grid graphs (which do not contain an induced 2×2 subgraph) with a given optimal cycle cover. They give a 2.5-approximation algorithm for obtaining a cycle cover, resulting in a 3.75-approximation algorithm for tours. The complexity of finding an optimal cycle cover in a 2-dimensional grid graph was established as *Problem 53* in *The Open Problems Project* [15].

Maurer [23] proves that a cycle *partition* with a minimum number of turns in grid graphs can be computed in polynomial time and performs practical experiments for optimal cycle covers. De Assis and de Souza [14] computed a provably optimal solution for an instance with 76 vertices. For the abstract version on graphs (in which "turns" correspond to weighted changes between edges), Fellows et al. [20] show that the problem is fixed-parameter tractable by the number of turns, tree-width, and maximum degree. Benbernou [11] considered milling with turn costs on the surface of polyhedrons in the 3-dimensional grid. She gives a corresponding 8/3-approximation algorithm for tours.

Note that the theoretical work presented in this paper has significant practical implications. As described in our forthcoming conference paper [18], the IP/LP-characterization presented in Sect. 3 can be modified and combined with additional algorithm engineering techniques to allow solving instances with more than 1000 pixels to provable optimality (thereby expanding the range of de Assis and de Souza [14] by a factor of 15), and computing solutions for instances with up to 300,000 pixels within a few percentage points (thereby showing that the practical performance of our approximation techniques is dramatically better than the established worst-case bounds).

For mowing problems, i.e., covering a given area with a moving object that may leave the region, Stein and Wagner [25] give a 2-approximation algorithm on the number of turns for the case of orthogonal movement. If only the traveled distance is considered, Arkin et al. [7] provide approximation algorithms for milling and mowing.

Angle and Curvature-Constrained Tours and Paths. If the instances are in the $\mathbb{R}^2$ plane and only the turning angles are measured, the problem is called the *Angular Metric Traveling Salesman Problem*. Aggarwal et al. [3] prove hardness and provide an $O(\log n)$ approximation algorithm for cycle covers and tours that works even for distance costs and higher dimensions. As shown by Aichholzer et al. [4], this problem seems to be very hard to solve optimally with integer programming. Fekete and Woeginger [19] consider the problem of connecting a point set with a tour for which the angles between the two successive edges are constrained. Finding a curvature-constrained shortest *path* with obstacles has been shown to be NP-hard by Lazard et al. [22]. Without obstacles, the problem is known as the *Dubins path* [16] that can be computed efficiently. With complexity depending on the types of obstacles, Boissonnat and Lazard [12], Agarwal et al. [1], and Agarwal and Wang [2] provide polynomial-time algorithms when possible or $1 + \epsilon$ approximation algorithms otherwise. Takei et al. [26] consider the solution of the problem from a practical perspective.

Related Combinatorial Problems. Goemans and Williamson [21] provide an approximation technique for constrained forest problems and similar problems that deal with penalties. In particular, they provide a 2-approximation algorithm for *Prize-Collecting Steiner Trees* in general symmetric graphs and the *Penalty Traveling Salesman Problem* in graphs that satisfy the triangle inequality. An introduction into approximation algorithms for prize-collecting/penalty problems, k-MST/TSP, and minimum latency problems is given by Ausiello et al. [9].

1.2   Preliminaries

The angular metric traveling salesman problem resp. cycle cover problem ask for a cycle resp. set of cycles such that a given set P of n points in $\mathbb{R}^d$ is covered and the sum of turn angles in minimized. A cycle is a closed chain of segments and covers the points of the segments' joints. A cycle has to cover at least two points. The turn angle of a joint is the angle difference to $180°$. In the presence of polygonal obstacles, cycles are not allowed to cross them. We consider three coverage variants: Full, subset, and penalty. In full coverage, every point has to be covered. In subset coverage, only points in a subset $S \subseteq P$ have to be covered (which is only interesting for grid graphs). In penalty coverage, no point has to be covered but every uncovered point $p \in P$ induces a penalty $c(p) \in \mathbb{Q}_0^+$ on the objective value. Optionally, the objective function can be a linear combination of distance and turn costs.

In the following, we introduce the *discretized angular metric*, by considering for every point $p \in P$ a set of ω possible orientations (and thus, 2ω possible directions) for a trajectory through p. We model this by considering for each $p \in P$ a set O_p of ω infinitely short segments, which we call *atomic strips*; a point is covered if one of its segments is part of the cycle, see Fig. 2. The corresponding selection of atomic strips is called *Atomic Strip Cover*, i.e., a selection of one $o \in O_p$ for every $p \in P$.

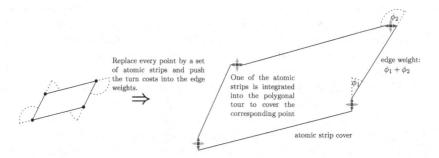

Fig. 2. Transforming an angular metric TSP instance and solution to an instance based on atomic strips, which can be considered infinitely small segments.

The atomic strips induce a weighted graph $G_O(V_O, E_O)$ with the endpoints of the atomic strips as vertices and the connections between the endpoints as edges. The weight of an edge in G_O equals the connection costs, in particular the turn costs on the two endpoints. Thus, the cycle cover problem turns into finding an Atomic Strip Cover with the minimum-weight perfect matching on its induced subgraph. As the cost of connections in it depends on *two* edges in the original graph, we call this generalized problem (in which the edge weights do not have to be induced by geometry) the *semi-quadratic cycle cover problem*.

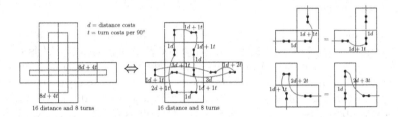

Fig. 3. (Left) From an optimal cycle cover (dotted) we can extract an Atomic Strip Cover (thick black), such that the matching (orange) induces an optimal solution. **(Right)** For turns it does not matter if we choose the horizontal or vertical atomic strip. (Color figure online)

It is important to note that the weights do not satisfy the triangle inequality; however, a direct connection is not more expensive than a connection that includes another atomic strip, giving rise to the following *pseudo-triangle inequalities*.

$$\forall v_1, v_2 \in V_O, w_1 w_2 \in O_p, p \in P : \begin{array}{l} \text{cost}(v_1 v_2) \le \text{cost}(v_1 w_1) + \text{cost}(w_2 v_2) \\ \text{cost}(v_1 v_2) \le \text{cost}(v_1 w_2) + \text{cost}(w_1 v_2) \end{array} \quad (1)$$

Our model allows the original objective function to be a linear combination of turn and distance costs, as it does not influence Eq. (1). Instances with polygonal obstacles for 2-dimensional geometric instances are also possible (however, for 3D, the corresponding edge weights can no longer be computed efficiently). A notable special case are *grid graphs* that arise as vertex-induced subgraphs of the infinite integer orthogonal grid. In this case, a point can only be covered straight, by a simple 90° turn, or by a 180° u-turn. We show grid graphs as polyominoes in which vertices are shown as *pixels*. We also speak of the number of *simple turns* (u-turns counting as two) instead of turn angles. More general grid graphs can be based on other grids, such as 3-dimensional integral or hexagonal grids.

Minimum turn cycle covers in grid graphs can be modeled as a semi-quadratic cycle cover problem with $\omega = 2$ and edge weights satisfying Eq. (1). One of the atomic strips represents being in a horizontal orientation (with an east and a west heading vertex) and the other being in a vertical orientation (with a north and a south heading vertex). The cost of an edge is as follows; see Fig. 3: Every vertex is connected to a position and a direction. The cost is the cheapest transition from the position and direction of the first vertex to the position and opposite heading of the second vertex (this is symmetric and can be computed efficiently). We can easily transform a cycle cover in a grid graph into one based on atomic strips and vice versa; see Fig. 3 (left). For each pixel we choose one of its transitions. If it is straight, we select the equally oriented strip; otherwise it does not matter, see Fig. 3 (right). With more atomic strips we can also model more general grid graphs such as hexagonal or 3-dimensional grid graphs with three atomic strips.

1.3 Our Contribution

We provide the following results.

- We resolve *Problem 53* in *The Open Problems Project* [15] by proving that finding a cycle cover of minimum turn cost is NP-hard, even in the restricted case of grid graphs. We also prove that finding a subset cycle cover of minimum turn cost is NP-hard, even in the restricted case of *thin* grid graphs, in which no induced 2 × 2 subgraph exists. This differs from the case of full coverage in thin grid graphs, which is known to be polynomially solvable [6].
- We provide a general IP/LP-based technique for obtaining $2 * \omega$ approximations for the semi-quadratic (penalty) cycle cover problem if Eq. (1) is satisfied, where ω is the maximum number of atomic strips per vertex.
- We show how to connect the cycle covers to minimum turn tours to obtain a 6 approximation for full coverage in regular grid graphs, 4ω approximations for full tours in general grid graphs, $4\omega + 2$ approximations for (subset) tours, and $4\omega + 4$ for penalty tours.

To the best of our knowledge, this is the first approximation algorithm for the subset and penalty variant with turn costs. For general grid graphs our techniques yields better guarantees than than the techniques of Arkin et al. who give a factor of $6 * \omega$ for cycle covers and $6 * \omega + 2$ for tours. In practice, our approach also yields better solutions for regular grid graphs, see [18].

2 Complexity

Problem 53 in *The Open Problems Project* asks for the complexity of finding a minimum-turn (full) cycle cover in a 2-dimensional grid graph. This is by no means obvious: large parts of a solution can usually be deduced by local information and matching techniques. In fact, it was shown by Arkin et al. [5,6] that the full coverage variant in *thin* grid graphs (which do not contain a 2 × 2 square, so every pixel is a boundary pixel) is solvable in polynomial time. In this section, we prove that finding a *full* cycle cover in 2-dimensional grid graphs with minimum turn cost is NP-hard, resolving *Problem 53*. We also show that *subset* coverage is NP-hard even for *thin* grid graphs, so the boundary techniques by Arkin et al. [5,6] do not provide a polynomial-time algorithm.

Theorem 1. *It is NP-hard to find a cycle cover with a minimum number of* $90°$ *turns* ($180°$ *turns counting as two) in a grid graph.*

The proof is based on a reduction from *One-in-three 3SAT* (1-in-3SAT), which was shown to be NP-hard by Schaefer [24]: for a Boolean formula in conjunctive normal form with only three literals per clause, decide whether there is a truth assignment that makes exactly one literal per clause `true` (and exactly two literals `false`). For example, $(x_1 \lor x_2 \lor x_3) \land (\overline{x_1} \lor \overline{x_2} \lor \overline{x_3})$ is not (1-in-3) satisfiable, whereas $(x_1 \lor x_2 \lor x_3) \land (\overline{x_1} \lor \overline{x_2} \lor \overline{x_4})$ is satisfiable.

See full version [17] for details, Fig. 5 for representing the one-clause formula $x_1 + x_2 + x_3 = 1$ with its three possible 1-in-3 solutions, and Fig. 4 for the instance $x_1 + x_2 + x_3 = 1 \land \overline{x_1} + \overline{x_2} + \overline{x_4} = 1 \land \overline{x_1} + x_2 + \overline{x_3} = 1$. For every variable we have a 𝕏 gadget consisting of a gray 🖽 gadget and a zig-zagging, high-cost path of blue pixels. A cheap solution traverses a blue path once and connect the ends through the remaining construction of gray and red pixels. Such *variable cycles* (highlighted in red) must either go through the upper (🖽) or lower (🖽) lane of the variable gadget; the former corresponds to a `true`, the later to a `false` assignment of the corresponding variable. A *clause gadget* modifies a lane of all three involved variable gadgets. This involves the gray pixels that are covered by the green cycles; we can show that they do not interfere with the cycles for covering the blue and red pixels, and cannot be modified to cover them. Thus, we only have to cover red and blue pixels, but can pass over gray pixels, too.

To this end, we must connect the ends of the blue paths; as it turns out, the formula is satisfiable if and only if we can perform this connection in a manner that also covers one corresponding red pixel with at most two extra turns.

For subset cover we can also show hardness for *thin* grid graphs. Arkin et al. [5,6] exploits the structure of these graphs to compute an optimal minimum-turn cycle cover in polynomial time. If we only have to cover a subset of the vertices, the problem becomes NP-hard again. The proof is inspired by the construction of Aggarwal et al. [3] for the angular-metric cycle cover problem and significantly simpler than the one for full coverage. See full version [17] for proof details.

Theorem 2. *The minimum-turn subset cycle cover problem is NP-hard, even in thin grid graphs.*

3   Approximation Algorithms

3.1   Cycle Cover

Now we describe a 2ω-approximation algorithm for the semi-quadratic (penalty) cycle cover problem with ω atomic strips per point if the edge weights satisfy Eq. (1). We focus on the full coverage version, as the penalty variant can be modeled in full coverage (with the same ω and while still satisfying Eq. (1)), by adding for every point $p \in P$ two further points that have a zero cost cycle only including themselves and a cycle that also includes p with the cost of the penalty.

Our approximation algorithm proceeds as follows. We first determine an atomic strip cover via linear programming. Computing an optimal atomic strip

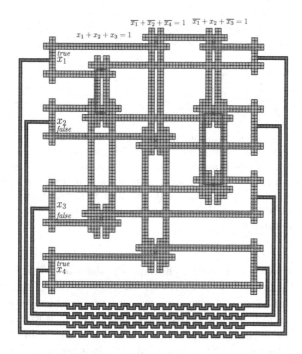

Fig. 4. Representing the *1-in-3SAT*-formula $x_1 + x_2 + x_3 = 1 \wedge \overline{x_1} + \overline{x_2} + \overline{x_4} = 1 \wedge \overline{x_1} + x_2 + \overline{x_3} = 1$. (Color figure online)

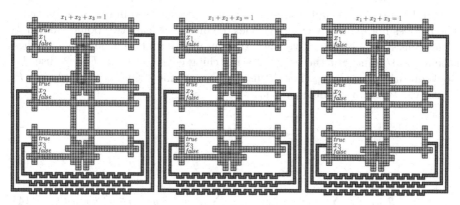

Fig. 5. Construction for the one-clause formula $x_1 + x_2 + x_3 = 1$ and three possible solutions. Every variable has a cycle traversing a zig-zagging path of blue pixels. A variable is **true** if its cycle uses the upper path (⊟) through green/red pixels, **false** if it takes the lower path (⊟). For covering the red pixels, we may use two additional turns. This results in three classes of optimal cycle covers, shown above. If we use the blue 4-turn cycle to cover the upper two red pixels, we are forced to cover the lower red pixel by the x_3 variable cycle, setting x_3 to **false**. The variable cycles of x_1 and x_2 take the cheapest paths, setting them to **true** or **false**, respectively. The alternative to a blue cycle is to cover all three red pixel by the variable cycles, as in the right solution. (Color figure online)

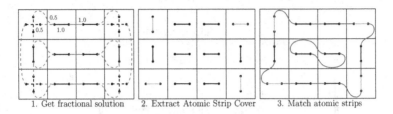

Fig. 6. Example of the approximation algorithm for a simple full cycle cover instance in a grid graph. First the fractional solution of the integer program (2)–(4) is computed. Strips and edges with value 0 are omitted, while dashed ones have value 0.5. Then the dominant (i.e., highest valued) atomic strips of this solution are selected. Finally, a minimum weight perfect matching on the ends of the atomic strips is computed. (Recall that atomic strips only have an but no length, so the curves in the corner indicate simple 90° turns.)

cover is NP-hard; we can show that choosing the *dominant* strips for each pixel in the fractional solution, i.e., those with the highest value, suffices to obtain provable good solutions. As a next step, we connect the atomic strips to a cycle cover, using a minimum-weight perfect matching. See Fig. 6 for an illustration.

We now describe the integer program whose linear programming relaxation is solved to select the dominant atomic strips. It searches for an optimal atomic strip cover that yields a perfect matching of minimum weight. To satisfy Eq. (1), transitive edges (connections implied by multiple explicitly given edges) may need to be added, especially loop-edges (which are not used in the final solution). The IP does not explicitly enforce cycles to contain at least two points: all small cycles consist only of transitive edges that implicitly contain at least one further atomic strip/point. For the usage of a matching edge $e = vw \in E_O$, we use the Boolean variable $x_e = x_{vw}$. For the usage of an atomic strip $o = vw \in O_p, p \in P$, we use the Boolean variable $y_o = y_{vw}$.

$$\min \quad \sum_{e \in E_O} \text{cost}(e) x_e \tag{2}$$

$$\text{s.t.} \quad \sum_{vw \in O_p} y_{vw} = 1 \qquad p \in P \tag{3}$$

$$2x_{vv} + \sum_{\substack{e \in E_O(v) \\ e \neq vv}} x_e = 2x_{ww} + \sum_{\substack{e \in E_O(w) \\ e \neq ww}} x_e = y_{vw} \; p \in P, vw \in O_p \tag{4}$$

We minimize the cost of the used edges, with Eq. (3) forcing the selection of one atomic strip per pixel (atomic strip cover) and Eq. (4) ensuring that exactly the vertices (endpoints) of the selected atomic strips are matched, with loop edges counting double due to their two ends.

Theorem 3. *Assuming edge weights that satisfy Eq. (1), there is a 2ω-approximation for semi-quadratic (penalty) cycle cover.*

Proof. Consider the described fractional atomic strip cover and matching of the integer program, which is a lower bound on the optimal cycle cover. We now show that we can transform this solution to a matching of the dominant strips with at most 2ω times the value. First we modify the solution such that exactly the dominant strips are used. In the current solution, the dominant strips are already used with at least $\frac{1}{\omega}$, so multiplying the solution by ω ensures a full usage of them. Now we can remove all superfluous strip usages by replacing two fractional matching edges that go through such a strip by a directly connecting matching edge without increasing the cost. This can create loop matching edges (assume these to have the same cost as the two edges they replace); these can easily be removed later. After this, we are left with a matching polytope that is half-integral (based on the same proof as for Theorem 6.13 in the book of Cook et al. [13]). Thus, we can assume our matching to be half-integral and double it to obtain an integral solution with double usages of strips. These double usages can be removed the same way as before while remaining integral. Whole redundant cycles may be removed on this way. We are now left with a feasible matching of the dominant strips that has at most 2ω times the cost of the original fractional solution, giving us the desired upper bound. More details on this proof can be found in the full version [17].

3.2   Tours

A given cycle cover approximation can be turned into a tour approximation at the expense of an additional constant factor. Because every cycle involves at least two points and a full rotation, we can use classic tree techniques known for TSP variants to connect the cycles and charge the necessary turns to the involved cycles. We sketch the basic ideas; see full version [17] for details.

Theorem 4. *Assuming validity of Eq. 1 we can establish the following approximation factors for tours.*

(i) *Full tours in regular grid graphs: 6-approximation.*
(ii) *Full tours in generalized grid graphs: 4ω-approximation.*
(iii) *Subset tours in (generalized) grid graphs: $(4\omega + 2)$-approximation.*
(iv) *Geometric full tours: $(4\omega + 2)$-approximation.*
(v) *Penalty tours (in grid graphs and geometric): $(4\omega + 4)$-approximation.*

These results also hold for objective functions that are linear combinations of length and turn costs.

Proof. It is crucial that (1) a cycle always has a turn cost of at least 360°, (2) two intersecting cycles can be merged with a cost of at most 360°, and (3) two cycles intersecting on a 180° turn can be merged without additional cost.

(i) For full tours in grid graphs, greedily connecting cycles provides a tour with at most 1.5 times the turn cost of the cycle cover, while a local optimization can be exploited to limit the length to 4 times the optimum, as shown by Arkin et al. [5].

(ii) In a cycle cover for (generalized) grid graphs, there are always at least two cycles with a distance of one, while every cycle has a length of at least 2; otherwise the cycle cover is already a tour. This allows iteratively merging cycles at cost at most as much as a cheapest cycle; the total number of merges is less than the number of cycles.

(iii) and (iv) For subset coverage in grid graphs or full coverage in the geometric case, we need to compute the cheapest paths between any two cycles, ignoring the orientations at the ends. First connect all intersecting cycles, charging the cost on the vanishing cycles. The minimum spanning tree on these edges is a lower bound on the cost of the tour. Doubling the MST connects all cycles with the cost of twice the MST, the cost of the cycle cover, and the turn costs at the end of the MST edges, which can be charged to the cycles.

(v) Penalty tours can be approximated in a similar manner. Instead of an MST, we use a Price-Collecting Steiner Tree, which is a lower bound on an optimal penalty tour. We use a 2-approximation for the PCST [21], as it is NP-hard. We achieve a cost of twice the 2-approximation of the PCST, the cost of the penalty cycle cover, and the cost of its cycles again for charging the connection costs. The penalties of the points not in the cycle cover are already paid by the penalty cycle cover.

□

4 Conclusions

We have presented a number of theoretical results on finding optimal tours and cycle covers with turn costs. In addition to resolving the long-standing open problem of complexity, we provided a generic framework to solve geometric (penalty) cycle cover and tours problems with turn costs.

As described in [10], the underlying problem is also of practical relevance. As it turns out, our approach does not only yield polynomial-time approximation algorithms; enhanced by an array of algorithm engineering techniques, they can be employed for actually computing optimal and near-optimal solutions for instances of considerable size in grid graphs. Further details on these algorithm engineering aspects will be provided in our forthcoming paper [18].

References

1. Agarwal, P.K., Biedl, T.C., Lazard, S., Robbins, S., Suri, S., Whitesides, S.: Curvature-constrained shortest paths in a convex polygon. SIAM J. Comp. 31(6), 1814–1851 (2002)
2. Agarwal, P.K., Wang, H.: Approximation algorithms for curvature-constrained shortest paths. SIAM J. Comp. 30(6), 1739–1772 (2000)
3. Aggarwal, A., Coppersmith, D., Khanna, S., Motwani, R., Schieber, B.: The angular-metric traveling salesman problem. SIAM J. Comp. 29(3), 697–711 (1999)

4. Aichholzer, O., Fischer, A., Fischer, F., Meier, J.F., Pferschy, U., Pilz, A., Staněk, R.: Minimization and maximization versions of the quadratic travelling salesman problem. Optimization **66**(4), 521–546 (2017)
5. Arkin, E.M., Bender, M.A., Demaine, E.D., Fekete, S.P., Mitchell, J.S.B., Sethia, S.: Optimal covering tours with turn costs. In: Proceedings of 12th ACM-SIAM Symposium Discrete Algorithms (SODA), pp. 138–147 (2001)
6. Arkin, E.M., Bender, M.A., Demaine, E.D., Fekete, S.P., Mitchell, J.S.B., Sethia, S.: Optimal covering tours with turn costs. SIAM J. Comp. **35**(3), 531–566 (2005)
7. Arkin, E.M., Fekete, S.P., Mitchell, J.S.B.: Approximation algorithms for lawn mowing and milling. Comp. Geom. **17**(1–2), 25–50 (2000)
8. Astroza, S., Patil, P.N., Smith, K.I., Bhat, C.R.: Transportation planning to accommodate needs of wind energy projects. In: Transportation Research Board-Annual Meeting (2017). Article 17–05309
9. Ausiello, G., Bonifaci, V., Leonardi, S., Marchetti-Spaccamela, A.: Prize-collecting traveling salesman and related problems. In: Gonzalez, T.F. (ed.) Handbook of Approximation Algorithms and Metaheuristics. Chapman and Hall/CRC, Boca Raton (2007)
10. Becker, A.T., Debboun, M., Fekete, S.P., Krupke, D., Nguyen, A.: Zapping Zika with a mosquito-managing drone: Computing optimal flight patterns with minimum turn cost. In: Proceedings of 33rd Symposium on Computational Geometry (SoCG), pp. 62:1–62:5 (2017). Video at https://www.youtube.com/watch?v=SFyOMDgdNao
11. Benbernou, N.M.: Geometric algorithms for reconfigurable structures. Ph.D. thesis, Massachusetts Institute of Technology (2011)
12. Boissonnat, J., Lazard, S.: A polynomial-time algorithm for computing a shortest path of bounded curvature amidst moderate obstacles. In: Proceedings of 12th Symposium on Computational Geometry (SoCG), pp. 242–251 (1996)
13. Cook, W., Cunningham, W., Pulleyblank, W., Schrijver, A.: Combinatorial Optimization. Wiley-Interscience, Hoboken (1997)
14. de Assis, I.R., de Souza, C.C.: Experimental evaluation of algorithms for the orthogonal milling problem with turn costs. In: Pardalos, P.M., Rebennack, S. (eds.) SEA 2011. LNCS, vol. 6630, pp. 304–314. Springer, Heidelberg (2011). https://doi.org/10.1007/978-3-642-20662-7_26
15. Demaine, E.D., Mitchell, J. S. B., O'Rourke, J.: The Open Problems Project. http://cs.smith.edu/~orourke/TOPP/
16. Dubins, L.E.: On curves of minimal length with a constraint on average curvature, and with prescribed initial and terminal positions and tangents. Am. J. Math. **79**(3), 497–516 (1957)
17. Fekete, S.P., Krupke, D.: Covering tours and cycle covers with turn costs: hardness and approximation. arXiv (2018). http://arxiv.org/abs/1808.04417
18. Fekete, S.P., Krupke, D.: Practical methods for computing large covering tours and cycle covers with turn cost. In: Proceedings of 21st SIAM Workshop on Algorithm Engineering and Experiments (ALENEX) (2019)
19. Fekete, S.P., Woeginger, G.J.: Angle-restricted tours in the plane. Comp. Geom. **8**, 195–218 (1997)
20. Fellows, M., Giannopoulos, P., Knauer, C., Paul, C., Rosamond, F.A., Whitesides, S., Yu, N.: Milling a graph with turn costs: a parameterized complexity perspective. In: Thilikos, D.M. (ed.) WG 2010. LNCS, vol. 6410, pp. 123–134. Springer, Heidelberg (2010). https://doi.org/10.1007/978-3-642-16926-7_13
21. Goemans, M.X., Williamson, D.P.: A general approximation technique for constrained forest problems. SIAM J. Comp. **24**(2), 296–317 (1995)

22. Lazard, S., Reif, J., Wang, H.: The complexity of the two dimensional curvature-constrained shortest-path problem. In: Proceedings of 3rd Workshop on the Algorithmic Foundations of Robotics (WAFR), pp. 49–57 (1998)

23. Maurer, O.: Winkelminimierung bei Überdeckungsproblemen in Graphen. Diplomarbeit, Technische Universität Berlin (2009)

24. Schaefer, T.J.: The complexity of satisfiability problems. In: Proceedings of 10th ACM symposium on Theory of computing (STOC), pp. 216–226 (1978)

25. Stein, C., Wagner, D.P.: Approximation algorithms for the minimum bends traveling salesman problem. In: Aardal, K., Gerards, B. (eds.) IPCO 2001. LNCS, vol. 2081, pp. 406–421. Springer, Heidelberg (2001). https://doi.org/10.1007/3-540-45535-3_32

26. Takei, R., Tsai, R., Shen, H., Landa, Y.: A practical path-planning algorithm for a simple car: a Hamilton-Jacobi approach. In: Proceedings of 29th American Control Conference (ACC), pp. 6175–6180 (2010)

The Parameterized Position Heap
of a Trie

Noriki Fujisato$^{(\boxtimes)}$, Yuto Nakashima, Shunsuke Inenaga, Hideo Bannai,
and Masayuki Takeda

Department of Informatics, Kyushu University, Fukuoka, Japan
{noriki.fujisato,yuto.nakashima,inenaga,bannai,takeda}@inf.kyushu-u.ac.jp

Abstract. Let Σ and Π be disjoint alphabets of respective size σ and π.
Two strings over $\Sigma \cup \Pi$ of equal length are said to *parameterized match*
(*p-match*) if there is a bijection $f : \Sigma \cup \Pi \to \Sigma \cup \Pi$ such that (1) f is
identity on Σ and (2) f maps the characters of one string to those of
the other string so that the two strings become identical. We consider
the p-matching problem on a (reversed) trie $\mathcal{T}$ and a string pattern P
such that every path that p-matches P has to be reported. Let N be
the size of the given trie $\mathcal{T}$. In this paper, we propose the *parameterized
position heap* for $\mathcal{T}$ that occupies $O(N)$ space and supports p-matching
queries in $O(m \log(\sigma + \pi) + m\pi + pocc))$ time, where m is the length of
a query pattern P and $pocc$ is the number of paths in $\mathcal{T}$ to report. We
also present an algorithm which constructs the parameterized position
heap for a given trie $\mathcal{T}$ in $O(N(\sigma + \pi))$ time and working space.

1 Introduction

The *parameterized matching problem* (*p-matching problem*), first introduced by
Baker [2], is a variant of pattern matching which looks for substrings of a text
that has "the same structure" as a given pattern. More formally, we consider
a parameterized string (p-string) that can contain static characters from an
alphabet Σ and parameter characters from another alphabet Π. Two equal
length p-strings x and y over the alphabet $\Sigma \cup \Pi$ are said to *parameterized
match* (*p-match*) if x can be transformed to y (and vice versa) by applying
a bijection which renames the parameter characters. The *p-matching problem*
is, given a text p-string w and pattern p-string p, to report the occurrences
of substrings of w that p-match p. Studying the p-matching problem is well
motivated by plagiarism detection, software maintenance, and RNA structural
pattern matching [2,15]. We refer readers to [11] for detailed descriptions about
these motivations.

Baker [2] proposed an indexing data structure for the p-matching problem,
called the *parameterized suffix tree* (*p-suffix tree*). The p-suffix tree supports
p-matching queries in $O(m \log(\sigma + \pi) + pocc)$ time, where m is the length of

YN, SI, HB, MT are respectively supported by JSPS KAKENHI Grant Numbers
JP18K18002, JP17H01697, JP16H02783, JP18H04098.

P. Heggernes (Ed.): CIAC 2019, LNCS 11485, pp. 237–248, 2019.
https://doi.org/10.1007/978-3-030-17402-6_20

pattern p, σ and π are respectively the sizes of the alphabets Σ and Π, and *pocc* is the number of occurrences to report [1]. She also showed an algorithm that builds the p-suffix tree for a given text S of length n in $O(n(\pi + \log \sigma))$ time with $O(n)$ space [2]. Later, Kosaraju [8] proposed an algorithm to build the p-suffix tree in $O(n \log(\sigma + \pi))$ time[1] with $O(n)$ space. Their algorithms are both based on McCreight's suffix tree construction algorithm [10], and hence are *offline* (namely, the whole text has to be known beforehand). Shibuya [15] gave an *left-to-right online* algorithm that builds the p-suffix tree in $O(n \log(\sigma + \pi))$ time with $O(n)$ space. His algorithm is based on Ukkonen's suffix tree construction algorithm [16] which scans the input text from left to right.

Diptarama et al. [5] proposed a new indexing structure called the *parameterized position heap* (*p-position heap*). They showed how to construct the p-position heap of a given p-string S of length n in $O(n \log(\sigma + \pi))$ time with $O(n)$ space in a *left-to-right online* manner. Their algorithm is based on Kucherov's position heap construction algorithm [9] which scans the input text from left to right. Recently, Fujisato et al. [7] presented another variant of the p-position heap that can be constructed in a *right-to-left online* manner, in $O(n \log(\sigma + \pi))$ time with $O(n)$ space. This algorithm is based on Ehrenfeucht et al.'s algorithm [6] which scans the input text from right to left. Both versions of p-positions heaps support p-matching queries in $O(m \log(\sigma + \pi) + m\pi + pocc)$ time.

This paper deals with indexing on *multiple* texts; in particular, we consider the case where those multiple texts are represented by a *trie*. It should be noted that our trie is a so-called *common suffix trie* (*CS trie*) where the common suffixes of the texts are merged and the edges are reversed (namely, each text is represented by a path from a leaf to the root). See also Fig. 1 for an example of a CS trie. There are two merits in representing multiple texts by a CS trie: Let N be the size of the CS trie of the multiple strings of total length Z. (1) N can be as small as $\Theta(\sqrt{Z})$ when the multiple texts share a lot of common long suffixes. (2) The number of distinct suffixes of the texts is equal to the number of the nodes in the CS trie, namely N. On the other hand, this is not the case with the ordinal common prefix trie (CP trie), namely, the number of distinct suffixes in the CP trie can be super-linear in the number of its nodes. Since most, if not all, indexing structures require space that is dependent of the number of distinct suffixes, the CS trie is a more space economical representation for indexing than its CP trie counterpart.

Let N be the size of a given CS trie. Due to Property (1) above, it is significant to construct an indexing structure *directly* from the CS trie. Note that if we expand all texts from the CS trie, then the total string length can blow up to $O(N^2)$. Breslauer [3] introduced the suffix tree for a CS trie which occupies $O(N)$ space, and proposed an algorithm which constructs it in $O(N\sigma)$ time and working space. Using the suffix tree of a CS trie, one can report all paths of the CS trie

---

[1] The original claimed time bounds in Kosaraju [8] and in Shibuya [15] are $O(n(\log \sigma + \log \pi))$. However, assuming by symmetry that $\sigma \geq \pi$, we have $\log \sigma + \log \pi = \log(\sigma\pi) \leq \log \sigma^2 = 2\log \sigma = O(\log \sigma)$ and $\log(\sigma + \pi) \leq \log(2\sigma) = \log 2 + \log \sigma = O(\log \sigma)$.

that exactly matches with a given pattern of length m in $O(m \log \sigma + occ)$ time, where occ is the number of such paths to report. Shibuya [14] gave an optimal $O(N)$-time construction for the suffix tree for a CS trie in the case of integer alphabets of size $N^{O(1)}$. Nakashima et al. [12] proposed the position heap for a CS trie, which can be built in $O(N\sigma)$ time and working space and supports exact pattern matching in $O(m \log \sigma + occ)$ time. Later, an optimal $O(N)$-time construction algorithm for the position heap for a CS trie in the case of integer alphabets of size $N^{O(1)}$ was presented [13].

In this paper, we propose the *parameterized position heap* for a CS trie $\mathcal{T}$, denoted by $\mathsf{PPH}(\mathcal{T})$, which is the *first* indexing structure for p-matching on a trie. We show that $\mathsf{PPH}(\mathcal{T})$ occupies $O(N)$ space, supports p-matching queries in $O(m \log(\sigma + \pi) + m\pi + pocc)$ time, and can be constructed in $O(N(\sigma + \pi))$ time and working space. Hence, we achieve optimal pattern matching and construction in the case of constant-sized alphabets. The proposed construction algorithm is fairly simple, yet uses a non-trivial idea that converts a given CS trie into a smaller trie based on the p-matching equivalence. The simplicity of our construction algorithm comes from the fact that each string stored in (p-)position heaps is represented by an explicit node, while it is not the case with (p-)suffix trees. This nice property makes it easier and natural to adopt the approaches by Brealauer [3] and by Fujisato et al. [7] that use *reversed suffix links* in order to process the texts from left to right. We also remark that all existing p-suffix tree construction algorithms [2,8,15] in the case of a single text require somewhat involved data structures due to non-monotonicity of parameterized suffix links [1,2], but our p-position heap does not need such a data structure even in the case of CS tries (this will also be discussed in the concluding section).

2 Preliminaries

Let Σ and Π be disjoint ordered sets called a *static alphabet* and a *parameterized alphabet*, respectively. Let $\sigma = |\Sigma|$ and $\pi = |\Pi|$. An element of Σ is called an *s-character*, and that of Π is called a *p-character*. In the sequel, both an s-character and a p-character are sometimes simply called a *character*. An element of Σ^* is called a *string*, and an element of $(\Sigma \cup \Pi)^*$ is called a *p-string*. The length of a (p-)string w is the number of characters contained in w. The empty string ε is a string of length 0, namely, $|\varepsilon| = 0$. For a (p-)string $w = xyz$, x, y and z are called a *prefix*, *substring*, and *suffix* of w, respectively. The set of prefixes of a (p-)string w is denoted by $\mathsf{Prefix}(w)$. The i-th character of a (p-)string w is denoted by $w[i]$ for $1 \le i \le |w|$, and the substring of a (p-)string w that begins at position i and ends at position j is denoted by $w[i..j]$ for $1 \le i \le j \le |w|$. For convenience, let $w[i..j] = \varepsilon$ if $j < i$. Also, let $w[i..] = w[i..|w|]$ for any $1 \le i \le |w|$. For any (p-)string w, let w^R denote the reversed string of w, i.e., $w^R = w[|w|] \cdots w[1]$.

Two p-strings x and y of length k each are said to *parameterized match* (*p-match*) iff there is a bijection f on $\Sigma \cup \Pi$ such that $f(a) = a$ for any $a \in \Sigma$ and $f(x[i]) = y[i]$ for all $1 \le i \le k$. For instance, let $\Sigma = \{\mathsf{a}, \mathsf{b}\}$ and $\Pi = \{\mathsf{x}, \mathsf{y}, \mathsf{z}\}$, and consider two p-strings $x = \mathsf{axbzzayx}$ and $y = \mathsf{azbyyaxz}$. These two strings

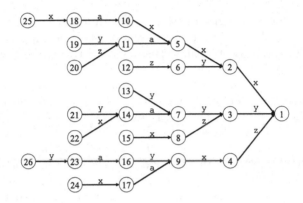

Fig. 1. The CS trie for a set $\{\texttt{xaxxx}, \texttt{yaxx}, \texttt{zaxx}, \texttt{zyx}, \texttt{yyy}, \texttt{yayy}, \texttt{xayy}, \texttt{xzy}, \texttt{yayxz}, \texttt{xaxz}\}$ of 10 p-strings over $\Sigma \cup \Pi$, where $\Sigma = \{\texttt{a}\}$ and $\Pi = \{\texttt{x}, \texttt{y}, \texttt{z}\}$.

p-match, since x can be transformed to y by applying a renaming bijection f such that $f(\texttt{a}) = \texttt{a}$, $f(\texttt{b}) = \texttt{b}$, $f(\texttt{x}) = \texttt{z}$, $f(\texttt{y}) = \texttt{x}$, and $f(\texttt{z}) = \texttt{y}$ to the characters in x. We write $x \approx y$ iff two p-strings x and y p-match. It is clear that $\approx$ is an equivalence relation on p-strings over $\Sigma \cup \Pi$. We denote by $[x]$ the equivalence class for p-string x w.r.t. $\approx$. The representative of $[x]$ is the lexicographically smallest p-string in $[x]$, which is denoted by $\mathsf{spe}(x)$. It is clear that two p-strings x and y p-match iff $\mathsf{spe}(x) = \mathsf{spe}(y)$. In the running example, $\mathsf{spe}(\texttt{axbzzayx}) = \mathsf{spe}(\texttt{azbyyaxz}) = \texttt{axbyyazx}$.

A *common suffix trie* (*CS trie*) $\mathcal{T}$ is a reversed trie such that (1) each edge is directed towards the root, (2) each edge is labeled with a character from $\Sigma \cup \Pi$, and (3) the labels of the in-coming edges to each node are mutually distinct. Each node of the trie represents the (p-)string obtained by concatenating the labels on the path from the node to the root. An example of a CS trie is illustrated in Fig. 1. $\mathsf{CST}(W)$ denotes the CS trie which represents a set W of (p-)strings.

3   Parameterized Position Heap of a Common Suffix Trie

In this section, we introduce the parameterized pattern matching (p-matching) problem on a common suffix trie that represents a set of p-strings, and propose an indexing data structure called a parameterized position heap of a trie.

3.1   p-Matching Problem on a Common Suffix Trie

We introduce the *p-matching problem* on a common suffix trie $\mathcal{T}$ and a pattern p. We will say that a node v in a common suffix trie p-matches with a pattern p-string p if the prefix of length $|p|$ of the p-string represented by v and p p-match. In this problem, we preprocess a given common suffix trie $\mathcal{T}$ so that later, given a query pattern p, we can quickly answer every node v of $\mathcal{T}$ whose prefix of

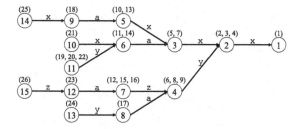

Fig. 2. Illustration of $\mathsf{pCST}(\mathcal{T})$ for $\mathcal{T}$ (where $\mathcal{T}$ is the common suffix trie illustrated in Fig. 1). Each node of $\mathsf{pCST}(\mathcal{T})$ corresponds to nodes of $\mathcal{T}$ which are labeled by elements in the tuple above the node of $\mathsf{pCST}(\mathcal{T})$. For example, the node of $\mathsf{pCST}(\mathcal{T})$ labeled 6 corresponds to the nodes of $\mathcal{T}$ labeled 11 and 14.

length $|p|$ and p p-match. For the common suffix trie in Fig. 1, when given query pattern $P = \mathsf{azy}$, then we answer the nodes 17 and 23.

Let $W_{\mathcal{T}}$ be the set of all p-strings represented by nodes of $\mathcal{T}$. By the definition of the common suffix trie, there may exist two or more nodes which represent different p-strings, but p-match. We consider the common suffix trie which merges such nodes into the same node by using the representative of the parameterized equivalent class of these strings. We define the set $\mathsf{pcs}(\mathcal{T})$ of p-strings as follows: $\mathsf{pcs}(\mathcal{T}) = \{\mathsf{spe}(w^R)^R \mid w \in W_{\mathcal{T}}\}$. Then, the reversed trie which we want to consider is $\mathsf{CST}(\mathsf{pcs}(\mathcal{T}))$. We refer to this reversed trie as the *parameterized-common suffix trie* of $\mathcal{T}$, and denote it by $\mathsf{pCST}(\mathcal{T})$ (i.e., $\mathsf{pCST}(\mathcal{T}) = \mathsf{CST}(\mathsf{pcs}(\mathcal{T}))$). Each node of $\mathsf{pCST}(\mathcal{T})$ stores pointers to its corresponding node(s) of $\mathcal{T}$. Then, by solving the p-matching problem on $\mathsf{pCST}(\mathcal{T})$, we can immediately answering p-matching queries on $\mathcal{T}$. Figure 2 shows an example of $\mathsf{pCST}(\mathcal{T})$. In the rest of this paper, N denotes the number of nodes of $\mathcal{T}$ and N_p denotes the number of nodes of $\mathsf{pCST}(\mathcal{T})$. Note that $N \geq N_p$ always holds.

3.2 Parameterized Position Heap of a Common Suffix Trie

Let $\mathcal{S} = \langle s_1, \ldots, s_k \rangle$ be a sequence of strings such that for any $1 < i \leq k$, $s_i \notin \mathsf{Prefix}(s_j)$ for any $1 \leq j < i$.

Definition 1 (Sequence hash trees [4]). *The sequence hash tree of a sequence* $\mathcal{S} = \langle s_1, \ldots, s_k \rangle$ *of strings, denoted* $\mathsf{SHT}(\mathcal{S}) = \mathsf{SHT}(\mathcal{S})^k$, *is a trie structure that is recursively defined as follows: Let* $\mathsf{SHT}(\mathcal{S})^i = (V_i, E_i)$. *Then*

$$\mathsf{SHT}(\mathcal{S})^i = \begin{cases} (\{\varepsilon\}, \emptyset) & \text{if } i = 1, \\ (V_{i-1} \cup \{u_i\}, E_{i-1} \cup \{(v_i, a, u_i)\}) & \text{if } 2 \leq i \leq k, \end{cases}$$

where v_i *is the longest prefix of* s_i *which satisfies* $v_i \in V_{i-1}$, $a = s_i[|v_i| + 1]$, *and* u_i *is the shortest prefix of* s_i *which satisfies* $u_i \notin V_{i-1}$.

Note that since we have assumed that each $s_i \in \mathcal{S}$ is not a prefix of s_j for any $1 \leq j < i$, the new node u_i and new edge (v_i, a, u_i) always exist for each $1 \leq i \leq k$. Clearly $\mathsf{SHT}(\mathcal{S})$ contains k nodes (including the root).

Let $\mathcal{W}_{\mathcal{T}} = \langle \mathsf{spe}(w_1), \ldots, \mathsf{spe}(w_{N_p}) \rangle$ be a sequence of p-strings such that $\{w_1, \ldots, w_{N_p}\} = \mathsf{pcs}(\mathcal{T})$ and $|w_i| \leq |w_{i+1}|$ for any $1 \leq i \leq N_p - 1$. $\mathcal{W}_{\mathcal{T}}(i)$ denote the sequence $\langle \mathsf{spe}(w_1), \ldots, \mathsf{spe}(w_i) \rangle$ for any $1 \leq i \leq N_p$, and $\mathsf{pCST}(\mathcal{T})^i$ denote the common suffix trie of $\{\mathsf{spe}(w_1), \ldots, \mathsf{spe}(w_i)\}$, namely, $\mathsf{pCST}(\mathcal{T})^i = \mathsf{CST}(\{\mathsf{spe}(w_1), \ldots, \mathsf{spe}(w_i)\})$. The node of $\mathsf{pCST}(\mathcal{T})$ which represents w_i is denoted by c_i. Then, our indexing data structure is defined as follows.

Definition 2 (Parameterized positions heaps of a CST). *The* parameterized position heap *(*p-position heap*) for a common suffix trie $\mathcal{T}$, denoted by* $\mathsf{PPH}(\mathcal{T})$, *is the sequence hash tree of $\mathcal{W}_{\mathcal{T}}$ i.e., $\mathsf{PPH}(\mathcal{T}) = \mathsf{SHT}(\mathcal{W}_{\mathcal{T}})$.*

Let $\mathsf{PPH}(\mathcal{T})^i = \mathsf{SHT}(\mathcal{W}_{\mathcal{T}}(i))$ for any $1 \leq i \leq N_p$ (i.e., $\mathsf{PPH}(\mathcal{T})^{N_p} = \mathsf{PPH}(\mathcal{T})$). The following lemma shows the exact size of $\mathsf{PPH}(\mathcal{T})$.

Lemma 1. *For any common suffix trie $\mathcal{T}$ such that the size of $\mathsf{pCST}(\mathcal{T})$ is N_p, $\mathsf{PPH}(\mathcal{T})$ consists of exactly N_p nodes. Also, there is a one-to-one correspondence between the nodes of $\mathsf{pCST}(\mathcal{T})$ and the nodes of $\mathsf{PPH}(\mathcal{T})$.*

Proof. Initially, $\mathsf{PPH}(\mathcal{T})^1$ consists only of the root that represents ε since $w_1 = \varepsilon$. Let i be an integer in $[1..N_p]$. Since w_i does not p-match with w_j and $|\mathsf{spe}(w_i)| \geq |\mathsf{spe}(w_j)|$ for any $1 \leq j < i$, there is a prefix of $\mathsf{spe}(w_i)$ that is not represented by any node of $\mathsf{PPH}(\mathcal{T})^{i-1}$. Therefore, when we construct $\mathsf{PPH}(\mathcal{T})^i$ from $\mathsf{PPH}(\mathcal{T})^{i-1}$, then exactly one node is inserted, which corresponds to the node representing w_i. $\qquad\square$

Let h_i be the node of $\mathsf{PPH}(\mathcal{T})$ which corresponds to w_i. For any p-string $p \in (\Sigma \cup \Pi)^+$, we say that p is *represented* by $\mathsf{PPH}(\mathcal{T})$ iff $\mathsf{PPH}(\mathcal{T})$ has a path which starts from the root and spells out p.

Ehrenfeucht et al. [6] introduced *maximal reach pointers*, which are used for efficient pattern matching queries on position heaps. Diptarama et al. [5] and Fujisato et al. [7] also introduced maximal reach pointers for their p-position heaps, and showed how efficient pattern matching queries can be done. We can naturally extend the notion of maximal reach pointers to our p-position heaps:

Definition 3 (Maximal reach pointers). *For each $1 \leq i \leq N_p$, the* maximal reach pointer *of the node h_i points to the deepest node v of $\mathsf{PPH}(\mathcal{T})$ such that v represents a prefix of $\mathsf{spe}(w_i)$.*

The node which is pointed by the maximal reach pointer of node h_i is denoted by $\mathsf{mrp}(i)$. The *augmented* $\mathsf{PPH}(\mathcal{T})$ is $\mathsf{PPH}(\mathcal{T})$ with the maximal reach pointers of all nodes. For simplicity, if $\mathsf{mrp}(i)$ is equal to h_i, then we omit this pointer. See Fig. 3 for an example of augmented $\mathsf{PPH}(\mathcal{T})$.

3.3   p-Matching with Augmented Parameterized Position Heap

It is straightforward that by applying Diptarama et al.'s pattern matching algorithm to our $\mathsf{PPH}(\mathcal{T})$ augmented with maximal reach pointers, parameterized pattern matching can be done in $O(m \log(\sigma + \pi) + m\pi + \mathit{pocc}')$ time where pocc'

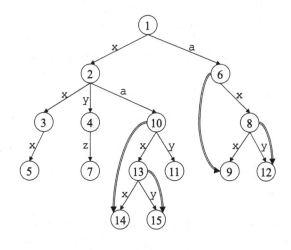

spe(w_1)	ε
spe(w_2)	x
spe(w_3)	xx
spe(w_4)	xy
spe(w_5)	xxx
spe(w_6)	axx
spe(w_7)	xyz
spe(w_8)	axy
spe(w_9)	axxx
spe(w_{10})	xaxx
spe(w_{11})	xayy
spe(w_{12})	axyz
spe(w_{13})	xaxy
spe(w_{14})	xaxxx
spe(w_{15})	xaxyx

Fig. 3. To the left is the list of spe(w_i) for p-strings represented by pCST($\mathcal{T}$) of Fig. 2, where $\Sigma = \{a\}$ and $\Pi = \{x, y, z\}$. To the right is an illustration for augmented PPH($\mathcal{T}$) where the maximal reach pointers are indicated by the double-lined arrows. The underlined prefix of each spe(w_i) in the left list denotes the longest prefix of spe(w_i) that was represented in PPH($\mathcal{T}$) and hence, the maximal reach pointer of the node with label i points to the node which represents this underlined prefix of spe(w_i).

is the number of nodes in pCST($\mathcal{T}$) that p-match with the pattern. Since each node in pCST($\mathcal{T}$) stores the pointers to the corresponding nodes in $\mathcal{T}$, then we can answer all the nodes that p-match with the pattern.

Diptarama et al.'s algorithm stands on Lemmas 13 and 14 of [5]. These lemmas can be extended to our PPH($\mathcal{T}$) as follows:

Lemma 2. *Suppose* spe(p) *is represented by a node u of augmented* PPH($\mathcal{T}$). *Then p p-matches with the prefix of length $|p|$ of w_i iff* mrp(i) *is u or a descendant of u.*

Lemma 3. *Suppose that* spe(p) *is not represented in augmented* PPH($\mathcal{T}$). *There is a factorization $q_1, \ldots, q_k$ of p s.t. q_j is the longest prefix of* spe($p[|q_1 \cdots q_{j-1}| + 1..|p|]$) *that is represented in augmented* PPH($\mathcal{T}$). *If p p-matches with the prefix of length $|p|$ of w_i, then* mrp($i + |q_1 \cdots q_{j-1}|$) *is the node which represents* spe(q_j) *for any $1 \leq j < k$ and* mrp($i + |q_1 \cdots q_{k-1}|$) *is the node which represents* spe(q_k) *or a descendant of* mrp($i + |q_1 \cdots q_{k-1}|$).

Theorem 1. *Using our augmented* PPH($\mathcal{T}$), *one can perform parameterized pattern matching queries in $O(m \log(\sigma + \pi) + m\pi + pocc)$ time.*

4   Construction of Parameterized Position Heaps

In this section, we show how to construct the augmented PPH($\mathcal{T}$) of a given common suffix trie $\mathcal{T}$ of size N. For convenience, we will sometimes identify each node v of $\mathsf{PPH}(\mathcal{T})$ with the string which is represented by v. In Sect. 4.1, we show how to compute $\mathsf{pCST}(\mathcal{T})$ from a given common suffix trie $\mathcal{T}$. In Sect. 4.2, we propose how to construct $\mathsf{PPH}(\mathcal{T})$ from $\mathsf{pCST}(\mathcal{T})$.

4.1   Computing $\mathsf{pCST}(\mathcal{T})$ from $\mathcal{T}$

Here, we show how to construct $\mathsf{pCST}(\mathcal{T})$ of a given $\mathcal{T}$ of size N.

Lemma 4. *For any common suffix trie $\mathcal{T}$ of size N, $\mathsf{pCST}(\mathcal{T})$ can be computed in $O(N\pi)$ time and space.*

Proof. We process every node of $\mathcal{T}$ in a breadth first manner. Let x_j be the p-string which is represented by j-the node of $\mathcal{T}$. Suppose that we have processed the first k nodes and have computed $\mathsf{pCST}(\mathcal{T})^i$ $(i \leq k)$. We assume that the j-th node of $\mathcal{T}$, for any $1 \leq j \leq k$, holds the resulting substitutions from x_j to $\mathsf{spe}((x_j)^R)^R$ (i.e., $x_j[\alpha]$ is mapped to $\mathsf{spe}((x_j)^R)^R[\alpha]$), and also a pointer to the corresponding node of $\mathsf{pCST}(\mathcal{T})^i$ (i.e., pointer to the node representing $\mathsf{spe}((x_j)^R)^R$). We consider processing the $(k+1)$-th node of $\mathcal{T}$. Since x_{k+1} is encoded from right to left, we can determine a character $\mathsf{spe}((x_{k+1})^R)^R[1]$ in $O(\pi)$ time. Then, we can insert a new node that represents $\mathsf{spe}((x_{k+1})^R)^R$ as a parent of the node which represents $\mathsf{spe}((x_{k+1}[2..|x_{k+1}|])^R)^R$ if there does not exist such a node in $\mathsf{pCST}(\mathcal{T})^i$. Therefore, we can compute $\mathsf{pCST}(\mathcal{T})$ in $O(N\pi)$ time and space.   □

4.2   Computing $\mathsf{PPH}(\mathcal{T})$ from $\mathsf{pCST}(\mathcal{T})$

For efficient construction of our $\mathsf{PPH}(\mathcal{T})$, we use *reversed suffix links* defined as follows.

Definition 4 (Reversed suffix links). *For any node v of $\mathsf{PPH}(\mathcal{T})$ and a character $a \in \Sigma \cup \Pi$, let*

$$\mathsf{rsl}(a, v) = \begin{cases} \mathsf{spe}(av) & \text{if } \mathsf{spe}(av) \text{ is represented by } \mathsf{PPH}(\mathcal{T}), \\ undefined & otherwise. \end{cases}$$

See Fig. 4 for an example of $\mathsf{PPH}(\mathcal{T})$ with reversed suffix links. In our algorithm, firstly, we insert a new node h_i of $\mathsf{PPH}(\mathcal{T})^i$ to $\mathsf{PPH}(\mathcal{T})^{i-1}$. After that, we add new suffix links which point to h_i. When we have computed $\mathsf{PPH}(\mathcal{T})$, then we compute all maximal reach pointers of $\mathsf{PPH}(\mathcal{T})$.

Inserting a New Node. Assume that c_j (i.e., j-th node of $\mathsf{pCST}(\mathcal{T})$) is the child of c_i for any $2 \leq i \leq N_p$. Consider to insert h_i (i.e., the node of $\mathsf{PPH}(\mathcal{T})$ which corresponds to c_i) to $\mathsf{PPH}(\mathcal{T})^{i-1}$. We show how to find the parent of h_i by starting from h_j. There are 3 cases based on $w_i[1]$ as follows:

- $w_i[1] \in \Pi$ and $w_i[1]$ appears in $w_j[1..|h_j|]$ (Lemma 5),

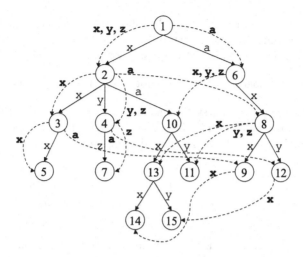

Fig. 4. PPH($\mathcal{T}$) with all reversed suffix links is illustrated in this figure. Each dashed arrow shows a reversed suffix link. The label of a suffix link is drawn by a bold character.

- $w_i[1] \in \Pi$ and $w_i[1]$ does not appear in $w_j[1..|h_j|]$ (Lemma 6),
- $w_i[1] \in \Sigma$ (Lemma 7).

Lemma 5. *Assume that $w_i[1] \in \Pi$ appears in $w_j[1..|h_j|]$, and a is the character in Π such that $a = \mathsf{spe}(w_j[1..|h_j|])[\alpha]$ and $w_i[1] = w_j[\alpha]$ for some $1 \leq \alpha \leq |w_j[1..|h_j|]|$. Let h_k be the node of PPH($\mathcal{T}$)$^{i-1}$ which is the lowest ancestor of h_j that has a reversed suffix link labeled with a. Then, h_i is a child of the node representing $\mathsf{rsl}(a, h_k)$.*

Proof. Let ℓ be the length of $\mathsf{rsl}(a, h_k)$. To prove this lemma, we show that

1. $\mathsf{rsl}(a, h_k) = \mathsf{spe}(w_i)[1..\ell]$, and
2. There does not exist a node which represents $\mathsf{spe}(w_i)[1..\ell + 1]$ in PPH($\mathcal{T}$)$^{i-1}$.

By the definition of reversed suffix links and spe, we have

$$\mathsf{rsl}(a, h_k) = \mathsf{spe}(a \cdot \mathsf{spe}(w_j[1..\ell - 1])) = \mathsf{spe}(w_i[1] \cdot w_j[1..\ell - 1])$$
$$= \mathsf{spe}(w_i[1] \cdot w_i[2..\ell]) = \mathsf{spe}(w_i[1..\ell]).$$

Thus, we have proved the first statement.

By a similar argument, we also have $\mathsf{spe}(a \cdot \mathsf{spe}(w_j[1..\ell])) = \mathsf{spe}(w_i[1..\ell + 1])$. Thus, if $\mathsf{spe}(w_i)[1..\ell + 1]$ is represented in PPH($\mathcal{T}$)$^{i-1}$, the node representing $\mathsf{spe}(w_j[1..\ell])$ must have a reversed suffix link labeled with a. This contradicts the fact that h_k is the lowest ancestor of h_j which has a reversed suffix link labeled with a. $\square$

Lemma 6. *Assume that $w_i[1] \in \Pi$ does not appear in $w_j[1..|h_j|]$. Let h_k be the node of PPH($\mathcal{T}$)$^{i-1}$ which is the lowest ancestor of h_j that has a reversed suffix link labeled with $a \in \Pi \setminus \{h_j[\alpha] \mid 1 \leq \alpha \leq |h_j|\}$. Then, h_i is a child of the node representing $\mathsf{rsl}(a, h_k)$.*

Proof. Let ℓ be the length of $\mathsf{rsl}(a, h_k)$. We show similar statements to the proof of the previous lemma hold, but with different assumptions on $w_i[1]$ and a. By the definition of reversed suffix links and spe, we have

$$\mathsf{rsl}(a, h_k) = \mathsf{spe}(a \cdot \mathsf{spe}(w_j[1..\ell - 1])) = \mathsf{spe}(a \cdot w_j[1..\ell - 1]) = \mathsf{spe}(w_i[1..\ell]).$$

Thus, we have proved the first statement.

By a similar argument, we also have $\mathsf{spe}(a \cdot \mathsf{spe}(w_j[1..\ell])) = \mathsf{spe}(w_i[1..\ell + 1])$. This implies that the second statement holds (similar to the proof of the previous lemma).   □

Lemma 7. *Assume that $w_i[1] \in \Sigma$. Let h_k be the node in $\mathsf{PPH}(\mathcal{T})^{i-1}$ which is the lowest ancestor of h_j that has a reversed suffix link labeled with $w_i[1]$. Then, h_i is a child of the node representing $\mathsf{rsl}(w_i[1], h_k)$.*

Proof. Since $w_i[1] \in \Sigma$, we can show the lemma in a similar way to the above proofs.   □

Inserting New Reversed Suffix Links. In our algorithm, we will add reversed suffix links which point to h_i after inserting a new node h_i. The following lemma shows the number of nodes which point to h_i by reversed suffix links is at most one.

Lemma 8. *For any node v of $\mathsf{PPH}(\mathcal{T})$, the number of nodes which point to v by reversed suffix links is at most one.*

Proof. Let v_1, v_2 be nodes of $\mathsf{PPH}(\mathcal{T})$. Assume that $\mathsf{rsl}(a_1, v_1) = \mathsf{rsl}(a_2, v_2)$ for some $a_1, a_2 \in \Sigma \cup \Pi$ and $v_1 \neq v_2$ hold. By the definition of reversed suffix links, $\mathsf{spe}(a_1 \cdot v_1) = \mathsf{spe}(a_2 \cdot v_2)$. Namely, $a_1 \cdot v_1 \approx a_2 \cdot v_2$ holds. This implies that $v_1 \approx v_2$, i.e., $\mathsf{spe}(v_1) = \mathsf{spe}(v_2)$. Since v_1 and v_2 are node of $\mathsf{PPH}(\mathcal{T})$, $\mathsf{spe}(v_1) = v_1$ and $\mathsf{spe}(v_2) = v_2$ hold. This contradicts the fact that $v_1 \neq v_2$.   □

By the above lemma and arguments of insertion, the node which points to the new node h_i by reversed suffix links is only a child of h_k which is an ancestor of h_j.

Construction Algorithm. Finally, we explain our algorithm of constructing our position heap. From the above lemmas, we can use similar techniques to Nakashima et al. [12] which construct the position heap of a trie of normal strings. One main difference is the computation of the label of inserted edges/reversed suffix links. In so doing, each node h_α holds the resulting substitutions from $w_\alpha[1..|h_\alpha|]$ to $\mathsf{spe}(w_\alpha[1..|h_\alpha|])$. By using these substitutions, we can compute the corresponding label in $O(\pi)$ time. Thus, we can insert new nodes and new suffix links in $O(\pi)$ time for each node of $\mathsf{pCST}(\mathcal{T})$. In fact, since we need to use $(\sigma + \pi)$-copies of the position heap for nearest marked ancestor queries on each character, we use $O(\sigma + \pi)$ time to update the data structures needed for each node of $\mathsf{pCST}(\mathcal{T})$. Therefore, we have the following lemma.

Lemma 9. *We can compute* $\mathsf{PPH}(\mathcal{T})$ *from* $\mathsf{pCST}(\mathcal{T})$ *of size* N_p *in* $O(N_p(\sigma + \pi))$ *time and space.*

Therefore, we can obtain the following result by Lemmas 4 and 9.

Theorem 2. *We can compute* $\mathsf{PPH}(\mathcal{T})$ *of a given common suffix trie* $\mathcal{T}$ *of size* N *in* $O(N(\sigma + \pi))$ *time and space.*

Since we can also compute all maximal reach pointers of $\mathsf{PPH}(\mathcal{T})$ efficiently in a similar way to [12] (this algorithm is also similar to suffix link construction), we also have the following lemma.

Lemma 10. *We can compute all the maximal reach pointers for* $\mathsf{PPH}(\mathcal{T})$ *in* $O(N_p(\sigma + \pi))$ *time and space.*

Hence, we can get the following result.

Theorem 3. *We can compute the augmented* $\mathsf{PPH}(\mathcal{T})$ *of a given common suffix trie* $\mathcal{T}$ *of size* N *in* $O(N(\sigma + \pi))$ *time and space.*

5 Conclusions and Open Problems

This paper proposed the p-position heap for a CS trie $\mathcal{T}$, denoted $\mathsf{PPH}(\mathcal{T})$, which is the first indexing structure for the p-matching problem on a trie. The key idea is to transform the input CS trie $\mathcal{T}$ into a parameterized CS trie $\mathsf{pCST}(\mathcal{T})$ where p-matching suffixes are merged. We showed that the p-matching problem on the CS trie $\mathcal{T}$ can be reduced to the p-matching problem on the parameterized CS trie $\mathsf{pCST}(\mathcal{T})$. We proposed an algorithm which constructs $\mathsf{PPH}(\mathcal{T})$ in $O(N(\sigma + \pi))$ time and working space, where N is the size of the CS trie $\mathcal{T}$. We also showed that using $\mathsf{PPH}(\mathcal{P})$ one can solve the p-matching problem on the CS trie $\mathcal{T}$ in $O(m \log(\sigma + \pi) + m\pi + pocc)$ time, where m is the length of a query pattern and $pocc$ is the number of occurrences to report.

Examples of open problems regarding this work are the following:

- Would it be possible to shave the $m\pi$ term in the pattern matching time using p-position heaps? This $m\pi$ term is introduced when the depth of the corresponding path of $\mathsf{PPH}(\mathcal{T})$ is shorter the pattern length m and thus the pattern needs to be partitioned into $O(\pi)$ blocks in the current pattern matching algorithm [5].
- Can we efficiently build the p-suffix tree for a CS trie? It is noted by Baker [1, 2] that the *destination* of a parameterized suffix link (p-suffix link) of the p-suffix tree can be an *implicit node* that lies on an edge, and hence there is no monotonicity in the chain of p-suffix links. If we follow the approach by Breslauer [3] which is based on Weiner's algorithm [17], then we need to use the reversed p-suffix link. It is, however, unclear whether one can adopt this approach since the *origin* of a reversed p-suffix link may be an implicit node. Recall that in each step of construction we need to find the nearest (implicit) ancestor that has a reversed p-suffix link labeled with a given character. Since there can be $\Theta(N^2)$ implicit nodes, we cannot afford to explicitly maintain information about the reversed p-suffix links for all implicit nodes.

References

1. Baker, B.S.: A theory of parameterized pattern matching: algorithms and applications. In: STOC 1993, pp. 71–80 (1993)
2. Baker, B.S.: Parameterized pattern matching: algorithms and applications. J. Comput. Syst. Sci. **52**(1), 28–42 (1996)
3. Breslauer, D.: The suffix tree of a tree and minimizing sequential transducers. Theor. Comput. Sci. **191**(1–2), 131–144 (1998)
4. Coffman, E., Eve, J.: File structures using hashing functions. Commun. ACM **13**, 427–432 (1970)
5. Diptarama, Katsura, T., Otomo, Y., Narisawa, K., Shinohara, A.: Position heaps for parameterized strings. In: Proceedings of CPM 2017, pp. 8:1–8:13 (2017)
6. Ehrenfeucht, A., McConnell, R.M., Osheim, N., Woo, S.W.: Position heaps: a simple and dynamic text indexing data structure. J. Discrete Algorithms **9**(1), 100–121 (2011)
7. Fujisato, N., Nakashima, Y., Inenaga, S., Bannai, H., Takeda, M.: Right-to-left online construction of parameterized position heaps. In: Proceedings of PSC, vol. 2018, pp. 91–102 (2018)
8. Kosaraju, S.R.: Faster algorithms for the construction of parameterized suffix trees (preliminary version). In: FOCS 1995, pp. 631–637 (1995)
9. Kucherov, G.: On-line construction of position heaps. J. Discrete Algorithms **20**, 3–11 (2013)
10. McCreight, E.M.: A space-economical suffix tree construction algorithm. J. ACM **23**(2), 262–272 (1976)
11. Mendivelso, J., Pinzón, Y.: Parameterized matching: solutions and extensions. In: Proceedings of PSC, pp. 118–131 (2015)
12. Nakashima, Y., Tomohiro, I., Inenaga, S., Bannai, H., Takeda, M.: The position heap of a trie. In: Calderón-Benavides, L., González-Caro, C., Chávez, E., Ziviani, N. (eds.) SPIRE 2012. LNCS, vol. 7608, pp. 360–371. Springer, Heidelberg (2012). https://doi.org/10.1007/978-3-642-34109-0_38
13. Nakashima, Y., Tomohiro, I., Inenaga, S., Bannai, H., Takeda, M.: Constructing LZ78 tries and position heaps in linear time for large alphabets. Inf. Process. Lett. **115**(9), 655–659 (2015)
14. Shibuya, T.: Constructing the suffix tree of a tree with a large alphabet. IEICE Trans. Fundam. Electron. **E86–A**(5), 1061–1066 (2003)
15. Shibuya, T.: Generalization of a suffix tree for RNA structural pattern matching. Algorithmica **39**(1), 1–19 (2004)
16. Ukkonen, E.: On-line construction of suffix trees. Algorithmica **14**(3), 249–260 (1995)
17. Weiner, P.: Linear pattern-matching algorithms. In: Proceedings of 14th IEEE Annual Symposium on Switching and Automata Theory, pp. 1–11 (1973)

Parameterized Algorithms
for Generalizations of Directed
Feedback Vertex Set

Alexander Göke[1]($\boxtimes$), Dániel Marx[2], and Matthias Mnich[1]

[1] Universität Bonn, Bonn, Germany
{alexander.goeke,mmnich}@uni-bonn.de
[2] SZTAKI, Budapest, Hungary
dmarx@cs.bme.hu

Abstract. The DIRECTED FEEDBACK VERTEX SET (DFVS) problem takes as input a directed graph G and seeks a smallest vertex set S that hits all cycles in G. This is one of Karp's 21 NP-complete problems. Resolving the parameterized complexity status of DFVS was a long-standing open problem until Chen et al. in 2008 showed its fixed-parameter tractability via a $4^k k! n^{\mathcal{O}(1)}$-time algorithm, where $k = |S|$.

Here we show fixed-parameter tractability of two generalizations of DFVS:

- Find a smallest vertex set S such that every strong component of $G - S$ has size at most s: we give an algorithm solving this problem in time $4^k (ks + k + s)! \cdot n^{\mathcal{O}(1)}$.
- Find a smallest vertex set S such that every non-trivial strong component of $G - S$ is 1-out-regular: we give an algorithm solving this problem in time $2^{\mathcal{O}(k^3)} \cdot n^{\mathcal{O}(1)}$.

We also solve the corresponding arc versions of these problems by fixed-parameter algorithms.

1 Introduction

The DIRECTED FEEDBACK VERTEX SET (DFVS) problem is that of finding a smallest vertex set S in a given digraph G such that $G - S$ is a directed acyclic graph. This problem is among the most classical problems in algorithmic graph theory. It is one of the 21 NP-complete problems on Karp's famous list [12].

Consequently, the DFVS problem has long attracted researchers in approximation algorithms. The current best known approximation factor achievable in polynomial time for n-vertex graphs with optimal fractional solution value[1] τ^* is $\mathcal{O}(\min\{\log \tau^* \log \log \tau^*, \log n \log \log n\})$ due to Seymour [17], Even et al. [8] and Even et al. [7]. On the negative side, Karp's NP-hardness reduction shows

---

[1] In unweighted digraphs, $\tau^* \leq n$; in weighted digraphs we assume all weights are at least 1.

© Springer Nature Switzerland AG 2019
P. Heggernes (Ed.): CIAC 2019, LNCS 11485, pp. 249–261, 2019.
https://doi.org/10.1007/978-3-030-17402-6_21

the problem to be APX-hard, which rules out the existence of a polynomial-time approximation scheme (PTAS) assuming P $\neq$ NP. Assuming the Unique Games Conjecture, the DFVS problem does not admit a polynomial-time $\mathcal{O}(1)$-approximation [10,11,18].

The DFVS problem has also received a significant amount of attention from the perspective of parameterized complexity. The main parameter of interest there is the optimal solution size $k = |S|$. The problem can easily be solved in time $n^{\mathcal{O}(k)}$ by enumerating all k-sized vertex subsets $S \subseteq V(G)$ and then seeking a topological order of $G - S$. The interesting question is thus whether the DFVS problem is *fixed-parameter tractable* with respect to k, which is to devise an algorithm with running time $f(k) \cdot n^{\mathcal{O}(1)}$ for some computable function f depending only on k. It was a long-standing open problem whether DFVS admits such an algorithm. The question was finally resolved by Chen et al. who gave a $4^k k! k^4 \cdot \mathcal{O}(nm)$-time algorithm for graphs with n vertices and m arcs. Recently, an algorithm for DFVS with run time $4^k k! k^5 \cdot \mathcal{O}(n+m)$ was given by Lokshtanov et al. [14]. It is well-known that the *arc* deletion variant is parameter-equivalent to the *vertex* deletion variant and hence DIRECTED FEEDBACK ARC SET (DFAS) can also be solved in time $4^k k! k^5 \cdot \mathcal{O}(n + m)$.

Once the breakthrough result for DFVS was obtained, the natural question arose how much further one can push the boundary of (fixed-parameter) tractability. On the one hand, Chitnis et al. [4] showed that the generalization of DFVS where one only wishes to hit cycles going through a specified subset of nodes of a given digraph is still fixed-parameter tractable when parameterized by solution size. On the other hand, Lokshtanov et al. [15] show that finding a smallest set of vertices of hitting only the *odd* directed cycles of a given digraph is W[1]-hard, and hence not fixed-parameter tractable unless FPT = W[1].

Our Contributions. For another generalization the parameterized complexity is still open: In the EULERIAN STRONG COMPONENT ARC (VERTEX) DELETION problem, one is given a directed multigraph G, and asks for a set S of at most k vertices such that every strong component of $G - S$ is Eulerian, that is, every vertex has the same in-degree and out-degree within its strong component. The arc version of this problem was suggested by Cechlárová and Schlotter [2] in the context of housing markets. Marx [16] explicitly posed determining the parameterized complexity of EULERIAN STRONG COMPONENT VERTEX DELETION as an open problem. Notice that these problems generalize the DFAS/DFVS problems, where each strong component of $G - S$ has size one and thus is Eulerian.

Theorem 1. EULERIAN STRONG COMPONENT VERTEX DELETION *is* W[1]-*hard parameterized by solution size* k, *even for* $(k + 1)$-*strong digraphs.*

Alas, we are unable to determine the parameterized complexity of EULERIAN STRONG COMPONENT ARC DELETION, which appears to be more challenging. Hence, we consider two natural generalizations of DFAS which may help to gain better insight into the parameterized complexity of that problem.

First, we consider the problem of deleting a set of k arcs or vertices from a given digraph such that every strong component has size at most s. Thus, the

DFAS/DFVS problems corresponds to the special case when $s = 1$. Formally, the problem BOUNDED SIZE STRONG COMPONENT ARC (VERTEX) DELETION takes as input a multi-digraph G and integers k, s, and seeks a set S of at most k arcs or vertices such that every strong component of $G - S$ has size at most s.

The *undirected* case of BOUNDED SIZE STRONG COMPONENT ARC (VERTEX) DELETION was studied recently. There, one wishes to delete at most k vertices of an undirected n-vertex graph such that each connected component of the remaining graph has size at most s. For s being constant, Kumar and Lokshtanov [13] obtained a kernel of size $2sk$ that can be computed in $n^{\mathcal{O}(s)}$ time; note that the degree of the run time in the input size n depends on s and is thus not a fixed-parameter algorithm. For general s, there is a $9sk$-sized kernel computable in time $\mathcal{O}(n^4m)$ by Xiao [19]. The directed case—which we consider here—generalizes the undirected case by replacing each edge by arcs in both directions.

Our main result here is to solve the directed case of the problem by a fixed-parameter algorithm:

Theorem 2. *There is an algorithm that solves* BOUNDED SIZE STRONG COMPONENT ARC (VERTEX) DELETION *in time* $4^k(ks + k + s)! \cdot n^{\mathcal{O}(1)}$ *for n-vertex multi-digraphs G and parameters $k, s \in \mathbb{N}$.*

In particular, our algorithm exhibits the same asymptotic dependence on k as does the algorithm by Chen et al. [3] for the DFVS/DFAS problem, which corresponds to the special case $s = 1$.

Another motivation for this problem comes from the k-linkage problem, which asks for k pairs of terminal vertices in a digraph if they can be connected by k mutually arc-disjoint paths. The k-linkage problem is NP-complete already for $k = 2$ [9]. Recently, Bang-Jensen and Larsen [1] solved the k-linkage problem in digraphs where strong components have size at most s. Thus, finding induced subgraphs with strong components of size at most s can be of interest in computing k-linkages.

Our second problem is that of deleting a set of k arcs or vertices from a given digraph such that each remaining non-trivial strong component is *1-out-regular*, meaning that every vertex has out-degree exactly 1 in its strong component. (A strong component is *non-trivial* if it has at least two vertices.) So in particular, every strong component is Eulerian, as in the EULERIAN STRONG COMPONENT ARC DELETION problem. Observe that in the DFAS/DFVS problem we delete k arcs or vertices from a given directed graph such that each remaining strong component is 0-out-regular (trivial). Formally, we consider the 1-OUT-REGULAR ARC (VERTEX) DELETION problem in which for a given multi-digraph G and integer k, we seek a set S of at most k arcs (vertices) such that every non-trivial component of $G - S$ is 1-out-regular. Note that this problem is equivalent to deleting a set S of at most k arcs (vertices) such that every non-trivial strong component of $G - S$ is an induced directed cycle. In contrast to EULERIAN STRONG COMPONENT VERTEX DELETION, the 1-OUT-REGULAR ARC (VERTEX) DELETION problem *is* monotone, in that every superset of a solution is again a solution: if we delete an additional arc or vertex that breaks a strong component that is an induced cycle into several strong components, then each of these newly created strong components is trivial.

Our result for this problem reads as follows.

Theorem 3. *There is an algorithm solving* 1-OUT-REGULAR ARC (VERTEX) DELETION *in time* $2^{\mathcal{O}(k^3)} \cdot \mathcal{O}(n^4)$ *for n-vertex digraphs G and parameter* $k \in \mathbb{N}$.

Notice that for BOUNDED SIZE STRONG COMPONENT ARC (VERTEX) DELETION and 1-OUT-REGULAR ARC (VERTEX) DELETION, there are infinitely many instances for which solutions are arbitrarily smaller than those for DFAS (DFVS), and for any instance they are never larger. Therefore, our algorithms strictly generalize the one by Chen et al. [3] for DFAS (DFVS). As a possible next step towards resolving the parameterized complexity of EULERIAN STRONG COMPONENT ARC DELETION, one may generalize our algorithm for 1-OUT-REGULAR ARC DELETION to r-OUT-REGULAR ARC DELETION for arbitrary r.

We give algorithms for vertex deletion variants only, and defer algorithms for arc deletion variants and proofs marked by ⋆ to the full version of this paper.

2   Notions and Notations

We consider finite directed graphs (or digraphs) G with vertex set $V(G)$ and arc set $A(G)$. We allow multiple arcs and arcs in both directions between the same pairs of vertices. For each vertex $v \in V(G)$, its *out-degree* in G is the number $d_G^+(v)$ of arcs of the form (v, w) for some $w \in V(G)$, and its *in-degree* in G is the number $d_G^-(v)$ of arcs of the form (w, v) for some $w \in V(G)$. A vertex v is *balanced* if $d_G^+(v) = d_G^-(v)$. A digraph G is *balanced* if every vertex $v \in V(G)$ is balanced.

For each subset $V' \subseteq V(G)$, the subgraph induced by V' is the graph $G[V']$ with vertex set V' and arc set $\{(u, v) \in A(G) \mid u, v \in V'\}$. For any set X of arcs or vertices of G, let $G - X$ denote the subgraph of G obtained by deleting the elements of X from G. For subgraphs G' of G and vertex sets $X \subseteq V(G)$ let $R_{G'}^+(X)$ denote the set of vertices that are *reachable* from X in G', i.e. vertices to which there is a path from some vertex in X. For an s-t-walk P and a t-q-walk R we denote by $P \circ R$ the *concatenation* of these paths, i.e. the s-q-walk resulting from first traversing P and then R.

Let G be a digraph. Then G is *1-out-regular* if every vertex has out-degree exactly 1. Further, G is called *strong* if either G consists of a single vertex (then G is called *trivial*), or for any distinct $u, v \in V(G)$ there is a directed path from u to v. A *strong component* of G is an inclusion-maximal strong induced subgraph of G. Also, G is *t-strong* for some $t \in \mathbb{N}$ if for any $X \subseteq V(G)$ with $|X| < t$, $G - X$ is strong. We say that G is *weakly connected* if its underlying undirected graph $\langle G \rangle$ is connected. Finally, G is *Eulerian* if there is a closed walk in G using each arc exactly once.

Definition 4. *For disjoint non-empty vertex sets* X, Y *of a digraph* G, *a set* S *is an* $X - Y$ *separator if* S *is disjoint from* $X \cup Y$ *and there is no path from* X

to Y in $G - S$. An $X - Y$ separator S is minimal if no proper subset of S is an $X - Y$ separator. An $X - Y$ separator S is important if there is no $X - Y$ separator S' with $|S'| \leq |S|$ and $R_{G-S}^+(X) \subset R_{G-S'}^+(X)$.

Notice that S can be either a vertex set or an arc set.

Proposition 5 ([5]). *Let G be a digraph and let $X, Y \subseteq V(G)$ be disjoint non-empty vertex sets. For every $p \geq 0$ there are at most 4^p important $X - Y$ separators of size at most p, all of which can be enumerated in time $4^p \cdot n^{\mathcal{O}(1)}$.*

3 Tools for Generalized DFVS/DFAS Problems

Iterative Compression. We use the standard technique of iterative compression. For this, we label the vertices of the input digraph G arbitrarily by $v_1, \ldots, v_n$, and set $G_i = G[\{v_1, \ldots, v_i\}]$. We start with G_1 and the solution $S_1 = \{v_1\}$. As long as $|S_i| < k$, we can set $S_{i+1} = S_i \cup \{v_{i+1}\}$ and continue. As soon as $|S_i| = k$, the set $T_{i+1} = S_i \cup \{v_{i+1}\}$ is a solution for G_{i+1} of size $k + 1$. The *compression variant* of our problem then takes as input a digraph G and a solution T of size $k + 1$, and seeks a solution S of size at most k for G or decides that none exists.

We call an algorithm for the compression variant on (G_{i+1}, T_{i+1}) to obtain a solution S_{i+1} or find out that G_{i+1} does not have a solution of size k, but then neither has G. By at most n calls to this algorithm we can deduce a solution for the original instance $(G_n = G, k)$.

Disjoint Solution. Given an input (G, T) to the compression variant, the next step is to ask for a solution S for G of size at most k that is disjoint from the given solution T of size $k + 1$. This assumption can be made by guessing the intersection $T' = S \cap T$, and deleting those vertices from G. Since T has $k + 1$ elements, this step creates 2^{k+1} candidates T'. The *disjoint compression variant* of our problem then takes as input a graph $G - T'$, a solution $T \setminus T'$ of size $k + 1 - |T'|$, and seeks a solution S' of size at most $k - |T'|$ disjoint from $T \setminus T'$.

Covering the Shadow of a Solution. The "shadow" of a solution S is the set of those vertices that are disconnected from T (in either direction) after the removal of S. A common idea of several fixed-parameter algorithms on digraphs is to first ensure that there is a solution whose shadow is empty, as finding such a shadowless solution can be a significantly easier task. A generic framework by Chitnis et al. [4] shows that for special types of problems as defined below, one can invoke the random sampling of important separators technique and obtain a set Z which is disjoint from a minimum solution and covers its shadow, i.e. the shadow is contained in Z. What one does with this set, however, is problem-specific. Typically, given such a set, one can use (some problem-specific variant of) the "torso operation" to find an equivalent instance that has a shadowless solution. Therefore, one can focus on the simpler task of finding a shadowless solution or more precisely, finding any solution under the guarantee that a shadowless solution exists.

Definition 6 (shadow). *Let G be a digraph and let $T, S \subseteq V(G)$. A vertex $v \in V(G)$ is in the forward shadow $f_{G,T}(S)$ of S (with respect to T) if S is a $T - \{v\}$-separator in G, and v is in the reverse shadow $r_{G,T}(S)$ of S (with respect to T) if S is a $\{v\} - T$-separator in G.*

A vertex is in the shadow of S if it is in the forward or reverse shadow of S.

Note that S itself is not in the shadow of S by definition of separators.

Definition 7 (T-connected and $\mathcal{F}$-transversal). *Let G be a digraph, let $T \subseteq V(G)$ and let $\mathcal{F}$ be a set of subgraphs of G. We say that $\mathcal{F}$ is T-connected if for every $F \in \mathcal{F}$, each vertex of F can reach some and is reachable by some (maybe different) vertex of T by a walk completely contained in F. For a set $\mathcal{F}$ of subgraphs of G, an $\mathcal{F}$-transversal is a set of vertices that intersects the vertex set of every subgraph in $\mathcal{F}$.*

Chitnis et al. [4] show how to deterministically cover the shadow of $\mathcal{F}$-transversals:

Proposition 8 (deterministic covering of the shadow, [4]). *Let $T \subseteq V(G)$. In time $2^{\mathcal{O}(k^2)} \cdot n^{\mathcal{O}(1)}$ one can construct $t \leq 2^{\mathcal{O}(k^2)} \log^2 n$ sets $Z_1, \ldots, Z_t$ such that for any set of subgraphs $\mathcal{F}$ which is T-connected, if there exists an $\mathcal{F}$-transversal of size at most k then there is an $\mathcal{F}$-transversal S of size at most k that is disjoint from Z_i and Z_i covers the shadow of S, for some $i \leq t$.*

4   Hardness of Vertex Deletion

In this section we prove Theorem 1, by showing NP-hardness and W[1]-hardness of the EULERIAN STRONG COMPONENTS VERTEX DELETION problem. Before the hardness proof we recall an equivalent characterization of Eulerian digraphs:

Lemma 9 (folklore). *Let G be a weakly connected digraph. Then G is Eulerian if and only if G is balanced.*

We can now state the hardness reduction, which relies on the hardness of the following problem introduced by Cygan et al. [6]. In DIRECTED BALANCED VERTEX DELETION, one is given a directed multigraph G and an integer $k \in \mathbb{N}$, and seeks a set S of at most k vertices such that $G - S$ is balanced.

Proposition 10 ([6]). DIRECTED BALANCED VERTEX DELETION *is NP-hard and W[1]-hard with parameter k.*

We will prove the hardness of EULERIAN STRONG COMPONENT VERTEX DELETION for $(k + 1)$-strong digraphs by adding vertices ensuring this connectivity. The proof of Theorem 1 is deferred to the full version of this paper.

5 Bounded Size Strong Component Arc (Vertex) Deletion

In this section we show a fixed-parameter algorithm for the vertex deletion variant of BOUNDED SIZE STRONG COMPONENT VERTEX DELETION.

We give an algorithm that, given an n-vertex digraph G and integers k, s, decides in time $4^k(ks + k + s)! \cdot n^{\mathcal{O}(1)}$ if G has a set S of at most k vertices such that every strong component of $G - S$ has size at most s. Such a set S will be called a *solution* of the instance (G, k, s).

The algorithm first executes the general steps "Iterative Compression" and "Disjoint Solution"; it continues with a reduction to a skew separator problem.

Reduction to Skew Separator Problem. Now the goal is, given a digraph G, integers $k, s \in \mathbb{N}$, and a solution T of $(G, k+1, s)$, to decide if (G, k, s) has a solution S that is disjoint from T. We solve this problem—which we call DISJOINT BOUNDED SIZE STRONG COMPONENT VERTEX DELETION REDUCTION—by reducing it to finding a small "skew separator" in one of a bounded number of reduced instances.

Definition 11. *Let G be a digraph, and let $\mathcal{X} = (X_1, \ldots, X_t), \mathcal{Y} = (Y_1, \ldots, Y_t)$ be two ordered collections of $t \geq 1$ vertex subsets of G. A skew separator S for $(G, \mathcal{X}, \mathcal{Y})$ is a vertex subset of $V(G) \setminus \bigcup_{i=1}^{t}(X_i \cup Y_i)$ such that for any index pair (i, j) with $t \geq i \geq j \geq 1$, there is no path from X_i to Y_j in the graph $G - S$.*

This definition gives rise to the SKEW SEPARATOR problem, which for a digraph G, ordered collections $\mathcal{X}, \mathcal{Y}$ of vertex subsets of G, and an integer $k \in \mathbb{N}$ asks for a skew separator for $(G, \mathcal{X}, \mathcal{Y})$ of size at most k. Chen et al. [3] showed:

Proposition 12 ([3, Theorem 3.5]). *There is an algorithm solving SKEW SEPARATOR in time $4^k k \cdot \mathcal{O}(n^3)$ for n-vertex digraphs G.*

The reduction from DISJOINT BOUNDED SIZE STRONG COMPONENT VERTEX DELETION REDUCTION to SKEW SEPARATOR is as follows. As T is a solution of $(G, k+1, s)$, we can assume that every strong component of $G - T$ has size at most s. Similarly, we can assume that every strong component of $G[T]$ has size at most s, as otherwise there is no solution S of (G, k, s) that is disjoint from T. Let $\{t_1, \ldots, t_{k+1}\}$ be a labeling of the vertices in T.

Lemma 13 (*). *There is an algorithm that, given an n-vertex digraph G, integers $k, s \in \mathbb{N}$, and a solution T of $(G, k+1, s)$, in time $\mathcal{O}((ks + s - 1)!) \cdot n^{\mathcal{O}(1)}$ computes a collection $\mathcal{C}$ of at most $(ks + s - 1)!$ vectors $C = (C_1, \ldots, C_{k+1})$ of length $k + 1$, where $t_h \in C_h \subseteq V(G)$ for $h = 1, \ldots, k+1$, such that for some solution S of (G, k, s) disjoint from T, there is a vector $C \in \mathcal{C}$ such that the strong component of $G - S$ containing t_h is exactly $G[C_h]$ for $h = 1, \ldots, k+1$.*

Armed with Lemma 13, we can hence restrict our search for a solution S of (G, k, s) disjoint from T to those S that additionally are "compatible" with a vector in $\mathcal{C}$. Formally, a solution S of (G, k, s) is *compatible* with a vector $C = (C_1, \ldots, C_{k+1}) \in \mathcal{C}$ if the strong component of $G - S$ containing t_h is exactly C_h for $h = 1, \ldots, k + 1$. For a given vector $C = (C_1, \ldots, C_{k+1})$, to determine whether a solution S of (G, k, s) disjoint from T and compatible with C exists, we create several instances of the SKEW SEPARATOR problem. To this end, note that if two sets $C_h, C_{h'}$ for distinct $t_h, t'_h \in T$ overlap, then actually $C_h = C_{h'}$ (and $t_h, t'_h \in C_h$). So for each set C_h we choose exactly one (arbitrary) *representative T-vertex* among all T-vertices in C_h with consistent choice over overlapping and thus equal C_h's. Let $T' \subseteq T$ be the set of these representative vertices. Now we generate precisely one instance $(G', \mathcal{X}_{\sigma'}, \mathcal{Y}_{\sigma'}, k)$ of SKEW SEPARATOR for each permutation σ' of T'. The graph G' is the same in all these instances, and is obtained from G by replacing each unique set C_h by two vertices t_h^+, t_h^- (where t_h is the representative of C_h), and connecting all vertices incoming to C_h in G by an in-arc to t_h^+ and all vertices outgoing from C_h in G by an arc outgoing from t_h^-. This way also arcs of the type (t_j^-, t_h^+) are added but none of type (t_j^-, t_h^-), (t_j^+, t_h^-) or (t_j^+, t_h^+). Notice that this operation is well-defined and yields a simple digraph G', even if $t_{h'} \in C_h$ for some distinct h, h'. The sets $\mathcal{X}_{\sigma'}$ and $\mathcal{Y}_{\sigma'}$ of "sources" and "sinks" depend on the permutation σ' with elements $\sigma'(1), \ldots, \sigma'(|T'|)$: let $\mathcal{X}_{\sigma'} = (t_{\sigma'(1)}^-, \ldots, t_{\sigma'(|T'|)}^-)$ and let $\mathcal{Y}_{\sigma'} = (t_{\sigma'(1)}^+, \ldots, t_{\sigma'(|T'|)}^+)$.

Thus, per triple $((G, k, s), T, C)$ we generate at most $|T'|! \leq |T|! = (k+1)!$ instances $(G', \mathcal{X}_{\sigma'}, \mathcal{Y}_{\sigma'}, k)$, the number of permutations of T'.

We now establish the correctness of this reduction, in the next two lemmas:

Lemma 14 ($\star$). *If an instance (G, k, s) admits a solution S disjoint from T, compatible with C and for which $(t_{\sigma'(1)}, \ldots, t_{\sigma'(|T'|)})$ is a topological order of the connected components of $G' - S$, then S forms a skew separator of size k for $(G, \mathcal{X}_{\sigma'}, \mathcal{Y}_{\sigma'})$.*

Lemma 15 ($\star$). *Conversely, if S is a skew separator of $(G', \mathcal{X}_{\sigma'}, \mathcal{Y}_{\sigma'})$ with size at most k, then S is a solution of (G, k, s) disjoint from S and compatible with C.*

In summary, we have reduced a single instance to the compression problem DISJOINT BOUNDED SIZE STRONG COMPONENT VERTEX DELETION REDUCTION to at most $|\mathcal{C}| \cdot |T'|!$ instances $(G', \mathcal{X}_{\sigma'}, \mathcal{Y}_{\sigma'}, k)$ of the SKEW SEPARATOR problem, where each such instance corresponds to a permutation σ' of T'. The reduction just described implies that:

Lemma 16. *An input (G, k, s, T) to the DISJOINT BOUNDED SIZE STRONG COMPONENT VERTEX DELETION problem is a "yes"-instance if and only if at least one of the instances $(G', \mathcal{X}_{\sigma'}, \mathcal{Y}_{\sigma'}, k)$ is a "yes"-instance for the SKEW SEPARATOR problem.*

So we invoke the algorithm of Proposition 12 for each of the instances $(G', \mathcal{X}_{\sigma'}, \mathcal{Y}_{\sigma'}, k)$. If at least one of them is a "yes"-instance then so is (G, k, s, T), otherwise (G, k, s, T) is a "no"-instance. Hence, we conclude that DISJOINT BOUNDED SIZE STRONG COMPONENT VERTEX DELETION REDUCTION is

fixed-parameter tractable with respect to the joint parameter (k, s), and so is
BOUNDED SIZE STRONG COMPONENT VERTEX DELETION. The overall run time
of the algorithm is thus bounded by $|\mathcal{C}| \cdot |T'|! \cdot n^{\mathcal{O}(1)} \cdot 4^k k n^3 = (ks + s - 1)! \cdot (k + 1)! \cdot 4^k \cdot n^{\mathcal{O}(1)} = 4^k (ks + k + s)! \cdot n^{\mathcal{O}(1)}$. This completes the proof of Theorem 2.

6 1-Out-Regular Arc (Vertex) Deletion

In this section we give a fixed-parameter algorithm for the vertex deletion variant
of Theorem 3. Let G be a digraph and let $k \in \mathbb{N}$. A *solution* for (G, k) is a set S
of at most k vertices of G such that every non-trivial strong component of $G - S$
is 1-out-regular.

We first apply the steps "Iterative Compression" and "Disjoint Solution"
from Sect. 3. This yields the DISJOINT 1-OUT-REGULAR VERTEX DELETION
REDUCTION problem, where we seek a solution S of (G, k) that is disjoint from
and smaller than a solution T of $(G, k + 1)$.

Then we continue with the technique of covering of shadows, as described in
Sect. 3. In our setting, let $\mathcal{F}$ be the collection of vertex sets of G that induce a
strongly connected graph different from a simple directed cycle. Then clearly $\mathcal{F}$
is T-connected and any solution S must intersect every such induced subgraph.

So we can use Proposition 8 to construct sets $Z_1, \ldots, Z_t$ with $t \leq 2^{\mathcal{O}(k^2)} \log^2 n$
such that one of these sets covers the shadow of our hypothetical solution S
with respect to T. For each Z_i we construct an instance, where we assume that
$Z = Z_i \setminus T$ covers the shadow. Note that a vertex of T is never in the shadow. As
we assume that $Z \cup T$ is disjoint of a solution we reject an instance if $G[Z \cup T]$
contains a member of $\mathcal{F}$ as a subgraph.

Observation 17. *$G[Z \cup T]$ has no subgraph in $\mathcal{F}$.*

Normally, one would give a "torso" operation which transforms (G, k) with
the use of Z into an instance (G', k') of the same problem which has a shadowless
solution if and only if the original instance has any solution. Instead, our torso
operation reduces to a similar problem while maintaining solution equivalence.

Reducing the Instance by the Torso Operation. Our torso operation works
directly on the graph. It reduces the original instance to one of a new problem
called DISJOINT SHADOW-LESS GOOD 1-OUT-REGULAR VERTEX DELETION
REDUCTION; afterwards we show the solution equivalence.

Definition 18. *Let (G, T, k) be an instance of DISJOINT 1-OUT-REGULAR
VERTEX DELETION REDUCTION and let $Z \subseteq V(G)$. Then $\mathsf{torso}(G, Z)$ defines
the digraph with vertex set $V(G) \setminus Z$ and good and bad arcs. An arc (u, v) for
$u, v \notin Z$ is introduced whenever there is an $u \to v$ path in G (of length at least 1)
whose internal vertices are all in Z. We mark (u, v) as good if this path P is
unique and there is no cycle O in $G[Z]$ with $O \cap P \neq \emptyset$. Otherwise we mark it
as a bad arc.*

Note that every arc between vertices not in Z also forms a path as above. Therefore $G[V(G) \setminus Z]$ is a subdigraph of $\mathsf{torso}(G, Z)$. Also, $\mathsf{torso}(G, Z)$ may contain self loops at vertices v from cycles with only the vertex v outside of Z. In $\mathsf{torso}(G, Z)$, we call a cycle *good* if it consists of only good arcs. (A non-good cycle in $\mathsf{torso}(G, Z)$ can contain both good arcs and bad arcs.)

Now we want to compute a vertex set of size k whose deletion from $G' = \mathsf{torso}(G, Z)$ yields a digraph whose every non-trivial strong component is a cycle of good arcs. We call this problem DISJOINT SHADOW-LESS GOOD 1-OUT-REGULAR VERTEX DELETION REDUCTION. To simplify notation we construct a set $\mathcal{F}_{\mathsf{bad}}$ which contains all strong subdigraphs of G that are not trivial or good cycles. Then S is a solution to G' if and only if $G' - S$ contains no subdigraph in $\mathcal{F}_{\mathsf{bad}}$. In the next lemma we verify that our new problem is indeed equivalent to the original problem, assuming that there is a solution disjoint from Z.

Lemma 19 ($\star$, torso preserves obstructions). *Let G be a digraph, $T, Z \subseteq V(G)$ as above and $G' = \mathsf{torso}(G, Z)$. For any $S \subseteq V(G) \setminus (Z \cup T)$ it holds that $G - S$ contains a subdigraph in $\mathcal{F}$ if and only if $G' - S$ contains a subdigraph in $\mathcal{F}_{\mathsf{bad}}$.*

The above lemma shows that S is a solution of an instance (G, T, k) for DISJOINT 1-OUT-REGULAR VERTEX DELETION REDUCTION disjoint of Z if and only if it is a solution of $(\mathsf{torso}(G, Z), T, k)$ for DISJOINT SHADOW-LESS GOOD 1-OUT-REGULAR VERTEX DELETION REDUCTION. As connections between vertices are preserved by the torso operation and the torso graph contains no vertices in Z, we can reduce our search for $(\mathsf{torso}(G, Z), T, k)$ to shadow-less solutions (justifying the name).

Finding a Shadowless Solution. Consider an instance (G, T, k) of DISJOINT SHADOW-LESS GOOD 1-OUT-REGULAR VERTEX DELETION REDUCTION. Normally, after the torso operation a pushing argument is applied. However, we give an algorithm that recovers the last connected component of G. As T is already a solution, but disjoint of the new solution S, we take it as a starting point of our recovery. Observe that, without loss of generality, each vertex t in T has out-degree at least one in $G - T \setminus \{t\}$, for otherwise already $T - t$ is a solution.

Consider a topological order of the strong components of $G - S$, say $C_1, \ldots, C_\ell$, i.e., there can be an arc from C_i to C_j only if $i < j$. We claim that the last strong component C_ℓ in the topological ordering of $G - S$ contains a non-empty subset T_0 of T. For if C_ℓ did not contain any vertex from T, then the vertices of C_ℓ cannot reach any vertex of T, contradicting that S is a shadowless solution of (G, k).

Since T_0 is the subset of T present in C_ℓ and arcs between strong components can only be from earlier to later components, we have that there are no outgoing arcs from C_ℓ in $G - S$.

We guess a vertex t inside T_0. This gives $|T| \le k + 1$ choices for t. For each guess of t we try to find the component C_ℓ, similarly to the bounded-size case. The component C_ℓ will either be trivial or not.

If C_ℓ is a trivial component, then $V(C_\ell) = \{t\}$, and so we delete all out-neighbors of t in $G - T$ and place them into the new set S. Hence, we must decrease the parameter k by the number of out-neighbors of t in $G - T$, which by assumption is at least one.

Else, if the component C_ℓ is non-trivial, define $v_0 = t$ and notice that exactly one out-neighbor v_1 of v_0 belongs to C_ℓ. Set $i = 0$ and notice that every out-neighbor of v_i other than v_{i+1} must be removed from the graph G as C_ℓ is the last component in the topological ordering of $G - T'$, there is no later component where those out-neighbors could go. This observation gives rise to a natural branching procedure: we guess the out-neighbor v_{i+1} of v_i that belongs to C_ℓ and remove all other out-neighbors of v_i from the graph. We then repeat this branching step with $i \mapsto i + 1$ until we get back to the vertex t of T_0 we started with. This way, we obtain exactly the last component C_ℓ, forming a cycle. This branching results in at least one deletion as long as v_i has out-degree at least two. If the out-degree of v_i is exactly one, then we simple proceed by setting $v_i := v_{i+1}$ (and increment i). In any case we stop early if (v_i, v_{i+1}) is a bad arc, as this arc may not be contained in a strong component.

Recall that the vertices $t = v_0, v_1, \ldots$ must *not* belong to S, whereas the deleted out-neighbors of v_i must belong to S. From another perspective, the deleted out-neighbors of v_i must *not* belong to T. So once we reached back at the vertex $v_j = t$ for some $j \geq 1$, we have indeed found the component C_ℓ that we were looking for.

Let us shortly analyze the run time of the branching step. As for each vertex v_i, we have to remove all its out-neighbors from G except one and include them into the hypothetical solution S of size at most k, we immediately know that the degree of v_i in G can be at most $k + 1$. Otherwise, we have to include v_0 into S. Therefore, there are at most $k + 1$ branches to consider to identify the unique out-neighbor v_{i+1} of v_i in C_ℓ. So for each vertex v_i with out-degree at least two we branch into at most $k + 1$ ways, and do so for at most k vertices, yielding a run time of $O((k + 1)^k)$ for the entire branching.

Once we recovered the last strong component C_ℓ of $G - S$, we remove the set $V(C_\ell)$ from G and repeat: we then recover $C_{\ell-1}$ as the last strong component, and so on until C_1.

Algorithm for Disjoint 1-Out-Regular Vertex Deletion Reduction.
Lemma 19 and the branching procedure combined give a bounded search tree algorithm for DISJOINT 1-OUT-REGULAR VERTEX DELETION REDUCTION:

Step 1. For a given instance $I = (G, T, k)$, use Proposition 8 to obtain a set of instances $\{Z_1, \ldots, Z_t\}$ where $t \leq 2^{\mathcal{O}(k^2)} \log^2 n$, and Lemma 19 implies
- If I is a "no"-instance then all reduced instances I/Z_j are "no"-instances, for $j = 1, \ldots, t$.
- If I is a "yes"-instance then there is at least one $i \in \{1, \ldots, t\}$ such that there is a solution $T^\star$ for I which is a shadowless solution for the reduced instance I/Z_i.

So at this step we branch into $t \leq 2^{\mathcal{O}(k^2)} \log^2 n$ directions.

Step 2. For each of the instances obtained from Step 1, recover the component C_ℓ by guessing the vertex $t = v_0$. Afterwards, recover $C_{\ell-1}, \ldots, C_1$ in this order.

So at this step we branch into at most $\mathcal{O}(k \cdot (k+1)^k)$ directions.

We then repeatedly perform Step 1 and Step 2. Note that for every instance, one execution of Step 1 and Step 2 gives rise to $2^{\mathcal{O}(k^2)} \log^2 n$ instances such that for each instance, we either know that the answer is "no" or the budget k has decreased, because each important separator is non-empty. Therefore, considering a level as an execution of Step 1 followed by Step 2, the height of the search tree is at most k. Each time we branch into at most $2^{\mathcal{O}(k^2)} \log^2 n \cdot \mathcal{O}(k \cdot (k+1)^k)$ directions. Hence the total number of nodes in the search tree is

$$\left(2^{\mathcal{O}(k^2)} \log^2 n \right)^k \cdot \mathcal{O}\left(k \cdot (k+1)^k \right) = \left(2^{\mathcal{O}(k^2)} \right)^k \left(\log^2 n \right)^k \cdot \mathcal{O}(k) \cdot \mathcal{O}((k+1)^k)$$

$$= 2^{\mathcal{O}(k^3)} \left(\log^2 n \right)^k = 2^{\mathcal{O}(k^3)} \cdot \mathcal{O}\left(((2k \log k)^k + n/2^k)^3 \right) = 2^{\mathcal{O}(k^3)} \cdot \mathcal{O}(n^3).$$

We then check the leaf nodes of the search tree and see if there are any strong components other than cycles left after the budget k has become zero. If for at least one of the leaf nodes the corresponding graph only has strong components that are cycles then the given instance is a "yes"-instance. Otherwise, it is a "no"-instance. This gives an $2^{\mathcal{O}(k^3)} \cdot n^{\mathcal{O}(1)}$-time algorithm for DISJOINT 1-OUT-REGULAR VERTEX DELETION REDUCTION. So overall, we have an $2^{\mathcal{O}(k^3)} \cdot n^{\mathcal{O}(1)}$-time algorithm for the 1-OUT-REGULAR VERTEX DELETION problem.

References

1. Bang-Jensen, J., Larsen, T.M.: DAG-width and circumference of digraphs. J. Graph Theory **82**(2), 194–206 (2016)
2. Cechlárová, K., Schlotter, I.: Computing the deficiency of housing markets with duplicate houses. In: Raman, V., Saurabh, S. (eds.) IPEC 2010. LNCS, vol. 6478, pp. 72–83. Springer, Heidelberg (2010). https://doi.org/10.1007/978-3-642-17493-3_9
3. Chen, J., Liu, Y., Lu, S., O'Sullivan, B., Razgon, I.: A fixed-parameter algorithm for the directed feedback vertex set problem. J. ACM **55**(5), 19 (2008). Article No. 21
4. Chitnis, R., Cygan, M., Hajiaghayi, M., Marx, D.: Directed subset feedback vertex set is fixed-parameter tractable. ACM Trans. Algorithms **11**(4), 28 (2015). Article No. 28
5. Chitnis, R., Hajiaghayi, M., Marx, D.: Fixed-parameter tractability of directed multiway cut parameterized by the size of the cutset. SIAM J. Comput. **42**, 1674–1696 (2013)
6. Cygan, M., Marx, D., Pilipczuk, M., Pilipczuk, M., Schlotter, I.: Parameterized complexity of Eulerian deletion problems. Algorithmica **68**(1), 41–61 (2014)
7. Even, G., Naor, J., Rao, S., Schieber, B.: Divide-and-conquer approximation algorithms via spreading metrics. J. ACM **47**(4), 585–616 (2000)
8. Even, G., Naor, J., Schieber, B., Sudan, M.: Approximating minimum feedback sets and multicuts in directed graphs. Algorithmica **20**(2), 151–174 (1998)

9. Fortune, S., Hopcroft, J., Wyllie, J.: The directed subgraph homeomorphism problem. Theoret. Comput. Sci. **10**(2), 111–121 (1980)
10. Guruswami, V., Håstad, J., Manokaran, R., Raghavendra, P., Charikar, M.: Beating the random ordering is hard: every ordering CSP is approximation resistant. SIAM J. Comput. **40**(3), 878–914 (2011)
11. Guruswami, V., Lee, E.: Simple proof of hardness of feedback vertex set. Theory Comput. **12**, 11 (2016). Article No. 6
12. Karp, R.M.: Reducibility among combinatorial problems. In: Miller, R.E., Thatcher, J.W., Bohlinger, J.D. (eds.) Complexity of Computer Computations, pp. 85–103. Springer, Boston (1972). https://doi.org/10.1007/978-1-4684-2001-2_9
13. Kumar, M., Lokshtanov, D.: A $2\ell k$ kernel for ℓ-component order connectivity. In: Proceedings of the IPEC 2016. Leibniz International Proceedings in Informatics, vol. 63, pp. 20:1–20:14 (2017)
14. Lokshtanov, D., Ramanujan, M.S., Saurabh, S.: When recursion is better than iteration: a linear-time algorithm for acyclicity with few error vertices. In: Proceedings of the SODA 2018, pp. 1916–1933 (2018)
15. Lokshtanov, D., Ramanujan, M., Saurabh, S.: Parameterized complexity and approximability of directed odd cycle transversal (2017). https://arxiv.org/abs/1704.04249
16. Marx, D.: What's next? Future directions in parameterized complexity. In: Bodlaender, H.L., Downey, R., Fomin, F.V., Marx, D. (eds.) The Multivariate Algorithmic Revolution and Beyond. LNCS, vol. 7370, pp. 469–496. Springer, Heidelberg (2012). https://doi.org/10.1007/978-3-642-30891-8_20
17. Seymour, P.D.: Packing directed circuits fractionally. Combinatorica **15**(2), 281–288 (1995)
18. Svensson, O.: Hardness of vertex deletion and project scheduling. Theory Comput. **9**, 759–781 (2013)
19. Xiao, M.: Linear kernels for separating a graph into components of bounded size. J. Comput. Syst. Sci. **88**, 260–270 (2017)

Shortest Reconfiguration Sequence
for Sliding Tokens on Spiders

Duc A. Hoang[1]([⊠])[iD], Amanj Khorramian[2][iD], and Ryuhei Uehara[1][iD]

[1] School of Information Science, JAIST,
1-1 Asahidai, Nomi, Ishikawa 923-1292, Japan
{hoanganhduc,uehara}@jaist.ac.jp
[2] Department of Electrical and Computer Engineering, University of Kurdistan,
Sanandaj, Iran
khorramian@gmail.com

Abstract. Suppose that two independent sets I and J of a graph with $|I| = |J|$ are given, and a token is placed on each vertex in I. The SLIDING TOKEN problem is to determine whether there exists a sequence of independent sets which transforms I into J so that each independent set in the sequence results from the previous one by sliding exactly one token along an edge in the graph. It is one of the representative reconfiguration problems that attract the attention from the viewpoint of theoretical computer science. For a yes-instance of a reconfiguration problem, finding a shortest reconfiguration sequence has a different aspect. In general, even if it is polynomial time solvable to decide whether two instances are reconfigured with each other, it can be NP-hard to find a shortest sequence between them. In this paper, we show that the problem for finding a shortest sequence between two independent sets is polynomial time solvable for spiders (i.e., trees having exactly one vertex of degree at least three).

Keywords: Sliding token · Shortest reconfiguration ·
Independent set · Spider tree · Polynomial-time algorithm

1 Introduction

Recently, the *reconfiguration problems* attracted the attention from the viewpoint of theoretical computer science. These problem arise when we like to find a step-by-step transformation between two feasible solutions of a problem such that all intermediate results are also feasible and each step abides by a fixed reconfiguration rule, that is, an adjacency relation defined on feasible solutions of the original problem. The reconfiguration problems have been studied extensively for several well-known problems, including INDEPENDENT SET [10,15,16,18], SATISFIABILITY [9,17], SET COVER, CLIQUE, MATCHING [15], and so on.

R. Uehara was partially supported by JSPS KAKENHI Grant Number JP17H06287 and 18H04091.

P. Heggernes (Ed.): CIAC 2019, LNCS 11485, pp. 262–273, 2019.
https://doi.org/10.1007/978-3-030-17402-6_22

A reconfiguration problem can be seen as a natural "puzzle" from the viewpoint of recreational mathematics. The *15-puzzle* is one of the most famous classic puzzles, that had the greatest impact on American and European societies (see [22] for its rich history). It is well known that the 15-puzzle has a parity, and one can solve the problem in linear time just by checking whether the parity of one placement coincides with the other or not. Moreover, the distance between any two reconfigurable placements is $O(n^3)$, that is, we can reconfigure from one to the other in $O(n^3)$ sliding pieces when the size of the board is $n \times n$. However, surprisingly, for these two reconfigurable placements, finding a shortest path is NP-complete in general [5, 20]. Namely, although we know that there is a path of length in $O(n^3)$, finding a shortest one is NP-complete. While every piece is a unit square in the 15-puzzle, we obtain the other famous classic puzzle when we allow to have rectangular pieces, which is called "Dad puzzle" and its variants can be found in the whole world (e.g., it is called "hako-iri-musume" in Japanese). Gardner said that "these puzzles are very much in want of a theory" in 1964 [8], and Hearn and Demaine gave the theory after 40 years [10]; they are PSPACE-complete in general [11].

Summarizing up, these sliding block puzzles characterize representative computational complexity classes; the decision problem for unit squares can be solved in linear time just by checking parities, finding a shortest reconfiguration for the unit squares is NP-complete, and the decision problem becomes PSPACE-complete for rectangular pieces. That is, this simple reconfiguration problem gives us a new sight of these representative computational complexity classes.

In general, the reconfiguration problems tend to be PSPACE-complete, and some polynomial time algorithms are shown in restricted cases. Finding a shortest sequence in the context of the reconfiguration problems is a new trend in theoretical computer science because it has a great potential to characterize the class NP from a different viewpoint from the classic ones.

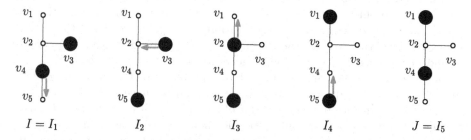

Fig. 1. A sequence $\langle I_1, I_2, \ldots, I_5 \rangle$ of independent sets of the same graph, where the vertices in independent sets are depicted by small black circles (tokens).

One of the important NP-complete problems is the INDEPENDENT SET problem. For this notion, a natural reconfiguration problem called SLIDING TOKEN was introduced by Hearn and Demaine [10]. (See [16] for an overview on different reconfiguration variants of INDEPENDENT SET.) Suppose that we are given two independent sets I and J of a graph $G = (V, E)$ such that $|I| = |J|$, and imagine that a *token* (coin) is placed on each vertex in I. For convenience, sometimes we identify the token with the vertex it is placed on and simply say "a token in an independent set." Then, the SLIDING TOKEN problem is to determine whether there exists a sequence $S = \langle I_1, I_2, \ldots, I_\ell \rangle$ of independent sets of G such that

(a) $I_1 = I$, $I_\ell = J$, and $|I_i| = |I| = |J|$ for all i, $1 \le i \le \ell$; and
(b) for each i, $2 \le i \le \ell$, there is an edge xy in G such that $I_{i-1} \setminus I_i = \{x\}$ and $I_i \setminus I_{i-1} = \{y\}$.

That is, I_i can be obtained from I_{i-1} by sliding exactly one token on a vertex $x \in I_{i-1}$ to its adjacent vertex $y \in I_i$ along an edge $xy \in E(G)$. Such a sequence S, if exists, is called a TS-*sequence* in G between I and J. We denote by a 3-tuple (G, I, J) an instance of SLIDING TOKEN problem. If a TS-sequence S in G between I and J exists, we say that I is *reconfigurable* to J (and vice versa), and write $I \overset{G}{\leftrightsquigarrow} J$. The sets I and J are the *initial* and *target* independent sets, respectively. For a TS-sequence S, the *length* $\mathsf{len}(S)$ of S is defined as the number of independent sets in S minus one. In other words, $\mathsf{len}(S)$ is the number of token-slides described in S. Figure 1 illustrates a TS-sequence of length 4 between two independent sets $I = I_1$ and $J = I_5$.

For the SLIDING TOKEN problem, linear-time algorithms have been shown for cographs (also known as P_4-free graphs) [16] and trees [4]. Polynomial-time algorithms are shown for bipartite permutation graphs [7], claw-free graphs [3], cacti [14], and interval graphs [2][1]. On the other hand, PSPACE-completeness is also shown for graphs of bounded tree-width [19], planar graphs [10,11], planar graphs with bounded bandwidth [26], and split graphs [1].

In this context, for a given YES-instance (G, I, J) of SLIDING TOKEN, we aim to find a shortest TS-sequence between I and J. Such a problem is called the SHORTEST SLIDING TOKEN problem. As seen for the 15-puzzle, the SHORTEST SLIDING TOKEN problem can be intractable even for these graph classes which the decision problem can be solved in polynomial time. Moreover, in the 15-puzzle, we already know that it has a solution of polynomial length for two configurations. However, in the SLIDING TOKEN problem, we have no upper bound of the length of a solution in general. To deal with this delicate issue, we have to distinguish two variants of this problem. In the *decision variant*, an integer ℓ is also given as a part of input, and we have to decide whether there exists a sequence between I and J of length at most ℓ. In the *non-decision variant*, we are asked to output a specific shortest TS-sequence. The length ℓ is not necessarily polynomial in $|V(G)|$ in general. When ℓ is super-polynomial, we may have that the decision variant is in P, while the non-decision one is not in P since it takes super-polynomial time to output the sequence. On the other hand, even when G is a perfect graph and ℓ is polynomial in $|V(G)|$, the decision variant of SHORTEST SLIDING TOKEN is NP-complete (see [16, Theorem 5]). In short, in the decision variant, we focus on the *length* of a shortest TS-sequence,

---

[1]   We note that the algorithm for a block graph in [12] has a flaw, and hence it is not yet settled [21].

while in the non-decision variant, we focus on the *construction* of a shortest TS-sequence itself.

From this viewpoint, the length of a token sliding is a key feature of the SHORTEST SLIDING TOKEN problem. If the length is super-polynomial in total, there exists at least one token that slides super-polynomial times. That is, the token visits the same vertex many times in its slides. That is, some tokens make *detours* in the sequence (the notion of detour is important and precisely defined later). In general, it seems to be more difficult to analyze "detours of tokens" for graphs containing cycle(s). As a result, one may first consider the problem for trees. The SLIDING TOKEN problem on a tree can be solved in linear time [4]. Polynomial-time algorithms for the SHORTEST SLIDING TOKEN problem were first investigated in [25]. In [25], the authors gave polynomial-time algorithms for solving SHORTEST SLIDING TOKEN when the input graph is either a proper interval graph, a trivially perfect graph, or a caterpillar. We note that caterpillars is the first graph class that required detours to solve the SHORTEST SLIDING TOKEN problem. A caterpillar is a tree that consists of a "backbone" called a *spine* with many *pendants*, or leaves attached to the spine. Each pendant can be used to escape a token, however, the other tokens cannot pass through it. Therefore, the ordering of tokens on the spine is fixed. In this paper, we consider the SHORTEST SLIDING TOKEN problem on a spider, which is a tree with one central vertex of degree more than 2. On this graph, we can use each "leg" as a stack and exchange tokens using these stacks. Therefore, we have many ways to handle the tokens, and hence we need more analyses to find a shortest sequence. In this paper, we give an $O(n^2)$ time algorithms for the SHORTEST SLIDING TOKEN problem on a spider, where n is the number of vertices. The algorithm is constructive, and the sequence itself can be output in $O(n^2)$ time. As mentioned in [25], the number of required token-slides in a sequence can be $\Omega(n^2)$, hence our algorithm is optimal for the number of token-slides. Due to space restriction, several details are omitted; they can be found in the full version of this paper [13].

Note: Recently, it is announced that the SHORTEST SLIDING TOKEN problem on a tree can be solved in polynomial time by Sugimori [23]. His algorithm is based on a dynamic programming on a tree [24]: though it runs in polynomial time, it seems to have much larger degree comparing to our case-analysis based algorithm.

2   Preliminaries

For common graph theoretic definitions, we refer the readers to the textbook [6]. Throughout this paper, we denote by $V(G)$ and $E(G)$ the vertex-set and edge-set of a graph G, respectively. We always use n for denoting $|V(G)|$. For a vertex $x \in V(G)$, we denote by $N_G(x)$ the set $\{y \in V(G) : xy \in E(G)\}$ of *neighbors* of x, and by $N_G[x]$ the set $N_G(x) \cup \{x\}$ of *closed neighbors* of x. In a similar manner, for an induced subgraph H of G, the set $N_G[H]$ is defined as $\bigcup_{x \in V(H)} N_G[x]$. The *degree* of x, denoted by $\deg_G(x)$, is the size of $N_G(x)$. For

$x, y \in V(G)$, the *distance* $\mathsf{dist}_G(x, y)$ between x and y is simply the length (i.e., the number of edges) of a shortest xy-path in G.

For a tree T, we denote by P_{xy} the (unique) shortest xy-path in T, and by T_y^x the subtree of T induced by y and its descendants when regarding T as the tree rooted at x. A *spider graph* (or *starlike tree*) is a tree having exactly one vertex (called its *body*) of degree at least 3. For a spider G with body v and a vertex $w \in N_G(v)$, the path G_w^v is called a *leg* of G. By definition, it is not hard to see that two different legs of G have no common vertex. For example, the graph in Fig. 1 is a spider with body $v = v_2$ and $\deg_G(v) = 3$ legs attached to v.

Let (G, I, J) be an instance of SHORTEST SLIDING TOKEN. A *target assignment* from I to J is simply a bijective mapping $f : I \to J$. A target assignment f is called *proper* if there exists a TS-sequence in G between I and J that moves the token on w to $f(w)$ for every $w \in I$. Given a target assignment $f : I \to J$ from I to J, one can also define the target assignment $f^{-1} : J \to I$ from J to I as follows: for every $x \in J$, $f^{-1}(x) = \{y \in I : f(y) = x\}$. Let $\mathcal{F}$ be the set of all target assignments from I to J. We define $M^*(G, I, J) = \min_{f \in \mathcal{F}} \sum_{w \in I} \mathsf{dist}_G(w, f(w))$. Intuitively, observe that any TS-sequence between I and J in G (if exists) uses at least $M^*(G, I, J)$ token-slides.

Let $S = \langle I_1, I_2, \ldots, I_\ell \rangle$ be a TS-sequence between two independent sets $I = I_1$ and $J = I_\ell$ of a graph G. Indeed, one can describe S in term of token-slides as follows: $S = \langle x_1 \to y_1, x_2 \to y_2, \ldots, x_{\ell-1} \to y_{\ell-1} \rangle$, where x_i and y_i ($i \in \{1, 2, \ldots, \ell-1\}$) satisfy $x_i y_i \in E(G)$, $I_i \setminus I_{i+1} = \{x_i\}$, and $I_{i+1} \setminus I_i = \{y_i\}$. The *reverse* of S (which reconfigures J to I), denoted by $\mathsf{rev}(S)$, is defined by $\mathsf{rev}(S) = \langle I_\ell, \ldots, I_2, I_1 \rangle$. One can also describe $\mathsf{rev}(S)$ in term of token-slides: $\mathsf{rev}(S) = \langle y_{\ell-1} \to x_{\ell-1}, \ldots, y_2 \to x_2, y_1 \to x_1 \rangle$. For example, the TS-sequence $S = \langle I_1, \ldots, I_5 \rangle$ described in Fig. 1 can also be written as $S = \langle v_4 \to v_5, v_3 \to v_2, v_2 \to v_1, v_5 \to v_4 \rangle$. Similarly, $\mathsf{rev}(S) = \langle I_5, \ldots, I_1 \rangle = \langle v_4 \to v_5, v_1 \to v_2, v_2 \to v_3, v_5 \to v_4 \rangle$.

For an edge $e = xy \in E(G)$, we say that S *makes detour over* e if both $x \to y$ and $y \to x$ are members of S. We emphasize that the steps $x \to y$ and $y \to x$ is *not* necessarily made by the same token. The *number of detours* S *makes over* e, denoted by $D_G(S, e)$, is defined to be twice the minimum between the number of appearances of $x \to y$ and the number of appearances of $y \to x$. The *total number of detours* S *makes in* G, denoted by $D_G(S)$, is defined to be $\sum_{e \in E(G)} D_G(S, e)$. As an example, one can verify that the TS-sequence S described in Fig. 1 satisfies $D_G(S, v_4 v_5) = 2$ and $D_G(S) = 2$. Let $\mathcal{S}$ be the set of all TS-sequences in G between two independent sets I, J. We define by $D^*(G, I, J) = \min_{S \in \mathcal{S}} D_G(S)$ the smallest number of detours that a TS-sequence between I and J in G can possibly make.

3   SHORTEST SLIDING TOKEN for Spiders

In this section, we claim that

Theorem 1. *Given an instance* (G, I, J) *of* SHORTEST SLIDING TOKEN *for spiders, one can construct a shortest* TS-*sequence between* I *and* J *in* $O(n^2)$ *time, where* n *denotes the number of vertices of the given spider* G.

First of all, from the linear-time algorithm for solving SLIDING TOKEN for trees (which also applies for spiders as well) presented in [4], we can simplify our problem as follows. For an independent set I of a tree T, the token on $u \in I$ is said to be (T, I)-*rigid* if for any I' with $I \overset{T}{\longleftrightarrow} I'$, $u \in I'$. Intuitively, a (T, I)-rigid token cannot be moved by any TS-sequence in T. One can find all (T, I)-rigid tokens in a given tree T in linear time. Moreover, a TS-sequence between I and J in T exists if and only if the (T, I)-rigid tokens and (T, J)-rigid tokens are the same, and for any component F of the forest obtained from T by removing all vertices where (T, I)-rigid tokens are placed and their neighbors, $|I \cap F| = |J \cap F|$. Thus, for an instance (G, I, J) of SHORTEST SLIDING TOKEN for spiders, we can assume without loss of generality that $I \overset{G}{\longleftrightarrow} J$ and there are no (G, I)-rigid and (G, J)-rigid tokens.

3.1 Our Approach

We now give a brief overview of our approach. For convenience, from now on, let (G, I, J) be an instance of SHORTEST SLIDING TOKEN for spiders satisfying the above assumption. Rough speaking, we aim to construct a TS-sequence in G between I and J of *minimum* length $M^*(G, I, J) + D^*(G, I, J)$, where $M^*(G, I, J)$ and $D^*(G, I, J)$ are respectively the smallest number of token-slides and the smallest number of detours that a TS-sequence between I and J in G can possibly perform, as defined in the previous section. Indeed, the following lemma implies that any TS-sequence in G between I and J must be of length at least $M^*(G, I, J) + D^*(G, I, J)$.

Lemma 1. *Let* I, J *be two independent sets of a tree* T *such that* $I \overset{T}{\longleftrightarrow} J$. *Then, for every* TS-*sequence* S *between* I *and* J, $\mathsf{len}(S) \geq M^*(T, I, J) + D^*(T, I, J)$.

As a result, it remains to show that any TS-sequence in G between I and J must be of length at most $M^*(G, I, J) + D^*(G, I, J)$, and there exists a specific TS-sequence S in G between I and J whose length is exactly $M^*(G, I, J) + D^*(G, I, J)$. To this end, we shall analyze the following cases.

- **Case 1:** $\max\{|I \cap N_G(v)|, |J \cap N_G(v)|\} = 0$.
- **Case 2:** $0 < \max\{|I \cap N_G(v)|, |J \cap N_G(v)|\} \leq 1$.
- **Case 3:** $\max\{|I \cap N_G(v)|, |J \cap N_G(v)|\} \geq 2$.

In each case, we claim that it is possible to simultaneously determine $D^*(G, I, J)$ and construct a TS-sequence in G between I and J whose length is minimum. More precisely, in **Case 1**, we show that it is always possible to construct a TS-sequence between I and J of length $M^*(G, I, J)$, that is, no detours are required. (Note that, no TS-sequence can use less than $M^*(G, I, J)$ token-slides.) However, this does not hold in **Case 2**. In this case, we show that in certain conditions, detours cannot be avoided, that is, any TS-sequence must make detours at least one time at some edge of G. More precisely, in such situations, we show that it is possible to construct a TS-sequence between I and J of length $M^*(G, I, J) + 2$, that is, the sequence makes detour at exactly one edge. Finally, in **Case 3**, we show that detours cannot be avoided at all, and it is possible to construct a TS-sequence between I and J of minimum length, without even knowing exactly how many detours it performs. As a by-product, we also describe how one can calculate this (smallest) number of detours precisely. Due to space restriction, in this paper, we consider only **Case 1**. For more details on **Case 2** and **Case 3**, please see the full version of this paper [13].

3.2 Case 1: $\max\{|I \cap N_G(v)|, |J \cap N_G(v)|\} = 0$

As mentioned before, in this case, we will describe how to construct a TS-sequence S in G between I and J whose length $\mathsf{len}(S)$ equals $M^*(G, I, J) + D^*(G, I, J)$. In general, to construct any TS-sequence, we need: (1) a target assignment f that tells us the final position a token should be moved to (say, a token on v should finally be moved to $f(v)$); and (2) an ordering of tokens that tells us which token should move first. From the definition of $M^*(G, I, J)$, it is natural to require that our target assignment f satisfies $M^*(G, I, J) = \sum_{w \in I} \mathsf{dist}_G(w, f(w))$. As you will see later, such a target assignment exists, and we can always construct one in polynomial time. We also claim that one can efficiently define a total ordering $\prec$ of vertices in I such that if $x, y \in I$ and $x \prec y$, then the token on x will be moved before the token on y in our desired TS-sequence. Combining these results, our desired TS-sequence will finally be constructed (in polynomial time).

Target Assignment. We now describe how to construct a target assignment f such that $M^*(G, I, J) = \sum_{w \in I} \mathsf{dist}_G(w, f(w))$. For convenience, we always assume that the given spider G has body v and $\deg_G(v)$ legs $L_1, \ldots, L_{\deg_G(v)}$. Moreover, we assume without loss of generality that these legs are labeled such that $|I \cap V(L_i)| - |J \cap V(L_i)| \leq |I \cap V(L_j)| - |J \cap V(L_j)|$ for $1 \leq i \leq j \leq \deg_G(v)$; otherwise, we simply re-label them. For each leg L_i ($i \in \{1, 2, \ldots, \deg_G(v)\}$), we define the corresponding independent sets I_{L_i} and J_{L_i} as follows: $I_{L_1} = (I \cap V(L_1)) \cup (I \cap \{v\})$; $J_{L_1} = (J \cap V(L_1)) \cup (J \cap \{v\})$; and for $i \in \{2, \ldots, d\}$, we define $I_{L_i} = I \cap V(L_i)$ and $J_{L_i} = J \cap V(L_i)$. In this way, we always have $v \in I_{L_1}$ (resp. $v \in J_{L_1}$) if $v \in I$ (resp. $v \in J$). This definition will be helpful when considering tokens placed at the body vertex v.

Under the above assumptions, we design Algorithm 1 for constructing f. We note that Algorithm 1 works even when the legs are labeled arbitrarily. However, our labeling of the legs of G will be useful when we use the produced target assignment for constructing a TS-sequence of length $M^*(G, I, J)$ between I and J in G.

Token Ordering. Intuitively, we want to have a total ordering $\prec$ of vertices in I such that if $x \prec y$, the token placed at x should be moved before the token placed at y. Ideally, once the token is moved to its final destination, it will never be moved again. From Algorithm 1, the following natural total ordering of vertices in I can be derived: for $x, y \in I$, set $x < y$ if x is assigned before y. Unfortunately, such an ordering does not always satisfy our requirement. However, we can use it as a basis for constructing our desired total ordering of vertices in I.

Algorithm 1. Find a target assignment between two independent sets I, J of a spider G such that $M^*(G, I, J) = \sum_{w \in I} \mathrm{dist}_G(w, f(w))$.

Input: Two independent sets I, J of a spider G with body v.
Output: A target assignment $f : I \to J$ such that $M^*(G, I, J) = \sum_{w \in I} \mathrm{dist}_G(w, f(w))$.

1: **for** $i = 1$ to $\deg_G(v)$ **do**
2:     **while** $I_{L_i} \neq \emptyset$ and $J_{L_i} \neq \emptyset$ **do**
3:         Let $x \in I_{L_i}$ be such that $\mathrm{dist}_G(x, v) = \max_{x' \in I_{L_i}} \mathrm{dist}_G(x', v)$.
4:         Let $y \in J_{L_i}$ be such that $\mathrm{dist}_G(y, v) = \max_{y' \in J_{L_i}} \mathrm{dist}_G(y', v)$.
5:         $f(x) \leftarrow y$; $I_{L_i} \leftarrow I_{L_i} \setminus \{x\}$; $J_{L_i} \leftarrow J_{L_i} \setminus \{y\}$.
6:     **end while**
7: **end for**
8: **while** $\bigcup_{i=1}^{\deg_G(v)} I_{L_i} \neq \emptyset$ and $\bigcup_{i=1}^{\deg_G(v)} J_{L_i} \neq \emptyset$ **do** ▷ From this point, for any leg L, either $I_L = \emptyset$ or $J_L = \emptyset$.
9:     Take a leg L_i such that there exists $x \in I_{L_i}$ satisfying $\mathrm{dist}_G(x, v) = \min_{x' \in \bigcup_{i=1}^{\deg_G(v)} I_{L_i}} \mathrm{dist}_G(x', v)$.
10:     Take a leg L_j such that there exists $y \in J_{L_j}$ satisfying $\mathrm{dist}_G(y, v) = \max_{y' \in \bigcup_{i=1}^{\deg_G(v)} J_{L_i}} \mathrm{dist}_G(y', v)$.
11:     $f(x) \leftarrow y$; $I_{L_i} \leftarrow I_{L_i} \setminus \{x\}$; $J_{L_j} \leftarrow J_{L_j} \setminus \{y\}$.
12: **end while**
13: **return** f.

Before showing how to construct $\prec$, we define some useful notation. Let $f : I \to J$ be a target assignment produced from Algorithm 1. For a leg L of G and a vertex $x \in I_L \cup J_L$, we say that the leg L *contains* x, and x is *inside* L. For each leg L of G, we define $I_L^1 = \{w \in I_L : f(w) \notin J_L\}$ and $I_L^2 = \{w \in I_L : f(w) \in J_L\}$. Roughly speaking, a token in I_L^1 (resp. I_L^2) must finally be moved to a target outside (resp. inside) the leg L. Given a total ordering $\lhd$ on vertices of I and a vertex $x \in I$, we define $K(x, \lhd) = N_G[P_{xf(x)}] \cap \{y \in I : x \lhd y\}$. Intuitively, if $y \in K(x, \lhd)$, then in order to move the token on x to its final target $f(x)$, one should move the token on y beforehand. In some sense, the token on y is an "obstacle" that forbids moving the token on x to its final target $f(x)$. If $x \in I_L$ for some leg L of G, we define $K^1(x, \lhd) = K(x, \lhd) \cap I_L^1$ and $K^2(x, \lhd) = K(x, \lhd) \cap I_L^2$. As before, a token in $K^1(x, \lhd)$ (resp. $K^2(x, \lhd)$) must finally be moved to a target outside (resp. inside) the leg L containing x. By definition, it is not hard to see that I_L^1 and I_L^2 (resp. $K^1(x, \lhd)$ and $K^2(x, \lhd)$) form a partition of I_L (resp. $K(x, \lhd)$).

Ideally, in our desired total ordering $\prec$, for any $w \in I$, we must have $K(w, \prec) = \emptyset$. This enables us to move tokens in a way that any token placed at $w \in I$ is moved directly to its final target $f(w)$ through the (unique) shortest path $P_{wf(w)}$ between them; and once a token is moved to its final target, it will never be moved again. As this does not always hold for the total ordering $<$ defined from Algorithm 1, a natural approach is to construct $\prec$ from $<$ by looking at all $w \in I$ with $K(w, <) \neq \emptyset$ and reversing the ordering of any pair of vertices that makes our desired moving strategy impossible. A formal description of this procedure is in Algorithm 2 below.

To provide a better explanation of Algorithm 2, we briefly introduce the cases that require changing the ordering $<$. Assume that $w \in I$ is such that $K(w, <) \neq \emptyset$.

- **Ordering between w and vertices in $K(w, <)$.** For each $x \in K(w, <)$, originally $w < x$, but in the new ordering, $x \prec w$. That is, to move the token on w, one should move any "obstacle" (which belongs to $K(w, <)$) beforehand;
- **Ordering between vertices in $K^2(w, <)$.** If $K^2(w, <) \neq \emptyset$, the token on w and any token in $K^2(w, <)$ must be moved to targets inside the leg L containing w. (If $f(w) \notin J_L$ then any "obstacle" between w and $f(w)$ must be moved to targets outside w, which means $K^2(w, <)$ is empty.) Consequently, for x, y in $K^2(w, <)$, if $x < y$, the token on x should move after the token on y, that is, we should define $x \succ y$.
- **Ordering of vertices between $K^1(w, <)$ and $K^2(w, <)$.** If both $K^1(w, <)$ and $K^2(w, <)$ are non-empty, then it is better (but not strictly required) if we move the tokens in $K^1(w, <)$ before moving any token in $K^2(w, <)$. Originally, vertices in $K^1(w, <)$ (whose targets is outside L) is assigned after those in $K^2(w, <)$ (whose targets is inside L) in Algorithm 1. Intuitively, this is because tokens in $K^1(w, <)$ is "closer" to the body vertex v than those in $K^2(w, <)$, and moving tokens in $K^1(w, <)$ creates "empty space" in L for moving tokens in $K^2(w, <)$ later.
  Note that when changing the ordering of vertices between $K^1(w, <)$ and $K^2(w, <)$, we also affect the ordering between vertices in $I_L^1 \supseteq K^1(w, <)$. However, the ordering of vertices in I_L^1 should remain unchanged, since Algorithm 1 always assign vertices in I_L^1 whose distance is closest to the body vertex v first. Thus, for each $x \in I_L^1 \setminus K^1(w, <)$ and $y \in K^1(w, <) \cup K^2(w, <) \cup \{w\}$, we need to set $x \prec y$.

The next lemma (Lemma 2) says that Algorithm 2 correctly produces a total ordering $\prec$ on vertices of I such that $K(w, \prec) = \emptyset$ for every $w \in I$. Intuitively, Lemma 2(i) and (ii) say that if $w_i \in I_L$ is the "chosen" vertex in line 2 of Algorithm 2 for some leg L of G, then only a subset $K(w_i, <) \cup I_L^1 \cup \{w_i\}$ of I_L contains "candidates" for "re-ordering". That is, the process of changing the ordering of tokens in each iteration of Algorithm 2 will not affect the ordering between tokens inside and outside L. Lemma 2(iii) guarantees that after "re-ordering", w_i will never be chosen again[2], and the next iteration of the main **while** loop can be initiated. As Algorithm 2 can "choose" at most $|I|$ vertices, and each iteration involving the "re-ordering" of at most $O(|I|)$ vertices, it will finally stop and produce the desired ordering in $O(|I|^2)$ time.

---

[2] $K(w_i, \prec) = \emptyset$ always holds, since none of the members of $K(w_i, <)$ will ever be larger than w_i in the new orderings $\prec$ produced in the next iterations.

Algorithm 2. Construct a total ordering $\prec$ of vertices in I.

Input: The natural ordering $<$ on vertices of I derived from Algorithm 1.
Output: A total ordering $\prec$ of vertices in I.

1: **while** there exists w such that $K(w, <) \neq \emptyset$ **do**
2:     Let w be the smallest element of I with respect to $<$ such that $K(w, <) \neq \emptyset$.
3:     Let L be the leg of G such that $w \in I_L$.
4:     **for** $x \in K(w, <)$ **do**
5:         Set $x \prec w$.
6:     **end for**
7:     **if** $\left|K^2(w, <)\right| \geq 2$ **then**
8:         For $x, y \in K^2(w, <)$, if $x < y$, then set $x \succ y$.
9:     **end if**
10:    **if** $\min\{\left|K^1(w, <)\right|, \left|K^2(w, <)\right|\} \geq 1$ **then**
11:        For $x \in K^1(w, <)$ and $y \in K^2(w, <)$, set $x \prec y$.
12:    **end if**
13:    **if** $\min\{\left|K^1(w, <)\right|, \left|I_L^1 \setminus K^1(w, <)\right|\} \geq 1$ **then**
14:        For $x \in I_L^1 \setminus K^1(w, <)$ and $y \in K^1(w, <) \cup K^2(w, <) \cup \{w\}$, set $x \prec y$.
15:    **end if**
16:    For $x, y \in I$ whose ordering has not been defined, if $x < y$ then set $x \prec y$.
17:    Re-define $<$ to use in the next iteration by setting $x < y$ if $x \prec y$ for every $x, y \in I$.
18: **end while**
19: **return** The total ordering $\prec$ of vertices in I.

Lemma 2. *Let (G, I, J) be an instance of* SHORTEST SLIDING TOKEN *for spiders, where the body v of G satisfies $\max\{|I \cap N_G(v)|, |J \cap N_G(v)|\} = 0$. Let $f : I \to J$ be a target assignment produced from Algorithm 1, and $<$ be the corresponding natural total ordering on vertices of I. Assume that $I = \{w_1, w_2, \ldots, w_{|I|}\}$ is such that $w_1 < w_2 < \cdots < w_{|I|}$. Let w_i be the smallest element in I (with respect to the ordering $<$) such that $K(w_i, <) \neq \emptyset$, and L be the leg of G such that $w_i \in I_L$. Then,*

(i) $K(w_i, <) \subseteq I_L^1$. Additionally, $w_i \in I_L^2$.
(ii) Let $\prec$ be the total ordering of vertices in I defined as in lines 2–17 of Algorithm 2, where the corresponding vertex w is replaced by w_i. Then,
(ii-1) If $x \in K(w_i, <)$, then $x > w_i$ and $x \prec w_i$.
(ii-2) If $x, y \in K^1(w_i, <)$, then $x < y$ if and only if $x \prec y$.
(ii-3) If $x, y \in K^2(w_i, <)$, then $x < y$ if and only if $x \succ y$.
(ii-4) If $x \in K^1(w_i, <)$ and $y \in K^2(w_i, <)$, then $x > y$ and $x \prec y$.
(ii-5) If $x \in I_L^1 \setminus K^1(w_i, <)$ and $y \in K^1(w_i, <)$, then $w_i < x < y$ and $x \prec y \prec w_i$.
(ii-6) If $x \in K(w_i, <) \cup I_L^1 \cup \{w_i\}$ and $y \in I \setminus (K(w_i, <) \cup I_L^1 \cup \{w_i\})$, then $x < y$ if and only if $x \prec y$.
(ii-7) If $x, y \in I \setminus (K(w_i, <) \cup I_L^1 \cup \{w_i\})$, then $x < y$ if and only if $x \prec y$.
(iii) Let $\prec$ be the total ordering of vertices in I described in (ii). Then, $K(w_i, \prec) = \emptyset$. Moreover, if w_j is the smallest element in I (with respect to the ordering $\prec$) such that $K(w_j, \prec) \neq \emptyset$, then $K(w_j, \prec) = K(w_j, <)$.

Now, we have

Lemma 3. *Let (G, I, J) be an instance of* SHORTEST SLIDING TOKEN *for spiders where the body v of G satisfies* $\max\{|I \cap N_G(v)|, |J \cap N_G(v)|\} = 0$. *Assume that there exists a leg L of G with* $|I_L| \neq |J_L|$. *Then, in $O(n^2)$ time, one can construct a* TS-*sequence S between I and J such that* $\mathsf{len}(S) = M^*(G, I, J)$.

In Lemma 3, we assumed that there is some leg L of G with $|I_L| \neq |J_L|$. In the next lemma, we consider the case $|I_L| = |J_L|$ for every leg L of G (regardless of whether $\max\{|I \cap N_G(v)|, |J \cap N_G(v)|\} = 0$).

Lemma 4. *Let (G, I, J) be an instance of* SHORTEST SLIDING TOKEN *for spiders. Let v be the body of G. Assume that* $|I_L| = |J_L|$ *for every leg L of G. Then, in $O(n^2)$ time, one can construct a* TS-*sequence S between I and J such that* $\mathsf{len}(S) = M^*(G, I, J)$.

4   Conclusion

In this paper, we have shown that one can indeed construct a TS-sequence of shortest length between two given independent sets of a spider graph (if exists). We hope that the ideas and approaches described in this paper will provide a useful framework for improving the polynomial-time algorithm for SHORTEST SLIDING TOKEN for trees [23].

References

1. Belmonte, R., Kim, E.J., Lampis, M., Mitsou, V., Otachi, Y., Sikora, F.: Token sliding on split graphs. arXiv preprint (2018)
2. Bonamy, M., Bousquet, N.: Token sliding on chordal graphs. In: Bodlaender, H.L., Woeginger, G.J. (eds.) WG 2017. LNCS, vol. 10520, pp. 127–139. Springer, Cham (2017). https://doi.org/10.1007/978-3-319-68705-6_10
3. Bonsma, P., Kamiński, M., Wrochna, M.: Reconfiguring independent sets in claw-free graphs. In: Ravi, R., Gørtz, I.L. (eds.) SWAT 2014. LNCS, vol. 8503, pp. 86–97. Springer, Cham (2014). https://doi.org/10.1007/978-3-319-08404-6_8
4. Demaine, E.D., et al.: Linear-time algorithm for sliding tokens on trees. Theor. Comput. Sci. **600**, 132–142 (2015). https://doi.org/10.1016/j.tcs.2015.07.037
5. Demaine, E.D., Rudoy, M.: A simple proof that the $(n^2 - 1)$-puzzle is hard. Theor. Comput. Sci. **732**, 80–84 (2018). https://doi.org/10.1016/j.tcs.2018.04.031
6. Diestel, R.: Graph Theory. Graduate Texts in Mathematics, vol. 173, 4th edn. Springer, Heidelberg (2017). https://doi.org/10.1007/978-3-662-53622-3
7. Fox-Epstein, E., Hoang, D.A., Otachi, Y., Uehara, R.: Sliding token on bipartite permutation graphs. In: Elbassioni, K., Makino, K. (eds.) ISAAC 2015. LNCS, vol. 9472, pp. 237–247. Springer, Heidelberg (2015). https://doi.org/10.1007/978-3-662-48971-0_21
8. Gardner, M.: The hypnotic fascination of sliding-block puzzles. Sci. Am. **210**, 122–130 (1964)

9. Gopalan, P., Kolaitis, P.G., Maneva, E.N., Papadimitriou, C.H.: The connectivity of boolean satisfiability: computational and structural dichotomies. SIAM J. Comput. **38**(6), 2330–2355 (2009). https://doi.org/10.1137/07070440X

10. Hearn, R.A., Demaine, E.D.: PSPACE-completeness of sliding-block puzzles and other problems through the nondeterministic constraint logic model of computation. Theor. Comput. Sci. **343**(1–2), 72–96 (2005). https://doi.org/10.1016/j.tcs.2005.05.008

11. Hearn, R.A., Demaine, E.D.: Games, Puzzles, and Computation. A K Peters (2009)

12. Hoang, D.A., Fox-Epstein, E., Uehara, R.: Sliding tokens on block graphs. In: Poon, S.-H., Rahman, M.S., Yen, H.-C. (eds.) WALCOM 2017. LNCS, vol. 10167, pp. 460–471. Springer, Cham (2017). https://doi.org/10.1007/978-3-319-53925-6_36

13. Hoang, D.A., Khorramian, A., Uehara, R.: Shortest reconfiguration sequence for sliding tokens on spider. arXiv preprint (2018). arXiv:1806.08291

14. Hoang, D.A., Uehara, R.: Sliding tokens on a cactus. In: Proceedings of ISAAC 2016, LIPIcs, vol. 64, pp. 37:1–37:26. Schloss Dagstuhl - Leibniz-Zentrum fuer Informatik (2016). https://doi.org/10.4230/LIPIcs.ISAAC.2016.37

15. Ito, T., et al.: On the complexity of reconfiguration problems. Theor. Comput. Sci. **412**(12–14), 1054–1065 (2011). https://doi.org/10.1016/j.tcs.2010.12.005

16. Kamiński, M., Medvedev, P., Milanič, M.: Complexity of independent set reconfigurability problems. Theor. Comput. Sci. **439**, 9–15 (2012). https://doi.org/10.1016/j.tcs.2012.03.004

17. Makino, K., Tamaki, S., Yamamoto, M.: An exact algorithm for the boolean connectivity problem for k-CNF. Theor. Comput. Sci. **412**(35), 4613–4618 (2011). https://doi.org/10.1016/j.tcs.2011.04.041

18. Mouawad, A.E., Nishimura, N., Raman, V., Simjour, N., Suzuki, A.: On the parameterized complexity of reconfiguration problems. In: Gutin, G., Szeider, S. (eds.) IPEC 2013. LNCS, vol. 8246, pp. 281–294. Springer, Cham (2013). https://doi.org/10.1007/978-3-319-03898-8_24

19. Mouawad, A.E., Nishimura, N., Raman, V., Wrochna, M.: Reconfiguration over tree decompositions. In: Cygan, M., Heggernes, P. (eds.) IPEC 2014. LNCS, vol. 8894, pp. 246–257. Springer, Cham (2014). https://doi.org/10.1007/978-3-319-13524-3_21

20. Ratner, D., Warmuth, M.: Finding a shortest solution for the $N \times N$-extension of the 15-puzzle is intractable. J. Symb. Comput. **10**, 111–137 (1990)

21. Ribeiro, M.T., dos Santos, V.F.: Personal communications, March 2018

22. Slocum, J., Sonneveld, D.: The 15 Puzzle Book: How It Drove the World Crazy. Socum Puzzle Foundations (2006)

23. Sugimori, K.: Shortest reconfiguration of sliding tokens on a tree. In: AAAC 2018, May 2018

24. Sugimori, K.: Personal communications, May 2018

25. Yamada, T., Uehara, R.: Shortest reconfiguration of sliding tokens on a caterpillar. In: Kaykobad, M., Petreschi, R. (eds.) WALCOM 2016. LNCS, vol. 9627, pp. 236–248. Springer, Cham (2016). https://doi.org/10.1007/978-3-319-30139-6_19

26. van der Zanden, T.C.: Parameterized complexity of graph constraint logic. In: Proceedings of IPEC 2015, LIPIcs, vol. 43, pp. 282–293. Schloss Dagstuhl - Leibniz-Zentrum fuer Informatik (2015). https://doi.org/10.4230/LIPIcs.IPEC.2015.282

Turing Tumble Is P(SPACE)-Complete

Matthew P. Johnson

Department of Computer Science, Lehman College
Ph.D. Program in Computer Science, The Graduate Center
City University of New York, New York, NY, USA

Abstract. Turing Tumble is a toy gravity-fed mechanical computer (similar to the classic Digi-Comp II, but including additional types of pieces such as *gears*), in which marbles roll down a board, along paths determined by the *locations* of ramps, toggles and gears, which are placed by the "programmer," and by their current *states*, which are altered by the passing marbles. Aaronson proved that a Digi-Comp II decision problem (viz., will any marbles reach the sink?) is CC-Complete, i.e., equivalent to evaluating *comparator circuits*, and posed the question of what additional functionality would raise the machine's computational power beyond CC, speculating that a capability for toggles to affect one another's states (which Turing Tumble's gears happen to provide) might suffice. This turns out to be so: we show, though a simple reduction from a variant of the circuit value problem (CVP), that the Turing Tumble decision problem is P-Complete. The two models also differ in complexity when exponentially (or unboundedly) many marbles are permitted: while Digi-Comp II remains in P, Turing Tumble becomes PSPACE-Complete.

1 Introduction

Turing Tumble is a toy gravity-fed mechanical computer, which was recently introduced commercially following a $400k Kickstarter campaign.[1] It consists of a 2D square grid pegboard, which stands upright; a collection of pieces including ramps, toggles (or "bits"), gears, crossovers, and interceptors; and a set of marbles (of two colors). The user "programs" the board through the placement of the pieces (and the choice of their initial states). The program is run by pressing a button to release a marble from the top, causing it to roll down the board, passing through some *diagonally contiguous* sequence of pieces (see Fig. 1), perhaps altering their states, until it reaches a paddle on the bottom-left or -right, which releases a second marble (from the upper-left or -right, respectively), and so on. The path traversed by a given marble depends both on the placement of the pieces, and on their *current* state.

A toggle can be in one of two states (0 or 1), which determine the direction the next marble rolls when exiting it (southwest or southeast, respectively). (The behavior is the same regardless of whether the marble *enters* the toggle from the

---

[1] https://www.turingtumble.com; emulations: https://jessecrossen.github.io/ttsim/ and https://www.lodev.org/jstumble/

P. Heggernes (Ed.): CIAC 2019, LNCS 11485, pp. 274–285, 2019.
https://doi.org/10.1007/978-3-030-17402-6_23

northwest or the northeast.) Crucially, each marble's visit to a toggle *flips its state*, toggling it back and forth. Moreover, a collection of *gear bit* toggles can be connected together by *gears, entangling* them, so that *flipping the value of any one of these connected gear toggles flips them all.*

The final state of (some distinguished subset of) the toggles can be interpreted as the program's output. Alternatively, with two marble colors, two sources, and two paddles, the color pattern of the marbles collected at the bottom of board can also be interpreted as output. Examples from the Turing Tumble website and instruction manual include programs implementing arithmetic operations and counting in binary, as well as for generating various color patterns.

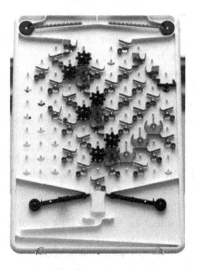

Fig. 1. Image from the Turing Tumble website.

Turing Tumble extends the Digi-Comp II,[2] a toy mechanical computer first sold in the 1960s, whose behavior is more restricted in a number of ways: rather than a general grid, its collection of paths is hardcoded (*in wood*), with its ramps and toggles in fixed locations; it has no gears, crossovers or interceptors; and it has only one color (and source) of marbles. (Programming it consists entirely in initializing the toggles' states.) Yet it also can perform interesting computations such as arithmetic and binary counting. Aaronson [1] investigated the computational power of a kind of generalized Digi-Comp II formalization in which the structure of paths and toggle locations is specified by an arbitrary directed acyclic graph (DAG) with a unique *source* and designated *target sink*, and whose internal nodes represent either toggles or merge points. He defined DIGICOMP as the problem of deciding, for a given instance (i.e., the DAG, the toggles' initial states, and the number of marbles available at the source, encoded in unary) whether any marbles released will eventually reach the target.

Aaronson showed that DIGICOMP is not circuit-universal, i.e., it cannot solve the circuit value problem (CVP) and is not P-Complete [11], but is instead merely CC-Complete [5]. That is, deciding DIGICOMP is equivalent (under logspace reductions) to solving the easier *comparator* circuit value problem (CCV). One of the open questions Aaronson raised was to ask what additional potential features to the DIGICOMP model might result in P-Completeness. In particular, he speculated that the addition of direct causal interaction between the pieces ("toggles and switches controlled by other toggles") might suffice.[3] In this paper we show that it does.

---

[2] https://digi-compii.com; emulation: https://museum.syssrc.com/joda/

[3] Aaronson describes the actual Digi-Comp II as including some such interaction, but this appears (based on the instruction manual [8]) to be in reference to the effects of the Clear switch and the CF1 toggle, which send the marble to an underground passageway, affecting the pieces above. Since the marbles that Clear and CF1 send underground only affect the pieces they pass directly under, however, this arguably should not count as interaction between pieces.

Contributions. We show that the TURINGTUMBLE decision problem (viz., deciding whether any marbles ever reach the designated target sink when an n-piece board configuration is run with m marbles) is P-Complete (under logspace reductions) when $m = \text{poly}(n)$ (and even when $m = n$), and that it is PSPACE-Complete when m is unbounded (and even when $m = 2^n$). The same results hold: (1) if we take n to be just the number of toggles and gear bits, and (2) already in the planar special case, i.e., without use of Turing Tumble's crossover pieces. Many proofs, figures, and details are omitted due to space limitations.

Related Work and Related Models. Aaronson [1] showed in a blogpost that DIGICOMP, which is equivalent to TURINGTUMBLE (on DAGs) without gear pieces, is CC-Complete, i.e., equivalent to the comparator circuit value (CCV) problem [5]. Meiburg [16], in a comment on Aaronson's blogpost, extended this argument to the special case of planar DAGs. Bickford [4] showed that a more general, *non*-acyclic setting, equivalent to TURINGTUMBLE on arbitrary directed graphs, can emulate a fixed-length-tape Turing Machine, i.e., a deterministic linearly bounded automaton, and hence is PSPACE-Complete [4]. We show that this complexity is reached already on DAGs *with exponentially many marbles* or equivalently (see Sect. 2.3) on graphs consisting of a DAG plus one back-arc, i.e., directed graphs with *feedback vertex set number* [9] equal to 1.

In the language of algorithmic combinatorial game theory [12], simulating or predicting the behavior of the Digi-Comp II and the Turing Tumble are classified as *zero-player* games. Although our reductions are not from constraint logic, our results are consistent the pattern commonly observed in connection with it [7], of zero-player games being either P-Complete or PSPACE-Complete, depending on whether they are bounded or unbounded, which here translates into the possible boundedness of the number of marbles. Note that in a sense this distinction collapses in case of the non-acyclic models cases mentioned above: with cycles allowed, a *single* marble, which can be made to cycle around the graph repeatedly, suffices. That is, the potential bound on number of marbles becomes a potential bound on execution time.

The toggles of Digi-Comp II and Turing Tumble are equivalent to 2-state periodic *rotors* in rotor-routing networks [10]. (One of the gadgets we construct, using toggles and ramps, is equivalent to a k-state periodic rotor, with k a power of 2.) Therefore the DAG of a Digi-Comp II instance is equivalent to (a special case of) an *acyclic* rotor network. (Due to their gears, Turing Tumble instances appear to transcend the rotor network model.)

Toggles are also effectively a *nonreversible* version of the *deterministic forks* or *3-spinners* from Demaine et al. [6]'s theory of single-robot motion planning gadgets: in their terminology, when the robot (i.e., the marble) enters at the center location (i.e., from the north) it exits from the left location (to the southwest) or the right location (to the southeast), depending on the 3-spinner's (the toggle's) current state, but (because the graph is a DAG) it can never enter at the left or right location and exit from the center location.

Collisions are central to the *billiard ball model* of computation [15], whereas in our setting there is only one active marble, which always rolls downhill. In dexterity games like Tilt, Labyrinth, and Pigs in Clover the challenge is to find (and successfully execute) a sequence of moves tilting the board in different directions in order to maneuver the ball towards the target location, whereas in our setting there is only one tilt direction, viz., downward (unless we count the release of the next marble as *tilting the board upward and then back down*, as discussed in Sect. 2.3), and the task is simply to predict the machine's behavior— a zero-player game.

Finally, closely related to the Turing Tumble/Digi-Comp II setting is Becker et al. [3]'s *particle computation* model. (Our work is analogous to the "internal computation" problems they study.) The two settings differ, first, in the mechanics of the board pieces involved (with toggles and gear bits, in Turing Tumble, which have state, versus obstacles in particle computation, which do not); second, Turing Tumble involves only one active marble and only the one fixed tilt direction versus multiple active marbles and multiple tilt directions in particle computation. Comparing with the *Single Instruction, Multiple Data* model of parallel algorithms [14], Becker et al. describe their model as "more extreme," and tantamount to *Single Instruction, Single Data, Multiple Locations*. In this sense our setting could be described as more extreme still, and analogous to *Single Instruction, Single Data, Single Location*. But of course, this simpler character of board and marble movement in our setting is complemented by the aforementioned more complicated nature of its individual board pieces.

2   Model

2.1   Hardware, Rules, and Dynamics

In terms of hardware components, the Turing Tumble consists of: the *game board*, which is covered by a square grid pattern of pegs; the *pieces*, which the user places on the pegs; the *marbles*; and a marble release mechanism involving two paddles located at the bottom of the board and two corresponding sources of marbles at the top of the board.

There are two types of pegs, *gear pegs* and *regular pegs*, which alternate along each grid row *and* along each grid column, thus partitioning the square grid into two coarser-grained square grids which are dual to one another (see Fig. 1). Gears can only be placed on gear pegs; other types of pieces (described below) can only be placed on regular pegs.[4] Multiple pieces cannot be placed on the same peg.

Only one marble is in motion on the board at a time. Execution is begun when the user manually causes the release of a marble, and it halts when the currently moving marble reaches one of the two paddles *and that paddle's corresponding marble source is empty* (or when the marble reaches an interceptor, but as already noted our constructions will not use these).

---

[4] Technically speaking, in the actual Turing Tumble product, nothing physically prevents gears from being placed on regular pegs as well, but there is no loss of generality in assuming that *gear bits*, rather than gears, will be placed on those pegs.

Upon a marble's release from a source at the top, it is deposited on a (regular) piece, from which it traverses a downward *path*, coming into direct physical contact with, or *visiting*, a sequence of (regular) pieces, until it exits the grid at the bottom of the board, and finally presses one of the two paddles, depending on whether its final horizontal location is left- or right-of-center. The sequence of (regular) pieces it visits is required to be *diagonally contiguous*, meaning that the ith piece visited must be on a peg located either to the direct southwest or the direct southeast of the $i-1$st piece visited. (Any non-diagonal *free-fall* is forbidden.) That is, the directed graph formed by the regular pegs and the paths marbles legally can potentially traverse is a -135-degree-rotated *layered grid graph* [2], viz., a -135-degree rotation of a directed graph whose nodes are members of $\mathbb{N} \times \mathbb{N}$ and whose arcs are all of the form $(i, j) \rightarrow (i + 1, j)$ or $(i, j) \rightarrow (i, j + 1)$. *Which* path a given marble actually follows will depend on the types of pieces it visits and on their current states (which may be affected *by* its visit).

With the exception of crossover pieces, the behavior that occurs when a marble visits a piece does not depend on whether it arrived *from* the northeast or the northwest. The behavior depends on the type of the piece as follows:

- **toggle:** In the 0 state (which we define as the one in which it appears to be pointing to the upper-left), it changes to the 1 state and the marble exits to the lower-right (and vice versa).
- **left ramp:** The marble exits to the lower-left.
- **right ramp:** The marble exits to the lower-right.
- **crossover:** The marble entering from the upper-left exits to the lower-right, and vice versa.
- **interceptor:** The marble does not exit.
- **gear bit:** The behavior is the same as the ordinary toggle's, except that when its state is flipped, by means of the gear bit's rotation, this causes the rotation (in the *opposite* direction) of any *adjacent* gears (i.e., any gears located directly above, below, left, or right), which in turn causes the same change in state (i.e., rotation in the same direction as the first gear bit) in any gear bits adjacent to those gears, and so on. Any two gear bits that are connected by such a *rectilinearly contiguous* alternating path of gear bits and gears (in which case we say the gear bits are *entangled*) will always have the same state. Note that an ordinary toggle adjacent to a gear is not affected.

More generally, gears behave as follows:

- **gears:** A maximal set of entangled gear bits is called a *gear component*.[5] A gear component need not be "convex" and can even have holes (although those in our constructions will not have holes). Let a *visit* to a gear component be a maximal subpath in the intersection of the gear component and a marble's

---

[5] More formally, gear components can be defined equivalently as follows. Consider the induced subgraph of the *solid* (undirected) grid graph that is induced by the gears and gear bits, i.e., each edge corresponds to a gear and a gear bit that are rectilinearly adjacent. Then a gear component is a component of this graph.

path, with the *length* of the visit referring to the length of this subpath. Note that a marble may visit a gear component multiple times, interspersed with visits to other pieces (potentially including toggles and/or other gear components' gear bits).

In the spirit of the mechanical interactions being modeled, where the gears and gear bits will not be perfectly friction-free and the marbles will have some finite weight, we assume that the size of any gear component (i.e., its number of gear bits) is bounded by some small global constant C, such as 8.

Observation 1. *The visit of a marble to a gear component flips the component's state iff the visit has odd length (i.e., consists of visits to an odd number of the component's gear bits).*

Crossover pieces can be interpreted as permitting nonplanarity in the board configuration; as already mentioned, our constructions will avoid the use of them. One way to do so is by designing a crossover gadget that simulates a crossover piece. An alternative approach, which will be more convenient in some situations in terms of computation of grid locations, is to carefully schedule marble arrivals in such a way that a *toggle* piece placed in a certain location will behave the same as a crossover piece in that location would have.

Observation 2. *If (a) all marbles visiting a given toggle piece alternate between arriving from the northwest and from the northeast, (b) the first one arrives from the northwest (respectively, from the northeast), and (c) the toggle is initialized to the 0 state (respectively, to the 1 state), then each such marble will exit in the same direction (viz., either to the southeast or to the southwest) as it would have if a crossover piece had been placed in the toggle's location instead.*

2.2   Problem Formulation

It is clear that Turing Tumble can be simulated in polynomial time, i.e., that the execution of an n-piece board configuration running on m marbles can be simulated in time polynomial in nm. This is so even if we permit the crossover pieces and interceptor pieces, and indeed even if we abstract away from the board's grid structure to an arbitrary DAG.

For proving hardness, we make a number of simplifying restrictions. We assume that there is only one marble color and only one *source*, but that there are *two* sinks (out-degree-0 nodes), one triggering the release of the next marble and the other being the designated *target sink*. (This is equivalent to the actual two-source/one-sink Turing Tumble in the special case that there is only one marble at the right-hand source and ramps are placed forming a path from that source to the right-hand sink.) Although our DAG will, unlike Turing Tumble pieces, have nodes with in-degree and/or out-degree greater than 2, such nodes will later translate into **bit** gadgets or 2^h-**rotor** gadgets (see below), both of which will be constructed out of Turing Tumble pieces, all of them having both in-degree and out-degree at most 2.

Now we formally define two variants of the decision problem. In the DAG variant, each node will be of one of the following types:

- **source** node: in-degree 0 and out-degree 1, stateless;
- **sink** node: in-degree arbitrary and out-degree 0, stateless;
- **merge** node: out-degree 1, in-degree > 1, stateless;
- 2^h-**rotor** node: out-degree 2^h (with outgoing arcs labeled 0 to $2^h - 1$), in-degree 1, state (in $\{0, ..., 2^h-1\}$) indicates the outgoing arc taken by the most recent marble entering the rotor, which would have cyclically incremented its state; or
- **bit** node: in-degree ≤ 4 (with each incoming arc labeled read, set to 1, set to 0, or flip), out-degree ≤ 8 (with each arc labeled with the operation performed, or equivalently the incoming arc used, *and* 0 or 1, indicating the current state *after* the current marble's visit), state (in $\{0,1\}$), which is altered by incoming marbles as per the labels of the incoming arcs used.

TuringTumble$_{\text{DAG}}^{m=m(n)}$ ($\mathbf{TT}_{\text{D}}^{m(n)}$)

INSTANCE: An n-node DAG with one source and two sinks (one designated as the *target*), and each other node either a **merge**, a 2^h-**rotor**, or a **bit**; the initial state (0 or 1) of each **bit** node; and a function specifying the number of marbles $m(n)$.

QUESTION: When the DAG in its initial state is run with $m = m(n)$ marbles at its source, do any marbles reach the target sink?

TuringTumble$_{\text{GRID};C}^{m=m(n)}$ ($\mathbf{TT}_{\text{G};C}^{m(n)}$)

INSTANCE: An n-piece board configuration placing at most one piece—either a **toggle**, a (**left** and **right**) **ramp**, a **gear bit**, or a **gear**—at each grid point in an $n \times n$ square grid graph, one point designated as the marble source, and two points designated as sinks (one designated as the *target*), such that every induced gear component has size at most C; the initial state (0 or 1) of each **toggle** and of **gear** component; and a function specifying the number of marbles $m(n)$.

QUESTION: When the board configuration in its initial state is run with $m = m(n)$ marbles at its source, do any marbles reach the target sink?

For the value of C, we can choose a small constant such as 8.

Remark 1. Note that while the DAG version abstracts away from Turing Tumble game board, the only way in which the grid version generalizes the actual physical product is in parameterizing the numbers of pieces and marbles (and thus the board size). All other respects in which the formulation differs from the actual product (e.g., having a strict bound on gear component size, and avoiding the use of crossover pieces or a second marble color) are *restrictions* from the product's behavior, which therefore only strengthen the hardness results.

Remark 2. In terms of an instance's encoding size, specifying an instance of the DAG version by the placement of all the individual pieces on the grid can be seen as analogous to writing a graph's edge weights in unary (or in a closer analogy, splitting each (integer) weight-w edge of a graph into a path of w unit-weight edges). In the eventual reduction, however, the number of ramp pieces used will be at most $O(n_b^2)$, where n_b is the number of toggles and gear bits used, and so the results are not affected if we represent the instance abstractly as a DAG and have n denote the DAG's number of nodes.

2.3   Single-Marble Non-acyclic Graph Interpretation

An alternative interpretation of the model is one where there is an additional arc, from the non-target sink to the source, and only a single marble. The back-arc means the model is no longer justifiable purely in terms of a unidirectional gravitational pull, but in a sense it is a minimal extension beyond this.

If we permit such a back-arc, then the "exponentially many marbles" situation perhaps becomes easier to motivate: the assumption of the machine being loaded with an exponential number of *physical objects* translates into the more conventional notion of a machine executing a procedure having exponential running time. Moreover, the problem can be interpreted as a *prediction* problem rather than a *simulation* problem: we can rephrase the decision problem as asking whether the marble *would* ever reach the target sink *if* we let it run without a time limit. We emphasize that a DAG plus a single back-arc, i.e., a directed graph with *feedback vertex set number* equal to 1, is highly restricted subclass of directed graphs generally.

3   Strategy: circuit $\mapsto \mathrm{TT}_{\mathrm{D}}^{m(n)}$ instance $\mapsto \mathrm{TT}_{\mathrm{G};C}^{m(n)}$ instance

We prove two main hardness results, P-hardness in the bounded m case and PSPACE-hardness in the unbounded m case. The two proofs share the same high-level structure, with the reduction's transformation procedure divided into two steps executed in succession:

1. constructing an instance of DAG-based TURINGTUMBLE, based on a given boolean circuit of the appropriate type, and
2. constructing an instance of grid-based TURINGTUMBLE, *based on the DAG-based* TURINGTUMBLE *instance just constructed.*

For simplicity we do not prove the two TURINGTUMBLE formulations are computationally equivalent (neither is a simple generalization of the other), though they likely are, even via relatively low-complexity transformation procedures. Indeed, we do not even prove a reduction in the direction that would appear most relevant here, i.e., from the DAG version to the grid version.

What we prove is a from a *special case of* the DAG version to the grid version. Specifically, we reduce from *the class of all DAG version instances that can be produced in step 1*. This is because the overall transformation from the formula to the grid version instance is the *composition* of the transformations performed by steps 1 and 2. The reductions for the bounded and unbounded settings will differ in the nature of the boolean circuit they start with, and in the complexity constraints the transformation procedure must obey.

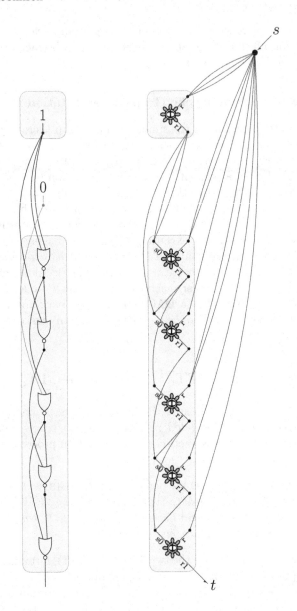

Fig. 2. Reduction from an example FIXEDINTOPNOR2CVP instance (left) to
TURINGTUMBLE instance (right). Each shaded circle in the TURINGTUMBLE instance
with multiple outgoing (downward) arcs indicates a **rotor** node whose arcs are taken
(by its incoming marbles) in left-to-right (counter-clockwise) order; each bolded gear
indicates a **bit** node (implemented with the gadget shown in Fig. 3), with its read and
set to 0 input arcs and its read 1 output arc labeled (and other output arcs, which are
directed to the non-sink terminal, not shown).

To prove the bounded setting P-hard, we give a logarithmic-space reduction from a variant of the Circuit Value Problem (CVP); to prove the unbounded setting PSPACE-hard, we will give a polynomial-time reduction from an analogous variant of the True Quantified Circuit problem.

4   Reducing CVP to TuringTumble$_{\text{DAG}}^{m=\text{poly}(n)}$

We reduce from a P-Complete variant of the Circuit Value Problem (CVP) that we will call FIXEDIN TOPOLOGICALLY ORDERED NOR2CVP (FIXEDINTOPNOR2CVP). NORCVP (equivalent to NANDCVP, Problem A.1.5 in [11]) is the CVP variant in which all gates are NOR gates, and is P-Complete [11]. Let NOR2CVP be the restriction of NORCVP in which all gates have fanin and fanout at most 2,[6] which remains P-Complete. Let TOPNOR2CVP be the restriction of NOR2CVP to instances having nodes listed in a topologically sorted order in the DAG's string encoding, which again remains P-Complete [11]. Finally, let FIXEDINTOPNOR2CVP be the P-Complete result of merging all the input's 1 bits into a single 1 bit and all its 0 bits into a single 0 bit (see [11], Sect. 6.1).[7]

Let such a circuit be specified by a sequence of $g + 2$ gates having values $G_{-1} = 1$, $G_1 = 0$, and for all $i \in [g]$, $G_i = (\overline{G}_{p_1(i)} \wedge \overline{G}_{p_2(i)})$ where $p_1(i), p_2(i) < i$. The task will be to evaluate the last gate G_g. Define $\text{ch}(G_j) = p_1^{-1}(j) \uplus p_2^{-1}(j)$, where $\uplus$ denotes multiset union, and out-deg$(G_j) = |\text{ch}(G_j)|$, which is at most 2 for all $i \in [g]$, i.e., fanout. For $i \in [g]$, let R_i refer to G_i's read input arc, and let R_i^1 refer to the output arc of G_i's that indicates value 1 was read.

Theorem 1. TURINGTUMBLE$_{\text{DAG}}^{m=n}$ *is P-hard under logspace reductions.*

Proof. Given the FIXEDINTOPNOR2CVP formula, we construct a Turing Tumble instance as follows. We create a **bit** node for each gate G_i, $i \in \{-1, 1, 2, ..., g-1\}$, initializing G_{-1} to 1, G_0 to 0, and G_i to 1 for $i \in [g]$.

We create a **rotor** node with one input arc coming from the source and with out-deg$(G_{-1}) + 2g - 1$ output arcs, where the first out-deg(G_{-1}) of these are directed to R_{-1}; then, for each $i \in [g-1]$, two arcs are directed to R_i; finally, one arc is directed to R_g. (Because the NOR gates are all initialized to 1, there is no need to create a **bit** node for G_0.)

For each $i \in \{-1, 1, 2, ..., g-1\}$, we create a **toggle** node with one input arc coming from R_i^1 and with out-deg(G_i) output arcs, directed (in arbitrary order) to the members of $\{R_j : G_j \in \text{ch}(G_i)\}$. Finally, we create an arc from R_g^1 to the target sink. (Any missing arcs at exit points with the potential to emit marbles should be assumed to go to the *non*-target sink.)

It is clear that the construction can be performed in logspace.

---

[6] Although NOR2CVP's P-hardness is not stated explicitly by [11], this is implicit in the reduction from AM2CVP to NORCVP ([11], Theorem 6.2.4).

[7] Our FIXEDINTOPNOR2CVP is somewhat in the spirit of [17]'s SEQUENTIAL NORCVP.

When run, for each $i \in \{-1, 1, 2, ..., g-1\}$, out-deg($G_i$) many marbles will travel from the source to G_i, and thence, if $G_i = 1$, to G_i's children, setting their states to 0. Finally, a marble travels from the source to G_g, and thence, if $G_g = 1$, to the target sink. Note that the total number of marbles used is at most $2g + 1$. We used a toggle for the output of every **bit** node except the last, and each of the $g+1$ **bit** nodes will itself certainly contain at least one **toggle/gear bit**, and so n is (conservatively) at least m.

Clearly TURINGTUMBLE is in P when $m = \text{poly}(n)$. Thus we conclude:

Corollary 1. TURINGTUMBLE$_{\text{DAG}}^{m=\text{poly}(n)}$ *is P-Complete under logspace reductions.*

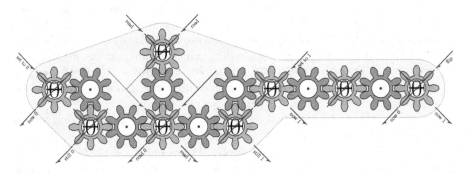

Fig. 3. Bit node gadget with operations: set to 1, set to 0, and read, in addition to flip, consisting of $C = 8$ gear bits, 7 gears, and 15 ramps.

5   Discussion

Due to space limits, we omit the details of how the DAG problem instance is embedded in the grid board, other than to say that the procedure involves the definition of three gadgets (which will be the only occurrences of gear components): a **bit** node gadget (illustrated in Fig. 3), a 2^h-**rotor** node gadget (which is a perfect height-h binary tree of **toggle** pieces, with its leaves' outputs bit-reversal permuted), a crossover gadget. (A **merge** node is simple a **toggle** piece.) Eventually we obtain:

Theorem 2. TURINGTUMBLE$_{\text{GRID};C}^{m=\text{poly}(n)}$ *is P-Complete under logspace reductions.*

Also omitted are the proofs for large number of marbles m:

Theorem 3. TURINGTUMBLE$_{\text{DAG}}^{m \geq 2^n}$ *and* TURINGTUMBLE$_{\text{GRID};C}^{m \geq 2^n}$ *are both PSPACE-Complete.*

Besides asking what potential extensions to the model might raise its power to P-Completeness, Aaronson presented as a possibly more interesting open problem the question of what extensions might raise its power beyond CC-Completeness but *not* all the way to P-Completeness. Our CVP reduction could perhaps be interpreted as discouraging in terms of that goal, since even a very limited use of interaction (i.e., gadgets that identify the states of four *adjacent* gear bits, using two gears) sufficed for P-Completeness. One potential direction, however, would be to investigate the computational power of Turing Tumble equipped with marble colors and sources/paddles, but *without* gears.

It would also be interesting to characterize Turing Tumble's computational power (and Digi-Comp II's) in relation to Demaine et al. [6]'s gadget primitives and to Holroyd et al. [13]'s Abelian logic gates.

Acknowledgements. This work was supported in part by NSF award INSPIRE-1547205, and by the Sloan Foundation via a CUNY Junior Faculty Research Award.

References

1. Aaronson, S.: The power of the Digi-Comp II: my first conscious paperlet, blogpost on Shtetl-Optimized (2010). https://www.scottaaronson.com/blog/?p=1902
2. Allender, E., Barrington, D.A.M., Chakraborty, T., Datta, S., Roy, S.: Planar and grid graph reachability problems. Theory Comput. Syst. **45**(4), 675–723 (2009)
3. Becker, A.T., Demaine, E.D., Fekete, S.P., Lonsford, J., Morris-Wright, R.: Particle computation: complexity, algorithms, and logic. Nat. Comput. **18**(1), 1–21 (2019)
4. Bickford, N.: Marble runs and Turing machines. Talk at Gathering 4 Gardner 11, March 2014. https://nbickford.wordpress.com/2014/03/25/images-from-marble-runs-and-turing-machines/
5. Cook, S.A., Filmus, Y., Lê, D.T.M.: The complexity of the comparator circuit value problem. ACM Trans. Comput. Theory **6**(4), 15 (2014)
6. Demaine, E.D., Grosof, I., Lynch, J., Rudoy, M.: Computational complexity of motion planning of a robot through simple gadgets. In: FUN, pp. 18:1–18:21 (2018)
7. Demaine, E.D., Hearn, R.A.: Constraint logic: a uniform framework for modeling computation as games. In: CCC, pp. 149–162 (2008)
8. Education Science Research, Incorporated: DIGI-COMP II instruction manual. https://cdn.evilmadscientist.com/KitInstrux/DCII-manual.pdf
9. Ganian, R., Hliněný, P., Kneis, J., Langer, A., Obdržálek, J., Rossmanith, P.: Digraph width measures in parameterized algorithmics. Discrete Appl. Math. **168**, 88–107 (2014)
10. Giacaglia, G.P., Levine, L., Propp, J., Zayas-Palmer, L.: Local-to-global principles for the hitting sequence of a rotor walk. Electron. J. Comb. **19**(1), 5 (2012)
11. Greenlaw, R., Hoover, H.J., Ruzzo, W.L., et al.: Limits to Parallel Computation: P-Completeness Theory. Oxford University Press, Oxford (1995)
12. Hearn, R.A., Demaine, E.D.: Games, Puzzles, and Computation. CRC Press, Boca Raton (2009)
13. Holroyd, A.E., Levine, L., Winkler, P.: Abelian logic gates. arXiv:1511.00422 (2015)
14. Leighton, F.T.: Introduction to Parallel Algorithms and Architectures: Arrays, Trees Hypercubes. Morgan Kaufmann, San Francisco (1992)
15. Margolus, N.: Physics-like models of computation. Phys. D: Nonlinear Phenom. **10**(1–2), 81–95 (1984)
16. Meiburg, A.: Comment #24 on Aaronson [1] (2010). https://www.scottaaronson.com/blog/?p=1902#comment-111711
17. Okhotin, A.: A simple P-complete problem and its language-theoretic representations. Theoret. Comput. Sci. **412**(1–2), 68–82 (2011)

Linear-Time In-Place DFS and BFS on the Word RAM

Frank Kammer[(✉)] and Andrej Sajenko

THM, University of Applied Sciences Mittelhessen, Giessen, Germany
{frank.kammer,andrej.sajenko}@mni.thm.de

Abstract. We present an in-place depth first search (DFS) and an in-place breadth first search (BFS) that runs on a word RAM in linear time such that, if the adjacency arrays of the input graph are given in a sorted order, the input is restored after running the algorithm. To obtain our results we use properties of the representation used to store the given graph and show several linear-time in-place graph transformations from one representation into another.

Keywords: Space efficient · Depth first search · Breadth first search · Restore model

1 Introduction

Motivated by the rapid growth of the data sizes in nowadays applications, algorithms that are designed to efficiently utilize both time and space are becoming more and more important. Another reason for the need of such algorithms is the limitation in the memory sizes of the tiniest devices.

To measure the total amount of memory that an algorithm requires we distinguish two types of memory. The memory that stores the input is called the *input memory*. The memory that an algorithm additionally occupies during the computation is called the *working memory*.

Several models of computation have been considered for the case when writing in the input memory is restricted. In the *multi-pass streaming* model [21] the input is assumed to be held in a read-only sequentially-accessible media, and the optimization target is the number of passes an algorithm makes over the input. In the *word RAM* [15] the memory is partitioned into randomly-accessible *words*, each of size w, the input is in the first $N \in I\!N$ words and reading/writing a word as well as the arithmetic operations (addition, subtraction, multiplication and bit-shift) take constant time if applied on inputs that fit into a word. As usual, we assume $w = \Omega(\log N)$. In the *read-only word RAM* [15] the input memory is assumed to be read-only. Another model allows data in the input memory to be permuted, but not destroyed [5]. A variant of the latter model is called the *restore model* [8] where the input memory is allowed to be modified during the process of answering a query, but it has to be restored to its original state afterwards.

© Springer Nature Switzerland AG 2019
P. Heggernes (Ed.): CIAC 2019, LNCS 11485, pp. 286–298, 2019.
https://doi.org/10.1007/978-3-030-17402-6_24

There are several algorithms for the read-only word RAM, e.g., for sorting [4, 22], geometric problems [1,3], or graph algorithms [2,7,10,12,17]. Unfortunately, most of the algorithms on n-vertex graphs (including *depth first search* (DFS) and *breadth first search* (BFS)) have to use roughly $\Omega(n)$ bits of working memory in the read-only RAM model since there is a lower bound for the *reachability problem*, i.e., the problem to find out if two given vertices of a given graph are in the same connected component. The lower bound essentially says that we can solve reachability in polynomial time only if we have roughly $\Theta(n)$ bits of working memory [11].

Our focus is to find *space-efficient algorithms*, i.e., algorithms that (1) run (almost) as fast as the best known algorithms for the problem without any space limitations and that (2) use space economically. To bypass the lower bound we consider *in-place* algorithms. An in-place algorithm [9] can use the input memory and the working memory for writing, and the result of the algorithm may be written to the input or can be sent to an output stream. Moreover, the working memory size is restricted to $O(1)$ words. Sorting algorithms like heapsort and bubblesort are classic examples of in-place algorithms.

Usually, one runs several computations on a given graph. To allow the input to be reused we want to run our algorithms on the *weak restore word RAM*, i.e., given the input in a specific representation, as for example the sorted representation in the next section, it can be restored.

Graph algorithms usually do not specify the input format of a given graph since linear time is sufficient to convert between any two reasonable adjacency-list representations—e.g., reorder the adjacency arrays with radix sort. However, since we focus on linear-time in-place algorithms for DFS and BFS in the weak restore word RAM, we have to be more specific about the input format. Implementing an in-place algorithm on the weak restore word RAM model where the working memory is limited and the input memory must be restored, a trick is to use the redundancy in the input representation. Thus, the size of the input representation is very crucial. In the following, let n and m be the number of vertices and edges, respectively, of the given graph.

We are not aware of a linear-time DFS or BFS that runs in-place or uses this model. However, Chakraborty et al. [6] introduced another model where the adjacency arrays of a graph can be only rotated, but a restoration is not required. In their model, they recently showed that one can run an in-place DFS and a BFS in $O(n^3 \log n)$ time on an arbitrary graph. The space required to represent the graph is not mentioned explicitly, but based on their description they require at least $(n + 2m + \min\{n, m/w\})$ words for undirected graphs since each undirected edge is stored at both endpoints and since an adjacency array is used for each vertex where the size of the array must be known. Moreover, their representation for directed graphs uses at least $(2n + 2m + 2\min\{n, m/w\})$ words since adjacency arrays for in- and out-edges are stored for each vertex.

We use the weak restore word RAM to show linear-time, in-place algorithms for both DFS and BFS that runs on a graph with a representation consisting of only $(n + m + 2)$ words on directed graphs and $(n + 2m + 2)$ words on undirected graphs (each undirected edge occurs at both endpoints). To operate efficiently on that compact representation and to have also some kind of redundancy, we assume that the order and the content of the adjacency arrays are sorted as defined more precisely in the next section.

2   Representation

To show our results we use different representations of the given n-vertex graph $G = (V, E)$ with $V = \{1, \ldots, n\}$ that all need the same amount of memory. We next present different graph representations.

In our *sorted standard representation* (Fig. 1), we first store the number of vertices and a *table of pointers* T with one pointer per vertex that points to the adjacency array of the vertex. Subsequently, we store the total length of the adjacency arrays. We additionally assume for the sorted standard representation that the adjacency array of vertex i is stored before the adjacency array of vertex $i + 1$ for all $i = 1, \ldots, n - 1$ and that all vertices inside an adjacency array are also stored in ascending order. If the adjacency array of a vertex is not given in ascending order, then it can be sorted using an in-place linear-time radix sort [14]. However, in this case, we cannot restore to the representation of the given graph.

This representation is economical in space and implicitly contains the information to compute the degree of each vertex $v \in V$. The degree $\deg(v)$ of a vertex v equals the length of its adjacency array, and since the adjacency array of a vertex v is written directly before the adjacency array of a vertex $v + 1$, the degree of v equals the pointer differences of $T[v]$ and $T[v + 1]$ for all $v \in V \setminus \{n\}$. For the last vertex $v = n$ the degree equals the difference of the pointer $T[v]$ and the total length of the array $n + m + 2$ with $n = A[0]$ and $m = A[n + 1]$. If a vertex $v \in V \setminus \{n\}$ has degree zero, then its adjacency array is empty and therefore $T[v]$ and $T[v + 1]$ point at the same position.

For our DFS described subsequently, we require to encode information like the state of visited and unvisited vertices. To be able to do this we transform the sorted standard representation first into a so-called *adjacency-array begin-pointer representation* or short the *begin-pointer representation* and finally into a so-called *swapped begin-pointer representation*.

We obtain the *begin-pointer representation* (Fig. 2) by taking the sorted standard representation and replacing each vertex name v in the adjacency arrays by a pointer to the beginning of the adjacency array of vertex v. Since a vertex of degree zero does not have an adjacency array, we cannot create a pointer into it. In this case we keep the vertex name, but we mark such a vertex by replacing its pointer in the table T by a self reference, i.e., set $T[v] = v$.

In the begin-pointer representation we can jump from one adjacency array into another, but lack the ability to find out the vertex name of the adjacency array in constant time if we jump into it using some edge. To resolve this issue we use the swapped begin-pointer representation (Fig. 3) where we swap the first adjacency pointer of a vertex v by v and move the pointer stored there into the table T of position v. In this representation we are still able to access the moved pointer by a lookup at $T[v]$, and know immediately to which vertex the adjacency belongs to.

In our full version of the paper [18], we show that in-place linear-time transformations between all kinds of representation exist.

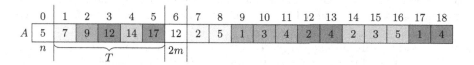

Fig. 1. Sorted standard representation of a graph with m undirected or $2m$ directed edges.

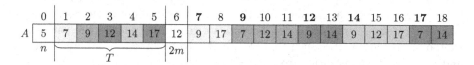

Fig. 2. Begin-pointer representation of the graph from Fig. 1. Every adjacency array entry v is replaced with the pointer $p = T[v]$ to the first position of v's adjacency array.

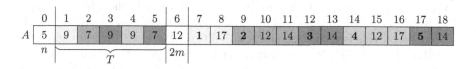

Fig. 3. Swapped begin-pointer representation of the graph in Fig. 1.

3 Depth-First Search

Usually a DFS is only an algorithmic scheme how a graph can be explored step by step and does nothing useful. Its usefulness comes in combination with additional computational steps that are defined by a user for a specific application. These steps can be encapsulated in functions that we call *user-implemented functions*.

To introduce the user-implemented functions `pre-` and `postprocess` as well as `pre-` and `postexplore` we start to sketch their usage in a standard DFS. Initially all vertices of a graph are unvisited, also called *white*. The algorithm starts by visiting a start vertex u. Whenever a DFS visits a vertex u for the first time it colors u *gray* to mark it as visited and executes `preprocess(u)`. For each outgoing edge (u, v) of u, it first calls `preexplore(u, v)` and second visits vertex v if v is white. When finally v has no outgoing white neighbors, it marks v as done by coloring it *black* and calls `postprocess(v)` and backtracks to the parent u. After backtracking from v to u the algorithm calls `postexplore(u, v)`.

By using suitable implementations for the four user-implemented functions, the user knows exactly how the exploration takes place and can easily output, e.g., the vertices in pre-, post-, or inorder with respect to the constructed DFS tree. Not every DFS algorithm supports all these functions. Thus, we can also measure the usefulness of a DFS implementation by the number of supported functions.

To obtain a linear-time in-place DFS on directed graphs, we cannot support calls of the functions `preexplore` and `postexplore`, which are often not necessary, i.e., to compute pre- and post-order.

We now start the description of our DFS algorithm where we expect the graph being given in the swapped begin-pointer representation. Our goal is to encode two information in the representation, but with the knowledge that we have to restore the representation later. First, we need to encode the color of each vertex. Instead of encoding all three colors we use only the colors white and *gray-black* (as gray or black). Second, we require to encode the path that we took to reach a vertex such that we are able to backtrack to a parent vertex and continue the exploration from there.

For simplicity, we assume that every vertex of the directed graph has at least two neighbors, and we so can conclude that every pointer in the adjacency arrays points at a position storing a vertex name $v \in V = \{1, \ldots, n\}$. In our full version of the paper [18], we describe how to handle vertices of degree one and zero.

Our idea is to store the colors of the vertices implicitly by using the following invariant: A vertex v is white exactly if the first pointer p in the adjacency array of v, which is stored in $T[v]$, points at a value at most n, i.e., $A[p] \leq n$. By our conclusion this is initially true for all vertices. We next want to enable the algorithm to backtrack from a visited vertex to its parent. Whenever a DFS takes a path from a vertex u to a vertex v it has to return to the vertex u from v, i.e., backtrack from v to u, if all white neighbors of v are visited. Our idea is to reverse the path from vertex u to the vertex v whenever we visit a white vertex v by using so-called *reverse pointers*. In other words, the idea is to turn the pointer to v in u's adjacency array to a pointer to u in v's adjacency array.

Now we describe the construction of a reserve pointer in detail. See also Fig. 4. Assume that our DFS currently visits a vertex u, and we iterate through u's adjacency array. Iterating over u's adjacency array, e.g., at a position p, we find a pointer q pointing into an adjacency array of a white vertex $v = A[q]$. Inside v's adjacency array the first pointer that we have to inspect is $q' = T[v]$. Because we know that we left from position p to q to reach v, we want to store a pointer to p as a reverse pointer from v to u. (Returning to u, the algorithm can continue exploring u's adjacency array from $p + 1$.) We store p inside $T[v]$. The pointer p is now the reverse pointer from v to u. Naively doing so we overwrite the pointer q'. This would cause an information loss. Therefore, we have to find a new location for q'. What we can observe is that when using the reverse pointer,

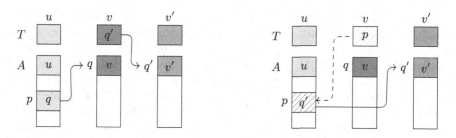

Fig. 4. The figure shows the state before (left) and after (right) creating a reverse pointer. The squares at the top are locations in T and the array bellow of each square of T is the adjacency array of the vertex written on the top. The variables u, v, v' are vertices and $p, q, q' \geq n + 2$ are array positions/pointers. Normal arrows are pointers from an adjacency array into another and dashed arrows are reverse pointers.

we can restore the original pointer from u to v such that we do not need to keep the pointer q in $A[p]$ (part of u's adjacency array) as long as we have the reverse pointer. Hence, we use $A[p]$ as a temporary location to store q'. Note that q' is still accessible from v by following the reverse pointer stored in $T[v]$.

In the example above we showed how to visit a vertex from a position p. If p is not the first position of u's adjacency array the creation of a reverse pointer that points at p has a nice side-effect: The vertex v becomes gray-black since the value stored in $T[v]$ points at a value larger than n.

What if p is the first position in u's adjacency array? Then we encounter two problems. To handle the problems, recall that a reverse pointer of a vertex v is always stored in $T[v]$. In this scenario the reverse pointer $p = T[v]$ points to the first position of an adjacency array that stores a vertex name $u = A[p]$. The first problem is that v is no longer white because p is the position of a value at most n. The second problem arises when we try to temporary store the pointer $q' = T[v]$ to $A[p]$, which stores the vertex name u in our swapped representation. Alternatively, storing the pointer q' in $T[u]$ overwrites the reverse pointer of vertex u, unless u is the start vertex.

We avoid both problems by never leaving a vertex from the first position of its adjacency array. If we have to visit a vertex by following the first pointer stored at the first position p, i.e, stored in $T[u]$ with $u = A[p]$, then we first swap the pointers in $T[u]$ and $A[p+1]$ and follow afterwards the pointer stored at the second position $p + 1$. Since the pointers in our adjacency arrays are stored in ascending order, we can check if we have swapped pointers. Whenever we return to a vertex that we left from a second position p in its adjacency array and the value stored at p is smaller than the value in $T[u]$ with $u = A[p-1] \wedge 1 \leq u \leq n$, we swap the pointers in $A[p]$ and $T[u]$ back, and follow the pointer at position p to the second vertex. This ensures that we never leave from the first adjacency position of a vertex and thus never have to store a reverse pointer pointing to a first adjacency position.

We have shown how to create reverse pointers; now it remains to describe how to remove them again. After exploring every neighbor of a vertex v, our algorithm finds the start of the adjacency array of vertex v'', i.e., we find a position q'' with $1 \le A[q''] \le n$ (or q'' is the end of the whole array A). Note that $v'' = v + 1$, but we do not know v at this point and thus, we cannot search for $v + 1$. Now, we need to backtrack and thus find the reverse pointer of v. We find the reverse pointer $p = T[v]$ by iterating backwards until we find a position q with $A[q] \le n$. In fact, then $A[q] = v$. Now we move the temporary stored pointer $q' = A[p]$ into $T[v]$ again, and restore the original pointer to v at position p by setting $A[p] = q$. However, this turns v into a white vertex again, which we solve by incrementing the first pointer $q' = T[v]$ of v by one such that the pointer points to a position storing a value larger than n. Since we assume a degree of at least two for all vertices the incrementation has the effect that the pointer points at a value strictly greater than n. The incrementation is easily reversible such that the restoration is trivial.

Before we present the remaining details of our algorithm, we summarize the possible modifications in T and the adjacency arrays of the vertices in the following three invariants that hold before and after each call of FOLLOW and BACKTRACK. Before, note that the only other operation that changes values is NEXTNEIGHBOR, which only swaps adjacency pointers, but does not change colors of vertices and the invariants are not affected.

1. A vertex v is white exactly if v is not a start vertex and $1 \le A[T[v]] \le n$.
2. Every gray-black vertex v on a current DFS path, except the start vertex, stores the reverse pointer at $T[v]$ that points into its parent adjacency array at a position $p = T[v]$ with $A[p] \ge n$. Moreover, p is the position where the parent of v originally stored the pointer to v.
3. The first pointer $q = T[v]$ in the adjacency array of a gray-black vertex v that is not on the current DFS path points with its first pointer $q = T[v]$ to the second position q' of another vertex adjacency array, i.e., $1 \le A[q' - 1] \le n$.

In detail, our DFS runs as follows. If a start-vertex $1 \le v_s \le n$ is given, we search for the first position p with $v_s = A[p]$ of its adjacency array in $O(m)$ time. Alternatively, we search for a position p with $v_s = A[p] \wedge 1 \le v_s \le n$. Then, we call VISIT($p$) that is described now.

- VISIT(p): (Visit the vertex whose adjacency array starts at position p.) In the swapped begin pointer representation, $v = A[p]$ is always the vertex name. First, call preprocess(v). Finally, start iterating through the neighbors starting from position p by executing NEXTNEIGHBOR(p, TRUE).
- NEXTNEIGHBOR(p, IGNORECHECK): (Follows the edge at position p if the opposite endpoint of the edge is white. Otherwise, it tries the position $p + 1$.) First of all, we test if p is the first position in the current adjacency array or two position after it by determining if ($\neg$IGNORECHECK $\wedge (1 \le A[p] \le n)$) or if $1 \le A[p - 2] \le n$, respectively. If so, define p' (and p'') such that p' is the first (p'' is the second) position in the adjacency array and check additionally

if the first pointer (which is temporary stored in a parent vertex in $A[r]$ with $r = T[u], u = A[p']$), and the second pointer in $A[p'']$ are swapped, which means that the first is larger than the second pointer. Use the information computed above and proceed with Substep 1.

Substep 1. If p is the first entry, increment p by one, swap the two pointers in $A[r]$ and $A[p'']$ as well as proceed with Substep 3 to visit the first neighbor (if white) from the second position of the adjacency array.

If p is two positions after the first entry and the two pointers are swapped, (i.e., we just returned from the first neighbor), decrement p by one, swap the two pointers as described above and also proceed with Substep 3 to visit the second neighbor (if white) from the second position of the adjacency array.

Otherwise, we just returned from the second, third, etc. neighbor. Then, we go to Substep 2 to test if we reached the end of the current adjacency array and then proceed with Substep 3.

Substep 2. We check if we require to backtrack, i.e, we reached the next adjacency array or are out of index in array A. Hence, check if $(1 \leq A[p] \leq n) \vee (p > n + m + 2)$. If we have to backtrack, search for the largest position $q < p$ such that $1 \leq A[q] \leq n$ and call BACKTRACK(q) unless $A[q] = v_s$. In that case color v_s gray-black by incrementing its firs adjacency pointer $T[v_s]$ by one. We now have to explored everything reachable from v_s. If wanted, start a new DFS with a next white vertex.

Substep 3. Check if the edge at p points to a white vertex $v = A[q]$ with $q = A[p]$ by running the non-recursive procedure ISWHITE(v). If p does, call FOLLOW(p). Otherwise, call NEXTNEIGBOR($p + 1$, FALSE).

- ISWHITE(v): (Return TRUE exactly if the vertex v is white.) We check the first invariant, i.e., return $v \neq v_s \wedge 1 \leq A[T[v]] \leq n$.

- FOLLOW(p): (Discover a new child via an edge e stored at position p and color the new discovered vertex implicitly gray-black.) First we determine the position $q = A[p]$ and the vertex $v = A[q]$ where e points to. Second, we are going to create a reverse pointer in $T[v]$ to backtrack later. To not lose the pointer previously stored in $T[v]$ we store it in $A[p]$. In detail, remember the first pointer $x = T[v]$ of the neighbor. Now, store the pointer inside $A[p] = x$ and create a reverse pointer from the neighbors first adjacency entry into its parent's adjacency array by setting $T[v] = p$. Finally, visit the neighbor by executing VISIT(q).

- BACKTRACK(q): (From a child v go to its parent where q is the beginning of v's adjacency array and $p = T[v]$ with $v = A[q]$ is a reverse pointer to the adjacency array of the parent.) Before going to the parent, we have to restore the edges that we modified by visiting v such that we fulfill the third invariant. In detail, we first restore the child's edge that was temporarily stored in the parent's adjacency array, but let it point one edge further to guarantee the third invariant. Thus, we set $T[v] = A[p] + 1$ and $A[p] = q$ with $v = A[q]$ and $p = T[v]$. Finally, we call postprocess(v) and subsequently NEXTNEIGHBOR($p + 1$, FALSE).

Concerning the running time on n-vertex m-edge directed graphs, we can observe that all functions of our in-place DFS run in constant time per call. Moreover, VISIT and BACKTRACK are called $O(n)$ times whereas all other functions are called $O(m)$ times. Thus, our in-place DFS runs in $O(n + m)$ time. Ignoring the calls for the user-defined functions as well as for ISWHITE, which is not recursive, we only make tail-calls and consequently require no recursion stack.

Theorem 1. *There is an in-place DFS for (un)directed graphs on the weak restore word RAM that runs in $O(n + m)$ time on n-vertex m-edge graphs on our sorted standard representation consisting of $n + m + 2$ words ($n + 2m + 2$ words) and supports calls of the user defined functions **pre-** and **postprocess**.*

If $O(n(n + m))$ time is allowed, we can support **pre-** and **postexplore**: Whenever backtracking from a vertex v to a vertex u we know v's name and return to a position p in u's adjacency entry. Thus, $O(n)$ time allows us to lookup the vertex name $u = A[q]$ by searching for the largest $q < p$ with $1 \leq A[q] \leq n$.

4   Breadth-First Search

As usual for a BFS, our algorithm runs in rounds and, in round $z - 1$ with $z \in \mathbb{N}$, all vertices of distance z from a start vertex are added into a new list. Then our algorithm can always iterate through a list of vertices and for each such vertex u, we iterate through u's adjacency array. For a simpler description, assume that all vertices are initially white. After adding u's white neighbors into a list for the next BFS round, the vertex turns black.

To implement our BFS we make use of the following observation. In the sorted standard representation all words in the table T are stored in ascending order. Our idea is to partition T in regions such that the most significant bits of the words are equal per region. We use this to create a *shifted representation* of T by ignoring the most significant bits and shifting the words in T together (Lemma 2) such that we have a linear number of bits free to store a c-color choice dictionary [16,17,19] as demonstrated in Fig. 5.

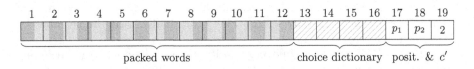

Fig. 5. Shifted representation with c-color choice dictionary.

Lemma 2. *Let $c > 0$ be a constant and $n \geq 2^{c+1}w$ be an integer. Having an array of n ordered words we can pack it in linear time with an in-place algorithm such that we have cn unused bits free and that we still can access all elements of the array in constant time. Afterwards, we can similarly unpack the words.*

Proof. The idea is to partition the array into parts such that each pair of words in a part has the same $c' = c + 1$ significant bits. Since the sequence is ordered, we iterate over all words and look for the positions where one of the most c' significant bits change.

Now, the most significant c' bits of each word are equal per region. We treat them as unused space. If we store the remaining $(w - c')$ bits of all words consecutively, they occupy $n(w - c')$ bits in total such that it leaves $c'n$ bits free to use. We use the last $2^{c'}$ words to store c' and all the positions. Thus, $c'n - 2^{c'}w \geq c'n - n = cn$ bits remain free.

For implementing a function READ($i \in \{1, \ldots, n\}$) that reads the ith original word, we have to identify its current position that can be distributed between two words, to cut its bits out of the two words and to use the remembered position to reconstruct its most significant bits. For the following description assume that the bits of a word are numbered from 0 (least significant) to $w - 1$ (most significant). In detail, the ith word in T originally stored at bit position $w(i - 1)$ was shifted exactly $c(i - 1)$ bits and now starts after $x = (w - c)(i - 1)$ bits, i.e., it starts with bit $y = (x \bmod w)$ in the word $((x \operatorname{div} w) - 1)$ and consists of the next $w - c$ bits. Using suitable shift operations we can get the ith word in constant time. To reconstruct its most significant bits, scan over the last c' words to determine the part to which i belongs. $\qquad\square$

Before we now obtain our linear-time BFS, we want to remark that the shifted representation cannot be used to run a standard DFS in-place since a stack for the DFS can require $\Theta(n \log n)$ bits on n-vertex graphs and that many bits are not free in the shifted representation.

We first prepare the shifted representation of our graph (Lemma 2). Then we can use the free bits to implement a c-color choice dictionary in which we store the colors of the vertices, and to iterate over colored vertices in constant time per vertex. The c-color choice dictionary provides the following functions.

- SETCOLOR(v, q): Colors an entry v with the color $q \in \{0, \ldots, c - 1\}$.
- COLOR(v): Returns the color of the entry v.
- CHOICE(q): Returns an (arbitrary) entry that has the color $q \in \{0, \ldots, c - 1\}$.

To start our BFS at vertex v, we first initialize a c-color choice dictionary D for four colors {WHITE, LIGHT-GRAY, DARK-GRAY, BLACK} with all vertices being initially white. Remember in a global variable a round counter $z = 0$ to output the round number for each vertex. Then, color the root vertex v light-gray by calling D.SETCOLOR(v, LIGHT-GRAY). Finally, we start to process the whole DFS-tree as follows.

Whenever the current round counter z is even, the idea is to iterate over the light-gray vertices and color their white neighbors dark-gray and if z is odd we do vice versa. We next explain the details for the case where z is even. For an odd z, simply switch the words light-gray and dark-gray below.

As long as there is a light-gray vertex $v = D$.CHOICE(LIGHT-GRAY), we output (v, z), color v's white neighbors dark-gray, and color v black. To color the neighbors we iterate over v's adjacency array starting at position $p = T[v]$ and ending at $q = \mathcal{T}[v+1] - 1$ where we define $\mathcal{T}[n+1] = n + m + 2$ as the end of our graph representation. For every neighbor $u = A[j]$ with $p \le j \le q$ we check if $D.\text{COLOR}(u) = \text{WHITE}$ and if so, we color u dark-gray by calling $D.\text{SETCOLOR}(u, \text{DARK-GRAY})$, otherwise we ignore it. After the iteration over v's adjacency array we call $D.\text{SETCOLOR}(v, \text{BLACK})$. Since v is now black, the next call of $D.\text{CHOICE}(\text{LIGHT-GRAY})$ returns the next light-gray vertex if one exists.

If we could color a vertex dark-gray during the current iteration over the light-gray vertices, then there are vertices left to process: We increase z by one and start a new round by iterating now over the dark-gray colored vertices as described. Otherwise, the BFS finishes.

By Lemma 2, we can restore to the sorted standard representation.

Theorem 3. *There is an in-place BFS for (un)directed graphs on the weak restore word RAM that runs in $O(n+m)$ time on n-vertex m-edge graphs on our sorted standard representation consisting of $n + m + 2$ words ($n + 2m + 2$ words).*

5   Conclusion

We showed linear-time in-place algorithms for DFS and BFS on the weak restore word RAM that have the same asymptotic running time as the standard algorithms. To evaluate the usability in practice we implemented the folklore and the linear-time in-place DFS. The implementations are published on GitHub [20].

Even if we consider our graph representation to be economical in its space requirement, Farzan and Munro [13] showed a succinct graph representation with constant access-time that requires only $(1+\epsilon) \log \binom{n^2}{m}$ bits for any constant $\epsilon > 0$. An interesting open question is if it is possible to implement a (linear-time) in-place algorithm for DFS or BFS by using the succinct graph representation of Farzan and Munro or one that requires a little more space.

Acknowledgments. Andrej Sajenko was funded by the Deutsche Forschungsgemeinschaft (DFG, German Research Foundation) – 379157101.

References

1. Asano, T., et al.: Reprint of: memory-constrained algorithms for simple polygons. Comput. Geom. Theory Appl. **47**((3, Part B)), 469–479 (2014). https://doi.org/10.1016/j.comgeo.2013.11.004
2. Asano, T., et al.: Depth-first search using $O(n)$ bits. In: Ahn, H.-K., Shin, C.-S. (eds.) ISAAC 2014. LNCS, vol. 8889, pp. 553–564. Springer, Cham (2014). https://doi.org/10.1007/978-3-319-13075-0_44
3. Barba, L., Korman, M., Langerman, S., Silveira, R.I., Sadakane, K.: Space-time trade-offs for stack-based algorithms. In Proceedings of the 30th International Symposium on Theoretical Aspects of Computer Science (STACS 2013). LIPIcs, vol. 20, pp. 281–292. Schloss Dagstuhl - Leibniz-Zentrum für Informatik (2013). https://doi.org/10.4230/LIPIcs.STACS.2013.281

4. Beame, P.: A general sequential time-space tradeoff for finding unique elements. SIAM J. Comput. **20**(2), 270–277 (1991). https://doi.org/10.1137/0220017
5. Brönnimann, H., Iacono, J., Katajainen, J., Morin, P., Morrison, J., Toussaint, G.: Space-efficient planar convex hull algorithms. Theor. Comput. Sci. **321**(1), 25–40 (2004). https://doi.org/10.1016/j.tcs.2003.05.004
6. Chakraborty, S., Mukherjee, A., Raman, V., Satti, S.R.: A framework for in-place graph algorithms. In: Proceedings 26th Annual European Symposium on Algorithms (ESA 2018), LNCS, pp. 13:1–13:16. Springer (2018). https://doi.org/10.4230/LIPIcs.ESA.2018.13
7. Chakraborty, S., Raman, V., Satti, S.R.: Biconnectivity, chain decomposition and st-numbering using $O(n)$ bits. In: Proceedings 27th International Symposium on Algorithms and Computation (ISAAC 2016). LIPIcs, vol. 64, pp. 22:1–22:13. Schloss Dagstuhl - Leibniz-Zentrum für Informatik (2016). https://doi.org/10.4230/LIPIcs.ISAAC.2016.22
8. Timothy, M.C., Munro, J.I., Raman, V.: Selection and sorting in the "restore" model. In: Proceedings 25th Annual ACM-SIAM Symposium on Discrete Algorithms (SODA 2014), pp. 995–1004. SIAM (2014). https://doi.org/10.1137/1.9781611973402.74
9. Timothy, M.C., Munro, J.I., Raman, V.: Selection and sorting in the restore model. ACM Trans. Algorithms **14**(2), 11:1–11:18 (2018). https://doi.org/10.1145/3168005
10. Datta, S., Kulkarni, R., Mukherjee, A.: Space-efficient approximation scheme for maximum matching in sparse graphs. In: Proceedings 41st International Symposium on Mathematical Foundations of Computer Science (MFCS 2016). LIPIcs, vol. 58, pp. 28:1–28:12. Schloss Dagstuhl - Leibniz-Zentrum für Informatik (2016). https://doi.org/10.4230/LIPIcs.MFCS.2016.28
11. Edmonds, J., Poon, C.K., Achlioptas, D.: Tight lower bounds for st-connectivity on the NNJAG model. SIAM J. Comput. **28**(6), 2257–2284 (1999). https://doi.org/10.1137/S0097539795295948
12. Elmasry, A., Hagerup, T., Kammer, F.: Space-efficient basic graph algorithms. In: Proceedings of the 32nd International Symposium on Theoretical Aspects of Computer Science (STACS 2015). LIPIcs, vol. 30, pp. 288–301. Schloss Dagstuhl - Leibniz-Zentrum für Informatik (2015). https://doi.org/10.4230/LIPIcs.STACS.2015.288
13. Farzan, A., Munro, J.I.: Succinct encoding of arbitrary graphs. Theor. Comput. Sci. **513**, 38–52 (2013). https://doi.org/10.1016/j.tcs.2013.09.031
14. Franceschini, G., Muthukrishnan, S., Pătraşcu, M.: Radix sorting with no extra space. In: Arge, L., Hoffmann, M., Welzl, E. (eds.) ESA 2007. LNCS, vol. 4698, pp. 194–205. Springer, Heidelberg (2007). https://doi.org/10.1007/978-3-540-75520-3_19
15. Frederickson, G.N.: Upper bounds for time-space trade-offs in sorting and selection. J. Comput. Syst. Sci. **34**(1), 19–26 (1987). http://dl.acm.org/citation.cfm?id=762350.762354
16. Hagerup, T.: Small uncolored and colored choice dictionaries. Computing Research Repository (CoRR), abs/1809.07661 (2018)
17. Hagerup, T., Kammer, F.: Succinct choice dictionaries. Computing Research Repository (CoRR), abs/1604.06058 (2016)
18. Kammer, F., Sajenko, A.: Linear-time in-place DFS and BFS on the word RAM. Computing Research Repository (CoRR), abs/1803.04282 (2018)

19. Kammer, F., Sajenko, A.: Simple 2^f-color choice dictionaries. In Proceedings of the 29th International Symposium on Algorithms and Computation (ISAAC 2018). LIPIcs, vol. 123, pp, 66:1–66:12. Schloss Dagstuhl - Leibniz-Zentrum für Informatik (2018). https://doi.org/10.4230/LIPIcs.ISAAC.2018.66
20. Kammer, F., Sajenko, A.: Space efficient (graph) algorithms (2018). https://github.com/thm-mni-ii/sea
21. Munro, J.I., Paterson, M.S.: Selection and sorting with limited storage. Theor. Comput. Sci. **12**(3), 315–323 (1980). https://doi.org/10.1016/0304-3975(80)90061-4
22. Pagter, J., Rauhe, T.: Optimal time-space trade-offs for sorting. In: Proceedings 39th Annual IEEE Symposium on Foundations of Computer Science (FOCS 1998), pp. 264–268. IEEE Computer Society (1998). https://doi.org/10.1109/SFCS.1998.743455

A Faster Algorithm for the Strongly Stable b-Matching Problem

Adam Kunysz[✉]

Institute of Computer Science, University of Wrocław, Wrocław, Poland
aku@cs.uni.wroc.pl

Abstract. We study a generalisation of the stable matching problem to the many-to-many variant in which vertices of the bipartite graph $G = (A \cup B, E)$ may involve ties in their preference lists. We investigate the notion of strong stability and give an $O(m \sum_{y \in B} b(y))$ algorithm for computing a strongly stable b-matching optimal for vertices of A, where we denote $m = |E|$. Our result improves on the previous algorithm by Chen and Ghosh [2]. The main technique allowing us to speed up the algorithm is a generalisation of the notion of level-maximal matchings [8] to the case of b-matchings.

As a byproduct of our results we obtain an $O(nm)$ algorithm for a many-to-one restriction of the problem also known as the hospitals-residents problem, where we denote $n = |A \cup B|$, $m = |E|$. The previous best algorithm had an $O(m \sum_{y \in B} b(y))$ runtime [8].

Keywords: Stable marriage · Strongly stable matching · b-matching

1 Introduction

An instance of the strongly stable b-matching problem is an undirected bipartite graph $G = (A \cup B, E)$, where each vertex $v \in A \cup B$ has a capacity $b(v)$. Additionally each vertex ranks its neighbours in order of preference with possible ties. Formally the preference list of a vertex v is a linearly ordered list of ties, which are vertices equally good for v. Ties are disjoint and may contain one vertex.

A set of edges $M \subseteq E$ is a b-matching if for each vertex $v \in A \cup B$ we have $|M(v)| \leq b(v)$, where $M(v)$ is the subset of edges of M incident to v.

Let M be a b-matching of G. We first introduce the notion of a *blocking edge*. Intuitively, an edge $(a, b) \in E \setminus M$ is a blocking edge with respect to M if by getting matched with each other neither of the vertices a and b would become worse off and at least one of them would become better off than in M. We say that a matching is *strongly stable* if there is no blocking edge with respect to it. We give a formal definition of the problem in Sect. 2.

Partly supported by Polish National Science Center grant 2018/29/B/ST6/02633.

P. Heggernes (Ed.): CIAC 2019, LNCS 11485, pp. 299–310, 2019.
https://doi.org/10.1007/978-3-030-17402-6_25

Our task is to find a strongly stable b-matching of G or report that none exists.

Motivation. The stable matching problem and its extensions have widespread applications to matching schemes [15].

The notion of strong stability considered in the paper allows us to prevent bribery amongst the agents. Let us consider the following scenario. Suppose that two agents, a and b are matched to respectively b' and a' in a matching M. Additionally we assume that there exists an edge (a, b) in the graph, the agent a strictly prefers b to its partner b' in M and the agent b is indifferent between a and a'. In order to improve their situation a may try to bribe b. Since b would not get worse after accepting a, they might yield to a and accept the bribe. Clearly such situations cannot happen if M is strongly stable.

Previous Results. Strongly stable matchings in the one-to-one variant have been extensively studied in the literature. Irving [5] gave an $O(n^4)$ algorithm for the version of the problem where the graph is complete and there is an equal number of vertices on both sides of the bipartition. In [12] Manlove extended this algorithm to incomplete bipartite graphs obtaining $O(m^2)$ complexity. Later on Kavitha et al. [8] introduced the notion of a level-maximal matching which allowed them to speed up the algorithm to $O(nm)$ time.

The one-to-many variant of the problem, also known as the hospital-residents problem, has also been studied in the literature. As is customary we call vertices of A and B respectively residents and hospitals, where we assume that residents have unit capacity. Irving et al. [6] gave an $O(m^2)$ algorithm for the problem, which was improved by Kavitha et al. [8] to $O(m \sum_{y \in B} b(y))$.

A generalisation of the stable matching problem to the many-to-many setting, known as the stable b-matching problem has been first studied by Baïou and Balinski [1]. They have shown that in instances with strict preferences the problem can be solved in $O(n^2)$ time. Malhotra [11] studied the problem under the notion of strong stability in instances with ties. He gave a polynomial-time algorithm for computing a strongly stable b-matching or reporting that none exists. Chen and Ghosh [2] showed that this algorithm is, in fact, incorrect and they described an $O(m^3 n)$ algorithm for the problem.

Our Results. Chen and Ghosh [2] asked whether it was possible to improve the complexity of their algorithm using techniques from [8]. Manlove posed this problem again in his recent book [14]. In this paper we show that such an improvement is indeed possible. We generalise the notion of level-maximal matchings to the case of b-matchings. This allows us to bound the cost of matching augmentations performed by the algorithm by $O(m \sum_{y \in B} b(y))$. We also show that the residual graph G_r can be maintained more efficiently than in the algorithm of Chen and Ghosh [2]. Their algorithm recomputes the residual graph from scratch in each phase. We prove that this graph can be efficiently updated after each phase of the algorithm. Combining the above two ideas we obtain an algorithm of time complexity $O(m \sum_{y \in B} b(y))$.

It is important to note that our algorithm calculates a strongly stable b-matching optimal for vertices of A however its runtime depends only on capacities of vertices belonging to B. We can reverse the roles of the vertices in the algorithm and obtain a strongly stable b-matching optimal for vertices of B in time $O(m \sum_{x \in A} b(x))$. This observation implies that we can in fact determine whether a strongly stable b-matching exists and find one in time $O(m \min\{\sum_{y \in B} b(y), \sum_{x \in A} b(x)\})$. As a byproduct of our algorithm we can solve the one-to-many restriction of our problem (also known as the hospitals-residents problem) in time $O(nm)$, thereby answering an open question given in [8].

Related Work. Stable and strongly stable matchings have been studied in so called stable roommate setting where the underlying graph is non-bipartite. Irving [4] gave a linear time algorithm for computing a stable matching in non-bipartite graphs with strict preferences or reporting that none exists. This algorithm has been later extended to the many-to-many setting by Irving and Scott [7]. Strongly stable matchings have also been studied in non-bipartite graphs. Scott [16] gave an $O(m^2)$ algorithm for computing a strongly stable matching in non-bipartite graphs or reporting that none exists. Kunysz [9] described a faster $O(nm)$ algorithm for this problem.

Many structural properties of the stable matching problem have been described over the years. In [3] Gusfield and Irving have proven that in the case of no ties the set of stable matching solutions forms a distributive lattice. They also have shown that despite its exponential size, the lattice can be represented as a set of closed sets of a certain partial order on $O(m)$ elements. Such a representation can be built in time $O(m)$ using the notion of a rotation. These results have been generalised to the case of strong stability. In [13] Manlove proved that strongly stable matchings form a distributive lattice. Recently, Kunysz et al. [10] described an $O(nm)$ algorithm for constructing a representation of the lattice and generalised the notion of a rotation to the case of strong stability.

2   Preliminaries

We first introduce some additional notation and formally define the problem.

Let b_1 and b_2 be two neighbours of a in G. If a strictly prefers b_1 to b_2 we denote it by $b_1 \succ_a b_2$. If a is indifferent between b_1 and b_2 we write $b_1 =_a b_2$. Similarly if a strictly prefers b_2 to b_1 we denote it by $b_1 \prec_a b_2$.

If a prefers b_1 to b_2 or is indifferent between them then we say that a *weakly prefers* b_1 to b_2 and denote it as $b_1 \succeq_a b_2$. Let M be a b-matching of G. We say that a vertex v is *free* in the b-matching M if there are less than $b(v)$ edges incident to v in M. Similarly we say that a vertex v is *full* in M if there are exactly $b(v)$ edges incident to v in M.

Definition 1. *An edge* $(v, w) \in E \backslash M$ *is blocking with respect to M if any of the following conditions hold:*

1. *Both v and w are free with respect to M.*

2. *The vertex v is free with respect to M and there exists a vertex v' such that $(v', w) \in M$ and $v \succeq_w v'$.*
3. *The vertex w is free with respect to M and there exists a vertex w' such that $(v, w') \in M$ and $w \succeq_v w'$.*
4. *There exist vertices v' and w' such that $(v', w) \in M$, $(v, w') \in M$ and additionally either $v \succ_w v'$ and $w \succeq_v w'$ or $v \succeq_w v'$ and $w \succ_v w'$.*

By $head_E(v)$ we denote the set of the most valued neighbours in E of v, formally $head_E(v) = \{w \in N_E(v) : (\forall w' \in N_E(v))w \succeq_v w'\})$, where $N_E(v)$ is the set of edges incident to v belonging to E. Similarly by $tail_E(v)$ we denote the set of the least valued neighbours of v in the graph spanned by E.

For $X \subseteq E$ by $d_X(v)$ we denote the number of edges incident to v in X.

Let $p = (v_1, v_2, \ldots, v_k)$ be a path in G. We say that p is an M-alternating path if $(v_{2i+1}, v_{2i+2}) \notin M$ and $(v_{2i}, v_{2i+1}) \in M$ for each i. If an alternating path p is of an odd length we say that p is an *augmenting path*.

3   Description of the Algorithm

In this section we give an overview of Algorithm 1.

The algorithm proceeds in phases corresponding to executions of the outer *while* loop (line 3). Throughout the execution we maintain two auxiliary graphs $G_c = (A \cup B, E_c)$ - current graph and $G_r = (A \cup B, E_r)$ - residual graph. We also denote $E' \subseteq E$ to be a subset of edges of E not considered by the algorithm yet. We initially set $E' = E$ and $E_c = \emptyset$, $E_r = \emptyset$. A matching M is maintained throught the execution. Each phase of the algorithm consists of three steps which we describe in Subsect. 3.1, 3.2 and 3.3. Additionally we describe some details regarding the maintenance of the graph G_r in Sect. 5.

3.1   Step 1 - Proposal-Rejection Sequence

In the first step we construct the auxiliary graph G_c. The pseudocode is contained in lines 4–10 of Algorithm 1.

Each free agent $v \in A$ first proposes to the most preferred neighbours on his preference list, amongst the ones not considered yet (i.e., to the vertices of the set $head_{E'}(v)$). Each agent of B upon receiving a proposal may accept or reject it. Edges corresponding to accepted proposals are added to G_c. Note that an agent $v \in A$ may propose to multiple agents belonging to B and similarly an agent $w \in B$ may accept proposals from multiple agents belonging to A.

Suppose that y receives a proposal from x. How do we determine if y should accept the proposal? Let us consider two cases:

1. $d_{E_c}(y) < b(y)$
2. $d_{E_c}(y) \geq b(y)$

In the first case y has leftover capacity and accepts the proposal. Let us now consider the second case. We say that a vertex x is *dominated* on the list of y in the graph G_c if $|\{(z, y) \in E_c : z \succ_y x\}| \geq b(y)$. If x is dominated on the list of y we call (x, y) dominated. Clearly y should reject a proposal from x if (x, y) is dominated as in this case y already has at least $b(y)$ better candidates. If (x, y) is not dominated then y accepts the proposal and (x, y) is added to G_c.

Once (x, y) is added to G_c some edges already present in G_c may become dominated. Additionally some edges belonging to the set E' may also become dominated. We remove all such edges from $E' \cup E_c$ in line 10. In Lemma 3 we prove that such edges do not belong to any strongly stable b-matching.

Note that once (x, y) is added to G_c some vertices belonging to A may become free due to the deletion in line 10. Agents belonging to A continue proposing until all of them are full or they run out of proposals. If at some point there are at least $b(w)$ vertices incident to w in E_c then such a vertex becomes *marked*. Lemma 4 implies that if a vertex w is marked then it must be matched to exactly $b(w)$ vertices in every strongly stable b-matching.

3.2   Step 2 - Construction of the Residual Graph G_r

The pseudocode of step 2 is described in lines 11–19 of Algorithm 1.

We first identify the set of *preferred edges* of G_c matched in every strongly stable b-matching contained in this graph. Then we construct graph G_r based on edges which are not preferred. The strongly stable b-matching M consists of preferred edges and a b_r-matching of G_r, where $b_r(v)$ is leftover capacity of v.

Suppose that $v \in A$. Similarly as in [2] we split edges incident to v in G_c into sets of preferred edges $P_{E_c}(v)$ and indifferent edges $I_{E_c}(v)$.

Definition 2. *Let $v \in A$. We divide the set $N_{E_c}(v)$ of neighbours of v into levels $L_1, L_2, ..., L_k$ according to his preference list. The vertex v is indifferent between all nodes in the same level L_i and strictly prefers each L_i to L_{i+1}. Let $p^* = \max\{p : \sum_{i=1}^{p-1} |L_i| \leq b(v)\}$. We denote:*

- $P_{E_c}(v) = \{(v, w)|(\exists i < p^*)w \in L_i\}$.
- $I_{E_c}(v) = \{(v, w)|w \in L_{p^*}\}$.

Note that the set $I_{E_c}(v)$ can be nonempty only if $d_{E_c}(v) > b(v)$. In this case $I_{E_c}(v)$ consists of edges belonging to the tail of v in G_c and $P_{E_c}(v)$ consists of the remaining edges incident to v in G_c. If we have $d_{E_c}(v) \leq b(v)$ then the set $I_{E_c}(v)$ is empty and all the edges incident to v in G_c belong to $P_{E_c}(v)$. We analogously define sets P_{E_c} and I_{E_c} for agents belonging to B. From the definition it follows that for any node v we have $|P_{E_c}(v)| \leq b(v)$.

Definition 3. *Let $(v, w) \in E_c$ be an edge. We say that (v, w) is of type X with respect to v if $(v, w) \in X_{E_c}(v)$ where $X \in \{P, I\}$. Similarly we say that (v, w) is of type X with respect to w if $(v, w) \in X_{E_c}(w)$ where $X \in \{P, I\}$. For $X, Y \in \{I, P\}$ we say that (v, w) if of type (X, Y) if (v, w) is of type X with respect to v and (v, w) is of type Y with respect to w. An edge is preferred if it is of type (I, P), (P, I) or (P, P).*

Let us consider a vertex $w \in B$. Recall that there are at most $b(w)$ edges of type P with respect to w. There may also exist preferred edges of type I with respect to w but of type P with respect to the other endpoint. It can happen that the set of preferred edges incident to w exceeds its capacity. In this case in lines 11–16 of Algorithm 1 we remove a subset of edges incident to w and claim that such edges do not belong to any strongly stable matching. After the removal there are no more than $b(w)$ preferred edges incident to each $w \in B$.

It may happen that there are more than $b(v)$ preferred edges incident to a vertex $v \in A$. We temporarily allow capacities of vertices belonging to A to be exceeded and deal with this problem later. Note that in line 15 we again mark a vertex w if at some point it is incident to more than $b(w)$ edges in E_c.

The residual graph G_r is a subgraph of G_c constructed from edges which are not preferred. We simply set $E_r \leftarrow \{(v, w) \in E_c : (v, w) \text{ is of type } (I, I)\}$ (line 17). Additionally we define a capacity function $b_r(v) = b(v) - |\{(v, w) : (v, w) \text{ is preferred}\}|$.

3.3   Step 3 - Augmentation of G_r

Let us now describe step 3 of the algorithm. The pseudocode of this step is presented in lines 20–30 of Algorithm 1. We distinguish two cases:

1. G_r contains a perfect b_r-matching M_r
2. a maximum matching M_r of G_r is not a perfect b_r-matching

In the case (1) our algorithm simply exits the while loop and step 3 is finished. In the case (2) we identify a subset of edges to be removed from the graph and restart the process. A maximum matching of G_r is computed using augmentation paths in lines 21–24. Note that in line 22 we search for an augmenting path of maximum level. This notion is used to speed up the algorithm and it is precisely defined in Sect. 6. For the correctness of the algorithm it suffices to find an arbitrary augmenting path.

Note that in line 29 we mark w if any edges incident to w are removed.

4   Correctness of the Algorithm

In this section we prove the correctness of our algorithm. We say that an *edge is strongly stable* if it belongs to some strongly stable b-matching.

In the following two lemmas we show that if the algorithm returns a b-matching, this b-matching is strongly stable.

Lemma 1. *If an edge $e = (v, w)$ is removed during the execution of Algorithm 1, it does not block a b-matching output by the algorithm.*

Lemma 2. *If Algorithm 1 returns a b-matching, this b-matching is strongly stable.*

In order to prove that the algorithm is correct we need to show that if a strongly stable b-matching exists, our algorithm is going to return one. We first show that no edge which belongs to a strongly stable b-matching is ever deleted.

Lemma 3. *No strongly stable edge is ever deleted during the execution of Algorithm 1.*

The following auxiliary lemma is needed to prove the correctness of the algorithm. Note that in particular this lemma implies that all the strongly stable b-matchings are of the same size and match the same sets of vertices.

Lemma 4. *Assume that there exists a strongly stable b-matching M_0 in G. Let M be the set computed by the algorithm when it exits the while loop (lines 3–30). Then:*

1. $|M| = |M_0|$
2. *For each $v \in A \cup B$ we have $d_M(v) = d_{M_0}(v)$ and $d_M(v) = \min(b(v), d_{E_c}(v))$*

We say that a strongly stable b-matching is *A-optimal* if every agent of A is matched to the best possible set of partners amongst all the possible strongly stable b-matchings.

Lemma 5. *If a strongly stable b-matching exists, then Algorithm 1 outputs an A-optimal strongly stable b-matching.*

5   Maintenance of G_r

In order to perform augmentations of M_r described in Subsect. 3.3 we need to be able to efficiently determine which edges belong to G_r. Each modification of G_c may result in some edges changing their types. As a result the graph G_r may change as well. We cannot afford to recalculate types from scratch after each modification of G_c as such an approach would significantly slow down the algorithm. We show that each modification of G_c only affects types of neighbouring vertices and show how to efficiently maintain G_r.

In the next lemma we discuss potential changes of G_r resulting from operations performed in step 1. We show that when an agent $v \in A$ makes his proposals then edges newly added to G_r are either of the form (v, w) where $w \in head_{E'}(v)$ or of the form (v', w) where $w \in head_{E'}(v)$ and (v', w) was a preferred edge before proposals made by v.

Lemma 6. *Consider an execution of Algorithm 1. Let us denote by G_r^j the residual graph after j iterations of the while loop in lines 4–10. Suppose that in iteration j an agent $v \in A$ proposed to agents belonging to $head_{E'}(v)$.*

Let (x, y) be an edge added to G_r^j during iteration j. Then exactly one of the following holds:

1. *We have $x = v$ and $y \in head_{E'}(v)$. The agent x proposed to y in the iteration j.*

2. We have $y \in head_{E'}(v)$. The edge (x, y) was of type (I, P) before the iteration j of the while loop.

We also show that as a result of steps 2 and 3 no new edges are added to G_r.

Lemma 7. *Let G_r^i be the residual graph during phase i of the algorithm when we exit the while loop in lines 4–10. Denote by H_r^i the residual graph obtained from G_r^i after the execution of steps 2 and 3. Then H_r^i is a subgraph of G_r^i.*

It can proven that once an edge is added or removed from G_c we can easily determine types of all the edges in the updated graph. In order to do so it suffices to carefully analyse proofs of the above two lemmas (see the full version of the paper).

In order to keep the pseudocode of Algorithm 1 easier to read we do not include instructions related to the maintenance of types. It is important to remember that after each modification of G_c we have to update types according to cases described in proofs of Lemmas 6 and 7.

Lemma 8. *Let (v, w) where $v \in A$ and $w \in B$ be an edge added to G_c at some point of the execution of Algorithm 1. Then the following hold:*

1. *If (v, w) is of type P with respect to v at some point of the execution then it remains of type P with respect to v as long as it belongs to G_c.*
2. *If (v, w) is of type I with respect to w at some point of the execution then it remains of type I with respect to w as long as it belongs to G_c.*

The correctness of the above lemma is straightforward from proofs of Lemmas 6 and 7. Note that the type of (v, w) with respect to $v \in A$ can change once during the execution of Algorithm 1. The only possible change is from I to P. Similarly the type of (v, w) with respect to $w \in B$ can only change once. In the full version of the paper we exploit this fact and present low-level details regarding data structures used to maintain the graph G_r. We show there that the cost of maintaining the graph G_r is at most $O(nm)$ throughout the execution.

Recall that throughout the execution of the algorithm we make sure that all the preferred edges are added to M (see line 18). Additionally we define M_r to be M restricted to the graph G_r. It may happen that during the execution of the algorithm a preferred edge e changes its type from (I, P) to (I, I). It is important to note that such an edge remains in M even though it is no longer preferred. Once e changes type to (I, I) it is added to E_r and M_r. This fact will be used in the analysis of level-maximality of M_r described in Sect. 6.

6   Level-Maximal Matchings

In this section we extend the notion of level-maximal matchings [8] to the case of b-matchings.

For a given edge e we define its *level* $l(e)$ to be the number of the phase in which e was added to G_r. Levels of edges not added to G_r remain undefined.

We say that the *level* of a vertex v is a minimal level of an edge incident to this vertex in G_r and we denote this value as $l(v)$. If v is isolated in G_r then the level of v remains undefined. Let M be a b-matching in G. We define the level of M as:

$$l(M) = \sum_{w \in B, d_M(w) > 0} l(w) d_M(w)$$

Note that since we sum over vertices w such that $d_M(w) > 0$ each term in the above sum is correctly defined.

Definition 4. *A matching M is said to be* level-maximal *in G if for every matching M' such that $\forall(v \in A) d_M(v) = d_{M'}(v)$ we have $l(M) \geq l(M')$.*

In the remainder of this section we show that the matching M_r maintained by the algorithm is level-maximal in G_r throughout the execution. To prove this we need two auxiliary lemmas. In Lemma 9 we give a necessary and sufficient condition for the matching to be level-maximal. In Lemma 10 we show that if we are given a level-maximal matching M, we can choose an augmenting path p such that $M \oplus p$ is level maximal as well.

Lemma 9. *Matching M is level-maximal if and only if there is no alternating path p from a vertex $w \in B$ to a vertex $w' \in B$ such that:*

1. *The edge of p incident to w does not belong to M*
2. *The vertex w is free with respect to M*
3. *The inequality $l(w') < l(w)$ holds*

Lemma 10. *Let M be level-maximal matching, $v \in A$ be free vertex with respect to M. Let $w \in B$ be a vertex of maximal level reachable by an augmenting path p from v. Then matching $N = M \oplus p$ is level-maximal.*

A path p as in the statement of Lemma 10 is called a *level-maximal augmenting path* with respect to M. It turns out that the removal of edges from G_r does not affect level-maximality of M_r.

Lemma 11. *Let M_r be a level-maximal matching in $G_r = (A \cup B, E_r)$ and let E'_r be a subset of E_r. Then $M'_r = M_r \cap E'_r$ is a level-maximal matching in $G'_r = (A \cup B, E'_r)$.*

It remains to show that the matching M_r maintained by the algorithm is level-maximal throughout the execution.

Lemma 12. *Matching M_r is level-maximal in G_r at all times of the execution of Algorithm 1.*

7 Level-Maximal Augmenting Paths

In this section we describe how to implement the search for level-maximal augmenting paths (see line 22 of Algorithm 1). Our algorithm is an extension of the procedure described in [8] for the many-to-one version of the problem.

Let us consider the graph G_r during phase i of the execution of Algorithm 1 and let $x \in A$ be a free node with respect to M_r. Our goal is to find a free vertex $w \in B$ reachable from x by an augmenting path from x, such that the level of w is maximal. If such a vertex w does not exist then we will find a set Z of vertices belonging to B reachable from x by alternating paths.

Lemma 13. *Let p be an augmenting path with respect to M_r from $x \in A$ to a vertex $w \in B$ such that w is of maximal possible level. Then the following hold:*

1. *For each vertex $w' \in B$ belonging to the path p we have $l(w') \geq l(w)$.*
2. *For each edge e belonging to p we have $l(e) \geq l(w)$.*

The search for augmenting paths is organized in rounds $i, i - 1, i - 2 \ldots$. We start in round i as this is the maximal possible level of an edge in the graph during phase i. In round k we are going to check whether there exists a level-maximal augmenting path ending in a vertex of level at least k. If we find such a path we return it, otherwise we proceed to the round $k - 1$. We continue until we either run out edges in the graph or find a desired path. From Lemma 13 it follows that when we check for the existence of a path ending in a vertex of level at least k it suffices to test edges of level k or higher.

Note that when we proceed from the round k to the round $k - 1$ we do not start the search from scratch. We reuse alternating paths explored in previous rounds and extend those using edges of level $k - 1$.

In order to implement this process we maintain an array o buckets X which implement a simple priority queue. A bucket X_j contains vertices of level j belonging to B which correspond to endpoints of currently explored alternating paths. We initialise this structure by adding all neighbours of x into the appropriate buckets. In round k we simply continue the search from vertices in the bucket X_k. There are three cases to consider:

1. X_k is empty and there exists a non-empty bucket.
2. All the remaining buckets are empty.
3. X_k is non-empty.

In the case (1) we simply proceed to the round $k - 1$. If (2) holds we finish the search and conclude that there exists no augmenting path starting at x. In the case (3) we expand current alternating paths. Let $w \in X_k$ be a vertex in the bucket. If w is free we found an augmenting path. From the construction of the algorithm it follows that the path from x to w is a level-maximal augmenting path. If w is full we explore alternating paths from this vertex. Vertices of level lower than k encountered during the search are added to their appropriate buckets. Once we explore all the paths from vertices of level at least k we proceed to the round $k - 1$ and continue the search.

Lemma 14. *Total cost of augmenting path searches throughout the execution of the algorithm is bounded by $O(m \sum_{y \in B} b(y))$.*

In order to prove that the algorithm runs in $O(m \sum_{y \in B} b(y))$ time, it remains to show that the additional overhead required to maintain the graph G_r does

---

Algorithm 1. For computing a strongly stable b-matching

1: $E' \leftarrow E$
2: $E_c, E_r, M \leftarrow \emptyset$
3: **while** $\exists(v \in A)v$ is free and $d_{E'}(v) > 0$ **do**
4:     **while** $\exists(v \in A)v$ is free and $d_{E'}(v) > 0$ **do**   ▷ Step 1
5:         $H \leftarrow head_{E'}(v)$
6:         **for** $w \in H$ **do**
7:             add (v, w) to E_c
8:             **if** $d_{E_c}(w) \geq b(w)$ **then**
9:                 marked(w) = true
10:             delete dominated edges belonging to $E_c \cup E'$ and incident to w
11:     **for** $w \in B$ **do**   ▷ Step 2
12:         **if** $|\{(v, w) : (v, w) \text{ is of type } (P, I), (P, P) \text{ or } (I, P)\}| > b(w)$ **then**
13:             **for** $(v', w) \in E_c \cup E'$ **do**
14:                 **if** $v' \preceq_w tail_{E_c}(w)$ **then**
15:                     marked(w) = true
16:                     delete (v', w)
17:     $E_r \leftarrow \{(v, w) \in E_c : (v, w) \text{ is of type } (I, I)\}$
18:     add all preferred edges to M
19:     $M_r \leftarrow M \cap E_r$
20:     **while** there exists a free $m \in A$ in G_r with respect to M_r **do**   ▷ Step 3
21:         **while** there exists an augmenting path p starting from m **do**
22:             $p \leftarrow$ an augmenting path from m to a free $w \in B$ [of maximum level]
23:             $M \leftarrow M \oplus p$
24:             $M_r \leftarrow M_r \oplus p$
25:         **if** m is free in G_r with respect to M_r **then**
26:             $Z \leftarrow$ the set of vertices belonging to B reachable from m by alternating paths
27:             **for** w, m' such that $w \in Z$ and $(m', w) \in E_c \cup E'$ **do**
28:                 **if** $m' \preceq_w tail_{E_c}(w)$ **then**
29:                     marked(w) = true
30:                     delete (m', w)
31: **if** $\exists(v \in A)d_M(v) > b(v)$ **then**
32:     **return** no strongly stable matching exists
33: **for** $w \in B$ **do**
34:     **if** marked(w) = true $\wedge d_M(w) \neq b(w)$ **then**
35:         **return** no strongly stable matching exists
36:     **if** marked(w) = false $\wedge d_M(w) \neq \min(d_{E_c}(w), b(w))$ **then**
37:         **return** no strongly stable matching exists
38: **return** M

---

not exceed the cost of augmenting path searches. Due to the space contraints we defer the details to the full version of the paper.

References

1. Baïou, M., Balinski, M.: Many-to-many matching: stable polyandrous polygamy (or polygamous polyandry). Discrete Appl. Math. **101**(1–3), 1–12 (2000)
2. Chen, N., Ghosh, A.: Strongly stable assignment. In: de Berg, M., Meyer, U. (eds.) ESA 2010. LNCS, vol. 6347, pp. 147–158. Springer, Heidelberg (2010). https://doi.org/10.1007/978-3-642-15781-3_13
3. Gusfield, D., Irving, R.W.: The Stable Marriage Problem - Structure and Algorithms. Foundations of Computing Series. MIT Press, Cambridge (1989)
4. Irving, R.W.: An efficient algorithm for the "stable roommates" problem. J. Algorithms **6**(4), 577–595 (1985)
5. Irving, R.W.: Stable marriage and indifference. Discrete Appl. Math. **48**(3), 261–272 (1994)
6. Irving, R.W., Manlove, D.F., Scott, S.: Strong stability in the hospitals/residents problem. In: Alt, H., Habib, M. (eds.) STACS 2003. LNCS, vol. 2607, pp. 439–450. Springer, Heidelberg (2003). https://doi.org/10.1007/3-540-36494-3_39
7. Irving, R.W., Scott, S.: The stable fixtures problem - a many-to-many extension of stable roommates. Discrete Appl. Math. **155**(16), 2118–2129 (2007)
8. Kavitha, T., Mehlhorn, K., Michail, D., Paluch, K.E.: Strongly stable matchings in time O(nm) and extension to the hospitals-residents problem. ACM Trans. Algorithms **3**(2), 15 (2007)
9. Kunysz, A.: The strongly stable roommates problem. In: Sankowski, P., Zaroliagis, C.D. (eds.) 24th Annual European Symposium on Algorithms, ESA 2016, Aarhus, Denmark, 22–24 August 2016, Volume 57 of LIPIcs, pp. 60:1–60:15. Schloss Dagstuhl - Leibniz-Zentrum fuer Informatik (2016)
10. Kunysz, A., Paluch, K.E., Ghosal, P.: Characterisation of strongly stable matchings. In: Proceedings of the Twenty-Seventh Annual ACM-SIAM Symposium on Discrete Algorithms, SODA 2016, Arlington, VA, USA, 10–12 January 2016, pp. 107–119 (2016)
11. Malhotra, V.S.: On the stability of multiple partner stable marriages with ties. In: Albers, S., Radzik, T. (eds.) ESA 2004. LNCS, vol. 3221, pp. 508–519. Springer, Heidelberg (2004). https://doi.org/10.1007/978-3-540-30140-0_46
12. Manlove, D.F.: Stable marriage with ties and unacceptable partners. Technical report, University of Glasgow (1999)
13. Manlove, D.F.: The structure of stable marriage with indifference. Discrete Appl. Math. **122**(1–3), 167–181 (2002)
14. Manlove, D.F.: Algorithmics of Matching Under Preferences, Volume 2 of Series on Theoretical Computer Science. World Scientific (2013)
15. Roth, A.E.: The evolution of the labor market for medical interns and residents: a case study in game theory. J. Polit. Econ. **92**(6), 991–2016 (1984)
16. Scott, S.: A study of stable marriage problems with ties. Ph.D. thesis, University of Glasgow (2005)

Eternal Domination in Grids

Fionn Mc Inerney$^{(\boxtimes)}$, Nicolas Nisse, and Stéphane Pérennes

Université Côte d'Azur, Inria, CNRS, I3S, Sophia Antipolis, France
{fionn.mc-inerney,nicolas.nisse,stephane.perennes}@inria.fr

Abstract. In the eternal domination game played on graphs, an attacker attacks a vertex at each turn and a team of guards must move a guard to the attacked vertex to defend it. The guards may only move to adjacent vertices on their turn. The goal is to determine the eternal domination number γ_{all}^{∞} of a graph which is the minimum number of guards required to defend against an infinite sequence of attacks.

This paper continues the study of the eternal domination game on strong grids $P_n \boxtimes P_m$. Cartesian grids $P_n \square P_m$ have been vastly studied with tight bounds existing for small grids such as $k \times n$ grids for $k \in \{2, 3, 4, 5\}$. It was recently proven that $\gamma_{all}^{\infty}(P_n \square P_m) = \gamma(P_n \square P_m) + O(n + m)$ where $\gamma(P_n \square P_m)$ is the domination number of $P_n \square P_m$ which lower bounds the eternal domination number [Lamprou et al. CIAC 2017]. We prove that, for all $n, m \in \mathbb{N}^*$ such that $m \geq n$, $\lfloor \frac{n}{3} \rfloor \lfloor \frac{m}{3} \rfloor + \Omega(n + m) = \gamma_{all}^{\infty}(P_n \boxtimes P_m) = \lceil \frac{n}{3} \rceil \lceil \frac{m}{3} \rceil + O(m\sqrt{n})$ (note that $\lceil \frac{n}{3} \rceil \lceil \frac{m}{3} \rceil$ is the domination number of $P_n \boxtimes P_m$). Our technique may be applied to other "grid-like" graphs.

Keywords: Eternal domination · Combinatorial games · Graphs · Grids

1 Introduction

The origins of the eternal domination game date back to the 1990's where the military strategy of Emperor Constantine for defending the Roman Empire was studied in a mathematical setting [1,21–23]. Roughly, a limited number of armies must be placed in such a way that an army can always move to defend against an attack by invaders.

Precisely, eternal domination is a 2-player game on graphs introduced in [6] and defined as follows. Initially, k guards are placed on some vertices of a graph $G = (V, E)$. Turn-by-turn, an *attacker* first chooses a vertex $v \in V$ to attack. Then, if no guard is occupying v or a vertex adjacent to v, then the attacker wins. Otherwise, one guard must move along an edge to occupy v if it is not already occupied, and the next turn starts. If the attacker never wins whatever

This work has been partially supported by ANR program "Investments for the Future" under reference ANR-11-LABX-0031-01, the Inria Associated Team AlDyNet. Due to a lack of space, several proofs have been omitted and can be found in [14].

P. Heggernes (Ed.): CIAC 2019, LNCS 11485, pp. 311–322, 2019.
https://doi.org/10.1007/978-3-030-17402-6_26

be its sequence of attacks, then the guards win. So, clearly, there is no point in the attacker attacking an occupied vertex. The aim in eternal domination is to minimize the number of guards that must be used in order to win. Hence, let $\gamma^{\infty}(G)$ be the minimum integer k such that there exists a strategy allowing k guards to win, regardless of what the attacker does [6].

In this paper, we consider the "all guards move" variant of eternal domination, proposed in [11], where, at their turn, every guard may move to a neighbour of its position (still satisfying that the attacked vertex is occupied by a guard at the end of the turn). Let $\gamma_{all}^{\infty}(G)$ be the minimum number of guards for which a winning strategy exists in this setting. By definition, $\gamma(G) \leq \gamma_{all}^{\infty}(G) \leq \gamma^{\infty}(G)$ for any graph G where $\gamma(G)$ is the minimum size of a dominating set in G^1.

Variants of the eternal domination game also differ in the fact that one or more guards may simultaneously occupy the same vertex. In the initial variant where a single guard is allowed to move each turn, this is not a strong constraint [6]. That is, imposing that a vertex cannot be occupied by more than one guard does not increase the number of guards required to win. In the case when multiple guards may move each turn, there are some graphs where this constraint increases the number of guards [18]. Let $\gamma_{all}^{*\infty}(G)$ be the minimum number of guards to win in G, moving several guards per turn, and in such a way that a vertex cannot be occupied by several guards.

Previous works mainly studied lower and upper bounds on $\gamma^{\infty}(G)$ and $\gamma_{all}^{\infty}(G)$ in function of other parameters of G, such as its domination number $\gamma(G)$ [11], independence number $\alpha(G)^2$ [6,11], and clique cover number $\theta(G)^3$ [6]. Notably, these results give the following inequalities $\gamma(G) \leq \gamma_{all}^{\infty}(G) \leq \alpha(G) \leq \gamma^{\infty}(G) \leq \theta(G)$ [6]. Particular graph classes have also been studied such as paths and cycles [11], trees [16], and proper interval graphs [5]. In particular, the class of grids and graph products has been widely studied [4,10,12,18–20,24].

In this paper, we focus on the class of *strong grids SG* and provide an almost tight asymptotical value for $\gamma_{all}^{\infty}(SG)$. Our result also holds for $\gamma_{all}^{*\infty}(SG)$. Our main result is a new technique to prove upper bounds that we believe can be generalized to many other "grid-like" graphs.

1.1   Related Work

The "all guards move" variant of eternal domination was shown to be NP-complete in Hamiltonian split graphs [3]. Note that it is not known whether the problem of deciding γ_{all}^{∞} is in NP in general graphs. Moreover, given a graph G and an integer k as inputs, the problem of deciding if $\gamma^{\infty}(G) \leq k$ is coNP-hard [2].

Several graph classes have been studied. For a path P_n on n vertices, $\gamma_{all}^{\infty}(P_n) = \lceil \frac{n}{2} \rceil$ and for a cycle C_n on n vertices, $\gamma_{all}^{\infty}(C_n) = \lceil \frac{n}{3} \rceil$ [11]. In [16], the authors present a linear-time algorithm to determine $\gamma_{all}^{\infty}(T)$ for all trees T.

---

[1] $D \subseteq V$ is a dominating set of G if every vertex is in D or adjacent to a vertex in D.
[2] $\alpha(G)$ is the maximum size of an independent set in G.
[3] $\theta(G)$ is the minimum number of complete subgraphs of G whose union covers $V(G)$.

It was proven that if G is a proper interval graph, then $\gamma_{all}^{\infty}(G) = \alpha(G)$ [5]. In the past few years, a lot of effort was put in by several authors to determine the eternal domination number of cartesian grids, $\gamma_{all}^{\infty}(P_n \,\square\, P_m)$. Exact values were determined for $2 \times n$ cartesian grids [12] and $4 \times n$ cartesian grids [4]. Asymptotical tight bounds for $3 \times n$ cartesian grids were obtained in [10] and improved in [20]. Finally, bounds for $5 \times n$ cartesian grids were given in [24]. The best known lower bound for $\gamma_{all}^{\infty}(P_n \,\square\, P_m)$ for values of n and m large enough, is the domination number with the latter only being recently determined in [13]. The best known upper bound for $\gamma_{all}^{\infty}(P_n \,\square\, P_m)$ was determined recently in [19], where it was shown that $\gamma_{all}^{\infty}(P_n \,\square\, P_m) = \gamma(P_n \,\square\, P_m) + O(n+m)$. Note that all the results discussed in this subsection also hold for $\gamma_{all}^{*\infty}$.

There are also many other variants of the game that exist and here we give a brief description and references for some of them. Recently, the eternal domination game and a variant have been studied in digraphs, including orientations of grids and toroidal strong grids [2]. Eternal total domination was studied in [17], where a total dominating set must be maintained by the guards each turn. The eviction model of eternal domination was studied in [15], where a vertex containing a guard is attacked each turn, which forces the guard to move to an adjacent empty vertex with the condition that the guards must maintain a dominating set each turn. The authors of the current paper studied a generalization of eternal domination, called the Spy game, in [7,8]. For more information and results on the original eternal domination game and its variants, see the survey [18].

1.2 Our Results

The main result of this paper is that, for all $n, m \in \mathbb{N}^*$ such that $m \geq n$,

$$\left\lfloor \frac{n}{3} \right\rfloor \left\lfloor \frac{m}{3} \right\rfloor + \Omega(n+m) = \gamma_{all}^{\infty}(P_n \boxtimes P_m) = \left\lceil \frac{n}{3} \right\rceil \left\lceil \frac{m}{3} \right\rceil + O(m\sqrt{n}).$$

In [14], we show that this result also holds in the case when at most one guard may occupy each vertex.

Note that, in toroidal strong grids $C_n \boxtimes C_m$, the problem becomes trivial and $\gamma_{all}^{\infty}(C_n \boxtimes C_m) = \lceil \frac{n}{3} \rceil \lceil \frac{m}{3} \rceil$ for any n and m. However, in strong grids, border-effects make the problem much harder. The upper bound is proven by defining a set of specific configurations that each dominate the grid and are "invariant" to the movements required by the defined strategy to defend against attacks. That is, the attacks are separated into three types of attacks: horizontal, vertical, and diagonal, and the strategy defined gives the movement of the guards based on the type of attack. It is shown that in each of the three cases of attacks, the guards are able to move from their current configuration to another configuration in the set of configurations (so, it does not matter which configuration was the initial one and which new configuration the guards reach after their moves) and hence, the guards can defend against an infinite sequence of attacks.

The lower bound is proven by showing that, in any winning configuration in eternal domination, there are some vertices that are dominated by more than one guard, and/or some guards dominate at most 6 vertices. By double counting, this leads to the necessity of having $\Omega(n + m)$ extra guards compared to the classical domination (when $n \equiv 0 \pmod 3$ and $m \equiv 0 \pmod 3$).

2   Preliminaries

We use classic graph-theory terminology [9]. Notably, given a graph $G = (V, E)$ and $S \subseteq V$, let $N(S) = \{v \in V \setminus S \mid \exists w \in S, \{v, w\} \in E\}$ denote the set of neighbours (not in S) of the vertices in S and let $N[S] = N(S) \cup S$ denote the *closed neighbourhood* of S. For $v \in V$, let $N(v) = N(\{v\})$ and $N[v] = N(v) \cup \{v\}$.

Let $n, m \in \mathbb{N}^*$ be such that $m \geq n$ and let the $n \times m$ strong grid, denoted by $SG_{n \times m}$, be the strong product $P_n \boxtimes P_m$ of an n-node path with an m-node path. Precisely, $SG_{n \times m}$ is the graph with the set of vertices $\{(i, j) \mid 1 \leq i \leq n, 1 \leq j \leq m\}$, and two vertices (i_1, j_1) and (i_2, j_2) are adjacent if and only if $\max\{|i_2 - i_1|, |j_2 - j_1|\} = 1$. That is, the vertices are identified by their Cartesian coordinates, *i.e.*, the vertex (i, j) is the vertex in *row i* and *column j*. The vertex $(1, 1)$ is in the *bottom-left corner* and the vertex (n, m) is in the *top-right corner*.

Definition 1. *The set of border vertices of $SG_{n \times m}$ is the set*

$$B = \bigcup_{1 \leq i \leq n, 1 \leq j \leq m} \{(1, j), (n, j), (i, 1), (i, m)\} \text{ of vertices of degree} \leq 5.$$

The set of pre-border vertices of $SG_{n \times m}$ is the set $PB = N(B)$.

Equivalently, PB is the set of border vertices of the strong grid induced by $V(SG_{n \times m}) \setminus B$.

We consider the turn-by-turn 2-player game in graphs called eternal domination. Each *turn*, each vertex of a graph $G = (V, E)$ may be occupied by one or more guards. Let $k \in \mathbb{N}^*$ be the total number of guards. The positions of the guards are formally defined by a multi-set C of vertices, called a *configuration*, where the number of occurrences of a vertex $v \in C$ corresponds to the number of guards at $v \in V$ and $k = |C|$. Each turn, given a current configuration $C = \{v_i \mid 1 \leq i \leq k\}$ of k guards, Player 1, the *attacker*, attacks a vertex $v \in V$. Then, Player 2 (the *defender*) may move each of its *guards* to a neighbour of their current position, thereby, achieving a new configuration $C' = \{w_i \mid 1 \leq i \leq k\}$ such that $w_i \in N[v_i]$ for every $1 \leq i \leq k$ (we then say that C' is *compatible* with C, which is clearly a symmetric relation). If $v \notin C'$, then the attacker *wins*, otherwise, the game goes on with a next turn (given the new configuration C').

A *strategy* for k guards is defined by an initial configuration of size k and by a function that, for every current configuration C and every attacked vertex $v \in V$, specifies a new configuration C' compatible with C. A strategy $\mathcal{S}$ for the guards is *winning* if, for every sequence of attacked vertices, the attacker never wins when the defender plays according to $\mathcal{S}$.

Our main contribution is the design of a winning strategy for $\gamma(SG_{n \times m}) + o(\gamma(SG_{n \times m}))$ guards in $SG_{n \times m}$, where $\gamma(SG_{n \times m}) = \lceil \frac{n}{3} \rceil \lceil \frac{m}{3} \rceil$ is the *domination number* of $SG_{n \times m}$. The next lemma is key for this winning strategy.

In our strategy, it will often be useful to move a guard from a node $u \in PB$ of the pre-border to another node $v \in PB$ such that u and v are not necessarily adjacent. For this purpose, the idea is to place a sufficient number of guards on the vertices of the border such that a "flow" of the guards on the border vertices will simulate the move of the guard from u to v in one turn.

Precisely, given a configuration C and $u, v \in V(SG_{n \times m})$ with $u \in C$, a guard is said to *jump* from u to v if the configuration $(C \setminus \{u\}) \cup \{v\}$ is compatible with C, *i.e.*, the guards, in one turn, can move to achieve the same configuration as C except that there is one guard less on u and one guard more on v. More generally, given $U \subset C$ and $W \subset V(SG_{n \times m})$, a set of guards is said to *jump* from U to W if the configuration $(C \setminus U) \cup W$ is compatible with the configuration C.

Lemma 1. *Let $\alpha, \beta \in \mathbb{N}^*$ such that $\beta \leq \alpha$. Let $U, W \subseteq PB$ be two subsets of pre-border vertices such that $|U| = |W| = \beta$. In any configuration C such that $U \subseteq C$ and C contains at least α occurrences of each vertex in B (i.e., each border vertex is occupied by at least α guards), then β guards may "jump" from U to W in one turn. Moreover, only guards in $U \cup B$ move.*

Proof. The proof is by induction on β. The inductive hypothesis is that if each vertex in B contains α guards, then $\beta \leq \alpha$ guards may "jump" from U to W in one turn such that at most β guards move off of each vertex $w \in B$ in this turn. For the base case, let us assume that $U = \{u\}$ and $W = \{w\}$. Let us show how 1 guard can "jump" from u to w in one turn. If $u = w$, the result trivially holds, so let $u \neq w$. Let $u' \in B$ (resp., w') be a neighbour of u (of w) that shares one coordinate with u (with w). Let $Q = (u' = v_0, v_1, \ldots, v_\ell = w')$ be a path from u' to w' induced by the border vertices. In one turn, a guard at u moves to u', for every $0 \leq i < \ell$, a guard at v_i moves to v_{i+1}, and a guard at v_ℓ moves to w.

Now, assume the inductive hypothesis holds for $\beta \geq 1$. If $\beta = \alpha$, we are done, so assume $\beta < \alpha$. Let $|U| = |W| = \beta + 1 \leq \alpha$ and let $u \in U$ and $w \in W$. By the inductive hypothesis, β guards may jump from $U \setminus \{u\}$ to $W \setminus \{w\}$ in one turn in such a way that, for every vertex $b \in B$, at most β guards move off of b during this turn. Since every vertex of B is occupied by $\alpha > \beta$ guards, at least one guard is unused on every vertex of B. Thus, it possible to use the same strategy as in the base case to make one guard jump from u to w on this same turn. $\square$

3   Upper Bound Strategy

This section is devoted to proving that for all $n, m \in \mathbb{N}^*$ such that $m \geq n$, $\gamma_{all}^\infty(SG_{n \times m}) = \lceil \frac{n}{3} \rceil \lceil \frac{m}{3} \rceil + O(m\sqrt{n})$.

Before considering the general case, let us first assume that $n - 2 \equiv 0 \pmod{3}$ and that there exists $k \in \mathbb{N}^*$ such that $k - 2 \equiv 0 \pmod{3}$, and $m \equiv 0 \pmod{k}$. The $n \times m$ strong grid will be partitioned into *blocks* which are subgrids of size $n \times k$. More precisely, for all $1 \leq q \leq \frac{m}{k}$, the qth block contains columns $(q-1)k + 1$ through qk of $SG_{n \times m}$.

3.1   Horizontal Attacks

In this section, we only consider one block of $SG_{n\times m}$. W.l.o.g., let us consider the block $SG_{n\times k}$ induced by $\{(i,j) \mid 1 \le i \le n, 1 \le j \le k\}$. Let us first define a family of parameterized configurations for this block.

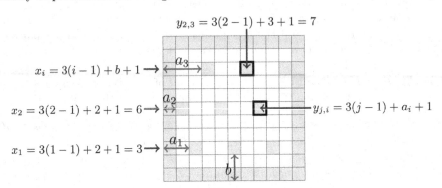

Fig. 1. $P_{11} \boxtimes P_{11}$ where the squares are vertices and two squares sharing a side and/or a corner are adjacent. Example of a configuration $C_H(X)$ where $X = (b = 2, a_1 = 2, a_2 = 1, a_3 = a_{\frac{n-2}{3}} = 3)$, there is one guard at each square in gray, and the white squares contain no guards.

Let $\mathcal{X} = \{(b, a_1, \ldots, a_{\frac{n-2}{3}}) \mid b \in \{1,2,3\}, a_i \in \{1,2,3\} \text{ for } i = 1, \ldots, \frac{n-2}{3}\}$.

Given $X = (b, a_1, \ldots, a_{\frac{n-2}{3}}) \in \mathcal{X}$, let $x_i(X) = 3(i-1) + b + 1$, and $y_{j,i}(X) = 3(j-1) + a_i + 1$ for every $1 \le i \le \frac{n-2}{3}$ and $1 \le j \le \frac{k-2}{3}$. We set $x_i = x_i(X)$ and $y_{j,i} = y_{j,i}(X)$ when there is no ambiguity. Intuitively, b will represent the *vertical shift* of the positions of the guards in configuration X. Similarly, for every $1 \le i \le \frac{n-2}{3}$, a_i represents the *horizontal shift* of the positions of the guards in row $x_i(X)$ in configuration X (see Fig. 1).

Horizontal Configurations. Let us define the set $\mathcal{C}_H$ of configurations as follows. For every $X \in \mathcal{X}$, let $C_H(X) = B \cup \{(x_i(X), y_{j,i}(X)) \mid 1 \le i \le \frac{n-2}{3}, 1 \le j \le \frac{k-2}{3}\}$ be the configuration where there is one guard at every vertex of B and one guard at each vertex $(x_i(X), y_{j,i}(X)) = (3(i-1) + b + 1, 3(j-1) + a_i + 1)$ for every $1 \le i \le \frac{n-2}{3}$ and $1 \le j \le \frac{k-2}{3}$. See an example in Fig. 1. Then,

$$\mathcal{C}_H = \{C_H(X) \mid X \in \mathcal{X}\}.$$

Note that $|C_H(X)| = \frac{(n-2)(k-2)}{9} + 2(n+k) - 4 = \kappa_H$ for every $X \in \mathcal{X}$. That is, any horizontal configuration uses κ_H guards.

Lemma 2. *Every configuration $C_H(X) \in \mathcal{C}_H$ is a dominating set of $SG_{n\times k}$.*

In this subsection, we limit the power of the attacker by allowing it to attack only some predefined vertices (this kind of attack will be referred to as a *horizontal attack*). For every configuration $C_H(X) \in \mathcal{C}_H$ and for any such attack, we show that the guards may be moved (in one turn) in such a way to defend the attacked vertex and reach a new configuration in $\mathcal{C}_H$.

Horizontal Attacks. Let $X = (b, a_1, \ldots, a_{\frac{n-2}{3}}) \in \mathcal{X}$ and $C_H(X) \in \mathcal{C}_H$. Let

$$A_H(X) = \{(x_i, y) \mid 1 \leq i \leq \frac{n-2}{3}, 1 \leq y \leq k\}.$$

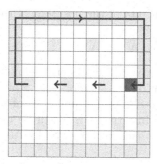

Fig. 2. Example of a horizontal attack at the red square. The arrows (in blue) show the movements of the guards in response to the attack. (Color figure online)

A *horizontal attack with respect to* X is an attack at any vertex in $A_H(X)$, *i.e.*, an attack at any vertex of a row where some non-border vertex is occupied by a guard. Note that, for every vertex $v \in A_H(X)$, either v is occupied by a guard or there is a guard on the vertex to the left or to the right of v.

The next lemma proves that, from any horizontal configuration and against any horizontal attack (with respect to this current configuration), there is a strategy for the guards that defends against this attack and leads to a (new) horizontal configuration. Therefore, starting from any horizontal configuration, there is a strategy of the guards that wins against any sequence of horizontal attacks. See Fig. 2 for a schematic representation of how the guards react to one of these attacks.

Lemma 3. *For any* $X \in \mathcal{X}$ *and any* $v \in A_H(X)$, *there exists* $X' \in \mathcal{X}$ *such that* $v \in C_H(X')$ *and configurations* $C_H(X)$ *and* $C_H(X')$ *are compatible. That is, in one turn, the guards may move from* $C_H(X)$ *to* $C_H(X')$ *and defend against an attack at* v.

3.2   Vertical Attacks

In this section, we consider the entire strong grid $SG_{n \times m}$ partitioned into $\frac{m}{k}$ blocks $SG_{n \times k}$ with block q, for $1 \leq q \leq \frac{m}{k}$, being induced by $\{(i, j + (q-1)k) \mid 1 \leq i \leq n, 1 \leq j \leq k\}$. We first define a family of parameterized configurations for this graph. A configuration for the whole grid will be defined as the union of some configurations for each of the q blocks. Formally, for every $1 \leq q \leq \frac{m}{k}$, let

$$\mathcal{X}^q = \{(b^q, a_1^q, \ldots, a_{\frac{n-2}{3}}^q) \mid b^q \in \{1,2,3\}, a_i^q \in \{1,2,3\} \text{ for } i = 1, \ldots, \frac{n-2}{3}$$

and $q = 1, \ldots, \frac{m}{k}\}$.

Given $X^q = (b^q, a_1^q, \ldots, a_{\frac{n-2}{3}}^q) \in \mathcal{X}^q$, let $x_i^q(X^q) = 3(i-1) + b^q + 1$, and $y_{j,i}^q(X^q) = (q-1)k + 3(j-1) + a_i^q + 1$ for every $1 \le i \le \frac{n-2}{3}$, $1 \le j \le \frac{k-2}{3}$, and $1 \le q \le \frac{m}{k}$. We set $x_i^q = x_i^q(X^q)$ and $y_{j,i}^q = y_{j,i}^q(X^q)$ when there is no ambiguity.

That is, intuitively, b^q will represent the *vertical shift* of the positions of the guards in configuration X^q in the qth block. Similarly, for every $1 \le i \le \frac{n-2}{3}$, a_i^q represents the *horizontal shift* of the positions of the guards in row $x_i(X)$ in configuration X^q in the qth block.

Finally, let $\mathcal{Y} = \{(X^1, \ldots, X^{\frac{m}{k}}) \mid X^q \in \mathcal{X}^q \text{ for } q = 1, \ldots, \frac{m}{k}\}$.

Vertical Congurations. In order to properly define the following set of configurations, the following notation is used. For a set S of vertices in a configuration $\mathcal{C}$ and an integer $x > 0$, let $S^{[x]}$ be the multi-set of vertices that consists of x copies of each vertex in S. Intuitively, $S^{[x]}$ will be used to define a configuration where x guards occupy each vertex of S. Let us now define the set $\mathcal{C}_V$ of configurations as follows.

For every $Y = (X^1, \ldots, X^{\frac{m}{k}}) \in \mathcal{Y}$, let $C_V(Y) = B^{[\frac{k-2}{3}]} \cup \bigcup\limits_{q=1}^{\frac{m}{k}} C_H(X^q)$ be the configuration obtained as follows. First, for any $1 \le q \le \frac{m}{k}$, guards are placed in configuration $C_H(X^q)$ in the qth block. Then, $\frac{k-2}{3}$ guards are added to every border vertex. Note that overall, there are $\frac{k-2}{3} + 1$ guards at each vertex of B. See an example in Fig. 3. Then, $\mathcal{C}_V = \{C_V(Y) \mid Y \in \mathcal{Y}\}$.

Note that $|C_V(Y)| = \frac{m}{k}\kappa_H + 2(\frac{k-2}{3})(n+m-2) = \kappa_V$ for every $Y \in \mathcal{Y}$. That is, any vertical configuration uses κ_V guards.

Lemma 4. *Every configuration $C_V(Y) \in \mathcal{C}_V$ is a dominating set of $SG_{n \times m}$.*

In this subsection, we limit the power of the attacker by allowing it to attack only some *vertical* vertices. For every configuration $C_V(X) \in \mathcal{C}_V$ and for any such attack, we show that the guards may be moved (in one turn) in such a way to defend the attacked vertex and reach a new configuration in $\mathcal{C}_V$.

Vertical Attacks. Let $Y = (X^1, \ldots, X^{\frac{m}{k}}) \in \mathcal{Y}$ and $C_V(Y) \in \mathcal{C}_V$. Let

$$A_V(Y) = \{(x_i^q - 1, y_{j,i}^q), (x_i^q + 1, y_{j,i}^q) \mid 1 \le i \le \frac{n-2}{3}, 1 \le j \le \frac{k-2}{3}, 1 \le q \le \frac{m}{k}\}$$

$$\cup \{(2, y_{j,n-1}^q) \mid 1 \le j \le \frac{k-2}{3}, 1 \le q \le \frac{m}{k} \text{ and } b^q = 3\}$$

$$\cup \{(n-1, y_{j,2}^q) \mid 1 \le j \le \frac{k-2}{3}, 1 \le q \le \frac{m}{k} \text{ and } b^q = 1\}$$

A *vertical attack with respect to Y* is an attack at any vertex in $A_V(Y)$, i.e., an attack at any non-border vertex above or below a guard not on a border vertex. Moreover, if the vertical shift b^q of the qth block equals 3, then some vertices of the second row of the qth block may also be attacked (depending on the horizontal shift a_{n-1}^q). Finally, if the vertical shift b^q of the qth block equals 1, then some vertices of the $(n-1)^{th}$ row of the qth block may also be attacked (depending on the horizontal shift a_2^q).

Note that $A_V(Y) \cap C_V(Y) = \emptyset$, and $A_V(Y) \cap A_H(X^q) = \emptyset$ for any $X^q \in Y$, *i.e.*, any vertical attack with respect to Y is not a horizontal attack with respect to $X^q \in Y$ and vice versa.

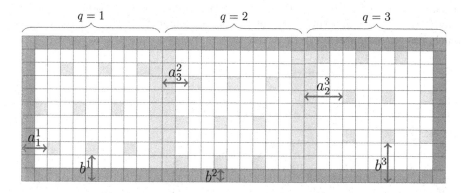

Fig. 3. A configuration $C_V(Y)$ where $k = 11$, $Y = (X^1, X^2, X^3)$, $X^1 = (2, 2, 1, 3)$, $X^2 = (1, 1, 1, 2)$, $X^3 = (3, 3, 3, 1)$, there are $(k-2)/3 + 1 = 4$ guards at each square in dark gray, 1 guard at each square in light gray, and the white squares contain no guards.

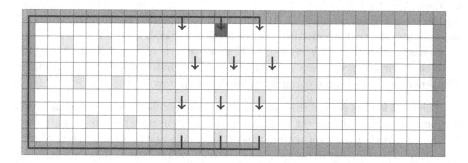

Fig. 4. Example of a vertical attack at the red square and how the guards react. (Color figure online)

The next lemma proves that, from any vertical configuration and against any vertical attack (with respect to this current configuration), there is a strategy for the guards that defends against this attack and leads to a (new) vertical configuration. Therefore, starting from any vertical configuration, there is a strategy of the guards that wins against any sequence of vertical attacks. See Fig. 4.

Lemma 5. *For any $Y \in \mathcal{Y}$ and any $v \in A_V(Y)$, there exists $Y' \in \mathcal{Y}$ such that $v \in C_V(Y')$ and configurations $C_V(Y)$ and $C_V(Y')$ are compatible. That is, in one turn, the guards may move from $C_V(Y)$ to $C_V(Y')$ and defend against an attack at v.*

3.3   Diagonal Attacks

The same $n \times m$ strong grid $SG_{n \times m}$, notations, and configurations for the guards used in Subsect. 3.2 will be used here. In this subsection, we limit the power of the attacker by allowing it to attack only some *diagonal* vertices. For every configuration $C_V(X) \in \mathcal{C}_V$ and for any such attack, we show that the guards may be moved (in one turn) in such a way to defend the attacked vertex and reach a new configuration in $\mathcal{C}_V$.

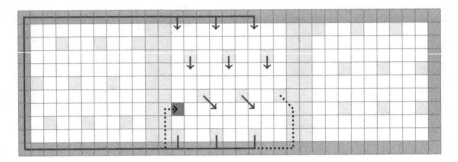

Fig. 5. Example of a diagonal attack at the red square. The dotted arrow in black is to differentiate between the different guards jumping. (Color figure online)

Diagonal Attacks. Let $Y = (X^1, \dots, X^{\frac{m}{k}}) \in \mathcal{Y}$ and $C_V(Y) \in \mathcal{C}_V$. Let $A_D(Y) = V(SG_{n \times m}) \setminus (B \cup A_H(Y) \cup A_V(Y))$. That is, $A_D(Y)$ covers all possible attacks that are neither horizontal nor vertical.

A *diagonal attack with respect to Y* is an attack at any vertex in $A_D(Y)$. Note that, for every vertex $v \in A_D(Y)$, there is a guard on a vertex adjacent to v and neither in the same column nor in the same row as v.

The next lemma proves that, from any vertical configuration and against any diagonal attack (with respect to this current configuration), there is a strategy for the guards that defends against this attack and leads to a (new) vertical configuration. Therefore, starting from any vertical configuration, there is a strategy of the guards that wins against any sequence of diagonal attacks. See Fig. 5.

Lemma 6. *For any $Y \in \mathcal{Y}$ and any $v \in A_D(Y)$, there exists $Y' \in \mathcal{Y}$ such that $v \in C_V(Y')$ and configurations $C_V(Y)$ and $C_V(Y')$ are compatible. That is, in one turn, the guards may move from $C_V(Y)$ to $C_V(Y')$ and defend against an attack at v.*

3.4   Upper Bound in Strong Grids

Note that, for any $Y = (X^1, \dots, X^{\frac{m}{k}}) \in \mathcal{Y}$, $A_D(Y) \cup A_V(Y) \cup \bigcup_{q=1}^{\frac{m}{k}} A_H(X^q) \cup B = V(SG_{n \times m})$. That is, any attack by the attacker in $SG_{n \times m}$ is either an attack at an occupied vertex or a horizontal, vertical or diagonal attack. Hence, Lemmas 3, 5, and 6 hold for any possible attack, which leads to our main theorem.

Theorem 1. *For all* $n, m \in \mathbb{N}^*$ *such that* $m \geq n$,

$$\gamma_{all}^{\infty}(SG_{n \times m}) = \left\lceil \frac{n}{3} \right\rceil \left\lceil \frac{m}{3} \right\rceil + O(m\sqrt{n}) = (1 + o(1))\gamma(SG_{n \times m}).$$

Sketch of Proof. Let k be the integer closest to $\sqrt{n}$ such that $k - 2 \equiv 0 \pmod 3$. $O(m + n\sqrt{n})$ guards suffice to place one guard at every vertex of some rows and columns so that it can be assumed that n and m satisfy $n - 2 \equiv 0 \pmod 3$ and $m \equiv 0 \pmod k$. Let $Y \in \mathcal{Y}$ be any configuration. The guards initially occupy the configuration $C_V(Y)$. By Lemma 4, the guards occupy a dominating set. We show that, for an attack at any vertex v, there is $Y' \in \mathcal{Y}$ such that $v \in C_V(Y')$ and $C_V(Y')$ is compatible with $C_V(Y)$. Indeed, the guards respond to attacks according to their type, *i.e.*, horizontal, vertical or diagonal. Since $k = \Theta(\sqrt{n})$, the strategy uses $\kappa_V = \lceil \frac{n}{3} \rceil \lceil \frac{m}{3} \rceil + O(m\sqrt{n})$ guards.   ◇

4 Lower Bound in Strong Grids

So far, the best lower bound for $\gamma_{all}^{\infty}(SG_{n \times m})$ was the trivial lower bound $\gamma(SG_{n \times m})$. In this section, we slightly increase this lower bound, reducing the gap with the new upper bound of the previous section.

Theorem 2. *For all* $n, m \in \mathbb{N}^*$, $\gamma_{all}^{\infty}(SG_{n \times m}) = \lfloor \frac{n}{3} \rfloor \lfloor \frac{m}{3} \rfloor + \Omega(n + m)$.

Sketch of Proof. If n and m are divisible by 3, there is a unique minimum dominating set of $SG_{n \times m}$ and each vertex is dominated by exactly one guard in this dominating set. The idea of the proof is that, in any winning configuration, some vertices are dominated by more than one guard, and/or some guards dominate at most 6 vertices. Indeed, this is because if there is a 4×5 subgrid that includes 5 border vertices with only one guard in it, then the attacker can win in at most two turns. By double counting, this leads to the necessity of having $\Omega(n + m)$ extra guards compared to the classical domination.   ◇

5 Further Work

Our results in the strong grid leave the open problem of tightening the bounds. Also, for which other grid graphs can our techniques used in obtaining the upper bound be applied? The technique of considering subgrids where only certain attacks are permitted and packing the borders of these subgrids as well as the entire grid with guards should allow to prove that $\gamma_{all}^{\infty}(G) = \gamma(G) + o(nm)$ for many types of $n \times m$ grids G. This should be true since, for all Cayley graphs H obtainable from abelian groups, $\gamma_{all}^{\infty}(H) = \gamma(H)$ [11], and many grid graphs can be represented as Cayley graphs obtained from abelian groups which are truncated. This truncation may increase the number of guards needed but our technique should permit the additional $o(nm)$ guards to suffice. Lastly, as mentioned in the introduction, it is known that given a graph G and an integer k as inputs and asking whether $\gamma_{all}^{\infty}(G) \leq k$ is NP-hard in general [3] but the exact complexity of the decision problem is open.

References

1. Arquilla, J., Fredricksen, H.: "Graphing" an optimal grand strategy. Mil. Oper. Res. **1**(3), 3–17 (1995)
2. Bagan, G., Joffard, A., Kheddouci, H.: Eternal dominating sets on digraphs and orientations of graphs. CoRR, abs/1805.09623 (2018)
3. Bard, S., Duffy, C., Edwards, M., Macgillivray, G., Yang, F.: Eternal domination in split graphs. J. Comb. Math. Comb. Comput. **101**, 121–130 (2017)
4. Beaton, I., Finbow, S., MacDonald, J.A.: Eternal domination numbers of $4 \times n$ grid graphs. J. Comb. Math. Comb. Comput. **85**, 33–48 (2013)
5. Braga, A., Souza, C., Lee, O.: The eternal dominating set problem for proper interval graphs. Inf. Process. Lett. **115**, 582–587 (2015)
6. Burger, A., Cockayne, E.J., Gründlingh, W.R., Mynhardt, C.M., van Vuuren, J.H., Winterbach, W.: Infinite order domination in graphs. J. Comb. Math. Comb. Comput. **50**, 179–194 (2004)
7. Cohen, N., Mc Inerney, F., Nisse, N., Pérennes, S.: Study of a combinatorial game in graphs through linear programming. Algorithmica (2018, to appear)
8. Cohen, N., Martins, N.A., Mc Inerney, F., Nisse, N., Pérennes, S., Sampaio, R.: Spy-game on graphs: complexity and simple topologies. Theor. Comput. Sci. **725**, 1–15 (2018)
9. Diestel, R.: Graph Theory. Graduate Texts in Mathematics, vol. 173, 4th edn. Springer, Heidelberg (2012)
10. Finbow, S., Messinger, M.E., van Bommel, M.F.: Eternal domination in $3 \times n$ grids. Australas. J. Comb. **61**, 156–174 (2015)
11. Goddard, W., Hedetniemi, S.M., Hedetniemi, S.T.: Eternal security in graphs. J. Comb. Math. Comb. Comput. **52**, 160–180 (2005)
12. Goldwasser, J.L., Klostermeyer, W.F., Mynhardt, C.M.: Eternal protection in grid graphs. Util. Math. **91**, 47–64 (2013)
13. Gonçalves, D., Pinlou, A., Rao, M., Thomassé, S.: The domination number of grids. SIAM J. Discrete Math. **25**(3), 1443–1453 (2011)
14. Mc Inerney, F., Nisse, N., Pérennes, S.: Eternal domination in grids. Technical report, INRIA (2018). RR, https://hal.archives-ouvertes.fr/hal-01790322
15. Klostermeyer, W.F., Lawrence, M., MacGillivray, G.: Dynamic dominating sets: the eviction model for eternal domination. Manuscript (2014)
16. Klostermeyer, W.F., MacGillivray, G.: Eternal dominating sets in graphs. J. Comb. Math. Comb. Comput. **68**, 97–111 (2009)
17. Klostermeyer, W.F., Mynhardt, C.M.: Eternal total domination in graphs. Ars Comb. **68**, 473–492 (2012)
18. Klostermeyer, W.F., Mynhardt, C.M.: Protecting a graph with mobile guards. Appl. Anal. Discrete Math. **10**, 1–29 (2014)
19. Lamprou, I., Martin, R., Schewe, S.: Perpetually dominating large grids. In: Fotakis, D., Pagourtzis, A., Paschos, V.T. (eds.) CIAC 2017. LNCS, vol. 10236, pp. 393–404. Springer, Cham (2017). https://doi.org/10.1007/978-3-319-57586-5_33
20. Messinger, M.E., Delaney, A.Z.: Closing the gap: eternal domination on $3 \times n$ grids. Contrib. Discrete Math. **12**(1), 47–61 (2017)
21. Revelle, C.S.: Can you protect the Roman Empire? Johns Hopkins Mag. **50**(2), 40 (1997)
22. Revelle, C.S., Rosing, K.E.: Defendens imperium romanum: a classical problem in military strategy. Am. Math. Mon. **107**, 585–594 (2000)
23. Stewart, I.: Defend the Roman Empire! Sci. Am. **281**, 136–138 (1999)
24. van Bommel, C.M., van Bommel, M.F.: Eternal domination numbers of $5 \times n$ grid graphs. J. Comb. Math. Comb. Comput. **97**, 83–102 (2016)

On the Necessary Memory to Compute the Plurality in Multi-agent Systems

Emanuele Natale[1,2] and Iliad Ramezani[3(✉)]

[1] Max Planck Institute for Informatics, Saarbrücken, Germany
[2] Université Côte d'Azur, CNRS, I3S, Inria, Sophia Antipolis, France
natale@i3s.unice.fr
[3] Sharif University of Technology, Tehran, Iran
iliramezani@ce.sharif.edu

Abstract. We consider the *Relative-Majority Problem* (also known as *Plurality*), in which, given a multi-agent system where each agent is initially provided an input value out of a set of k possible ones, each agent is required to eventually compute the input value with the highest frequency in the initial configuration. We consider the problem in the general Population Protocols model in which, given an underlying undirected connected graph whose nodes represent the agents, edges are selected by a *globally fair* scheduler.

The *state complexity* that is required for solving the Plurality Problem (i.e., the minimum number of memory states that each agent needs to have in order to solve the problem), has been a long-standing open problem. The best protocol so far for the general multi-valued case requires polynomial memory: Salehkaleybar et al. (2015) devised a protocol that solves the problem by employing $\mathcal{O}(k2^k)$ states per agent, and they conjectured their upper bound to be optimal. On the other hand, under the strong assumption that agents initially agree on a total ordering of the initial input values, Gąsieniec et al. (2017), provided an elegant logarithmic-memory plurality protocol.

In this work, we refute Salehkaleybar et al.'s conjecture, by providing a plurality protocol which employs $\mathcal{O}(k^{11})$ states per agent. Central to our result is an ordering protocol which allows to leverage on the plurality protocol by Gąsieniec et al., of independent interest. We also provide a $\Omega(k^2)$-state lower bound on the necessary memory to solve the problem, proving that the Plurality Problem cannot be solved within the mere memory necessary to encode the output.

1 Introduction

Consider a network of n people, where each person supports one opinion from a set of k possible opinions. There is also a *scheduler* who decides in each round which pair of neighbors can interact. The goal is to *eventually* reach an agreement on the opinion with the largest number of supporters, i.e. the *plurality* opinion (or *majority* when $k = 2$). Here, eventually means at an unspecified moment in

© Springer Nature Switzerland AG 2019
P. Heggernes (Ed.): CIAC 2019, LNCS 11485, pp. 323–338, 2019.
https://doi.org/10.1007/978-3-030-17402-6_27

time, which the agents are not necessarily aware of (i.e. *global termination* is not required [25]).

The main resource we are interested in minimizing is the *state complexity* of each node:

How many different states does each person need to go through during such computation?

This *voting* task is known as the *Plurality Problem* (or as the *Voting Problem*) in the asynchronous Population Protocols model [1,24]. For $k = 2$, the problem is well understood: each person needs to maintain two bits in order for the people to elect the opinion of the majority [9,21], regardless of the network size n, and the problem cannot be solved with a single bit [21].

However, the state complexity of the problem for general k has so far remained elusive: a clever protocol by Salehkaleybar et al. [24], called DMVR, shows how to solve the problem with $O(k2^k)$ states per person. They conjectured the DMVR protocol to be optimal:

"We conjecture that the DMVR protocol is an optimal solution for majority voting problem, i.e. at least $k2^{k-1}$ states are required for any possible solution."

On the other hand, under the assumption that agents initially agree on a total ordering of the initial input values, [17] provide an elegant plurality protocol which makes use of a polynomial number of states only. It remained however rather unclear whether the above assumption can be removed in order to achieve a polynomial number of states for the general Plurality Problem as well.

1.1 Related Work

Progress towards understanding the inherent computational complexity for a multi-agent system to achieve certain tasks has been largely empirical in nature. More recently, deeper insights have been offered by analytical studies with respect to some coordination problems [23]. In this regard, understanding the amount of memory necessary for a multi-agent system in order to solve a computational problem is a fundamental issue, as it constrains the simplicity of the individual agents which make up the system [22]. Several research areas such as Chemical Reaction Networks [13] and Programmable Matter [18] investigate the design of computing systems composed of elementary units; in this regard, a high memory requirement for a computational problem constitute a prohibitive barrier to its feasibility in such systems.

The Plurality Problem (also known as Plurality *Consensus* Problem in Distributed Computing), is an extensively studied problem in many areas of distributed computing, such as population protocols [1,8,9,21,24], fixed-volume Chemical Reaction Networks [13,27], asynchronous Gossip protocols [5,6,10,15, 16], Statistical Physics [12] and Mathematical Biology [7,11,20,26].

In the Population Protocols model, the memory is usually measured in terms of the number of *states* (state complexity) rather than the number of bits, following the convention for abstract automata [19]. In the context of the Plurality

Problem, for $k = 2$, the protocols of [9, 21] require 4 states per node, and in [21], they showed that the problem cannot be solved with 3 states. For general k, the protocol of [24] uses $O(2^{k-1} \cdot k)$ states per node, and the only lower bound known has been so far the trivial $\Omega(k)$, as each node/agent needs at least k distinct states to specify its own opinion (which is from a set of size k). Under the crucial assumption that agents initially agree on a representation of the input values as distinct integers, [17] provides an elegant solution to the Plurality Problem which employs $\mathcal{O}(k^6)$ states only.

1.2   Our Results

In this work we refute the conjecture of [24], by devising a general *ordering* protocol which allows the agents to agree on a mapping of the initial k input values to the integers $\{0, \cdots, k-1\}$, thus satisfying the assumption of the protocol by [17]. We further show how to adapt the plurality protocol by [17] in a way that allows to couple its execution in parallel with the ordering protocol such that, once the ordering protocol has converged to the aforementioned mapping, the execution of the plurality protocol is also eventually consistent with the provided ordering of colors. We emphasize that agents are not required to detect when the protocol terminates; this is indeed easily shown to be impossible under the general assumption of a fair scheduler. The resulting plurality protocol make use of $\mathcal{O}(k^{11})$ states per agent.

Theorem 1. *There is a population protocol $P_{general}$ which solves the Plurality Problem under a globally fair scheduler, by employing $\mathcal{O}(k^{11})$ states per agent.*

Furthermore, we prove that $k^2 - k$ states per node are necessary (Theorem 2).

Insights on the Ordering Problem. The main idea for solving the ordering problem is to have some agents form a linked list, where each node is a single agent representing one of the initial colors. The fairness property of the scheduler allows for an *adversarial* kind of asynchronicity in how agents' interactions take place. Because of this distributed nature of the problem, (temporary) creation of multiple linked lists cannot be avoided. Thus, it is necessary to devise a way to eliminate multiple linked lists, whenever more than one of them are detected. We achieve this goal by having agents from one of the linked lists *leave* it; also, as soon as these leaving agents interact with their successor or predecessor in their former list, they force them to leave the list as well, thus propagating the removal process until the entire list gets destroyed.

On the other hand, in order to form the linked list, the simple idea of having *removed* agents appending themselves to an existing linked list does not work. One of the issues with this naive approach is that a free agent u may interact with the last agent v of a list ℓ which is in the process of being destroyed, but the removal process in ℓ may still not have reached v. Our approach to resolve this latter issue consists of, firstly, forcing the destruction process of a linked lists to start from the first agents of the lists, and secondly, forcing free agents to attach to an existing list by *climbing it up* from its first agent and appending themselves to its end once they have traversed it all. This way, by the time that there is only one *first agent* r of a linked list (we call such agents *root* agents), we can be sure that all the free agents must follow the linked list starting by the agent r, thus avoid extending incomplete linked lists.

1.3   Model and Basic Definitions

Population Protocols. In this work, we consider the communication model of Populations Protocols [1]: the multi-agent system is represented by a connected graph $G = (V, E)$ of n nodes/agents, where each node implements a finite state machine with state space Σ. The communication in this model proceeds in discrete steps. We remark that, as for asynchronous continuous-time models with Poisson transition rates, they can always be mapped to a discrete-time model [14].

At each time step, an (oriented) edge is chosen by a certain *scheduler*, and the two endpoint nodes interact. Furthermore, there is a transition function $\Gamma : \Sigma \times \Sigma \to \Sigma \times \Sigma$ that, given an ordered pair of states $(\sigma_u, \sigma_v) \in \Sigma \times \Sigma$ for two interacting nodes u and v, returns their new states $\Gamma((\sigma_u, \sigma_v)) = (\sigma'_u, \sigma'_v)$. We call *configuration*, and denote it by $S^{(t)}$, the vector whose entry u corresponds to agent u's state after t time steps. We say that a configuration S'_1 is *reachable* from configuration S'_2 if there exists a sequence of edges s_{seq} such that if we start from S'_2 and we let the nodes interact according to s_{seq}, the resulting configuration is S'_1.

In recent works, the scheduler in this model is typically assumed to be probabilistic: the edge that is selected at each step is determined by a probability distribution on the edges. The most general studied scheduler is the *fair scheduler* [2], which guarantees the following *global fairness property* [3,4].

Definition 1. *A scheduler is said to be* globally fair, *iff whenever a configuration S appears infinitely often in an infinite execution $S^{(1)}, S^{(2)}, \cdots$, also any configuration S' reachable from S appears infinitely often.*

Some of our results hold for an even weaker[1] version of scheduler, which satisfies the *weak fairness property* [4,17].

Definition 2. *A scheduler is said to be* weakly fair, *iff any edge $e \in E$ appears infinitely often in the activation series $e_1, e_2, \ldots$.*

Note that any probabilistic scheduler which selects any edge with a positive probability, is a globally fair scheduler, in the sense that the global fairness property holds with probability 1. Indeed, the fairness condition for a scheduler may

---

[1] Formally, the globally fair scheduler is not a special case of the weak one since, if the activation of an edge does not lead to a different configuration, it can be ignored under a globally fair scheduler. However, if such *useless* activations are ignored, it is easy to see that the globally fair scheduler is a special case of the weak one.

be viewed as an attempt to capture, in a general way, useful probability-1 properties in a probability-free model [2]. This is crucially the case when correctness is required to be deterministic (i.e. the probability of failure should be 0) [21,24].

We emphasize that our theoretical results concern the *existence* of certain times in the execution of the protocols for which some given properties hold, but no general time upper bound is provided, since a fair scheduler can typically delay some edge activation arbitrarily.

k-Plurality Problem. Let $G = (V, E)$ be a network of n agents, such that each agent $v \in V$ initially supports a value in a set of possible values C of size k. We refer to the k input values as *colors*. For each color $c \in C$, denote by $supp(c)$ the set of agents supporting color c. We further denote $c(v)$ as the input color of $v \in V$. We say that a population protocol solves the k-plurality problem if it reaches any configuration $S^{(t)}$, such that for any $t' \geq t$ it holds that the agents agree on the color with the greatest number of supporters in the initial configuration $S^{(0)}$. More formally, there is an *output function* $\Phi : \Sigma \to C$ such that for any $t' \geq t$ and any agent u, $\Phi((S^{(t')})_u)$ equals the plurality color. If the relative majority is not unique, the agents should reach agreement on any of the plurality colors.

In this work, we focus on solving the k-Plurality Problem under a fair scheduler with the goal of optimizing the state complexity, which we denote by $\mathcal{M}_k$.

We emphasize that we do not assume any non-trivial lower bounds on the support of the initial majority compared to other colors, nor that the agents know the size of the network n, or that they know in advance the number of colors k. We do not make any assumption on the underlying graph other than connectedness. We remark that the analysis of our protocol $P_{general}$ in Theorem 1 holds for strongly connected directed graphs; however, for the sake of simplicity, we restrict ourselves to the original setting by [23].

Crucially, motivated by real-world scenarios such as DNA computing and biological protocols, we do not even assume that the nodes initially agree on a binary representations of the colors: they are only able to recognize whether two colors are equal and to memorize them. This latter assumption separates the polynomial state complexity of [17] from the exponential state complexity of [23].

2   Lower Bound on $\mathcal{M}_k$

Since the agents need at least to be able to distinguish their initial colors from each other, the trivial lower bound $\mathcal{M}_k \geq k$ follows. In this section, we show that $\mathcal{M}_k \in \Omega(k^2)$.

Theorem 2. *Any protocol for the k-Plurality Problem requires at least $k^2 - k$ memory states per agent.*

Proof. The high level idea is to employ an indistinguishability argument. That is, we show that for any protocol with less than $k^2 - k$ states, there must be two initial configurations, $S^{(0)}{}_1$ and $S^{(0)}{}_2$, with different plurality colors, such that a configuration is reachable from both $S^{(0)}{}_1$ and $S^{(0)}{}_2$. Therefore, the protocol must fail in at least one of these two initial configurations.

Let P be a protocol that solves the plurality consensus problem with k initial colors, and let $\Phi : \Sigma \to C$ be the output function of P. Define $\text{ENDSTATES}(c) = \{\sigma \in \Sigma \,|\, \Phi(\sigma) = c\}$. We start by observing that there must be some color $c^* \in C$, such that $|\text{ENDSTATES}(c^*)| \leq |\Sigma|/k$. For any initial configuration $S^{(0)}$ and color c, let $\Delta_c^{S^{(0)}}$ be the number of agents in $S^{(0)}$ with initial color c.

Definition 3. *For an odd integer $x > 0$, let $S_x^{c^*}$ be the set of all initial configurations $S^{(0)}$, such that $|S^{(0)}| = 2x - 1$, $\Delta_{c^*}^{S^{(0)}} = x$ and for any color $c \neq c^*$, $\Delta_c^{S^{(0)}}$ is an even number.*

Given that, for the sake of the lower bound, we can assume a complete topology, the number of configurations in $S_x^{c^*}$ is equal to the number of ways to put $(x - 1)/2$ pair of balls into $k - 1$ bins. Therefore, we have $|S_x^{c^*}| \geq (\frac{x-1}{2} + k - 2 / k - 2)^{k-2}$. For each $S^{(0)} \in S_x^{c^*}$, since the plurality color in $S^{(0)}$ is c^*, $S^{(0)}$ will reach a configuration that Φ maps all agents in the configuration to c^*. The number of such possible configurations is at most the number of ways to put $2x - 1$ balls into $|\text{ENDSTATES}(c^*)|$ bins. For a sufficiently large x, the number of such possible configurations is at most $((2x - 1 + \frac{|\Sigma|}{k} - 1)e / \frac{|\Sigma|}{k} - 1)^{\frac{|\Sigma|}{k} - 1}$. Observe that for $|\Sigma| < k^2 - k$ and sufficiently large x, the upper bound on the number of possible final configurations is less than the lower bound on $|S_x^{c^*}|$. Therefore, there must be two distinct initial configurations $S^{(0)}{}_1, S^{(0)}{}_2 \in S_x^{c^*}$ and a configuration S in which all agents are mapped to c^*, such that S is reachable from both $S^{(0)}{}_1$ and $S^{(0)}{}_2$, by some activation sequences T_1 and T_2 respectively. By definition of $S_x^{c^*}$, we have the following observation.

Observation 1. *For each $S^{(0)}, S^{(0)\prime} \in S_x^{c^*}$ where $S^{(0)} \neq S^{(0)\prime}$, there exists a color c such that $|\Delta_c^{S^{(0)}} - \Delta_c^{S^{(0)\prime}}| \geq 2$.*

Let c be the color obtained from Observation 2 when applied to $S^{(0)}{}_1$ and $S^{(0)}{}_2$. Without loss of generality, assume that $\Delta_c^{S^{(0)}{}_1} \geq \Delta_c^{S^{(0)}{}_2} + 2$. Let $S^{(0)}{}_3$ be an initial configuration with $x - \Delta_c^{S^{(0)}{}_1} + 1$ agents, all having initial color c. Let us consider the two initial configurations $S^{(0)}{}_4 = S^{(0)}{}_1 \cup S^{(0)}{}_3$ and $S^{(0)}{}_5 = S^{(0)}{}_2 \cup S^{(0)}{}_3$. Observe that the plurality color in $S^{(0)}{}_5$ is still c^*, while the plurality color in $S^{(0)}{}_4$ is now c. Since T_1 and T_2 are possible initial sequences of interactions in $S^{(0)}{}_4$ and $S^{(0)}{}_5$ respectively, both $S^{(0)}{}_4$ and $S^{(0)}{}_5$ can reach the configuration $S \cup S^{(0)}{}_3$. Therefore, a protocol P using only $k^2 - k - 1$ states can fail to distinguish between initial configurations $S^{(0)}{}_4$ and $S^{(0)}{}_5$. Hence, P fails to solve the problem on at least one initial configuration.

3   Upper Bound on $\mathcal{M}_k$

In the following, we present a protocol that solves the problem with polynomial state complexity; we prove that $\mathcal{M}_k \in O(k^{11})$. The protocol proposed by

Gąsieniec et al. [2] solves the problem using a polynomial number of states, under the hypothesis that agents agree on a way to represent each color with a m-bit label.

First, we present a protocol that constructs such a shared labeling for the input colors (Theorem 3). Then, we combine these two protocols to design a new protocol that solves the k-Plurality Problem (Theorem 1).

3.1   Protocol for the Ordering Problem

In the Ordering Problem, each agent $a \in V$ initially obtains its input color c_a, from a set of possible colors C of size k. The goal of the agent is to eventually agree on a bijection between the set of the possible input colors of size k, and the integers $\{0, ..., k-1\}$. In other words, each agent a eventually gets a *label* $d_a \in \{0, 1, ..., k-1\}$, such that for any two agents a and b, $d_a = d_b$ iff $c_a = c_b$. We want to solve the Ordering Problem by means of a protocol which uses as few states as possible.

A weakly fair scheduler activates pairs of agents to interact. We consider the underlying topology of possible interactions to be a complete directed graph. We show how to remove such assumption in General Graphs section.

In this section, we prove the following theorem.

Theorem 3. *There is a population protocol P_o which solves the Ordering Problem under a weakly fair scheduler, by employing $\mathcal{O}(k^4)$ states per agent.*

We refer the reader to the section Insights on the Difficulty in the Introduction for an overview of the main ideas behind protocol P_o.

Memory Organization. The state of each agent a, encodes the following information:

1. c_a, the initial color, which never changes.
2. d_a, the desired value, stored in $\lceil log_2 k \rceil$ bits.
3. l_a, a bit, indicating whether or not a is a leader.
4. r_a, a bit, indicating whether or not a is a root.
5. pre_a, a color from the set C. If $r_a = 0$ and a is on a linked list, then pre_a is the color of the agent preceding a on the linked list. Otherwise pre_a is set to be c_a.
6. suc_a, a color from the set C. If a is on a linked list, suc_a is the color of the agent succeeding a on the linked list (or c_a if a is the last agent in the linked list). Otherwise, suc_a is the color of the agent whom a is following on a linked list, to reach the end of that linked list, or c_a if a is not following a linked list yet.

Thus, the number of states used is at most $8k^4$.

Definitions. An agent a is called a *leader*, iff l_a is set. A leader a is called a *root*, iff r_a is set. A leader a is called *isolated*, iff a is not a root and $pre_a = c_a$.

A linked list of n links, is a sequence of leaders $a_0, a_1, .., a_n$, such that only a_0 is a root, and $\forall i, 0 < i \leq n : suc_{a_{i-1}} = c_{a_i} \wedge pre_{a_i} = c_{a_{i-1}}$. A linked list is said to be *consistent*, iff none of its agents' information change by any sequence of further activations, except possibly suc_a where a is the last agent on the linked list.

An isolated agent a is a *good* agent, iff suc_a is either c_a or the color of one of the agents of a consistent linked list.

Initialization. Before the execution of the protocol, each agent sets $d = 0$, $l = 1$, $r = 1$, $pre = c$ and $suc = c$.

Transition Function. Let us suppose two agents $a, b \in A$ interact, $a \neq b$. The transition function Γ_o that updates their states is given by the following Python code, where clear function is for *isolating* an agent.

```python
def clear(u):
    r_u = d_u = 0
    pre_u = suc_u = c_u


if c_a == c_b:
    if l_a and l_b and (not r_a or r_b):
        l_a = r_a = 0    #1
    elif l_a and not l_b:
        d_b = d_a    #2
elif c_a != c_b and l_a and l_b:
    if r_a and r_b:
        clear(a)    #3
    elif not r_a and pre_a == c_a and suc_a == c_b and pre_b == c_b:
        clear(a)    #4
    elif r_a and not r_b:
        if suc_a == c_a or (suc_a == c_b and (pre_b != c_a or d_b != 1)):
            d_b = 1    #5
            suc_a = suc_b = c_b
            pre_b = c_a
        elif pre_a == c_a:
            suc_b = suc_a    #6
    elif not r_a and not r_b:
        if pre_a != c_a and suc_a == c_b and pre_b != c_a:
            suc_a = c_a    #7
        elif pre_b == c_a and (pre_a == c_a or suc_a != c_b):
            clear(b)    #8
        elif pre_a != c_a and suc_a == c_b and pre_b == c_a and d_a + 1 != d_b:
            suc_a = c_a    #9
            clear(b)
        elif pre_a == c_a and suc_a == c_b:
            if suc_b != c_b:
                suc_a = suc_b    #10
            else:
                d_a = d_b + 1    #11
                suc_b = suc_a = c_a
                pre_a = c_b
```

Algorithm 1.1. Protocol P_o for the Ordering Problem.

As seen above, there are 11 rules. The rules are defined for directed pair interactions, but can easily be modified to handle the undirected-interaction case.

Proof of Theorem 3. We now prove the correctness of Protocol P_o (Algorithm 1). We have the following.

Lemma 1. *After some number of activations T, in each nonempty set $supp(c)$ of agents, only one is a leader, and among all leaders only one is a root. After such configuration is reached, the leader and root bits of all agents will never change.*

Proof. The protocol never changes a leader or root bit from False to True. When two leaders with the same color interact, one of them clears its leader bit, due to Rule 1 (notice that the direction of interaction is relevant here). Therefore, the number of leaders decreases until no two leaders have the same color, after which no leader bit of any agent ever changes. Afterwards, when two roots interact, they now have different colors and only one of them remains a root, due to Rule 3. Furthermore, note that when two leaders interact where one of them is a root, the one who remains a leader is also a root, due to Rule 1. Hence, we conclude that there is always a root, and after some number of interactions the root must be unique, after which no root bit of any agent ever changes.

Let T be the number of activations described in Lemma 1. Let L be the set of leaders after T activations and let $q \in L$ represent the unique root. We now prove, by using induction on n, that for any integer $n, 1 \le n < |L|$, after some number of activations $t_n \ge T$, there is a consistent linked list of n links whose agents belong to L.

From now on, we may refer to a leader $a \in L$ by its color c_a. Observe that, since there is only one root and no two leaders have the same color, any linked list that exists after T activations, is a consistent one.

Base case $n = 1$. If after T activations, q does not have a successor (i.e. $suc_q = c_q$), then as soon as q interacts with another leader, it makes the other one its successor, due to Rule 5. Otherwise, as soon as q interacts with suc_q, by Rule 5 we can be sure that they form a consistent linked list of 1 link.

Induction Step. Suppose that $n + 1 < |L|$ and after $t_n \ge T$ activations, a consistent linked list of n links exists. Let $v_1, v_2, ..., v_n$ denote the agents succeeding agent q on the linked list, respectively. Suppose $suc_{v_n} \ne c_{v_n}$, and let p denote suc_{v_n}. Consider the first interaction between v_n and p, after t_n activations. After such interaction, if $pre_p = c_{v_n}$ and $d_p = d_{v_n} + 1$, we have a consistent linked list of $n + 1$ links; otherwise, Rule 7 or Rule 9 executes and $suc_{v_n} = c_{v_n}$. We now assume $suc_{v_n} = c_{v_n}$.
    We prove the following.

Lemma 2. *Suppose some number of activation $T' \ge T$ has passed, and q, v_1, v_2, ..., v_n form a consistent linked list of n links where $n + 1 < |L|$, and also $suc_{v_n} = c_{v_n}$. After some more activations, a good agent exists.*

Proof. (Proof of Lemma 2). If a good agent already exists after T' activations, the claim is proved. Therefore, we assume that no good agent exists right after

T' activations. Define $M = L \setminus \{q, v_1, v_2, ..., v_n\}$. Let M_1 be the set of agents in M which are isolated and $M_2 = M \setminus M_1$. It follows from the hypothesis that $|M| > 0$. First, we prove the lemma assuming $|M_1| = |M|$. Then, we prove the other cases by induction on the size of the $|M_1|$.

Case $|M_1| = |M|$. From the definitions above, it follows that $|M_1| = |M|$ implies $|M_1| > 0$ and $|M_2| = 0$. Let $a \in M_1$ be an agent. Since a is not a good agent, we have $suc_a \neq c_a$. Let b denote suc_a. Consider the first moment after t_k activations in which a and b interact. If one of a or b became a good agent, we are done. Otherwise, Rule 4 executes and a is cleared. Thus, after some number of activations $t \geq T'$, a is a good agent.

We now use induction on $|M_1| = 0, 1, ..., |M| - 1$ to prove the remaining cases.

Base case $|M_1| = 0$. Consider an agent $a_0 \in M_2$. Let agent a_1 be $prev_{a_0}$. If $a_1 \in M_2$, let agent a_2 be $prev_{a_1}$. We repeat this process until we reach some agent a_i such that either $a_i \notin M_2$ or $a_i = a_j$ for some $j < i$. Since $|M_1| = 0$, if $a_i \notin M_2$ then $a_i \notin M$ and a_{i-1} is cleared by the time it and a_i interact, due to Rule 8. Note that the only way M_1 gets new members, is that an agent becomes cleared, which implies the existence of a good agent. Otherwise, $a_i = a_j$ for some $j < i$, which means that we incur in a cycle when we follow the *prev* values of agents. In particular, there will be a pair of agents on this cycle such that when they interact (if they are not already cleared by that time), Rule 8 or Rule 9 executes and an agent is cleared. Therefore, after some activations $t \geq t_n$, an agent is cleared and a good agent exists.

Induction Step. Suppose $h < |M|$ and the statement holds for all $|M_1| < h$. We show that it also holds for $|M_1| = h$. Again, we repeat the process described in the base case. This time, we stop at agent a_i if any of the following holds: *(i)* $a_i \notin M$, *(ii)* $a_i = a_j$ for some $j < i$, or *(iii)* $a_i \in M_1$.

The first two cases follow from the same argument as in the base case. In the third case, suppose that agents a_{i-1} and a_i interact at time t_{ind}. If an agent $a \in M$ has been cleared by time t_{ind}, then we have a good agent. Otherwise, if no agents has been cleared between t_n activations and t_{ind} and, by time t_{ind}, agent a_i is not in M_1 anymore, then the size of M_1 has been reduced by at least 1. The latter event implies that, by induction hypothesis, after some more activations either good agent exists or, by Rule 8, an interaction between a_{i-1} and a_i clears a_i. Thus, eventually a good agent exists.

Let a be a good agent, whose existence is guaranteed by Lemma 2. The only activation that changes the state of a, is an interaction with suc_a (or q when $suc_a = c_a$). If suc_a is not the last agent of the linked list, it will be updated to be its successor (or v_1 when $suc_a = c_a$). Therefore, after at most n such activations, a interacts with the last agent on the linked list, and since $suc_{v_n} = c_{v_n}$, it is added to the linked list (provided that the linked list have not already increased its size by attaching another good agent to it). Therefore, after some activations $t_{n+1} \geq t_n \geq T$, a consistent linked list of $n + 1$ links is formed, concluding the induction.

We have thus proved that, after some number of activations $t_{|L|-1}$, there is a consistent linked list that includes all agents from L. Let a be the last agent on the linked list. Rule 7 ensures that after some activations, $suc_a = c_a$. Also, after some activations all non-leader agents copy the assigned number of their leader. Afterwards, the whole system stabilizes and no agent changes its state, concluding the proof of the theorem.

As a final remark notice that, for an agent a, there may be sequences of edge activations that lead the assigned label d_a to reach a value which grows as a function of n before stabilizing. We thus assume that the variable d_a *overflows* when exceeding the largest number it can store, and gets set back to 0. Notice that d_a is guaranteed to be large enough to store k. It is straightforward to verify that this latter assumption does not affect our analysis above.   □

4   Plurality Protocol with $\mathcal{O}(k^{11})$ States

We now come back to the original problem by proving the following result.

Theorem 4. *There is a population protocol P_{cli} which solves the k-Plurality Problem under a weakly fair scheduler, when the underlying graph is complete, by employing $\mathcal{O}(k^{11})$ states per agent.*

Recall that, initially, each agent $a \in V$ obtains its initial color which we shall rename to ic_a, from a set of possible colors C of size k. Let m be $\lceil log_2 k \rceil$. For the sake of simplicity, in this section we consider the underlying topology of possible interactions to be a complete graph. We show how to remove such assumption in Section General Graphs, thus proving Theorem 1. A weak scheduler activates pairs of agents to interact. The goal is for all agents to agree on the plurality color, using as few states as possible.

Main Intuition behind P_{cli}. The protocol proposed by Gąsieniec et al. [17], which we shall call P_r, solves this problem under the hypothesis that each color is denoted by a never changing m-bit label, such that each bit is either -1 or 1, rather than the more standard 0 or 1. We adopt the same notation and assume that the ordering protocol P_o stores the d values in such format. The idea is to run both protocols, P_o and P_r, in parallel and, whenever for an agent a, l_a and d_a are not equal, we ensure that after some activations, $\forall i, 0 \le i < m : c_a[i] = w(s_a[i])$. When the latter condition holds, we can set l_a to be d_a and reinitialize c_a and s_a according to initialization of P_r.

Notice that, since every agent is required to eventually learn the label of the plurality color, each agent also stores a color that corresponds to that label.

Memory Organization. The state of each agent a, encodes the following information:

1. $ic_a, d_a, ld_a, rt_a, pre_a$ and suc_a, as described for P_o (where c_a, l_a and r_a in P_o are renamed to ic_a, ld_a and rt_a, respectively),
2. l_a, c_a, s_a, as described in P_r, and

3. ans_a, a color from the set C, which holds the relative majority color.

The number of states used is at most $8k^{11}$.

Definitions. An agent a is called *unstable*, iff $l_a \neq d_a$. For each $i, 0 \leq i < m$, and for each x that is an i-bit number with bit values either -1 or 1, let us define L_x to be the set of all agents a such that the first i bits of l_a are equal to x.

Initialization. Before the execution of the protocol, for each agent a, the variables d_a, ld_a, rt_a, pre_a and suc_a are initialized according to P_o. Note that, instead of all bits set to 0, d_a has all bits set to -1. Moreover, we set $l_a = d_a$ and initialize c_a and s_a according to P_r. ans_a is set to be ic_a.

Transition Function. We now define the transition function Γ_{cli}. Let us suppose that two agents a and b are activated, with $a \neq b$. Let Γ_o be the transition function of P_o, and Γ_r be the transition function of P_r.

First, the values related to P_o are updated according to Γ_o. If l_a or l_b is the label of the winning color in $P_2(0)$ (as described in P_r), let us set $ans_a = ic_a$ or $ans_a = ic_b$, respectively. Afterwards,

1. If $d_a = l_a$ and $d_b = l_b$, we update the values related to P_r according to Γ_r.
2. If $d_a \neq l_a$, let $L_a = \{i | 0 \leq i < m \wedge c_a[i] < w(s_a[i])\}$, and let $G_a = \{i | 0 \leq i < m \wedge c_a[i] > w(s_a[i])\}$. Let L_b and G_b be analogously defined. If $L_a \cup G_a = \emptyset$, we set $l_a = d_a$ and initialize c_a and s_a according to initialization rule of P_r. Otherwise, let M be $L_a \cup G_a$ if $d_b = l_b$, $(L_a \cup G_a) \cap (L_b \cup G_b)$ otherwise. For each $i \in M$, if l_a and l_b share the same i-bit prefix, we have that
   (a) If $i \in (L_a \cap G_b) \cup (G_a \cap L_b)$, set $s_a[i] = [c_a[i]]$ and $s_b[i] = [c_b[i]]$,
   (b) Otherwise, update $s_a[i]$ and $s_b[i]$ according to Γ_r.
   If $d_b = l_b$, we update the array c_b and, if needed, we propagate the changes as in Γ_r.

Proof of Theorem 4. First, we prove that for each $i, 0 \leq i < m$, and for each x that is an i-bit number with bit values equal to either -1 or 1, the two invariants of $P_2(i)$ hold for L_x, that is

1. $\sum_{a \in L_x} c_a[i] = \sum_{a \in L_x} w(s[i])$, and
2. $\forall a \in L_x, |w(s_a[i]) - c_a[i]| \leq 1$.

The interactions in which states are updated according to P_r satisfy the invariants due to the correctness of P_r. The other interactions can be divided into the following two cases.

1. For any agent a, if l_a changes and the change includes a bit from the i-bit prefix of l_a, we know that $c_a[i] = w(s[i])$ by definition of the protocol. Therefore, if a is in L_x before the change, the same value is subtracted from both sides of the first invariant. Otherwise, if a is in L_x after the change, the same value is added to both sides of the first invariant. Moreover, since after the reinitialization of a it is still the case that $c_a[i] = w(s[i])$, the invariants hold.

2. If $a \in L_x$ and b are two agents such that $s_a[i]$ and $s_b[i]$ changes simultaneously, then we know that $b \in L_x$ by definition of the protocol and, because of the second invariant, either $c_a[i] = w(s_a[i]) - 1$ and $c_b[i] = w(s_b[i]) + 1$, or $c_a[i] = w(s_a[i]) + 1$ and $c_b[i] = w(s_b[i]) - 1$. In both cases, $w(s_a[i]) + w(s_b[i])$ remains unchanged after $s_a[i]$ is set to $[c_a[i]]$ and $s_b[i]$ is set to $[c_b[i]]$, so the first invariant holds. It is immediate to check that the second invariant still holds as well.

We proved that the two invariants hold throughout the execution of the protocol. It follows from the correctness of P_o that after some number of activations T_o, for each agent a, d_a doesn't change anymore.

Lemma 3. *After some number of activations $T \geq T_o$, for each agent a, l_a equals d_a.*

By the definition of P_{cli}, since d_a remains unchanged after T activations, it is obvious that l_a remains unchanged as well. Thus, from the correctness of P_r, it follows that the whole system eventually stabilizes and every agent knows the label of the plurality color. Furthermore, when an agent a interacts with an agent b with the winning color label, P_{cli} sets ans_a to ic_b. Otherwise, if a has the winning color itself, as soon as it is activated it sets ans_a to ic_a (if it is not set already). Therefore, it only remains to prove Lemma 3.

Proof of Lemma 3. Suppose T_o activations have passed. Since after T_o activations, the d value of agents remains unchanged, by the definition P_{cli} it immediately follows that the number of unstable agents never increases.

Hence, to conclude the proof it suffices to prove the following fact.

Fact 1. *Suppose that, after some number of activations $T \geq T_o$ have passed, an unstable agent still exists. Then, after some additional number of activations, the number of unstable agents decreases.*

To see why Fact 1 holds, suppose that some number of activations $T \geq T_o$ have passed and a is an unstable agent. Let $I = \{i | 0 \leq i < m \wedge c_a[i] \neq w(s_a[i]))\}$. Since the protocol does not change $s_a[i]$ and $c_a[i]$ for all $i \notin I$, the size of I never increases. We prove Fact 1 by induction on $|I|$.

Base case $|I| = 0$. As soon as a is activated, it will set l_a to d_a and thus the number of unstable agents will decrease.

Induction Step. Suppose $|I| = n, 0 < n \leq |A|$, and for all $|I| < n$, after some activations either $I = \emptyset$ or the number of unstable agents decreases. Let $i \in I$ be an integer, and let x denote the i-bit prefix of l_a. Let U be the set of all agents $a \in L_x$ such that a is unstable and $c_a[i] = w(s_a[i])$. P_{cli} does not let agents in U interact in $P_2(i)$, but any two agents from $L_x \setminus U$ can interact with each other. It can easily be seen that the two invariants hold for agents in $L_x \setminus U$ in $P_2(i)$. After some interactions, we can distinguish the following cases: *(i)* an unstable agent in U becomes stable, *(ii)* an unstable agent becomes stable and is added to L_x, or *(iii)* the agents in $L_x \setminus U$ in $P_2(i)$ stabilize.

In the latter case, suppose without loss of generality that $c_a[i] < w(s_a[i])$. By the first invariant of $P_2(i)$ on $L_x \setminus U$, we know that there will be another agent $b \in L_x \setminus U$ such that $c_b[i] > w(s_b[i])$. As soon as a and b interact, the protocol ensures that after the interaction, $i \notin I$. Thus, after some number of activations, either the number of unstable agents decreases or the size of I decreases. Hence, Fact 1 follows by the induction hypothesis, and the proof of Lemma 3 is completed.   □

5   General Graphs

The protocol P_{cli} works on complete directed graphs, but it can be easily modified to work on complete undirected graphs. We now present a protocol P_{gen} which works on undirected connected graphs, under a globally fair scheduler, and finally prove our main result, Theorem 1.

Plurality Protocol on General Graphs. The idea is that, whenever a pair of agents is activated, the two agents can swap their updated states. This way, the agents effectively *travel* on the nodes of the underlying graph and possibly interact with other agents that were not initially adjacent.

Therefore, let us define the transition function $\Gamma_{gen}(p,q) = \Gamma_{cli}(q,p)$, where Γ_{cli} is the transition functions of modified P_{cli}. The initialization of P_{gen} is the same as that of P_{cli}

Proof (Proof of Theorem 1). Let G be any connected graph. Let $S^{(0)}$, $S^{(1)}$, ... any an infinite sequence of configurations obtained by running P_{gen} on G under a globally fair scheduler, where $S^{(0)}$ is the initial configuration. Since the number of possible states is finite, the number of possible configurations is also finite. Therefore, there exists a configuration S that appears infinitely often in the sequence.

For all distinct pairs $u, v \in V(G)$, let $Path_{u,v}$ be a series of edges forming a path from u to v, and suppose that the edges in $Path_{u,v}$ gets activated first in · the order in which they appear in the path, and then in reverse order. Let $\mathcal{E}_{u,v}$ be the concatenation of such edge activations. If we activate edges according to $\mathcal{E}_{u,v}$, then u travels along $Path_{u,v}$ (possibly interacting with some other agents), until it interacts with v, and then travels back to its position. Therefore, the sequence of activations $\mathcal{E}_{u,v}$ ensures that pair $\{u,v\}$ of agents interact with each other at least once. If we keep activating edges according to the sequences $\{\mathcal{E}_{u,v}\}_{u,v \in V}$, for each pair of agents $\{u,v\}$, then starting from S, each pair of agents interact infinitely often.

Remark that, a globally fair scheduler is also a weakly fair one. By correctness of P_{cli} under a weakly fair scheduler (Theorem 4), by repeating the mentioned edge activation sequence starting from S, a *stable* configuration S' will be reached (a configuration in which all agents know the initial plurality color, and their guess remains correct thereafter). Therefore, S' is reachable from S. By the definition of a globally fair scheduler, since S is infinitely reached, the stable configuration S' is eventually reached.

References

1. Angluin, D., Aspnes, J., Diamadi, Z., Fischer, M.J., Peralta, R.: Computation in networks of passively mobile finite-state sensors. Distrib. Comput. **18**(4), 235–253 (2006)
2. Angluin, D., Aspnes, J., Eisenstat, D., Ruppert, E.: The computational power of population protocols. Distrib. Comput. **20**(4), 279–304 (2007)
3. Aspnes, J., Beauquier, J., Burman, J., Sohier, D.: Time and Space optimal counting in population protocols. In: 20th International Conference on Principles of Distributed Systems (OPODIS 2016), Leibniz International Proceedings in Informatics (LIPIcs), vol. 70, pp. 13:1–13:17. Dagstuhl, Germany (2017)
4. Beauquier, J., Burman, J., Clavière, S., Sohier, D.: Space-optimal counting in population protocols. In: Moses, Y. (ed.) DISC 2015. LNCS, vol. 9363, pp. 631–646. Springer, Heidelberg (2015). https://doi.org/10.1007/978-3-662-48653-5_42
5. Becchetti, L., Clementi, A.E.F., Natale, E., Pasquale, F., Silvestri, R.: Plurality consensus in the gossip model. In: Proceedings of the 26th Annual ACM-SIAM Symposium on Discrete Algorithms, SODA 2015, pp. 371–390 (2015)
6. Becchetti, L., Clementi, A.E.F., Natale, E., Pasquale, F., Trevisan, L.: Stabilizing consensus with many opinions. In: Proceedings of the 27th Annual ACM-SIAM Symposium on Discrete Algorithms, SODA 2016, pp. 620–635 (2016)
7. Ben-Shahar, O., Dolev, S., Dolgin, A., Segal, M.: Direction election in flocking swarms. Ad Hoc Netw. **12**, 250–258 (2014)
8. Benezit, F., Thiran, P., Vetterli, M.: The distributed multiple voting problem. IEEE J. Sel. Top. Signal Process. **5**(4), 791–804 (2011)
9. Bénézit, F., Thiran, P., Vetterli, M.: Interval consensus: from quantized gossip to voting. In: Proceedings of the IEEE International Conference on Acoustics, Speech, and Signal Processing, ICASSP 2009, pp. 3661–3664 (2009)
10. Boyd, S., Ghosh, A., Prabhakar, B., Shah, D.: Randomized gossip algorithms. IEEE/ACM Trans. Netw. **14**(SI), 2508–2530 (2006)
11. Couzin, I.D., Krause, J., Franks, N.R., Levin, S.A.: Effective leadership and decision-making in animal groups on the move. Nature **433**(7025), 513–516 (2005)
12. Levin, D.A., Peres, Y.: Markov Chains and Mixing Times, 1st edn. American Mathematical Society, Providence (2008)
13. Doty, D.: Timing in chemical reaction networks. In: Proceedings of the 25th Annual ACM-SIAM Symposium on Discrete Algorithms, SODA 2014, pp. 772–784 (2014)
14. Elsässer, R., Friedetzky, T., Kaaser, D., Mallmann-Trenn, F., Trinker, H.: Brief announcement: rapid asynchronous plurality consensus. In: Proceedings of the ACM Symposium on Principles of Distributed Computing, PODC 2017, pp. 363–365. ACM, New York (2017)
15. Ghaffari, M., Lengler, J.: Tight analysis for the 3-majority consensus dynamics. CoRR, abs/1705.05583 (2017)
16. Ghaffari, M., Parter, M.: A polylogarithmic gossip algorithm for plurality consensus. In: Proceedings of the 2016 ACM Symposium on Principles of Distributed Computing, PODC 2016, pp. 117–126 (2016)
17. Gąsieniec, L., Hamilton, D., Martin, R., Spirakis, P.G., Stachowiak, G.: Deterministic population protocols for exact majority and plurality. In: LIPIcs-Leibniz International Proceedings in Informatics, vol. 70 (2016)

18. Gmyr, R., Hinnenthal, K., Kostitsyna, I., Kuhn, F., Rudolph, D., Scheideler, C.: Shape recognition by a finite automaton robot. In: 43rd International Symposium on Mathematical Foundations of Computer Science (MFCS 2018), Leibniz International Proceedings in Informatics (LIPIcs), vol. 117, pp. 52:1–52:15, Dagstuhl, Germany (2018)

19. Holzer, M., Kutrib, M.: Descriptional and computational complexity of finite automata-a survey. Inf. Comput. **209**(3), 456–470 (2011)

20. Ma, Q., Johansson, A., Tero, A., Nakagaki, T., Sumpter, D.J.T.: Current-reinforced random walks for constructing transport networks. J. R. Soc. Interface **10**(80), 20120864 (2012)

21. Mertzios, G.B., Nikoletseas, S.E., Raptopoulos, C.L., Spirakis, P.G.: Determining majority in networks with local interactions and very small local memory. Distrib. Comput. **30**(1), 1–16 (2017)

22. Pitoni, V.: Memory management with explicit time in resource-bounded agents. In: Proceedings of the 32nd AAAI Conference on Artificial Intelligence, New Orleans, Louisiana, USA, 2–7 February 2018

23. Ranjbar-Sahraei, B., Ammar, H.B., Bloembergen, D., Tuyls, K., Weiss, G.: Theory of cooperation in complex social networks. In: Proceedings of the 28th AAAI Conference on Artificial Intelligence, AAAI 2014, pp. 1471–1477. AAAI Press, Québec City (2014)

24. Salehkaleybar, S., Sharif-Nassab, A., Golestani, S.J.: Distributed voting/ranking with optimal number of states per node. IEEE Trans. Signal Inf. Process. Netw. **1**(4), 259–267 (2015)

25. Santoro, N.: Design and Analysis of Distributed Algorithms, 1st edn. Wiley, Hoboken (2006)

26. Sumpter, D.J.T., Krause, J., James, R., Couzin, I.D., Ward, A.J.W.: Consensus decision making by fish. Curr. Biol. **18**(22), 1773–1777 (2008)

27. Temkin, O.N., Zeigarnik, A.V., Bonchev, D.G.: Chemical Reaction Networks: A Graph-Theoretical Approach, 1st edn. CRC Press, Boca Raton (1996)

Complexity of Vertex Switching on Edge-Bicolored Graphs

Ho Lam Pang and Leizhen Cai[✉]

Department of Computer Science and Engineering,
The Chinese University of Hong Kong, Shatin, Hong Kong SAR, China
{hlpang,lcai}@cse.cuhk.edu.hk

Abstract. An edge-bicolored graph is an undirected graph where each edge is colored exclusively either blue or red. We consider the following switching operation at vertices of edge-bicolored graphs: switching at a vertex changes colors of all edges incident with the vertex from blue to red and vice versa.

We study the complexity of using vertex switching to transform an edge-bicolored graph into a graph that satisfies a given property for the blue graph alone or for both blue and red graphs, and obtain polynomial-time algorithms and NP-completeness proofs for several fundamental properties such as connected graphs, Eulerian graphs, and acyclic graphs.

Keywords: Graph algorithms · Edge-bicolored graphs · Vertex switching

1 Introduction

An *edge-bicolored graph* G is an undirected graph where each edge is colored exclusively either blue or red, and we use G_b and G_r respectively to denote the blue and red graphs of G. Edge-bicolored graphs arise naturally from many applications. For instance, we can represent a communication network by using blue edges for active lines and red edges for inactive lines.

For an edge-bicolored graph, the operation of *switching* at a vertex changes colors of all edges incident with the vertex from blue to red and vice versa. This switching operation contains the classical Seidal switching [10] as a special case with G being edge-bicolored complete graphs and, as pointed out by Zaslavsky [12], was first described by Abelson and Rosenberg [1] in their mathematical system for structural analysis of attitudinal cognitions.

In this paper, we are interested in using switching operation to transform an edge-bicolored graph G into a graph that satisfies a given property for the blue graph alone or for both blue and red graphs. We will focus on polynomial algorithms and NP-completeness of the following two types of problems in terms of a given graph property Π, i.e., a family Π of graphs, where $G \odot S$ denotes the resulting graph after switching at vertices S.

Partially supported by CUHK Direct Grant 4055069.

P. Heggernes (Ed.): CIAC 2019, LNCS 11485, pp. 339–351, 2019.
https://doi.org/10.1007/978-3-030-17402-6_28

Π-GRAPH SWITCHING: Does input graph G contain vertices S such that $(G \odot S)_b$ is a Π-graph?

DUAL Π-GRAPH SWITCHING: Does input graph G contain vertices S such that both $(G \odot S)_b$ and $(G \odot S)_r$ are Π-graphs?

We will consider Π-GRAPH SWITCHING and DUAL Π-GRAPH SWITCHING for the following classical and fundamental Π-graphs: connected graphs, Eulerian graphs, even graphs, cluster graphs, and acyclic graphs.

1.1 Our Contributions

Table 1 summarizes our main results for Π-GRAPH SWITCHING and DUAL Π-GRAPH SWITCHING. We note that our cubic algorithms for EVEN SWITCHING and DUALLY EVEN SWITCHING are based on a connection with the light-flipping game of Dodis and Winkler [2], and our cubic algorithm for DUAL CLUSTER SWITCHING uses a reduction to the classical 2SAT problem.

Table 1. Complexities of Π-GRAPH SWITCHING and DUAL Π-GRAPH SWITCHING.

Π	Π-GRAPH SWITCHING	DUAL Π-GRAPH SWITCHING
Connected	$O(m + n)$	NP-complete
Even	$O(n^3)$	$O(n^3)$
Eulerian	Open	NP-complete
Cluster	NP-complete	$O(n^3)$
Acyclic	NP-complete	NP-complete

1.2 Related Work

Switching on *signed graphs* (edge-bicolored graphs with signs "+" and "−" as two colors and $\{+, -\}$ forms a group under product of signs) has been well studied in connection with matroid theory [12,13]. However, there is little attention in the literature regarding algorithmic problems we study in this paper for general edge-bicolored graphs.

For Π-GRAPH SWITCHING on edge-bicolored complete graphs, the problem is equivalent to that of transforming an uncolored graph into a Π-graph by Seidel switching which swaps neighbors of a vertex with its non-neighbors. For this special case, polynomial-time algorithms have been obtained for various Π-graphs, e.g., Hamiltonian graphs [8], Eulerian graphs [6], bipartite graphs [6], and graphs of fixed degeneracy [3,7]. On the other hand, there are a few intractable cases such as regular graphs [7].

1.3   Definitions

An *edge-bicolored graph* $G = (V, E_b \cup E_r)$ consists of *blue graph* $G_b = (V, E_b)$ and *red graph* $G_r = (V, E_r)$ with $E_b \cap E_r = \phi$. We use m and n for numbers of edges and vertices in G. A vertex u is a *blue neighbor* (resp., *red neighbor*) of v if u is a neighbor of v in G_b (resp., G_r), and the *blue-degree* $d_b(v)$ (resp., *red-degree* $d_r(v)$) of vertex v in G is the degree of v in G_b (resp., G_r). Two vertices u and v in G are *dually connected* if G contains both blue and red paths between u and v, and a subgraph of G is *monochromatic* if all edges in G have the same color.

An uncolored graph is *even* if every vertex has an even degree, *Eulerian* if it is even and connected, *acyclic* if it contains no cycle, and a *cluster graph* if it is the disjoint union of complete graphs. For any property Π, an edge-bicolored graph G is a *dual Π-graph* if both G_b and G_r are Π-graphs.

In all illustrations, solid lines represent blue edges and dashed lines represent red edges.

1.4   Basic Properties

We use $G \odot v$ to denote the resulting graph after switching at a vertex v, and $G \odot S$ the resulting graph after switching at every vertex of a set S of vertices. The following two observations directly follow from the definition of vertex switching: (a) for any vertex v of G, $(G \odot v) \odot v = G$, and (b) for any edge e, the color of e is changed iff S contains exactly one end of e.

The above two observations lead us to the following elementary properties of vertex switching, which will be used throughout this paper.

Lemma 1. *Let S be a subset of vertices of an edge-bicolored graph G.*

1. *$G \odot S$ is the graph obtained by changing the color of each edge in the cut $[S, V - S]$ from blue to red and vice versa.*
2. *$G \odot S$ is uniquely determined regardless of the ordering we switch at vertices.*
3. *For any subset S' of vertices, $(G \odot S) \odot S' = G \odot (S \Delta S')$, where $S \Delta S'$ is the symmetric difference of S and S'.*

2   Connected Graphs

We start with the fundamental property of connectedness of graphs as property Π and consider CONNECTED SWITCHING (Is $(G \odot S)_b$ a connected graph?) and DUALLY CONNECTED SWITCHING (Are $(G \odot S)_b$ and $(G \odot S)_r$ both connected graphs?). We give a simple $O(m + n)$-time algorithm for CONNECTED SWITCHING following a result of Zaslavsky [12], and prove the NP-completeness of DUALLY CONNECTED SWITCHING. For DUALLY CONNECTED SWITCHING on edge-bicolored complete graphs, we completely determine the existence of the required S in terms of the structure of G_b and G_r.

For CONNECTED SWITCHING, we first note the following characterization as a special case of a result of Zaslavsky [12]. Only a hint of proof was given in that paper, and here we fill in details for completeness.

Theorem 1 (Zaslavsky). *For any edge-bicolored graph G, there is a subset S of vertices such that $(G \odot S)_b$ is connected iff the underlying uncolored graph of G is connected.*

Proof. We need only prove the sufficiency as the necessity is obvious. Let T be an arbitrary spanning tree of G, and make T a rooted tree by fixing a vertex as the root. We perform switching operations starting from children of the root level-by-level as follows: switch at the current vertex if it is connected to its parent by a red edge. It is clear that when we finish, all edges of T become blue and hence the resulting graph is blue-connected, which establishes the claim. □

Since we can find a spanning tree T of G using BFS and process T level-by-level in $O(m + n)$ time, the proof implies the following result.

Corollary 1. CONNECTED SWITCHING *is solvable in $O(m + n)$ time.*

On the other hand, once we require both $(G \odot S)_b$ and $(G \odot S)_r$ to be connected, it becomes intractable to find the required S. We will prove this fact by a reduction from the following NP-complete problem SET SPLITTING for triples [9].

INSTANCE: Set X and collection $\mathcal{T}$ of triples from X.
QUESTION: Is there a partition $(X', X - X')$ of X such that every triple in $\mathcal{T}$ has its elements in both X' and $X - X'$?

Note that for any instance $(X, \mathcal{T})$ of SET SPLITTING for triples, a partition $(X', X - X')$ forms a solution iff every triple in $\mathcal{T}$ contains exactly one or two elements of X'.

Theorem 2. DUALLY CONNECTED SWITCHING *is NP-complete.*

Proof. The problem is clearly in NP and we give a reduction from SET SPLITTING for triples with instance $(X, \mathcal{T})$. We first construct from $(X, \mathcal{T})$ in polynomial time an edge-bicolored multigraph G as follows (see Fig. 1a for an example):

1. Create a vertex v^*. For each element $x \in X$, create an *element-vertex* v_x and connect it with v^* by a blue edge and a red edge.
2. For each triple $T \in \mathcal{T}$, create a *triple-vertex* v_T and connect v_T with each corresponding element-vertex of T by a blue edge.

For any $X' \subseteq X$, let $V(X')$ denote its corresponding vertices in G. We show that $(X, \mathcal{T})$ admits a valid splitting $(X', X - X')$ iff $G \odot V(X')$ is dually connected. For this purpose, we first note that all element-vertices of G are dually connected to v^* and it remains so regardless of what vertices of G are switched at. Secondly for any triple-vertex v_T, the only way to make v_T dually connected to v^* is to switch at exactly one or two vertices of $V(T)$. It follows that $G \odot V(X')$ is dually connected iff $V(X')$ contains exactly one or two vertices of $V(T)$ for every triple $T \in \mathcal{T}$, which is equivalent to $(X', V - X')$ being a valid splitting.

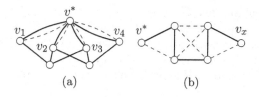

Fig. 1. (a) Example of reduction from SET SPLITTING where $X = \{1,2,3,4\}$ and $T = \{\{1,2,3\},\{2,3,4\}\}$, and (b) replacement gadget H for blue-red multiple edge.

Now we remove multiple edges in G by replacing, for each element-vertex v_x, the blue-red multiple edge connecting v_x and vertex v^* by gadget H in Fig. 1b. It is easily checked that all vertices in H are dually-connected to v^* and it remains so after switching at v_x. This ensures that this new graph is the required instance of DUALLY CONNECTED SWITCHING and hence the theorem follows. □

We now consider the special case of DUALLY CONNECTED SWITCHING with G being edge-bicolored complete graphs, and completely characterize the existence of required S in terms of the structure of G_b and G_r, which also leads to an $O(n^2)$-time algorithm. First we give two lemmas that enable us to switch at a single vertex without disconnecting a graph.

Lemma 2. *Let G be an edge-bicolored complete graph. For any connected G_b that is not a complete graph, there exists a vertex v such that $(G \odot v)_b$ remains connected.*

Proof. In order for a vertex v to possess the property in the lemma, it suffices for v to be a non-universal vertex of G_b such that $G_b - v$ remains connected. Indeed, if G_b contains a universal vertex, then we can take any non-universal vertex of G_b as v (G_b contains at least one such vertex). Otherwise, compute a spanning tree T of G_b and take any leaf of T as v. □

Lemma 3. *Let G be an edge-bicolored complete graph. If G_b is disconnected then for any non-universal vertex v of G_r, $(G \odot v)_r$ remains connected.*

Proof. First we note that G_r is connected as G is a complete graph. For any vertex u, if $G_r - u$ is disconnected, then $G_b - u$ is connected as $G - u$ is also a complete graph. By the assumption that G_b is disconnected, u is an isolated vertex of G_b, implying that u is a universal vertex of G_r. Therefore, $G_r - v$ remains connected for any non-universal vertex v, and hence $(G \odot v)_r$ remains connected for any such vertex. □

With the above two lemmas at hand, we now give a structural characterization for edge-bicolored complete graphs that can be transformed into dually connected ones by vertex switching.

Theorem 3. *An edge-bicolored complete graph G can be transformed into a dually connected graph by vertex switching iff neither G_b nor G_r is a complete graph or the union of two vertex-disjoint complete graphs.*

Proof. Suppose that neither G_b nor G_r is a complete graph or the union of two vertex-disjoint complete graphs. If both G_b and G_r are connected, then we are done. Otherwise, without loss of generality, we may assume that G_b is disconnected. Then G_r is connected as G is a complete graph. We consider two cases, which requires us to switch at one and two vertices respectively.

Case 1. G_b *contains one connected component* H *that is not a complete graph.* By applying Lemma 2 to $G[V(H)]$, we see that H contains a vertex v such that $(G[V(H)] \odot v)_b$ remains connected. It follows that $(G \odot v)_b$ is connected as v is connected to all vertices in $G - V(H)$ by blue edges after switching at vertex v. Since G_r is connected and v is not a universal vertex of G_r, by Lemma 3, $(G \odot v)_r$ remains connected.

Case 2. G_b *is the disjoint union of* $k \geq 3$ *complete graphs.* Since G_r is not a complete graph, it contains at least one non-universal vertex v. By Lemma 3, $(G \odot v)_r$ remains connected. Note that the connected component of $(G \odot v)_b$ containing v is not a complete graph. Therefore we are done if $(G \odot v)_b$ is connected, and otherwise Case 1 applies to $(G \odot v)_b$.

Conversely, we may assume, without loss of generality, that G_b violates the condition of the theorem. If G_b is a complete graph, then for any vertices S, $(G \odot S)_b$ is a union of two vertex-disjoint complete graphs, which is disconnected. Otherwise G_b consists of two vertex-disjoint complete graphs, and let H denote one of them. For any vertices S, we have $G \odot S = (G \odot V(H)) \odot (S \triangle V(H))$. Since $(G \odot V(H))_b$ is a complete graph, we see that $(G \odot S)_b$ consists of two vertex-disjoint complete graphs if $S \neq V(H)$, which is disconnected; and otherwise $(G \odot S)_b$ is a complete graph and hence $(G \odot S)_r$ is disconnected. □

Since it takes $O(n^2)$ time to check whether a graph is a complete graph or the union of two vertex-disjoint complete graphs, we have the following result.

Corollary 2. *For edge-bicolored complete graphs,* DUALLY CONNECTED SWITCHING *is solvable in* $O(n^2)$ *time.*

3   Eulerian Graphs

In connection with Eulerian graphs, we take even graphs and Eulerian graphs, respectively, as property Π and consider the following four problems:

- EVEN SWITCHING: Is $(G \odot S)_b$ an even graph?
- EULERIAN SWITCHING: Is $(G \odot S)_b$ a connected even graph, i.e., Eulerian graph?
- DUALLY EVEN SWITCHING: Are $(G \odot S)_b$ and $(G \odot S)_r$ both even graphs?
- DUALLY EULERIAN SWITCHING: Are $(G \odot S)_b$ and $(G \odot S)_r$ both Eulerian graphs?

We obtain $O(n^3)$-time algorithms for both EVEN SWITCHING and DUALLY EVEN SWITCHING through a connection with the light-flipping game of Dodis and Winkler [2], and establish NP-completeness for DUALLY EULERIAN SWITCHING. However, the complexity of EULERIAN SWITCHING remains open.

Let us start with two simple but crucial observations on degree parity once we switch at a vertex v in G:

1. Vertex v interchanges its blue-degree parity with its red-degree parity.
2. Every vertex u adjacent to v changes its blue-degree parity and red-degree parity, regardless of the color of edge uv.

The above observations lead us to the following light-flipping game of Dodis and Winkler [2], which contains Lights Out and Orbix games as special cases:

Let G^* be an undirected graph where each vertex has an indicator light with state either **On** or **Off**, and a button of type either **Excl** or **Incl**. Pressing an **Excl** button flips states of all its neighboring vertices, and pressing an **Incl** button flips state of the vertex and also states of all its neighboring vertices. Given an initial configuration of lights and buttons, the objective of the game is to press some buttons to turn off all lights, i.e., turn all vertices into **Off** states.

Indeed, we can express EVEN SWITCHING as a light-flipping game as follows:

1. Set G^* to be the underlying uncolored graph of edge-bicolored graph G.
2. The button type of a vertex is **Excl** if its blue- and red-degrees have the same parity and **Incl** otherwise.
3. Set the initial state of a vertex to be **Off** if its blue-degree is even and **On** otherwise.

With the above connection, we can adopt the ideas for the light-flipping game [2] to solve EVEN SWITCHING directly by solving a system of linear equations over GF(2) in polynomial time.

Theorem 4. EVEN SWITCHING *is solvable in* $O(n^3)$ *time.*

Proof. Let G be an edge-bicolored graph. For each vertex v_i of G, let $x_i \in \{0,1\}$ be a variable to indicate whether v_i is switched at ($x_i = 1$) or not ($x_i = 0$). Note that switching at a vertex v_i changes its blue-degree from $d_b(v_i)$ to $d_r(v_i)$, and also changes the blue-degree partity of every vertex adjacent to v_i. Therefore our problem is to find values for x_i's to satisfy the following system of linear equations over GF(2):

$$x_i d_r(v_i) + (1 - x_i)d_b(v_i) + \sum_{v_j \in N(v_i)} x_j \equiv 0 \pmod{2}, \qquad (1)$$

which can be solved by Gaussian elimination in $O(n^3)$ time. $\qquad \square$

For DUALLY EVEN SWITCHING, we observe that for an edge-bicolored graph G to be transformed into a dually even graph by vertex switching, the underlying graph of G must be an even graph. For such G, G_r is automatically an even graph whenever G_b becomes one. Therefore DUALLY EVEN SWITCHING is just EVEN SWITCHING on edge-bicolored graphs whose underlying graphs are even graphs, and the following result follows immediately from Theorem 4.

Corollary 3. DUALLY EVEN SWITCHING *is solvable in* $O(n^3)$ *time.*

Now we turn our attention to DUALLY EULERIAN SWITCHING. In spite of the polynomial solvability of DUALLY EVEN SWITCHING, the addition of connectivity requirement makes our problem intractable.

Theorem 5. DUALLY EULERIAN SWITCHING *is NP-complete.*

Proof. The problem is clearly in NP, and we give a reduction from DUALLY CONNECTED SWITCHING for the NP-hardness. For an arbitrary edge bicolored graph G, first we construct an edge bicolored multigraph G' by doubling each edge with the same color. It is easy to see that G' is a dually even graph, and remains so regardless of what vertices are switched at. Furthermore, for any subset S of vertices, $G \odot S$ is dually connected iff $G' \odot S$ is, implying that $G \odot S$ is dually connected iff $G' \odot S$ is dually Eulerian.

Now we remove multiple edges in G' by replacing each double edge in G' by the gadget H in Fig. 2 to obtain an edge-bicolored graph G^* without multiple edges. Note that the two 5-cliques in H correspond to the two end-vertices in the replaced double edge. For such a 5-clique K, the structure of H ensures that, for K to satisfy dully Eulerian property, we switched at either all or no vertices of K. It follows that $G' \odot S$ is dually Eulerian iff G^* becomes dually Eulerian after switching at all vertices in the 5-cliques corresponding to S, which implies the theorem.   □

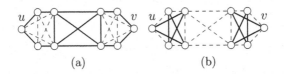

(a)   (b)

Fig. 2. Replacement gadget H for (a) blue double edge uv and (b) red double edge uv.

4   Cluster Graphs

We now turn our attention to cluster graphs as property Π and consider CLUSTER SWITCHING (Is $(G \odot S)_b$ a cluster graph?) and DUAL CLUSTER SWITCHING (Are $(G \odot S)_b$ and $(G \odot S)_r$ both cluster graphs?). We show that the former is NP-complete but the latter is solvable in $O(n^3)$ time. Recall that cluster graphs are exactly P_3-free graphs, i.e., graphs containing no induced paths P_3 on three vertices.

Theorem 6. CLUSTER SWITCHING *is NP-complete for edge-bicolored graphs of maximum degree at least 4, but is solvable in $O(n)$ time for edge-bicolored graphs of maximum degree at most 3.*

Proof. The problem is clearly in NP. For NP-hardness, we note that for any edge-bicolored graph G without red edges, G contains vertices S such that $(G \odot S)_b$ is a cluster graph iff both $G[S]$ and $G - S$ are cluster graphs. In other words, CLUSTER SWITCHING on edge-bicolored graphs without red edges is equivalent to the problem of vertex-partitioning an uncolored graph into two cluster graphs, which is known as the 2-SUBCOLORING problem and is NP-complete for graphs of maximum degree at least 4 [5].

We now discuss the linear-time case. We can always transform any edge-bicolored G of maximum degree 3 into a graph with maximum blue-degree at most 1 (i.e., a special cluster graph) by the following algorithm: *whenever the current graph contains a vertex v of blue-degree at least 2, switch at vertex v.*

Note that the above algorithm reduces at least one blue edge in each iteration, and hence terminates in at most $m \leq 3n/2$ iterations. We can implement it in linear time by maintaining a list of vertices of blue-degree at least 2. □

Moving on to DUAL CLUSTER SWITCHING, we will give an $O(n^3)$-time algorithm by a reduction to 2SAT, which has been a useful approach for Seidel switching (e.g. [6]). Recall that an instance (U, C) of 2SAT consists of a set U of Boolean variables and a collection C of binary clauses over U, and we want to determine if there is a truth assignment for U that satisfies every clause in C.

For our purpose, we regard each vertex u of G as a Boolean variable and express the existence of S that makes $G \odot S$ a dual cluster graph by the satisfaction of a collection of binary clauses, where the Boolean value of a vertex v is 1 iff $v \in S$. For convenience, we from now on use P_3-free graphs for cluster graphs and focus on destroying monochromatic induced P_3's by vertex switching. We start with two 3-vertex graphs: 3-path P_3 and triangle K_3.

Lemma 4. *If the underlying graph G^* of an edge-bicolored graph G is P_3 or K_3, then we can construct in $O(1)$ time a collection $C(G)$ of binary clauses such that G can be transformed into a dually P_3-free graph by vertex switching iff $C(G)$ is satisfiable.*

Proof. Let $\{u, v, w\}$ be vertices of G, and we denote the color of edge e by $c(e)$. For any $S \subseteq V(G)$, we say that S is *valid* if $G \odot S$ is dually P_3-free, and we express the validity of S by a logical formula $F(G)$ as follows:

Case 1. $G^* = P_3$ *with edges uv and vw.* If $c(uv) = c(vw)$, we set $F(G) = (u \in S \Leftrightarrow w \notin S)$ as S is valid iff it contains exactly one of u and w. Otherwise $c(uv) \neq c(vw)$, and we set $F(G) = (u \in S \Leftrightarrow w \in S)$ as S is valid iff it simultaneously includes or excludes u and w.

Case 2. $G^* = K_3$. If $c(uv) = c(vw) = c(wu)$, we set $F(G) = (u \in S \Leftrightarrow v \in S \Leftrightarrow w \in S)$ as we need to switch at either all vertices or no vertex to make G dually P_3-free. Otherwise we may assume $c(uv) = c(vw) \neq c(uw)$, and we set $F(G) = (u \in S \Leftrightarrow v \notin S \Leftrightarrow w \in S)$ as valid values of S are exactly $\{v\}$ and $\{u, w\}$.

It is straightforward to transform $F(G)$ into a collection $C(G)$ of binary clauses as formula $(u \in S \Leftrightarrow v \in S)$ is equivalent to clauses $\{\{u, \overline{v}\}, \{\overline{u}, v\}\}$ and formula $(u \in S \Leftrightarrow v \notin S)$ is equivalent to clauses $\{\{u, v\}, \{\overline{u}, \overline{v}\}\}$. The construction of $C(G)$ clearly takes $O(1)$ time, as each case contains at most 4 binary clauses.   □

Having Lemma 4 at hand, we now solve DUAL CLUSTER SWITCHING in $O(n^3)$ time by reducing it to 2SAT. To do so, we consider all $O(n^3)$ distinct vertex triples in the input graph G to construct the required binary clauses $C(G)$.

Theorem 7. DUAL CLUSTER SWITCHING *is solvable in* $O(n^3)$ *time.*

Proof. For an edge-bicolored graph G, let $G_1, G_2, \ldots, G_t$ with $t = \binom{n}{3}$, be subgraphs induced by all distinct vertex triples in G. Clearly, G is dually P_3-free iff every G_i is dually P_3-free. Therefore in order to find vertices S to make $G \odot S$ dually P_3-free, it suffices to find S such that $G_i \odot S$ is dually P_3-free for every i.

For this purpose, we consider each G_i. If the underlying graph G_i^* of G_i is either P_3 or K_3, then we use Lemma 4 to construct its binary clauses $C(G_i)$ in $O(1)$ time. Otherwise G_i is always dually P_3-free regardless of which vertices are switched at since G_i^* contains no P_3 at all, and hence we set $C(G_i) = \emptyset$. Finally we set $C(G) = \bigcup_{i=1}^{t} C(G_i)$. By Lemma 4, G has the required S iff $C(G)$ is satisfiable. Since $C(G)$ contains $O(n^3)$ binary clauses and 2SAT is solvable in linear time [4], we can solve DUAL CLUSTER SWITCHING in $O(n^3)$ time.   □

5   Acyclic Graphs

Finally we take acyclic graphs (i.e., forests) as property Π and will show the NP-completeness of both ACYCLIC SWITCHING (Is $(G \odot S)_b$ acyclic?) and DUALLY ACYCLIC SWITCHING (Are both $(G \odot S)_b$ and $(G \odot S)_r$ acyclic? Equivalently, does $G \odot S$ contain no monochromatic cycle?).

Theorem 8. ACYCLIC SWITCHING *is NP-complete for edge-bicolored graphs of maximum degree* 5.

Proof. The problem is clearly in NP. Similar to the proof of Theorem 6, ACYCLIC SWITCHING on edge-bicolored graphs without red edges is equivalent to the problem of vertex-partitioning an uncolored graph into two forests, which is NP-complete for graphs of maximum degree 5 [11].   □

For DUALLY ACYCLIC SWITCHING, we will establish its intractability by a reduction from SET SPLITTING for triples used in Sect. 2. We begin with a *consistency gadget* Γ in Fig. 3a with two terminal vertices x and x'. The gadget will be used in our reduction to connect a vertex v with another vertex v' representing same element by identifying v with x and v' with x'. The following lemma shows that we need to switch at both or none of x and x' to maintain the dual acyclicity of Γ.

Fig. 3. (a) Consistency gadget Γ, and (b) example of reduction from SET SPLITTING where $X = \{1, 2, 3, 4\}$ and $\mathcal{T} = \{\{1, 2, 3\}, \{2, 3, 4\}\}$.

Lemma 5. *For any vertices S in Γ, if S contains exactly one of x and x' then $\Gamma \odot S$ contains a monochromatic cycle. Otherwise $\Gamma \odot S$ contains no monochromatic (x, x')-path.*

Proof. If S contains exactly one of x and x', say x, then $\Gamma \odot x$ contains three blue (x, x')-paths. Thereafter, regardless of which degree-2 vertices of Γ are switched at, the resulting graph always contains a monochromatic cycle. Otherwise, we note that neither Γ nor $\Gamma \odot \{x, x'\}$ contains a monochromatic (x, x')-path, and switching at any degree-2 vertex of Γ creates no monochromatic (x, x')-path. $\square$

We now establish the intractability of DUALLY ACYCLIC SWITCHING. Recall that for any instance $(X, \mathcal{T})$ of SET SPLITTING for triples, a partition $(X', X - X')$ forms a solution iff every triple in $\mathcal{T}$ contains exactly one or two elements of X'.

Theorem 9. DUALLY ACYCLIC SWITCHING *is NP-complete.*

Proof. The problem is clearly in NP, and we give a reduction from SET SPLITTING for triples with instance $(X, \mathcal{T})$. We construct from $(X, \mathcal{T})$ in polynomial time an edge-bicolored graph G as follows (see Fig. 3b for an example):

1. For each triple $T \in \mathcal{T}$, construct a blue triangle B_T on three new vertices corresponding to the three elements of T.
2. For each element $x \in X$, create a vertex v_x and for each triple T containing element x, connect v_x with the corresponding vertex of x in blue triangle B_T by a new copy of the consistency gadget Γ.

We note that G_b consists of vertex-disjoint blue triangles and stars, and G_r is a forest consisting of vertex-disjoint stars. For an element $x \in X$, we use V_x to denote the set of vertices in G corresponding to x.

Suppose that $(X', X - X')$ is a solution of $(X, \mathcal{T})$, and let $S = \bigcup_{x \in X'} V_x$. Since every triple T of $\mathcal{T}$ contains exactly one or two elements of X', blue triangle B_T in G contains exactly one or two vertices of S and hence switching at S destroys all blue triangles of G. Furthermore Lemma 5 ensures that $G \odot S$ has no monochromatic path between any two vertices of V_x, implying that switching at S creates no monochromatic cycle and hence $G \odot S$ is dually acyclic.

Conversely, suppose that $G \odot S$ is dually acyclic for some vertices S of G. By Lemma 5, we see that for any $x \in X$, if $v_x \in S$ then S contains V_x and otherwise

S contains no vertex of V_x. Let X' be the subset of X consisting of all elements x with $v_x \in S$, and consider an arbitrary triple $T \in \mathcal{T}$. Since blue triangle B_T in G is destroyed by switching at S, S contains exactly one or two vertices of B_T. Therefore X' contains exactly one or two elements of T, and hence $(X', X - X')$ is a solution of $(X, \mathcal{T})$. $\qquad\square$

6   Concluding Remarks

We have studied computational complexity of some fundamental problems regarding vertex switching in edge-bicolored graphs, which has revealed interesting connections with Seidel's switching, light-flipping game of Dodis and Winkler, and graph 2-partition.

There are abundant of interesting problems for vertex switching in edge-bicolored graphs. For instance, most problems in this paper become NP-complete once we limit the number k of vertex switching operations, and it is natural to consider their parameterized complexity with respect to parameter k. We have identified quite some intractable cases in a forthcoming paper, but FPT cases seem uncommon and deserve attention and efforts.

From graph theoretical point of view, we note that for any hereditary property Π, the family $\mathcal{F}_\Pi$ of edge-bicolored graphs obtainable from Π-graphs by vertex switching is also hereditary. It is worthwhile to investigate characterizations of such families $\mathcal{F}_\Pi$ by obstruction sets, especially those of finite size.

References

1. Abelson, R.P., Rosenberg, M.J.: Symbolic psycho-logic: a model of attitudinal cognition. Behav. Sci. **3**(1), 1–13 (1958)
2. Dodis, Y., Winkler, P.: Universal configurations in light-flipping games. In: Proceedings of the 12th Annual ACM-SIAM Symposium on Discrete Algorithms, pp. 926–927 (2001)
3. Ehrenfeucht, A., Hage, J., Harju, T., Rozenberg, G.: Complexity issues in switching of graphs. In: Ehrig, H., Engels, G., Kreowski, H.-J., Rozenberg, G. (eds.) TAGT 1998. LNCS, vol. 1764, pp. 59–70. Springer, Heidelberg (2000). https://doi.org/10.1007/978-3-540-46464-8_5
4. Even, S., Itai, A., Shamir, A.: On the complexity of time table and multicommodity flow problems. SIAM J. Comput. **5**(4), 691–703 (1976)
5. Fiala, J., Jansen, K., Le, V.B., Seidel, E.: Graph subcolorings: complexity and algorithms. SIAM J. Discrete Math. **16**(4), 635–650 (2003)
6. Hage, J., Harju, T., Welzl, E.: Euler graphs, triangle-free graphs and bipartite graphs in switching classes. In: Corradini, A., Ehrig, H., Kreowski, H.-J., Rozenberg, G. (eds.) ICGT 2002. LNCS, vol. 2505, pp. 148–160. Springer, Heidelberg (2002). https://doi.org/10.1007/3-540-45832-8_13
7. Kratochvíl, J.: Complexity of hypergraph coloring and Seidel's switching. In: Bodlaender, H.L. (ed.) WG 2003. LNCS, vol. 2880, pp. 297–308. Springer, Heidelberg (2003). https://doi.org/10.1007/978-3-540-39890-5_26
8. Kratochvíl, J., Nešetril, J., Zýka, O.: On the computational complexity of Seidel's switching. Ann. Discrete Math. **51**, 161–166 (1992)

9. Lovász, L.: Coverings and colorings of hypergraphs. In: Proceedings of the 4th Southeastern Conference on Combinatorics, Graph Theory, and Computing, pp. 3–12 (1973)
10. Seidel, J.J.: A survey of two-graphs. In: Geometry and Combinatorics, pp. 146–176. Elsevier (1991)
11. Wu, Y., Yuan, J., Zhao, Y.: Partition a graph into two induced forests. J. Math. Study **29**(1), 1–6 (1996)
12. Zaslavsky, T.: Signed graphs. Discrete Appl. Math. **4**(1), 47–74 (1982)
13. Zaslavsky, T.: A mathematical bibliography of signed and gain graphs and allied areas. Electr. J. Comb. DS8 (2018)

Independent Lazy Better-Response Dynamics on Network Games

Paolo Penna[1(✉)] and Laurent Viennot[2]

[1] Department of Computer Science, ETH Zurich, Zurich, Switzerland
paolo.penna@inf.ethz.ch
[2] Inria – Université Paris Diderot, Paris, France
Laurent.Viennot@inria.fr

Abstract. We study an *independent* best-response dynamics on network games in which the nodes (players) decide to revise their strategies independently with some probability. We provide several bounds on the *convergence time* to an *equilibrium* as a function of this probability, the degree of the network, and the potential of the underlying games. These dynamics are somewhat more suitable for distributed environments than the classical better- and best-response dynamics where players revise their strategies "sequentially", i.e., no two players revise their strategies simultaneously.

1 Introduction

Complex and distributed systems are often modeled by means of *game dynamics* in which the participants (players) act spontaneously, typically striving to maximize their own payoff. Such selfish behavior often results in a so-called (pure Nash) *equilibrium* which, roughly speaking, corresponds to the situation in which no player has an incentive to change her current strategy.[1]

Consider the natural scenario in which people interact on a (social) network and take their decisions based on both their personal interests and also on what their friends decided. Situations of this sort are often modeled by means of *games* that are played locally by the nodes of some graph (see, e.g., [14] and [13, Chap. 19]). For example, players may have to choose between two alternatives (strategies), and each strategy becomes more valuable if other friends also choose it (perhaps it is easier to agree than to disagree, or it is better to adopt the same technology for working, rather than different ones).

A full version of this work is available online at [27].

Supported by IRIF (CNRS UMR 8243) and Inria project-team GANG.

---

[1] In this work we consider only pure Nash equilibria, which are the equilibria that occur in certain games when each player chooses one strategy out of the available ones. Other equilibrium concepts are also studied, most notably the mixed Nash equilibrium, where each player chooses a probability distribution over the available strategies.

© Springer Nature Switzerland AG 2019
P. Heggernes (Ed.): CIAC 2019, LNCS 11485, pp. 352–364, 2019.
https://doi.org/10.1007/978-3-030-17402-6_29

In many cases, an extremely simple procedure to convergence to an equilibrium is the so-called *best-response* dynamics in which at each step one player revises her strategy so to maximize her own payoff (and the others stay put). These dynamics work in more general settings (not only on network games), where convergence to an equilibrium is proven via a potential argument (every move reduces the value of a global function – called potential). Games of this nature are called potential games and they are used to model a variety of situations. Interestingly, this argument *fails* as soon as two or more players *move at the same time.*

In this work we study a natural variant of best-response dynamics in which we relax the requirement that one player at a time moves. That is, now players become active *independently* with some probability and all active players revise their strategy according to the best-response rule (or more generally any better-response rule). This is similar as before but allowing simultaneous moves. Specifically, we study the *convergence time* of these dynamics when players play on a network a "local" potential game: (1) each player interacts only with her neighbors, meaning that the strategies of the non-neighbors do not affect the payoff of this player, and (2) locally the game is a potential game (see Sect. 2 for more details and formal definitions).

Simple examples show that convergence is impossible if two players are always active (move all the time), or that the time to converge can be made arbitrarily long if they become active at almost every step. At the other extreme, if the probability of becoming active is too small, then the dynamics will also take a long time to converge since almost all the time nothing happens. The trade-off is between having sufficiently many active players and, at the same time, not too many neighboring players moving simultaneously.

1.1   Our Contribution

We investigate how the convergence time depends on the probabilities of becoming active and on the degree of the network. This is also motivated by the search for simple dynamics that the players can easily implement without global knowledge of the network (namely, they only need to known how many neighbors they have), nor without having complex reasoning (they still myopically better-respond). We first show that for the symmetric coordination game, the convergence time is polynomial whenever the probability of being active is slightly below the inverse of the maximum degree of the network (Theorem 2 and Corollary 1). This generalizes to *arbitrary* potential games on graphs, where every node plays a possibly *different* potential game with each of its neighbors, and the maximum degree is replaced by a *weighted maximum degree* (see Theorem 6). These results indeed hold whenever each active player uses a *better response* (not necessarily the best response). Finally, we prove a lower bound saying that, in general, the probabilities of becoming active *must* depend on the degree for otherwise the convergence time is *exponential* with high probability (Theorem 5 and Corollary 2). Note that this holds also for the simplest scenario of symmetric coordination games.

Our upper bounds can be seen as a probabilistic version of the potential argument (under certain conditions, the potential decreases in expectation *at every step* by some fixed amount). To the best of our knowledge, this is the first study on the convergence time of these natural variants of best-response dynamics. Prior studies (see next section) either focus on sufficient conditions to guarantee convergence to Nash equilibria, or they consider *noisy* best-response dynamics whose equilibria can be different from best-response.

We note that the general upper bound necessarily depends on the maximum value of the potential, as these games include *max-cut* games which are PLS-complete [30]: for such games, no centralized algorithm for computing a Nash equilibrium in time polynomial in the number of players is known, and these games are hard precisely when the potential can assume arbitrarily large values. Obviously, one cannot hope that simple distributed dynamics do better than the best centralized procedure.

1.2 Related Work

Several works study convergence to Nash equilibria for simple variants of best-response dynamics. A first line of research concerns the ability to converge to a Nash equilibrium when the strict schedule of the moves of the players (one player at a time) is relaxed [10]; they proved that any "separable" schedule guarantees convergence to a Nash equilibrium. Other works study the convergence time of specific dynamics with limited simultaneous moves: [19] introduce a "local" coordination mechanism for congestion games (which are equivalent to potential games [24]), while [15] shows that with limited simultaneous moves the dynamics reaches quickly a state whose cost is not too far from the worst Nash equilibrium [15]; Fast convergence can be achieved in certain linear congestion games if *approximate* equilibria are considered [9].

Another well-studied variant of best-response dynamics is that of *noisy* or *logit (response)* dynamics [1,6,7], where players' responses is *probabilistic* and determined by a noise parameter (as the noise tends to zero, players select almost surely best-responses, while for high noise they respond at random). These dynamics turn out to behave differently from "deterministic" best-response in many aspects. In the original logit dynamics by [6,7], where one randomly chosen player moves at a time, they essentially rest on a subset of potential minimizers. When the players' schedule is relaxed, this property is lost and additional conditions on the game are required [1,2,10,18,26]. Our *independent* better-response dynamics can be seen as an analog of the independent dynamics of [1] for logit response.

Potential games on graphs (a proper subclass of potential games) are well-studied because of their many applications. In physics, ferro-magnetic systems are modeled as *noisy* best-response dynamics on lattice graphs in which every player (node) plays a coordination game with each neighbor (see, for example, [23] and Chap. 15 of [22]). The version in which the coordination game is asymmetric (i.e., coordinating on one strategy is more profitable than another) is used to model the diffusion of new technologies [21,25] and opinions [17] in social networks. Finally, potential games on graphs (every node plays some potential game with each neighbor) characterize the class of potential games for which the equilibria of *noisy* best-response dynamics with *all* players updating simultaneously can be "easily" computed [3]. The convergence time of best-response dynamics for games on graphs is studied in [12,17]: Among other results, [12] showed that a polynomial number of steps are sufficient when the same game is played on all edges and the number of strategies is constant. Analogous results are proven for finite opinion games in [17]. Finally, [4] characterize the class of potential games which are also *graphical* games [20], where the potential can be decomposed into the sum of potentials of "maximal" cliques of an underlying graph. Graphical games have been studied in several works (see, e.g., [5,8,11,28]). The class of *local interaction* potential games [3] is the restriction in which the potential can be decomposed into pairwise (edge) potential games. In this work we deal precisely with this class of games. Since this class includes the so-called max-cut games, which are known to be PLS-complete [30], it is considered unlikely that an equilibrium can be computed efficiently, even by a centralized procedure.

Our dynamics are similar to the α-synchronous dynamics in cellular automata [16]. In particular, the case of symmetric coordination game corresponds to majority rule on general graphs [29] (where each cellular automaton tries to switch to the majority state of its neighbors, and stays put in case of ties). The present work can be seen as a first study of α-synchronous dynamics on general graphs for the rules that follow from best-response to some potential games with neighbors.

2 Model (Local Interaction Potential Games)

Intuitively speaking we consider a network (graph) where each node is a player who repeatedly plays with her neighbors. We assume that a two-player potential game (defined below) is associated to each edge of the graph. Each player must play the same strategy on all the games associated to its incident edges, and her payoff is the sum of the payoffs obtained in each of these games. We also assume finite strategies, i.e. each player chooses her strategy within a finite set.

Symmetric Coordination Game. One of the simplest (potential) games is the *symmetric coordination game* where each player chooses color B or W (for black or white) and her payoff is 1 if players agree on their strategies, and 0 otherwise (see Fig. 1a where the two numbers are the payoff for the row and the column player, respectively).

	B	W
B	1,1	0,0
W	0,0	1,1

	B	W
B	2,1	0,0
W	0,0	1,2

	B	W
B	-2	-1
W	0	-2

	B	W
B	2,2	1,1
W	0,0	2,2

(a) Symmetric   (b) Another   (c) Potential   (d) Game equivalent
Coordination Game. Coordination Game. for Game (b).   to Game (b).

Fig. 1. Examples of two-player games and potential function.

General Potential Games. In a general game, we have n players, and each of them can choose one color (strategy) and the combination $c = (c_1, \ldots, c_n)$ of all colors gives to each player u some payoff $PAY_u(c)$. In a *potential game*, when the change in the payoff of any player improves by some amount, some global function P called the *potential* will be decreased by the same amount: For any player u and any two configurations c and c' which differ only in u's strategy, it holds that

$$PAY_u(c') - PAY_u(c) = P(c) - P(c'). \tag{1}$$

A configuration c is a *(pure Nash) equilibrium* if no player u can improve her payoff, that is, the quantity above is negative or zero for all $c' = (c_1, \ldots, c'_u, \ldots, c_n)$. Conversely, c is not an equilibrium if there is a player u who can improve her payoff ($PAY_u(c') - PAY_u(c) > 0$) in which case c'_u is called a *better response* (to strategies c). A *best response* is a better response maximizing this improvement, over the possible strategies of the player. Potential games possess the following nice feature: A configuration c is an equilibrium if and only if no player can improve the potential function by changing her current strategy. In a general (two-player) potential game the payoff of the players is not the same, and the potential function is therefore not symmetric (see the example in Fig. 1c).

Local Interaction Potential Games [3]. In a local interaction potential game the potential function can be decomposed into the sum of two-player potential games, one for each edge of the network G:

$$P(c) = \sum_{uv \in E(G)} P_{uv}(c_u, c_v). \tag{2}$$

No edge exists if the strategies of the two players do not affect each others' payoff (the corresponding potential is constant and can be ignored). This definition captures the following natural class of games on networks: Each edge corresponds to some potential game, and the payoff of a player is the sum of the payoffs of the games with the neighbors. Note that a player chooses one strategy to be played on all these games.

(Independent) Better-Response Dynamics. A simple procedure for computing an equilibrium consists of repeatedly selecting *one* player who is currently not playing a best response and let her play a better or best response. Every step

reduces the potential by a finite amount, and therefore this procedure terminates into an equilibrium in $O(M)$ time steps, where M is the maximum value for the potential (w.l.o.g., we assume that the potential is always non-negative and takes integer values[2]). Here we consider the variant in which, at each time step, each player becomes *independently* active according to some probability, and those who can improve their payoff change strategy accordingly:

Definition 1. *In independent better-response dynamics, at each time step t players do the following:*

- *Each player (node) u becomes active with some probability p_u^t which can change over time (the case in which it is constant over time is a special case of this one).*
- *Every active player (node) revises her strategy according to a better (or best) response rule. If the current strategy is already a best response, then no change is made.*

Note that all players that are active at a certain time step may change their strategies *simultaneously*. So, for example, it may happen that on the symmetric coordination game in Fig. 1a the two players move from state BW to state WB and back if they are both active all the time.

Generic upper bound. To show that dynamics converge quickly, we show that the potential decreases in expectation at every step. To this end, we consider the probability space of all possible evolutions of the dynamics. A configuration c at a given time t is given by the colors chosen by players at the previous time step (strategy profile) and by the values p_u^t used by users for randomly deciding to be active at time t. The universe Ω is then defined as the set of all infinite sequences $c^0, c^1, \ldots$ of configurations.

Definition 2 (δ-improving dynamics). *Dynamics are δ-improving for a given (local interaction) potential game if in expectation the potential decreases by at least δ during each time step, unless the current configuration is an equilibrium. That is, for any configuration c which is not an equilibrium, and any event $F_c^t = \{c^0, c^1, \ldots \in \Omega \mid c^t = c\}$ where configuration c is reached at time t, we have*

$$E[P^{t+1} - P^t \mid F_c^t] \leq -\delta$$

where P^t denotes the potential at time t.

---

[2] As we assume that strategy sets are finite, the potential function is defined by a finite set of values. Rescaling the potential function so that different values are at least 1 apart, and then truncating the values to integers allows to obtain an equivalent game (with same dynamics). Additionally shifting the values allows to obtain a non-negative potential function for that game.

Standard Martingale arguments imply the following (see [27] for details):

Theorem 1. *The expected convergence time of any δ-improving dynamics is* $O\left(\frac{M_0}{\delta}\right)$ *where M_0 is the expected potential of the game at time 0.*

3   Networks with Symmetric Coordination Games

We first consider the scenario in which every edge of the network is the symmetric coordination game in Fig. 1a. The nodes of a graph G (players) can choose between two colors B and W and are rewarded according to the number of neighbors with same color. We are thus considering the dynamics in which nodes attempt to choose the *majority color of their neighbors* and every active node changes its color if more than half of its neighbors has the different color.

In order to analyze the convergence time of these dynamics, we shall relate the probabilities of being active to the number of neighbors having a different color. We say that u is *unstable* at time t if more than half of the neighbors has the other color, that is,

$$dc_u^t > \frac{1}{2}\delta_u$$

where δ_u is the degree of u and dc_u^t is the number of neighbors of u that have a color different from the color of u at time t. By definition, the dynamics converge if no node is unstable. Note that we have $dc_u^t \le \delta_u \le \Delta_u \le \Delta$ where $\Delta = \max_{u \in V(G)} \delta_u$ is the maximum degree of the graph, and $\Delta_u = \max_{uv \in E(G)} \delta_v$ is the local maximum degree in the neighborhood of u.

For the case of symmetric coordination games, the potential function of a configuration is the number of edges whose endpoints have different colors: An edge uv is said to be *conflicting* in configuration c if u and v have different colors. Therefore the potential is at most the number m of edges.

Theorem 2. *Fix some real values $p, q \in (0,1)$. If we have $p_u^t \in [\frac{p}{\Delta}, \frac{q}{\Delta_u}]$ for all u, t in a symmetric coordination game, then the expected convergence time is* $O\left(\frac{\Delta m_0}{p(1-q)}\right)$ *where m_0 is the initial number of conflicting edges, Δ is the maximum degree, and Δ_u is the maximum degree in the neighborhood of u.*

As an immediate corollary, we have the following result for the case in which all nodes are active with the *same* probability p.

Corollary 1. *If all unstable nodes are active with probability $p < \frac{1-\varepsilon}{\Delta}$ for $\varepsilon > 0$, then the dynamics converge to a stable state in $O(\frac{m_0}{p\varepsilon})$ expected time.*

Theorem 2 derives from the following lemma and Theorem 1.

Lemma 1. *Any dynamics satisfying the hypothesis of Theorem 2 are δ-improving for $\delta = p(1-q)/\Delta$.*

Proof. Consider the event F_c^t where a configuration c is reached at time t. Let C^t denote the number of conflicting edges in c, and U^t be the set of unstable nodes at time t respectively. Recall that the number of conflicting edges is equal to the potential, that is, $P^t = C^t$. We now express $E[C^{t+1} - C^t \mid F_c^t]$ as a function of the values $\{p_u^t \mid u \in V(G)\}$ associated to c.

For that purpose, we first analyze the probability that any given edge of c is conflicting after the random choices made at time t. We distinguish the following types of edges. Let S_1 (resp. S_2) denote the set of edges in c with the same color and one unstable extremity (resp. two). Similarly, let C_1 (resp. C_2) denote the set of edges in c with conflicting colors and one unstable extremity (resp. two). Note that $C^t = |C_1| + |C_2|$. A conflicting edge uv will become non-conflicting if only one extremity changes its color. Similarly, a non-conflicting edge uv will become conflicting if only one extremity changes its color. Due to independence of choices, this happens in both cases with probability $p_{uv}^t = p_u^t(1 - p_v^t) + (1 - p_u^t)p_v^t$ if both u and v are unstable, and with probability p_u^t if u is unstable and v is not. By linearity of expectation, we then obtain:

$$E[C^{t+1} - C^t \mid F_c^t] = \sum_{uv \in S_1} p_u^t + \sum_{uv \in S_2} p_{uv}^t - \sum_{uv \in C_1} p_u^t - \sum_{uv \in C_2} p_{uv}^t. \tag{3}$$

(When we note $uv \in C_1$ (resp. $uv \in S_1$), we assume that u is unstable and v is not.) By definition, each unstable node u sees more conflicting edges than non-conflicting ones, thus implying $1 + \sum_{v|uv \in S_1} 1 + \sum_{v|uv \in S_2} 1 \le \sum_{v|uv \in C_1} 1 + \sum_{v|uv \in C_2} 1$. By multiplying by p_u^t and then summing over all unstable nodes, we obtain:

$$\sum_{u \in U^t} p_u^t + \sum_{uv \in S_1} p_u^t + \sum_{uv \in S_2} (p_u^t + p_v^t) \le \sum_{uv \in C_1} p_u^t + \sum_{uv \in C_2} (p_u^t + p_v^t). \tag{4}$$

As $p_{uv}^t = p_u^t + p_v^t - 2p_u^t p_v^t$, we deduce from (3) and (4):

$$E[C^{t+1} - C^t \mid F_c^t] \le \sum_{uv \in C_2} 2p_u^t p_v^t - \sum_{u \in U^t} p_u^t. \tag{5}$$

Since every edge $uv \in C_2$ has both endpoints in U^t, we can rewrite (5) as

$$E[C^{t+1} - C^t \mid F_c^t] \le \sum_{u \in U^t} p_u^t \left(-1 + \sum_{v|uv \in C_2} p_v^t \right).$$

Using $p_v^t \le \frac{q}{\Delta_v} \le \frac{q}{\delta_u}$ and $p_u^t \ge \frac{p}{\Delta}$, we obtain the following inequality: $E[C^{t+1} - C^t \mid F_c^t] \le \sum_{u \in U^t} \frac{p}{\Delta}(-1 + q) = -p(1-q)\frac{|U^t|}{\Delta}$. This completes the proof. □

Adaptive Probabilities. The upper bound of Theorem 2 can be improved if nodes are aware of the number of neighbors that are willing to change strategy (unstable) and then set accordingly the probability of changing too. More precisely, one can think of active nodes announcing to their neighbors that they are unstable and that they would like to switch to the other color, before actually doing so. Then, each unstable node will switch with a probability inversely proportional to the number of unstable neighbors. The following theorem shows that this yields an improved upper bound on the convergence time.

Theorem 3. *Fix some real values $p, q \in \left(0, \frac{1}{2}\right)$. If we have $p_u^t \in [\frac{p}{d_u^t+1}, \frac{q}{d_u^t+1}]$ for all u, t in a symmetric coordination game, where d_u^t is the number of conflicting unstable neighbors of u, then the expected convergence time is $O\left(\frac{m_0}{p(1-2q)}\right)$ where m_0 is the initial number of conflicting edges.*

To prove this theorem we adapt the proof of Lemma 1 and show that these dynamics are δ-improving for $\delta = p(1 - 2q)$ (see [27] for details).

Fully Local Dynamics. Theorem 2 requires that each node is aware of a bound on the maximum degree, or the local maximum degree in her neighborhood for setting p_u^t. Theorem 3 requires knowledge of the number of conflicting unstable neighbors at each time step. We next consider dynamics that are fully local as each node u can set the probabilities p_u^t by only looking at its own degree.

Theorem 4. *Fix some real values $p, q \in \left(0, \frac{1}{2}\right)$. If we have $p_u^t \in [\frac{p}{\delta_u}, \frac{q}{\delta_u}]$ for all u, t in a symmetric coordination game, where δ_u is the degree of u, then the expected convergence time is $O\left(\frac{\Delta m_0}{p(1-2q)}\right)$ where m_0 is the initial number of conflicting edges.*

The proof of this theorem is similar to that of Theorems 2 and 3 (see [27]).

Tightness of the Results. Consider the following network composed of a clique and $r/2+1$ paths, for even r (see figure below). Each node in a path is connected to all nodes to the right and to the left path (or clique for the first path) as feature by demi-edges with degree indications w.r.t. the previous and the next part of the construction. Below each part, we indicate the number of nodes in the part.

Intuitively, the construction is such that the process proceeds from left to right, where nodes in certain path become unstable only after all nodes in the previous path became black; moreover, inside each path the process is also sequential, i.e., the path becomes black from extremities to center. These observations imply that any dynamics in which nodes become active with probability $p \simeq \alpha$, require $\Omega(r^2/\alpha) = \Omega(n/\alpha)$ steps.

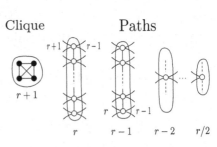

Since every node has degree $\Theta(r) = \Theta(\sqrt{n}) = \Theta(\Delta)$, and the initial configuration has $m_0 = \Theta(r^2) = \Theta(n)$ conflicting edges (those between the clique and the first path), non-adaptive dynamics take $\Theta(\Delta m_0) = \Theta(n^{3/2})$ time steps.

On the contrary, adaptive dynamics take $\Theta(m_0) = \Theta(n)$ steps since the number d_u^t of unstable conflicting neighbors of each node u is at most 1. Therefore, the analysis of Theorems 2, 3, and 4 is tight. Moreover, the adaptive dynamics are provably faster than non-adaptive ones.

4 An Exponential Lower Bound When the Degree Is Unbounded

In this section we prove a lower bound for the case of symmetric coordination game on each edge and dynamics with constant probabilities, that is, the case in which every node becomes active with some probability p which does not depend on the graph nor on the time, and which is the same over all nodes.

Theorem 5. *For every $p > 0$, there are starting configurations of the complete bipartite graph where the expected number of steps to converge to an equilibrium is exponential in the number of nodes.*

Proof Idea. Consider the *continuous* version of the problem in which, instead of a bipartite graph with n nodes on each side, we imagine L and R being two continuous intervals (see figure below). Start from a symmetric configuration in which a fraction $\alpha > 1/2$ of the players in L has color W and the same fraction in R has the other color B. Suppose that $\alpha = \frac{1}{2-p}$. Then after one step the system reaches the symmetric configuration, that is, a fraction α of nodes in L has color B and the same fraction in R has color W. Indeed, the fraction β of players with color B in L after one step is precisely $\beta = 1 - \alpha + p \cdot \alpha = \frac{1-p}{2-p} + \frac{p}{2-p} = \alpha$.

We next prove the theorem via Chernoff bounds. For $\epsilon = p/3$ consider the interval $around(\alpha) := [(1-\epsilon)\alpha, (1+\epsilon)\alpha]$, and let $CYCLE(t)$ be the following event:

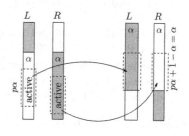

$CYCLE(t) := \{$At time t a fraction $\alpha_L \in around(\alpha)$ of the nodes in L has some color c, and a fraction $\alpha_R \in around(\alpha)$ of the nodes in R has the other color $\bar{c}$ (where $\overline{B} = W$ and $\overline{W} = B$).$\}$

We say that the configuration is *balanced* at time t when $CYCLE(t)$ holds. Since $\epsilon < p/2$ we have $(1-\epsilon)\alpha > 1/2$, and thus the best response of every (active) node in a balanced configuration is to switch color (since both $\alpha_L n$ and $\alpha_R n$ are strictly larger than $n/2$). Chernoff bounds guarantee that with high probability enough many nodes will be activated and therefore will switch to obtain a symmetric balanced configuration (see [27] for proof of next lemma):

Lemma 2. *For any t, it holds that $P[CYCLE(t+1)| \ CYCLE(t)] \geq 1 - 4\exp\left(-\frac{\delta^2}{3}\mu\right)$, where $\delta = \frac{\epsilon}{1+\epsilon}$ and $\mu = p(1+\epsilon)\alpha n$ with $\epsilon = p/3$.*

The above lemma implies that, starting from a balanced configuration, the probability of reaching an equilibrium in t steps is at least $(1 - q)^{t-1}$ where $q = 4\exp(-\frac{\delta^2}{3}\mu)$. The expected time to converge is thus at least $1/q^2$ which proves the theorem. Simple calculations lead to the following result (see [27] for details):

Corollary 2. *Starting from any balanced configuration, the expected number of steps to converge to an equilibrium in the complete bipartite graph is $e^{\Omega(n^{1-3c})}$, as long as $p \geq 1/n^c$ with $0 \leq c < 1/3$.*

5   General Local Interaction Potential Games

In this section we extend the upper bound of Theorem 2 to general local interaction potential games: each edge uv of G is associated with a (two-player) potential game with potential P_{uv}. Without loss of generality, we assume that the potential P_{uv} takes integer non-negative values. The upper bound is given in terms of the following quantity:

$$\Delta_P := \max_u \sum_{v \in N(u)} \Delta_{P_{uv}}, \tag{6}$$

where $\Delta_{P_{uv}}$ denotes the maximum value of P_{uv}. Note that for symmetric coordination games, Δ_P is simply the maximum degree Δ of the graph.

Theorem 6. *For any $p, q \in (0, 1/2)$, if we have $p_u^t \in [\frac{p}{\Delta_P}, \frac{q}{\Delta_P}]$ for all u and t and for Δ_P defined as in (6) in a general local interaction potential game, then the expected convergence time is $O\left(\frac{n\Delta_P^2}{p(1-2q)}\right)$.*

Since local interaction potential games include *max-cut games*, which are notoriously PLS-complete [30], one cannot hope to have convergence time polynomial independent of 'Δ_P' in general. Local interaction games also include finite *opinion games* [17] and, in particular, $16\Delta \leq \Delta_P \leq 16(\Delta + 1)$, where Δ is the maximum degree of the underlying graph (details in [27]). Theorem 6 implies:

Corollary 3. *In finite opinion games on networks of maximum degree Δ, the expected converge time of independent better-response dynamics is $O(n\Delta^2)$ whenever $p_u^t = \frac{\alpha}{\Delta}$ for some $\alpha \in [\frac{p}{16}, \frac{q}{16}]$ with $p, q \in (0, 1/2)$.*

6   Conclusion

This work provides bounds on the time to converge to a (pure Nash) equilibrium when players are active *independently* with some probability and they better or best respond to each others current strategies. Our study focuses on a natural (sub)class of potential games, namely, local interaction potential games. The bounds suggest that the time to converge to an equilibrium *must* depend on the *degree* of the nodes in the underlying network (cf. Theorems 2, 6 and Corollary 2).

Since our bounds hold for local interaction potential games, it would be interesting to investigate whether analogous results hold for *general* potential games. Here a relevant notion is that of graphical games [20] and the results in [4]. It would also be interesting to sharpen some of our bounds to show that $p \simeq 1/\Delta$ is essentially the threshold between fast and slow convergence, and to investigate the range $p \in [1/n, 1/n^{1/3}]$ (cf. Theorem 2 and Corollary 2).

Acknowledgments. We thank Damien Regnault and Nicolas Schabanel for inspiring discussions on closely related problems, and an anonymous reviewer for pointing out the last open question. Part of this work has been done while the first author was at LIAFA, Université Paris Diderot, supported by the French ANR Project DISPLEXITY.

References

1. Alós-Ferrer, C., Netzer, N.: The logit-response dynamics. Games Econ. Behav. **68**(2), 413–427 (2010)
2. Alós-Ferrer, C., Netzer, N.: On the convergence of logit-response to (strict) nash equilibria. Econ. Theory Bull. **5**(1), 1–8 (2017)
3. Auletta, V., Ferraioli, D., Pasquale, F., Penna, P., Persiano, G.: Logit dynamics with concurrent updates for local interaction potential games. Algorithmica **73**(3), 511–546 (2015)
4. Babichenko, Y., Tamuz, O.: Graphical potential games. J. Econ. Theory **163**, 889–899 (2016)
5. Bailey, J.P., Piliouras, G.: Multiplicative weights update in zero-sum games. In: ACM Conference on Economics and Computation (EC), pp. 321–338 (2018)
6. Blume, L.E.: The statistical mechanics of strategic interaction. Games Econ. Behav. **5**(3), 387–424 (1993)
7. Blume, L.E.: Population Games. Addison-Wesley, Boston (1998)
8. Cai, Y., Daskalakis, C.: On minmax theorems for multiplayer games. In: Annual ACM-SIAM Symposium on Discrete Algorithms (SODA), pp. 217–234 (2011)
9. Chien, S., Sinclair, A.: Convergence to approximate nash equilibria in congestion games. Games Econ. Behav. **71**(2), 315–327 (2011)
10. Coucheney, P., Durand, S., Gaujal, B., Touati, C.: General revision protocols in best response algorithms for potential games. In: Netwok Games, Control and OPtimization (NetGCoop) (2014)
11. Daskalakis, C., Papadimitriou, C.H.: On a network generalization of the minmax theorem. In: Albers, S., Marchetti-Spaccamela, A., Matias, Y., Nikoletseas, S., Thomas, W. (eds.) ICALP 2009. LNCS, vol. 5556, pp. 423–434. Springer, Heidelberg (2009). https://doi.org/10.1007/978-3-642-02930-1_35
12. Dyer, M., Mohanaraj, V.: Pairwise-interaction games. In: Aceto, L., Henzinger, M., Sgall, J. (eds.) ICALP 2011. LNCS, vol. 6755, pp. 159–170. Springer, Heidelberg (2011). https://doi.org/10.1007/978-3-642-22006-7_14
13. Easley, D., Kleinberg, J.: Networks, Crowds, and Markets. Cambridge University Press, Cambridge (2012)
14. Ellison, G.: Learning, local interaction, and coordination. Econometrica **61**(5), 1047–1071 (1993)
15. Fanelli, A., Moscardelli, L., Skopalik, A.: On the impact of fair best response dynamics. In: Rovan, B., Sassone, V., Widmayer, P. (eds.) MFCS 2012. LNCS, vol. 7464, pp. 360–371. Springer, Heidelberg (2012). https://doi.org/10.1007/978-3-642-32589-2_33

16. Fatès, N., Regnault, D., Schabanel, N., Thierry, É.: Asynchronous behavior of double-quiescent elementary cellular automata. In: Correa, J.R., Hevia, A., Kiwi, M. (eds.) LATIN 2006. LNCS, vol. 3887, pp. 455–466. Springer, Heidelberg (2006). https://doi.org/10.1007/11682462_43
17. Ferraioli, D., Goldberg, P.W., Ventre, C.: Decentralized dynamics for finite opinion games. Theor. Comput. Sci. **648**, 96–115 (2016)
18. Ferraioli, D., Penna, P.: Imperfect best-response mechanisms. Theory Comput. Syst. **57**(3), 681–710 (2015)
19. Fotakis, D., Kaporis, A.C., Spirakis, P.G.: Atomic congestion games: fast, myopic and concurrent. Theory Comput. Syst. **47**(1), 38–59 (2010)
20. Kearns, M., Littman, M.L., Singh, S.: Graphical models for game theory. In: Conference on Uncertainty in Artificial Intelligence (UAI), pp. 253–260 (2001)
21. Kreindler, G.E., Young, H.P.: Rapid innovation diffusion in social networks. Proc. Natl. Acad. Sci. **111**(Suppl. 3), 10881–10888 (2014)
22. Levin, D.A., Peres, Y., Wilmer, E.L.: Markov Chains and Mixing Times. American Mathematical Society, Providence (2009)
23. Martinelli, F.: Lectures on Glauber dynamics for discrete spin models. In: Bernard, P. (ed.) Lectures on Probability Theory and Statistics. LNM, vol. 1717, pp. 93–191. Springer, Heidelberg (1999). https://doi.org/10.1007/978-3-540-48115-7_2
24. Monderer, D., Shapley, L.S.: Potential games. Games Econ. Behav.or **14**(1), 124–143 (1996)
25. Montanari, A., Saberi, A.: The spread of innovations in social networks. Proc. Natl. Acad. Sci. **107**(47), 20196–20201 (2010)
26. Penna, P.: The price of anarchy and stability in general noisy best-response dynamics. Int. J. Game Theory **47**(3), 839–855 (2018)
27. Penna, P., Viennot, L.: Independent lazy better-response dynamics on network games. CoRR, abs/1609.08953 (2016)
28. Piliouras, G., Shamma, J.S.: Optimization despite chaos: convex relaxations to complex limit sets via poincaré recurrence. In: Annual ACM-SIAM Symposium on Discrete Algorithms (SODA), pp. 861–873 (2014)
29. Rouquier, J., Regnault, D., Thierry, E.: Stochastic minority on graphs. Theor. Comput. Sci. **412**(30), 3947–3963 (2011)
30. Schäffer, A.A., Yannakakis, M.: Simple local search problems that are hard to solve. SIAM J. Comput. **20**(1), 56–87 (1991)

Subset Feedback Vertex Set in Chordal and Split Graphs

Geevarghese Philip[1], Varun Rajan[2], Saket Saurabh[3,4],
and Prafullkumar Tale[3(✉)]

[1] Chennai Mathematical Institute and UMI ReLaX, Chennai, India
gphilip@cmi.ac.in
[2] Chennai Mathematical Institute, Chennai, India
varunrajan09@gmail.com
[3] The Institute of Mathematical Sciences, HBNI, Chennai, India
{saket,pptale}@imsc.res.in
[4] Department of Informatics, University of Bergen, Bergen, Norway

Abstract. In the SUBSET FEEDBACK VERTEX SET (SUBSET-FVS) problem the input consists of a graph G, a subset T of vertices of G called the "terminal" vertices, and an integer k. The task is to determine whether there exists a subset of vertices of cardinality at most k which together intersect all cycles which pass through the terminals. SUBSET-FVS generalizes several well studied problems including FEEDBACK VERTEX SET and MULTIWAY CUT. This problem is known to be NP-Complete even in split graphs. Cygan et al. proved that SUBSET-FVS is fixed parameter tractable (FPT) in general graphs when parameterized by k [SIAM J. Discrete Math (2013)]. In split graphs a simple observation reduces the problem to an equivalent instance of the 3-HITTING SET problem with same solution size. This directly implies, for SUBSET-FVS *restricted to split graphs*, (i) an FPT algorithm which solves the problem in $\mathcal{O}^*(2.076^k)$ time (The $\mathcal{O}^*()$ notation hides polynomial factors.) [Wahlström, Ph.D. Thesis], and (ii) a kernel of size $\mathcal{O}(k^3)$. We improve both these results for SUBSET-FVS on split graphs; we derive (i) a kernel of size $\mathcal{O}(k^2)$ which is the best possible unless NP $\subseteq$ coNP/poly, and (ii) an algorithm which solves the problem in time $\mathcal{O}^*(2^k)$. Our algorithm, in fact, solves SUBSET-FVS on the more general class of *chordal graphs*, also in $\mathcal{O}^*(2^k)$ time.

1 Introduction

In a *covering* or *transversal problem* we are given a universe of elements U, a family $\mathcal{F}$ ($\mathcal{F}$ could be given implicitly) of subsets of U, and an integer k and the objective is to check whether there exists a subset of U of size at most k which intersects all the elements of $\mathcal{F}$. Several natural problems on graphs can be framed in the form of such a problem. For instance, consider the classic FEEDBACK VERTEX SET (FVS) problem. Here, given a graph G and a positive integer k, the objective is to decide whether there exists a vertex subset X (also called a feedback vertex set) of size at most k which intersects all

© Springer Nature Switzerland AG 2019
P. Heggernes (Ed.): CIAC 2019, LNCS 11485, pp. 365–376, 2019.
https://doi.org/10.1007/978-3-030-17402-6_30

cycles, that is, such that $G - X$ is a forest. Other examples include ODD CYCLE TRANSVERSAL, DIRECTED FEEDBACK VERTEX SET and VERTEX COVER (VC). These problems have been particularly well studied in parameterized complexity [4,6,15,17,19,21].

In a natural generalization of covering problems, together with U, $\mathcal{F}$ and k we are given also a subset T of U and the objective is to decide whether there is a subset of U of size at most k that intersects all the sets in $\mathcal{F}$ that contain an element in T. This leads to the *subset variant* of classic covering problems; typical examples include SUBSET FEEDBACK VERTEX SET (SUBSET-FVS), SUBSET DIRECTED FEEDBACK VERTEX SET and SUBSET ODD CYCLE TRANSVERSAL. All these problems have received considerable attention and they have been shown to be *fixed-parameter tractable* (FPT) with k as the parameter [4,6,15].

In this paper we study SUBSET-FVS when the input belongs to certain restricted families of graphs. The (general) SUBSET-FVS problem was first introduced by Even et al. [9]. This problem generalizes several other well-studied problems like FVS, VC, and MULTIWAY CUT [10]. The question whether the SUBSET-FVS problem is fixed parameter tractable (FPT) when parameterized by the solution size was posed independently by Kawarabayashi and the third author in 2009. Cygan et al. [6] and Kawarabayashi and Kobayashi [15] independently answered this question positively in 2011. Wahlström [21] gave the first parameterized algorithm where the dependence on k is $2^{\mathcal{O}(k)}$. Lokshtanov et al. [17] presented a different FPT algorithm which has linear dependence on the input size. On the other hand, Fomin et al. presented a parameter preserving reduction from VC to SUBSET-FVS ([10, Theorem 2.1]) which rules out the possibility of an algorithm with subexponential dependence on k under the Exponential-Time Hypothesis. More recently, Hols and Kratsch showed—using matroid-based tools—that SUBSET-FVS has a randomized polynomial kernelization with $\mathcal{O}(k^9)$ vertices [14].

All the results we mentioned above hold for arbitrary input graphs. There has also been interest in studying SUBSET-FVS on various structured families of graphs, such as chordal graphs and split graphs. While FVS is polynomial-time solvable on split graphs, it turns out that SUBSET-FVS is NP-complete on these graphs [10]. Indeed, the known upper bounds on the number of minimal feedback vertex sets and of minimal *subset* feedback vertex sets in split graphs are n^2 and $3^{n/3}$, respectively [10]. Golovach et al. [12] initiated the algorithmic study of SUBSET-FVS on chordal graphs and presented an exact exponential time algorithm for the problem on these graphs. This algorithm was later improved by Chitnis et al. [3].

In this article we study SUBSET-FVS on chordal and split graphs in the realm of parameterized complexity. For a given set of vertices T, a T-*cycle* is a cycle which contains at least one vertex from T. Formally, the problem we study is as follows.

SUBSET FVS IN CHORDAL GRAPHS   **Parameter:** k
Input: A chordal graph $G = (V, E)$, a set of *terminal vertices* $T \subseteq V$, and
an integer k
Question: Does there exist a set $S \subseteq V$ of at most k vertices of G such that
the subgraph $G[V \setminus S]$ contains no T-cycle?

If the input graph in SUBSET FVS IN CHORDAL GRAPHS is restricted to split
graphs, then the problem will be called SUBSET FVS IN SPLIT GRAPHS.

A simple observation states that in chordal graphs it is enough to intersect all
T-triangles to hit all T-cycles. This provides a parameter preserving reduction
from SUBSET FVS IN SPLIT GRAPHS to 3-HITTING SET (3-HS). On the one
hand, this directly implies a polynomial compression (in fact, a polynomial kernel) of $\mathcal{O}(k^3)$ size for SUBSET FVS IN SPLIT GRAPHS [1], and an FPT algorithm
running in time $2.076^k n^{\mathcal{O}(1)}$ [20]. On the other hand, when we formulate the problem in terms of 3-HS, we lose structural properties of the input graph. These
structural properties can potentially be exploited to obtain better algorithms
and smaller kernels for the original problem. This was recently demonstrated by
Le et al. [16] who derived smaller-than-cubic kernels for several implicit 3-HS
problems. This article is written in the same spirit, of obtaining better results
for implicit 3-HITTING SET problems by exploiting structural properties of the
input graph.

Our Results and Methods: Our first main result is a quadratic kernel for
SUBSET FVS IN SPLIT GRAPHS. This is an improvement over the cubic-size
kernel obtained via the 3-HS route [1]. Formally, we obtain the following.

Theorem 1. SUBSET FVS IN SPLIT GRAPHS *has a quadratic-size kernel.*

We design the kernel for SUBSET FVS IN SPLIT GRAPHS using non-trivial applications of the expansion lemma—a combinatorial tool central to the design of
the first quadratic kernel for FVS [19]. Given an input $(G; T; k)$ where G is a
split graph, we first reduce the input to an instance $(G; T; k)$ where the terminal
set T is exactly the independent set I from a split partition (K, I) of G. Then
we show that if a (non-terminal) vertex $v \in K$ has at least $k + 1$ neighbours in
I then we can include v in a solution or safely delete one edge incident with v;
this leads to an instance where each $v \in K$ has at most k neighbours in I. We
apply the expansion lemma to this instance to bound the number of vertices in
K by $10k$; this gives the bound of $\mathcal{O}(k^2)$ on the number of vertices in I. We
also use the expansion lemma to identify an *irrelevant edge* incident to $v \in K$.
A simple parameter preserving reduction from VC to SUBSET FVS IN SPLIT
GRAPHS implies that this kernel size bound is tight: the problem has no kernel
of size $\mathcal{O}(k^{2-\epsilon})$ under the assumption that NP $\not\subseteq$ coNP/poly [7].

Our second main result is an FPT algorithm for SUBSET FVS IN CHORDAL
GRAPHS with a faster running time than the fastest known algorithm which was
based on solving 3-HS.

Theorem 2. SUBSET FVS IN CHORDAL GRAPHS *admits an algorithm with running time* $\mathcal{O}(2^k(n+m))$. *Here* n, m *are the number of vertices and the edges of the input graph* G, *respectively.*

Note that this running time is *linear* in the size of the input graph. To design our FPT algorithm we examine a *clique-tree* of the input chordal graph and find a useful vertex to branch on. The structure provided by this clique-tree plays a crucial part in obtaining the improved running time.

All missing proofs and the FPT algorithm can be found in the full version of the paper [18].

2   Preliminaries

Graphs. All our graphs are finite, undirected, and simple. We mostly conform to the graph-theoretic notation and terminology from the book of Diestel [8].

Let $S \subseteq V(G)$ and $F \subseteq E(G)$ be a vertex subset and an edge subset of a graph G, respectively. We use (i) $G[S]$ to denote the subgraph of G *induced* by S, (ii) $G - S$ to denote the graph $G[V \setminus S]$, and (ii) $G - F$ to denote the graph $(V(G), (E(G) \setminus F))$. A *triangle* is a cycle of length three. Set S is a *feedback vertex set* (FVS) of G if $G - S$ is a forest. A path P (or cycle C) *passes through* S if P (or C) contains a vertex from S. Let $T \subseteq V(G)$ be a specified set of vertices called *terminal* vertices (or *terminals*). A *T-cycle* ((*T*-triangle) is a cycle (triangle) which passes through T. Graph G is a *T-forest* if it contains no *T*-cycle. Vertex set S is a *subset feedback vertex set* (subset-FVS) of G with respect to terminal set T if the graph $G - S$ is a *T*-forest. Note that S may contain vertices from T, and that $G - S$ need not be a forest. Set S is a *subset triangle hitting set* (subset-THS) of G with respect to terminal set T if $G - S$ contains no *T*-triangle. More generally, we say that a vertex v *hits* a cycle C if C contains v. Vertex set S *hits* a set $\mathcal{C}$ of cycles if for each cycle $C \in \mathcal{C}$ there is a vertex $v \in S$ which hits C. We elide the phrase "with respect to T" when there is no ambiguity.

K_n is the complete graph on n vertices. A subset $S \subseteq V(G)$ of vertices of graph G is a *clique* if its vertices are all pairwise adjacent, and is an *independent set* if they are all pairwise non-adjacent. A clique C in G is a *maximal clique* if C is not a *proper* subset of some clique in G. A vertex v of G is a *simplicial vertex* (or is simplicial) in G if $N[v]$ is a clique. In this case we say that $N[v]$ is a *simplicial clique* in G and that v is a simplicial vertex *of* $N[v]$.

Fact 1 ([2], **Lemma 3**). *Vertex* v *is simplicial in graph* G *if and only* v *belongs to precisely one maximal clique of* G, *namely the set* $N[v]$.

Chordal Graphs. A graph G is *chordal* (or *triangulated*) if every induced cycle in G is a triangle; equivalently, if every cycle of length at least four has a chord. If G is a chordal graph then [13]: (i) every induced subgraph of G is chordal; (ii) G has a simplicial vertex, and if G is not a complete graph then G has two non-adjacent simplicial vertices. Whether a graph H is chordal or not can be found in time $\mathcal{O}(|V(H)| + |E(H)|)$, and if H is chordal then a simplicial vertex of H can be found within the same time bound [13].

Split Graphs. A graph G is a *split graph* if its vertex set can be partitioned into a clique and an independent set in G. Such a partition is called a *split partition* of G. We say that an edge uv in $G[K]$ is *highlighted* if there is a vertex x in I such that the vertices $\{x, u, v\}$ induce a triangle in G.

Lemma 1. *Let G be a chordal graph and let $T \subseteq V(G)$ be a specified set of terminal vertices. A vertex subset $S \subseteq V(G)$ is a subset-FVS of G with respect to T if and only if the graph $G - S$ contains no T-triangles.*

We use $(G; T; k)$ to denote an instance of SUBSET FVS IN CHORDAL GRAPHS or SUBSET FVS IN SPLIT GRAPHS where G is the input graph, T is the specified set of terminals, and k is the parameter.

Corollary 1. *An instance $(G; T; k)$ of SUBSET FVS IN CHORDAL GRAPHS (or of SUBSET FVS IN SPLIT GRAPHS) is a YES instance if and only if there is a vertex subset $S \subseteq V(G)$ of size at most k such that S is a T-THS of G.*

Let $(G; T; k)$ be an instance of SUBSET FVS IN CHORDAL GRAPHS. A subset $S \subseteq V(G)$ of vertices of G is a *solution* of this instance if S is a T-FVS (equivalently, a T-THS) of G.

Expansion Lemmas. Let t be a positive integer and G a bipartite graph with vertex bipartition (P, Q). A set of edges $M \subseteq E(G)$ is called a *t-expansion of P into Q* if (i) every vertex of P is incident with exactly t edges of M, and (ii) the number of vertices in Q which are incident with at least one edge in M is exactly $t|P|$. We say that M *saturates* the endvertices of its edges. Note that the set Q may contain vertices which are *not* saturated by M. We need the following generalizations of Hall's Matching Theorem known as *expansion lemmas*:

Lemma 2 ([5] **Lemma 2.18**). *Let t be a positive integer and G be a bipartite graph with vertex bipartition (P, Q) such that $|Q| \geq t|P|$ and there are no isolated vertices in Q. Then there exist nonempty vertex sets $X \subseteq P$ and $Y \subseteq Q$ such that (i) X has a t-expansion into Y, and (ii) no vertex in Y has a neighbour outside X. Furthermore the sets X and Y can be found in time polynomial in the size of G.*

Lemma 3 ([11]). *Let t be a positive integer and G be a bipartite graph with vertex bipartition (P, Q) such that $|Q| > \ell t$, where ℓ is the size of a maximum matching in G, and there are no isolated vertices in Q. Then there exist nonempty vertex sets $X \subseteq P$ and $Y \subseteq Q$ such that (i) X has a t-expansion into Y, and (ii) no vertex in Y has a neighbour outside X. Furthermore the sets X and Y can be found in time polynomial in the size of G.*

We need sets X, Y of Lemma 3 with an additional property:

Lemma 4. *If the premises of Lemma 3 are satisfied then we can find, in polynomial time, sets X, Y of the kind described in Lemma 3 and a vertex $w \in Y$ such that there exists a t-expansion M from X into Y which does not saturate w.*

3   Kernel Bounds for SUBSET FVS IN SPLIT GRAPHS

In this section we show that SUBSET FVS IN SPLIT GRAPHS has a quadratic-size kernel with a linear number of vertices on the clique side.

Theorem 3. SUBSET FVS IN SPLIT GRAPHS *has a quadratic-size kernel. More precisely: There is a polynomial-time algorithm which, given an instance* $(G;T;k)$ *of* SUBSET FVS IN SPLIT GRAPHS, *returns an instance* $(G';T';k')$ *of* SUBSET FVS IN SPLIT GRAPHS *such that (i)* $(G;T;k)$ *is a* **YES** *instance if and only if* $(G';T';k')$ *is a* **YES** *instance, and (ii)* $|V(G')| = \mathcal{O}(k^2)$, $|E(G')| = \mathcal{O}(k^2)$, *and* $k' \leq k$. *Moreover if* (K',I') *is a split partition of split graph* G' *then* $|K'| \leq 10k$.

Our algorithm works as follows. We first reduce the input to an instance $(G;T;k)$ where the terminal set T is exactly the independent set I from a split partition (K,I) of G. Then we show that if a (non-terminal) vertex $v \in K$ has at least $k+1$ neighbours in I then we can either include v in a solution or safely delete one edge incident with v; this leads to an instance where each $v \in K$ has at most k neighbours in I. We apply the expansion lemma (Lemma 3) to this instance to bound the number of vertices in K by $10k$; this gives the bound of $\mathcal{O}(k^2)$ on the number of vertices in I.

We now describe the reduction rules. Recall that we use $(G;T;k)$ and $(G';T';k')$ to represent the input and output instances of a reduction rule, respectively. We always apply the *first* rule—in the order in which they are described below—which applies to an instance. Thus we apply a rule to an instance *only if* the instance is reduced with respect to all previously specified reduction rules.

Recall that a split graph may have more than one split partition. To keep our presentation short we need to be able to refer to one split partition which "survives" throughout the application of these rules. Towards this we fix an arbitrary split partition $(K^\star, I^\star)$ of the original input graph. Whenever we say "the split partition (K,I) of graph G" we mean the ordered pair $((K^\star \cap V(G)), (I^\star \cap V(G)))$. The only ways in which our reduction rules modify the graph are: (i) delete a vertex, or (ii) delete an edge of the form uv ; $u \in K^\star, v \in I^\star$. So $((K^\star \cap V(G)), (I^\star \cap V(G)))$ remains a split partition of the "current" graph G at each stage during the algorithm.

Our first reduction rule deals with some easy instances.

Reduction Rule 1. *Recall that* (K,I) *is the split partition of graph* G. *Apply the first condition which matches* $(G;T;k)$:

1. *If* $T = \emptyset$ *then output* I_{YES}[1] *and stop.*
2. *If* $k < 0$, *or if* $k = 0$ *and there is a* T-*triangle in* G, *then output* I_{NO} *and stop.*
3. *If there is no* T-*triangle in* G *then output* I_{YES} *and stop.*
4. *If* $|K| \leq k+1$ *then output* I_{YES} *and stop.*
5. *If* $|K| = k+2$ *and there is an edge* uv *in* $G[K]$ *which is* not *highlighted then output* I_{YES} *and stop.*

---

[1] I_{YES} and I_{NO} are trivial **YES** and **NO** instances, respectively.

Each remaining rule deletes a vertex or an edge from the graph.

Reduction Rule 2. *If there is a vertex v of degree zero in G then delete v from G to get graph G'. Set $T' \leftarrow T \setminus \{v\}, k' \leftarrow k$. The reduced instance is $(G'; T'; k')$.*

Reduction Rule 3. *If there is a non-terminal vertex v in G which is not adjacent to a terminal vertex, then delete v from G to get graph G'. Set $T' \leftarrow T, k' \leftarrow k$. The reduced instance is $(G'; T'; k')$.*

Reduction Rule 4. *If there is a bridge e in G then delete edge e (not its end-vertices) to get graph G'. Set $T' \leftarrow T, k' \leftarrow k$. The reduced instance is $(G'; T'; k')$.*

Lemma 5. *Let $(G; T; k)$ be an instance of* SUBSET FVS IN SPLIT GRAPHS *which is reduced with respect to Reduction Rules 1, 2, 3, and 4. Then*

1. *Each vertex in G has degree at least two.*
2. *Every vertex in G is part of some T-triangle.*
3. *If $(G; T; k)$ is a* YES *instance then every terminal vertex on the clique side of G is present in* every *solution of $(G; T; k)$ of size at most k.*

It is thus safe to pick a terminal vertex from the clique side into the solution.

Reduction Rule 5. *If there is a terminal vertex t on the clique side then delete t to get graph G'. Set $T' \leftarrow T \setminus \{t\}, k' \leftarrow k-1$. The reduced instance is $(G'; T'; k')$.*

Observation 1. *Let $(G; T; k)$ be reduced with respect to Reduction Rules 1, 2, 3, 4 and 5. Let (K, I) be the split partition of G. Then $T = I$ and every vertex in K has a neighbour in I.*

Our kernelization algorithm can be thought of having two main parts: (i) bounding the number of vertices on the clique side by $\mathcal{O}(k)$, and (ii) bounding the number of independent set vertices in the neighbourhood of each clique-side vertex by k. We now describe the second part. We need some more notation. For a vertex $v \in K$ on the clique side of graph G we use (i) $N_1(v)$ for the set of neighbours $N(v) \cap I$ of v on the independent side I, and (ii) $N_2(v)$ to denote the set of all *other* clique vertices—than v—which are adjacent to some vertex in $N_1(v)$; that is, $N_2(v) = N(N_1(v)) \setminus \{v\}$. Informally, $N_2(v)$ is the second neighbourhood of v "going via I". We use $B(v)$ to denote the bipartite graph obtained from $G[N_1(v) \cup N_2(v)]$ by deleting every edge with both its endvertices in $N_2(v)$. Equivalently: Let H be the (bipartite) graph obtained by deleting, from G, every edge which has both its ends on the clique side of G. Then $B(v) = H[N_1(v) \cup N_2(v)]$. We call $B(v)$ the bipartite graph *corresponding* to vertex $v \in K$.

Bounding the Independent-Side Neighbourhood of a Vertex on the Clique Side. The first reduction rule of this part applies when there is a vertex $v \in K$ which is part of more than k T-triangles and these T-triangles are pairwise vertex-disjoint apart from the one common vertex v. In this case any solution of size at most k must contain v, so we delete v and reduce k.

Lemma 6. *Let $v \in K$ be a vertex on the clique side of graph G such that the bipartite graph $B(v)$ contains a matching of size at least $k + 1$. Then every T-FVS of G of size at most k contains v.*

Reduction Rule 6. *If there is a vertex v on the clique side K of graph G such that the bipartite graph $B(v)$ has a matching of size at least $k + 1$ then delete vertex v from G to get graph G'. Set $T' \leftarrow T, k' \leftarrow k - 1$. The reduced instance is $(G'; T'; k')$.*

Let $(G; T; k)$ be an instance which is reduced with respect to Reduction Rules 1 to 6. We show that if there is a vertex $v \in K$ on the clique side of G which has more than k neighbours in the independent side I, then we can find an *edge* of the form vw ; $w \in I$ which can safely be deleted from the graph. We get this by a careful application of the "matching" version (Lemma 3) of the Expansion Lemma together with Lemma 4. Let v be such a vertex and let $P = N_2(v), Q = N_1(v), t = 1$. Then (P, Q) is a bipartition of the graph $B(v)$ corresponding to vertex v. Let $\ell \leq k$ be the size of a maximum matching of $B(v)$. Note that $|Q| \geq (k + 1) > \ell t$ and that—by part (1) of Lemma 5—there are no isolated vertices in set $|Q|$. Thus Lemma 4 applies to graph $B(v)$ together with $P, Q, t = 1$. Since a 1-expansion from X into Y contains a matching between X and Y which saturates X we get

Corollary 2. *Let $(G; T; k)$ be an instance which is reduced with respect to Reduction Rule 6. Suppose there is a vertex $v \in K$ on the clique side of G which has more than k neighbours in the independent side I. Then we can find, in polynomial time, non-empty vertex sets $X \subseteq N_2(v) \subseteq K, Y \subseteq N_1(v) \subseteq I$ and a vertex $w \in Y$ such that (i) there is a matching M between X and Y which saturates every vertex of X and does not saturate w, and (ii) $N_G(Y) = X \cup \{v\}$.*

Lemma 7. *Let $(G; T; k)$ be an instance which is reduced with respect to Reduction Rule 6, and let $v \in K$ be a vertex on the clique side which has more than k neighbours in the independent side I. Let $X \subseteq K, w \in Y \subseteq I, M \subseteq E(G[X \cup Y])$ be as guaranteed to exist by Corollary 2. Let $G' = G - \{vw\}$, and let S' be a T-THS of G' of size at most k. If $v \notin S'$ then $(S' \setminus Y) \cup X$ is a T-THS of G' of size at most k.*

Reduction Rule 7. *If there is a vertex v on the clique side K of graph G such that v has more than k neighbours in the independent side I, then find a vertex $w \in I$ as described by Corollary 2 and delete the edge vw to get graph G'. Set $T' \leftarrow T, k' \leftarrow k$. The reduced instance is $(G'; T'; k')$.*

We now show how to bound the number of vertices on the clique side K of an instance $(G; T; k)$ which is reduced with respect to Reduction Rule 7.

Bounding the Size of the Clique Side. We partition the clique side K into three parts and bound the size of each part separately. To do this we first find a 3-approximate solution $\tilde{S}$ to $(G; T; k)$. For this we initialize $\tilde{S} \leftarrow \emptyset$ and iterate as follows: If there is a vertex v in the independent side I such that v is part of a triangle $\{v, x, y\}$ in the graph $G - \tilde{S}$—note that in this case $\{x, y\} \subseteq K$—then we set $\tilde{S} \leftarrow \tilde{S} \cup \{v, x, y\}$. We repeat this till there is no such vertex $v \in I$ or till $|\tilde{S}|$ becomes larger than $3k$, whichever happens first.

Reduction Rule 8. *Let $(G; T; k)$ be an instance which is reduced with respect to Reduction Rule 7 and let $\tilde{S}$ be the set constructed as described above. If $|\tilde{S}| > 3k$ then return I_{NO}.*

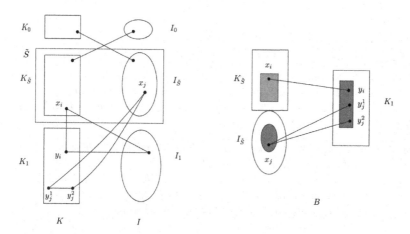

Fig. 1. The figure on the left shows the partition of $V(G)$ as described after Reduction Rule 8. On the right side, we have graph B as described before Corollary 4. Shaded regions in $\tilde{S}$ and K_1 represent the sets X, Y, respectively, as described in the same corollary.

At this point we have that the cardinality of the approximate solution $\tilde{S}$ is at most $3k$. We now partition the sets K, I into three parts each and bound each part separately (See Fig. 1):

- $K_{\tilde{S}}$ is the set of clique-side vertices included in $\tilde{S}$: $K_{\tilde{S}} = K \cap \tilde{S}$.
- $I_{\tilde{S}}$ is the set of independent-side vertices included in $\tilde{S}$: $I_{\tilde{S}} = I \cap \tilde{S}$.
- K_0 is the set of clique-side vertices not in $\tilde{S}$ whose neighbourhoods in the independent side I are all contained in $I_{\tilde{S}}$: $K_0 = \{u \in (K \setminus K_{\tilde{S}}) ; (N(u) \cap I \subseteq I_{\tilde{S}}\}$.
- I_0 is the set of independent-side vertices not in $\tilde{S}$ whose neighbourhoods are all contained in $K_{\tilde{S}}$: $I_0 = \{v \in I \setminus I_{\tilde{S}} ; N(v) \subseteq K_{\tilde{S}}\}$.
- K_1 is the set of clique-side vertices not in $\tilde{S}$ which have at least one neighbour in I outside of $I_{\tilde{S}} \cup I_0$. Equivalently, it is the set of clique-side vertices not in $K_{\tilde{S}} \cup K_0$: $K_1 = K \setminus (K_{\tilde{S}} \cup K_0)$.
- I_1 is the set of independent-side vertices which are not in $I_{\tilde{S}} \cup I_0$: $I_1 = I \setminus (I_{\tilde{S}} \cup I_0)$. Since $\tilde{S}$ is a solution each vertex in I_1—being a terminal—can have exactly one neighbour in K_1.

We list some simple properties of this partition.

Observation 2. $|K_{\tilde{S}}| \leq 2k$ and $|I_{\tilde{S}}| \leq k$. Each vertex in K_1 has (i) no neighbour in I_0 and (ii) at least one neighbour in I_1. Each vertex in I_1 has exactly one neighbour in K_1. The bipartite graph obtained from $G[K_1 \cup I_1]$ by deleting all the edges in $G[K_1]$ is a forest where each connected component is a star.

Let H be the bipartite graph obtained from $G[I_{\tilde{S}} \cup K_0]$ by deleting all the edges in $G[K_0]$. Since—Observation 1—every vertex in the set K_0 has at least one neighbour in the set I and since $(N(K_0) \cap I) \subseteq I_{\tilde{S}}$ by construction, we get that there are no isolated vertices in graph H. So if $|K_0| \geq 2|I_{\tilde{S}}|$ then Lemma 2 applies to graph H with $P \leftarrow I_{\tilde{S}}, Q \leftarrow K_0, t \leftarrow 2$ and we get

Corollary 3. Let $(G; T; k)$ be an instance which is reduced with respect to Reduction Rule 8, and let the sets $K_{\tilde{S}}, K_0, K_1, I_{\tilde{S}}, I_0, I_1$ be as described above. If $|K_0| \geq 2|I_S|$ then we can find, in polynomial time, non-empty vertex sets $X \subseteq I_{\tilde{S}} \subseteq I, Y \subseteq K_0 \subseteq K$ such that (i) X has a 2-expansion M into Y, and (ii) $N_G(Y) = X$.

Lemma 8. Let $(G; T; k)$ be an instance which is reduced with respect to Reduction Rule 8, and let the sets $K_{\tilde{S}}, K_0, K_1, I_{\tilde{S}}, I_0, I_1$ be as described above. Suppose $|K_0| \geq 2|I_{\tilde{S}}|$, and let $X \subseteq I_{\tilde{S}} \subseteq I, Y \subseteq K_0 \subseteq K, M \subseteq E(G[X \cup Y])$ be as guaranteed to exist by Corollary 3. If S is a T-THS of graph G of size at most k then $(S \setminus Y) \cup X$ is also a T-THS of G of size at most k.

Reduction Rule 9. If $|K_0| \geq 2|I_{\tilde{S}}|$ then find sets $X \subseteq I_{\tilde{S}}$ and $Y \subseteq K_0$ as described by Corollary 3. Set $G' \leftarrow G - X, T' \leftarrow T \setminus X, k' \leftarrow k - |X|$. The reduced instance is $(G'; T'; k')$.

At this point we have the bounds $|K_{\tilde{S}}| \leq 2k$ and $|K_0| < 2|I_{\tilde{S}}| = 2k$. We now use a more involved application of the Expansion Lemma to bound the size of the remaining part K_1 of the clique side. The general idea is that if K_1 is at least twice as large as the approximate solution $\tilde{S}$ then the 2-expansion which exists between subsets of these two sets will yield a non-empty set of "redundant" vertices in K_1.

Consider the bipartite graph B obtained from the induced subgraph $G[\tilde{S} \cup K_1]$ of G by (i) deleting all the edges in the two induced subgraphs $G[\tilde{S}]$ and $G[K_1]$, respectively, and (ii) deleting every edge uv ; $u \in K_{\tilde{S}}, v \in K_1$ if and only if there is no vertex $w \in I_1$ such that $\{u, v, w\}$ form a triangle in G. Consider a vertex $v \in K_1$. If v has a neighbour $w \in I_{\tilde{S}}$ then the edge vw is present in graph B and so v is not isolated in B. Now suppose v has no neighbour in $I_{\tilde{S}}$. From the construction we know that v has no neighbour in I_0 either. Then from Lemma 5 and Observation 1 we get that there is a triangle $\{v, x, y\}$ in G where $x \in (I \setminus (I_0 \cup I_{\tilde{S}})) = I_1$ and $y \in K$. Now by construction vertex $x \in I_1$ has no neighbour in the set K_0, and from Observation 2 we get that x has no neighbour other than v in the set K_1. Thus we get that $y \in K_{\tilde{S}}$, and hence that the edge vy survives in graph B. Hence v is not isolated in B in this case either. So if $|K_1| \geq 2|\tilde{S}|$ then Lemma 2 applies to the bipartite graph B with $P \leftarrow \tilde{S}, Q \leftarrow K_1, t \leftarrow 2$ and we get

Corollary 4. *Let $(G; T; k)$ be an instance which is reduced with respect to Reduction Rule 9, and let the sets $K_1, \tilde{S}, K_{\tilde{S}}, I_{\tilde{S}}, K$ and the bipartite graph B be as described above. If $|K_1| \geq 2|\tilde{S}|$ then we can find, in polynomial time, non-empty vertex sets $X \subseteq \tilde{S} = (K_{\tilde{S}} \cup I_{\tilde{S}}), Y \subseteq K_1 \subseteq K$ such that (i) X has a 2-expansion M into Y, and (ii) $N_B(Y) = X$.*

Lemma 9. *Let $(G; T; k)$ be an instance which is reduced with respect to Reduction Rule 8, and let the sets $K_1, \tilde{S}, K_{\tilde{S}}, I_{\tilde{S}}, K$ and the bipartite graph B be as described above. Suppose $|K_1| \geq 2|\tilde{S}|$, and let $X \subseteq \tilde{S} = (K_{\tilde{S}} \cup I_{\tilde{S}}), Y \subseteq K_1 \subseteq K, M \subseteq E(G[X \cup Y])$ be as guaranteed to exist by Corollary 3. If S is a T-THS of graph G of size at most k then $(S \setminus Y) \cup X$ is also a T-THS of G of size at most k.*

Reduction Rule 10. *If $|K_1| \geq 2|\tilde{S}|$ then find sets $X \subseteq \tilde{S}$ and $Y \subseteq K_1$ as described by Corollary 4. Set $G' \leftarrow G - X, T' \leftarrow T \setminus X, k' \leftarrow k - |X|$. The reduced instance is $(G'; T'; k')$.*

Proof Sketch of Theorem 3. On input $(G; T; k)$ the algorithm applies the various reduction rules exhaustively and in the given order, and outputs an equivalent reduced instance $(G'; T'; k')$. We assume without loss of generality that $(G'; T'; k')$ is not a trivial YES or NO instance. Since Reduction Rule 2 is not applicable to the reduced instance $(G'; T'; k)$, there is no isolated vertex in graph G'. Let (K', I') be the split partition of graph G'. Since Reduction Rules 6 and 7 are not applicable, every vertex in K' is adjacent with at most k vertices in I'. Since there is no isolated vertex in the graph, this implies $|I'| \leq k \cdot |K'|$. Since Reduction Rule 8 did not return I_{NO} the approximate solution $\tilde{S}$ is of size at most $3k$. By Observation 2, $|K' \cap \tilde{S}| = |K'_{\tilde{S}}| \leq 2k$ and $|I' \cap \tilde{S}| = |I'_{\tilde{S}}| \leq k$. Let K'_0, K'_1 be the partition as defined after Reduction Rule 8. Since Reduction Rule 9 is not applicable, this implies $|K_0| < 2|I'_{\tilde{S}}| < 2k$. Similarly, Reduction Rule 10 being not applicable implies that $|K_1| < 2|\tilde{S}| < 6k$. Since $K'_{\tilde{S}}, K_0$ and K_1 is a partition of K, we conclude that the cardinality of K is upper bounded by $10k$.

It is not difficult to verify that the kernelization algorithm runs in polynomial time.

References

1. Abu-Khzam, F.N.: A Kernelization algorithm for d-hitting set. J. Comput. Syst. Sci. **76**(7), 524–531 (2010)
2. Blair, J.R.S., Peyton, B.: An introduction to chordal graphs and clique trees. In: George, A., Gilbert, J.R., Liu, J.W.H. (eds.) Graph Theory and Sparse Matrix Computation. The IMA Volumes in Mathematics and its Applications, vol. 56, pp. 1–29. Springer, New York (1993)
3. Chitnis, R., Fomin, F.V., Lokshtanov, D., Misra, P., Ramanujan, M.S., Saurabh, S.: Faster exact algorithms for some terminal set problems. In: Gutin, G., Szeider, S. (eds.) IPEC 2013. LNCS, vol. 8246, pp. 150–162. Springer, Cham (2013)

4. Chitnis, R.H., Cygan, M., Hajiaghayi, M.T., Marx, D.: Directed subset feedback vertex set is fixed-parameter tractable. ACM Trans. Algorithms (TALG) **11**(4), 28:1–28:28 (2015)
5. Cygan, M., et al.: Parameterized Algorithms. Springer, Cham (2015)
6. Cygan, M., Pilipczuk, M., Pilipczuk, M., Wojtaszczyk, J.O.: Subset feedback vertex set is fixed-parameter tractable. SIAM J. Discrete Math. **27**(1), 290–309 (2013)
7. Dell, H., Van Melkebeek, D.: Satisfiability allows no nontrivial sparsification unless the polynomial-time hierarchy collapses. J. ACM **61**(4), 23:1–23:27 (2014)
8. Diestel, R.: Graph Theory, 5th edn. Springer, Heidelberg (2016)
9. Even, G., Naor, J., Zosin, L.: An 8-approximation algorithm for the subset feedback vertex set problem. SIAM J. Comput. **30**(4), 1231–1252 (2000)
10. Fomin, F.V., Heggernes, P., Kratsch, D., Papadopoulos, C., Villanger, Y.: Enumerating minimal subset feedback vertex sets. Algorithmica **69**(1), 216–231 (2014)
11. Fomin, F.V., Lokshtanov, D., Misra, N., Philip, G., Saurabh, S.: Hitting forbidden minors: approximation and Kernelization. SIAM J. Discrete Math. **30**(1), 383–410 (2016)
12. Golovach, P.A., Heggernes, P., Kratsch, D., Saei, R.: Subset feedback vertex sets in chordal graphs. J. Discrete Algorithms **26**, 7–15 (2014)
13. Golumbic, M.C.: Algorithmic Graph Theory and Perfect Graphs, vol. 57. Elsevier, Amsterdam (2004)
14. Hols, E.M.C., Kratsch, S.: A randomized polynomial kernel for subset feedback vertex set. Theory of Comput. Syst. **62**(1), 63–92 (2018)
15. Kawarabayashi, K., Kobayashi, Y.: Fixed-parameter tractability for the subset feedback set problem and the S-cycle packing problem. J. Comb. Theor. Ser. B **102**(4), 1020–1034 (2012)
16. Le, T.-N., Lokshtanov, D., Saurabh, S., Thomassé, S., Zehavi, M.: Subquadratic kernels for implicit 3-hitting set and 3-set packing problems. In: Proceedings of the Twenty-Ninth Annual ACM-SIAM Symposium on Discrete Algorithms, pp. 331–342. SIAM (2018)
17. Lokshtanov, D., Ramanujan, M.S., Saurabh, S.: Linear time parameterized algorithms for subset feedback vertex set. ACM Trans. Algorithms (TALG) **14**(1), 7 (2018)
18. Philip, G., Rajan, V., Saurabh, S., Tale, P.: Subset feedback vertex set in chordal and split graphs. arXiv e-prints, January 2019. https://arxiv.org/abs/1901.02209
19. Thomassé, S.: A $4k^2$ kernel for feedback vertex set. ACM Trans. Algorithms **6**(2), 3:21–3:28 (2010)
20. Wahlström, M.: Algorithms, measures and upper bounds for satisfiability and related problems. Ph.D. thesis, Department of Computer and Information Science, Linköpings universitet (2007)
21. Wahlström, M.: Half-integrality, LP-branching and FPT algorithms. In: Proceedings of the Twenty-fifth Annual ACM-SIAM Symposium on Discrete Algorithms, pp. 1762–1781. SIAM (2014)

Author Index

Printed in the United States
By Bookmasters